Lecture Notes in Physics

The Editorial Policy for Edited Volumes

The series *Lecture Notes in Physics* (LNP), founded in 1969, reports new developments in physics research and teaching - quickly, informally but with a high degree of quality. Manuscripts to be considered for publication are topical volumes consisting of a limited number of contributions, carefully edited and closely related to each other. Each contribution should contain at least partly original and previously unpublished material, be written in a clear, pedagogical style and aimed at a broader readership, especially graduate students and nonspecialist researchers wishing to familiarize themselves with the topic concerned. For this reason, traditional proceedings cannot be considered for this series though volumes to appear in this series are often based on material presented at conferences, workshops and schools.

Acceptance

A project can only be accepted tentatively for publication, by both the editorial board and the publisher, following thorough examination of the material submitted. The book proposal sent to the publisher should consist at least of a preliminary table of contents outlining the structure of the book together with abstracts of all contributions to be included. Final acceptance is issued by the series editor in charge, in consultation with the publisher, only after receiving the complete manuscript. Final acceptance, possibly requiring minor corrections, usually follows the tentative acceptance unless the final manuscript differs significantly from expectations (project outline). In particular, the series editors are entitled to reject individual contributions if they do not meet the high quality standards of this series. The final manuscript must be ready to print, and should include both an informative introduction and a sufficiently detailed subject index.

Contractual Aspects

Publication in LNP is free of charge. There is no formal contract, no royalties are paid, and no bulk orders are required, although special discounts are offered in this case. The volume editors receive jointly 30 free copies for their personal use and are entitled, as are the contributing authors, to purchase Springer books at a reduced rate. The publisher secures the copyright for each volume. As a rule, no reprints of individual contributions can be supplied.

Manuscript Submission

The manuscript in its final and approved version must be submitted in ready to print form. The corresponding electronic source files are also required for the production process, in particular the online version. Technical assistance in compiling the final manuscript can be provided by the publisher's production editor(s), especially with regard to the publisher's own LaTeX macro package which has been specially designed for this series.

LNP Homepage (springerlink.com)

On the LNP homepage you will find:
−The LNP online archive. It contains the full texts (PDF) of all volumes published since 2000. Abstracts, table of contents and prefaces are accessible free of charge to everyone. Information about the availability of printed volumes can be obtained.
−The subscription information. The online archive is free of charge to all subscribers of the printed volumes.
−The editorial contacts, with respect to both scientific and technical matters.
−The author's / editor's instructions.

E. Ben-Naim H. Frauenfelder Z. Toroczkai (Eds.)

Complex Networks

 Springer

Editors

Eli Ben-Naim
Hans Frauenfelder
Zoltan Toroczkai
Los Alamos National Laboratory
Complex Systems Group
Theoretical Division
87545 Los Alamos, NM
USA

E. Ben-Naim H. Frauenfelder Z. Toroczkai (Eds.), *Complex Networks*, Lect. Notes Phys. **650**
(Springer, Berlin Heidelberg 2004), DOI 10.1007/b98716

ISSN 0075-8450
ISBN 978-3-642-06110-3   e-ISBN 978-3-540-44485-5

Springer is a part of Springer Science+Business Media

springeronline.com

The use of general descriptive names, registered names, trademarks, etc. in this publication does not imply, even in the absence of a specific statement, that such names are exempt from the relevant protective laws and regulations and therefore free for general use.

Cover design: *design & production*, Heidelberg

Printed on acid-free paper
54/3141/ts - 5 4 3 2 1 0

Preface

The study of interacting particle systems has traditionally focused on cases where the underlying topology can be described by simple structures such as regular crystalline lattices or by a continuum medium. The emerging science of complex networks addresses complementary situations where the underlying topology is a graph whose structure is complex, irregular, and dynamically evolving. Complex networks are ubiquitous in nature. Natural networks include biological networks (metabolic networks, gene regulatory networks, protein interaction networks, signaling networks, epidemic networks), and ecological networks (food webs). Man-made networks include communications networks (WWW, Internet, phone, wireless), transportation infrastructures (power grid, waterways, natural gas, roadways, airlines), and social interactions (acquaintance networks, scientific collaboration networks, terrorist networks).

Network science dates back to Leonhard Euler who initiated graph theory by his solution in 1736 to the famous Königsberg bridges problem. For the next 200 years graph theory dealt with regular or small structures. Network science was reborn with the introduction of random graph theory, through the seminal works of Ray Solomonoff and A. Rapoport in 1951, and separately, by the works of Pál Erdős and Alfréd Rényi in 1959-1960 who introduced probabilistic methods to graph theory.

Currently, a third revolution is underway. It has been motivated by the emergence of communication networks and the need to characterize biological networks and facilitated by the availability of large data sets and the explosive growth in computing power. Based on characteristics of real-world networks, the small-world network model by Duncan Watts and Steven Strogatz and the preferential attachment model of scale-free networks by Albert-László Barabási and Réka Albert have reshaped the way we think of networks.

These contributions showed that the structure of many real-world large-scale complex networks are far from those of the traditional random graphs, and they opened up many avenues for future research. They demonstrated that complex networks is an intellectually deep and ripe area, relevant to many scientific disciplines including physics, biology, engineering, and social science, far beyond the traditional fields of mathematics and computer science.

Now, the research front turns to networks dynamics. Most networks have the role and function to transport or transfer entities (information, energy, etc.) along the links. Optimizing transport efficiency and quantifying network vulnerabilities and robustness constitute the next open questions. Predicting the dynamical evolution of the network structure and its coupling with the transport processes are the ultimate challenge for complex networks science.

This volume of the Lecture Notes in Physics series focuses on the application of techniques from statistical physics to characterization and modeling of complex networks. There is a deep connection between statistical physics and statistical graph theory as both aim to characterize macroscopic observables based on a probabilistic treatment of all microstates of the system. As a concrete example, the polymerization process proposed by Paul Flory and used by chemical physicists to model gelation is equivalent to the growth of a random graph. This natural connection between statistical mechanics and statistical graph theory is currently being exploited by many physicists and the present volume presents the state-of-the-art in the application of statistical physics methods to complex networks research.

This volume consists of four parts. The first two parts concern theory and modeling of networks while the last two parts involve applications to real-world networks. Part I deals with theoretical characterization of structural properties of networks including spectral and extremal properties and structural robustness. Part II addresses dynamical aspects of networks including evolving networks, dynamical processes and transport on networks, and synchronization of networks. Part III focuses on information and social networks including publication networks, collaboration networks, email communication, and board membership networks. Part IV starts with an overview of networks in biological systems, followed by applications to genetic and neural networks.

The articles in this volume were written by speakers at the conference "Complex Networks: Structure, Dynamics, and Function", the 23rd annual conference of the Center for Nonlinear Studies at Los Alamos National Laboratory, held from May 12–16, 2003 in Santa Fe, New Mexico, USA. The papers in this volume are review articles by experts in network science, many of whom made seminal contributions to the foundations of this novel field. As a collection, this volume covers a large fraction of the state-of-the art of complex network research. The articles are aimed at students, newcomers to the field, as well as experts. All articles have been carefully peer-reviewed not only for scientific content but also for self-consistency and readability.

The editors thank the authors for their contributions and the referees, whose comments improved the articles in a significant way. The editors also wish to thank the conference organizers Benjamin McMahon, Paul Fenimore, and Pieter Swart, as well as the conference coordinator Roderick Garcia.

Los Alamos, New Mexico, USA   *Eli Ben-Naim*
February 2004   *Hans Frauenfelder*
                                                     Zoltan Toroczkai

Contents

Part I   Network Structure

Tomography and Stability of Complex Networks
Tomer Kalisky, Reuven Cohen, Daniel ben-Avraham, Shlomo Havlin...... 3
1   Introduction 3
2   General Results 4
3   Scale-Free Networks 8
4   Tomography of Scale Free Networks 11
5   Random Breakdown 18
6   Intentional Attack 19
7   Critical Exponents 23
8   Conclusions... 31

Spectral Analysis of Random Networks
Sergei N. Dorogovtsev, Alexander V. Goltsev, José F.F. Mendes,
Alexander N. Samukhin... 35
1   Introduction 35
2   Random Walk on a Tree 36
3   General Theory 37
4   Spectra of Uncorrelated Graphs 39
5   Effective Medium Approximation 40
6   Tail Behavior and Finite-Size Effects....................... 40
7   Spectrum of a Transition Matrix 42
8   Spectra of Different Topological Graphs 43
9   Conclusions... 48

A Tractable Complex Network Model
Based on the Stochastic Mean-Field Model of Distance
David J. Aldous 51
1   Introduction 51
2   Formulas 53
3   The Model 59
4   Calculations 67
5   Further Calculations 77
6   Comparison with Other Models 84

The Small World Phenomenon in Hybrid Power Law Graphs
Fan Chung, Linyuan Lu 89
1   Introduction 89
2   Preliminaries 91
3   Local Graphs 93
4   The Hybrid Power Law Model 95
5   Several Facts Concerning Random Power Law Graphs............. 97
6   The Diameter of the Hybrid Model 99
7   Concluding Remarks.. 101

Classes of the Shortest Pathway Structures
in Scale Free Networks
Kwang-Il Goh, Eulsik Oh, Chul-Min Ghim, Byungnam Kahng,
Doochul Kim 105
1   Introduction 105
2   Load or Betweenness Centrality 107
3   Load-Load Correlation 115
4   Diameter Change Distribution 118
5   Conclusions and Discussion 123

The Optimal Path in an Erdős-Rényi Random Graph
Lidia A. Braunstein, Sergey V. Buldyrev, Sameet Sreenivasan,
Reuven Cohen, Shlomo Havlin, H. Eugene Stanley 127
1   Introduction 127
2   Theoretical Arguments...................................... 128
3   Numerical Analysis 129
4   Probability Distribution of the Maximal Weight
     on the Optimal Path.. 132

Clustering in Complex Networks
Gábor Szabó, Mikko Alava, János Kertész......................... 139
1   Introduction 139
2   Examples of Clustering..................................... 141
3   Models That Create Clustering.............................. 143
4   Rate-Equation Approach 151
5   Conclusions.. 159

Equilibrium Statistical Mechanics of Network Structures
Illés Farkas, Imre Derényi, Gergely Palla, Tamás Vicsek.............. 163
1   Introduction 163
2   Preliminaries 165
3   Graph Ensembles... 166
4   Main Features of Equilibrium Graphs: Local and Global Properties ... 176
5   Topological Phase Transitions in Equilibrium Network Ensembles 178
6   Summary.. 184

Information Theory of Complex Networks:
On Evolution and Architectural Constraints
Ricard V. Solé, Sergi Valverde 189
1   Introduction 189
2   Measuring Correlations 191
3   Entropy and Information 194
4   Model Networks... 196
5   Real Networks 198
6   Simulated Annealing Search 202
7   Discussion... 204

Part II   Network Dynamics

Extremal Properties of Random Structures
Eli Ben-Naim, Paul L. Krapivsky, Sidney Redner 211
1   Introduction 211
2   Random Trees 213
3   Random Graphs... 223
4   Random Networks... 225
5   Summary and Discussion 231

On the Analysis of Backtrack Procedures
for the Colouring of Random Graphs
Rémi Monasson 235
1   Introduction 235
2   Colouring in the Absence of Backtracking 239
3   Colouring in the Presence of Massive Backtracking 244
4   Conclusions: What Is Missing? 251

Small-World Synchronized Computing Networks
for Scalable Parallel Discrete-Event Simulations
Hasan Guclu, György Korniss, Zoltán Toroczkai, Mark A. Novotny 255
1   Introduction 255
2   The Basic Conservative Scheme 256
3   The Small-World Synchronized Conservative PDES Scheme.......... 261
4   Summary... 272

Critical Phenomena in a Small World
Matthew B. Hastings, Balázs Kozma 277
1   Introduction 277
2   Long-Range Versus Small-World 280
3   Edwards-Wilkinson Equation: An Example 288
4   Discussion 296

Attacks and Cascades in Complex Networks
Ying-Cheng Lai, Adilson E. Motter, Takashi Nishikawa 299
1   Introduction 299
2   Conceptual Network of Language 301
3   Attack-Induced Cascades in Complex Networks 302
4   Range-Based Attacks on Links in Complex Networks 305
5   Discussion 308

Part III   Information Networks & Social Networks

Scholarly Information Network
Paul Ginsparg 313
1   arXiv Background and Lessons 313
2   New Scholarly Publication Models 318
3   Novel Corpus Navigation Tools 322
4   Text Classification and Support Vector Machines 326
5   arXiv q-bio Extraction 329
6   Conclusion 334

Who Is the Best Connected Scientist?
A Study of Scientific Coauthorship Networks
Mark E.J. Newman 337
1   Introduction 337
2   Coauthorship Networks 339
3   Basic Results 341
4   Distances and Centrality 352
5   Weighted Collaboration Networks 361
6   Conclusions 366

Information Dynamics in the Networked World
Bernardo A. Huberman, Lada A. Adamic 371
1   Introduction 371
2   Email as Spectroscopy 372
3   Information Flow in Social Groups 379
4   Small World Search 386
5   Conclusion 395

Emergence of Complexity in Financial Networks
Guido Caldarelli, Stefano Battiston, Diego Garlaschelli,
Michele Catanzaro 399
1   Introduction 399
2   The Board and Director Networks 400
3   Network of Price Correlations 406
4   The Stock Investment Network 412

Topology, Hierarchy, and Correlations in Internet Graphs
Romualdo Pastor-Satorras, Alexei Vázquez, Alessandro Vespignani 425
1   Introduction 425
2   Internet Maps 427
3   Average Properties 428
4   Scale-Free Properties 430
5   Hierarchy and Correlations 434
6   Conclusions.. 438

Part IV   Biological Networks

Characteristics of Biological Networks
Albert-László Barabási, Zoltán N. Oltvai, Stefan Wuchty 443
1   Introduction 443
2   Basic Network Features 444
3   Network Models.. 445
4   Conclusions.. 453

Boolean Modeling of Genetic Regulatory Networks
Réka Albert 459
1   Introduction 459
2   The Segment Polarity Gene Network........................... 463
3   Description of the Model 465
4   Modeling the Wild Type Segment Polarity Genes................ 467
5   The Functional Topology of the Segment Polarity Network 469
6   Gene Mutations 472
7   Determination of the Steady States
    and Their Domains of Attraction 473
8   Possible Changes in the Assumptions 476
9   Conclusions.. 479

Theoretical Neuroanatomy: Analyzing the Structure, Dynamics, and Function of Neuronal Networks
Anil K. Seth, Gerald M. Edelman 483
1   Introduction 483
2   Structure.. 484
3   Dynamics 488
4   Function 493
5   General Discussion 504
Appendix A: Implementation Details 506

Index 513

List of Contributors

Lada A. Adamic
HP Labs
1501 Page Mill Road
Palo Alto, CA 94304
USA
ladamic@exch.hpl.hp.com

Mikko Alava
Laboratory of Physics
Helsinki University of Technology
P. O. Box 1100, FIN-02015 HUT
Finland
mja@fyslab.hut.fi

Réka Albert
Department of Physics
Pennsylvania State University
University Park, PA 16802
USA
ralbert@phys.psu.edu

David J. Aldous
Department of Statistics
367 Evans Hall
University of California
Berkeley, CA 94720
USA
aldous@stat.berkeley.edu

Albert-László Barabási
Department of Physics
University of Notre Dame
Notre Dame, IN 46556
USA
alb@nd.edu

Stefano Battiston
Laboratoire de Physique Statistique
ENS, 24 rue Lhomond
75005 Paris
France
battiston@ens.fr

Daniel ben-Avraham
Department of Physics
Clarkson University
Potsdam, NY 13699
USA
qd00@clarkson.edu

Eli Ben-Naim
Theoretical Division
Los Alamos National Laboratory
Los Alamos, NM 87545
USA
ebn@lanl.gov

Lidia A. Braunstein
Department of Physics
Boston University
Boston, MA 02215
USA
lidia@meta.bu.edu

Sergey V. Buldyrev
Department of Physics
Boston University
Boston, MA 02215
USA
sergey@meta.bu.edu

Guido Caldarelli
Dipartimento di Fisica
Università La Sapienza
P.le Moro 5
00185 Roma
Italy.
gcalda@pil.phys.uniroma1.it

Michele Catanzaro
Dipartimento di Fisica
Università La Sapienza
P.le Moro 5
00185 Roma
Italy
michele@pil.phys.uniroma1.it

Fan Chung
Department of Mathematics
University of California, San Diego
La Jolla, CA 92093
USA
fan@math.ucsd.edu

Reuven Cohen
Department of Computer Science &
Applied Mathematics
Weizmann Institute of Science
Rehovot
Israel
cohenr@shoshi.ph.biu.ac.il

Imre Derényi
Department of Biological Physics
Eötvös University
Pázmány P. stny. 1A,
1117 Budapest
Hungary
derenyi@angel.elte.hu

Sergei N. Dorogovtsev
Departamento de Física
Universidade de Aveiro
Campus Universitário de Santiago
3810-193 Aveiro
Portugal
sdorogov@fis.ua.pt

Gerald M. Edelman
The Neurosciences Institute
10640 John Jay Hopkins Drive
San Diego, CA 92121
USA
edelman@nsi.edu

Illés Farkas
Biological Physics Research Group
Eötvös University
Pázmány P. stny. 1A,
1117 Budapest
Hungary
fij@angel.elte.hu

Diego Garlaschelli
Dipartimento di Fisica
Università di Siena
Via Roma 56
53100 Siena
Italy
diego@pil.phys.uniroma1.it

Chul-Min Ghim
School of Physics
Seoul National University
Seoul 151-747
Korea
cmghim@phya.snu.ac.kr

Paul Ginsparg
Departments of Physics
Cornell University
Ithaca, NY 14853
USA
ginsparg@cornell.edu

Kwang-Il Goe
School of Physics
Seoul National University
Seoul 151-747
Korea
vrooom@soeh.snu.ac.kr

Alexander V. Goltsev
Departamento de Física
Universidade de Aveiro
Campus Universitário de Santiago
3810-193 Aveiro
Portugal
goltsev@fis.ua.pt

Hasan Guclu
Department of Physics
Rensselaer Polytechnic Institute
110 8th Street
Troy, NY 12180
USA
gucluh@rpi.edu

Matthew B. Hastings
Theoretical Division
Los Alamos National Laboratory
Los Alamos, NM 87545
USA
hastings@lanl.gov

Shlomo Havlin
Department of Physics
Bar-Ilan University
Ramat-Gan
Israel
havlin@ophir.ph.biu.ac.il

Bernardo A. Huberman
HP Labs
1501 Page Mill Road
Palo Alto, CA 94304
USA
huberman@hpl.hp.com

Byungnam Kahng
School of Physics
Seoul National University
Seoul 151-747
Korea
kahng@phya.snu.ac.kr

Tomer Kalisky
Department of Physics
Bar-Ilan University
Ramat-Gan
Israel
kaliskt@mail.biu.ac.il

János Kertész
Department of Theoretical Physics
Budapest University of Technology
8 Budafoki út, 1111
Hungary
kertesz@phy.bme.hu

Doochul Kim
School of Physics
Seoul National University
Seoul 151-747
Korea
dkim@snu.ac.kr

György Korniss
Department of Physics
Rensselaer Polytechnic Institute
110 8th Street
Troy, NY 12180
USA
korniss@rpi.edu

Balázs Kozma
Department of Physics
Rensselaer Polytechnic Institute
110 8th Street
Troy, NY 12180
USA
kozmab@rpi.edu

Paul L. Krapivsky
Department of Physics
Boston University
Boston, MA 02215
USA
paulk@bu.edu

Ying-C. Lai
Department of Mathematics
Arizona State University
Tempe, AZ 85287
USA
yclai@chaos1.la.asu.edu

Linyuan Lu
Department of Mathematics
University of California, San Diego
La Jolla, CA 92093
USA
llu@math.ucsd.edu

José F. F. Mendes
Departamento de Física
Universidade de Aveiro
Campus Universitário de Santiago
3810-193 Aveiro
Portugal
jfmendes@fis.ua.pt

Rémi Monasson
Laboratoire de Physique Théorique
ENS, 24 rue Lhomond
75005 Paris
France
monasson@lpt.ens.fr

Adilson E. Motter
Max Planck Institute for
the Physics of Complex Systems
Nöthnitzer Strasse 38
01187 Dresden
Germany
motter@mpipks-dresden.mpg.de

Mark E. J. Newman
Department of Physics
University of Michigan
Ann Arbor, MI 48109
USA
mejn@umich.edu

Takashi Nishikawa
Department of Mathematics
Southern Methodist University
Dallas, TX 75275
USA
tnishi@chaos6.la.asu.edu

Mark A. Novotny
Department of Physics
Mississippi State University
Mississippi State, MS 39762
USA
man40@ra.msstate.edu

Eulsik Oh
School of Physics
Seoul National University
Seoul 151-747
Korea
oes@phya.snu.ac.kr

Zoltán N. Oltvai
Department of Pathology
Northwestern University
Chicago, IL 60611
USA
zno008@northwestern.edu

Gergely Palla
Department of Biological Physics
Eötvös University
Pázmány P. stny. 1A
1117 Budapest
Hungary
pallag@angel.elte.hu

Romualdo Pastor-Satorras
Department de Física
Universitat Politècnica de Catalunya
Campus Nord
08034 Barcelona
Spain
romu@sinera.upc.es

Sidney Redner
Department of Physics
Boston University
Boston, MA 02215
USA
redner@bu.edu

Alexander N. Samukhin
Departamento de Física
Universidade de Aveiro
Campus Universitário de Santiago
3810-193 Aveiro
Portugal
samukhin@fis.ua.pt

Anil K. Seth
The Neurosciences Institute
10640 John Jay Hopkins Drive
San Diego, CA 92121
USA
seth@nsi.edu

Ricard V. Solé
Complex Systems Lab-ICREA
Universitat Pompeu Fabra (GRIB)
Dr Aiguader 80
08003 Barcelona
Spain
ricard.sole@upf.edu

Sameet Sreenivasan
Department of Physics
Boston University
Boston, MA 02215
USA
sameet@buphy.bu.edu

H. Eugene Stanley
Department of Physics
Boston University
Boston, MA 02215
USA
hes@bu.edu

Gábor Szabó
Department of Theoretical Physics
Budapest University of Technology
8 Budafoki út, 1111
Hungary
szabo@phy.bme.hu

Zoltán Toroczkai
Theoretical Division
Los Alamos National Laboratory
Los Alamos, NM 87545
USA
toro@lanl.gov

Sergi Valverde
Complex Systems Lab-ICREA
Universitat Pompeu Fabra (GRIB)
Dr Aiguader 80
08003 Barcelona
Spain

Alexei Vázquez
Department of Physics
University of Notre Dame
NotreDame, IN 46556
USA
avazque1@nd.edu

Alessandro Vespignani
Laboratoire de Physique Théorique
Bâtiment 210
Université de Paris-Sud
91405 ORSAY Cedex
France
Alessandro.Vespignani
@th.u-psud.fr

Tamás Vicsek
Department of Biological Physics
Eötvös University
Pázmány P. stny. 1A
1117 Budapest
Hungary
vicsek@angel.elte.hu

Stefan Wuchty
Department of Physics
University of Notre Dame
Notre Dame, IN 46556
USA
swuchty@nd.edu

Part I

Network Structure

Tomography and Stability of Complex Networks

Tomer Kalisky[1], Reuven Cohen[1,2], Daniel ben-Avraham[3], and Shlomo Havlin[1]

[1] Minerva Center and Department of Physics, Bar-Ilan University, Ramat-Gan, Israel
[2] Department of Computer Science and Applied Mathematics, Weizmann Institute of Science, Rehovot, Israel
[3] Department of Physics, Clarkson University, Potsdam, NY 13699, USA

Abstract. We study the structure of generalized random graphs with a given degree distribution $P(k)$, and review studies on their behavior under both random breakdown of nodes and intentional attack on the most highly connected nodes. We focus on scale free networks, where $P(k) \propto k^{-\lambda}$, for $m < k < K$. We first examine the "Tomography" of these networks, i.e. the structure of layers around a network node. It is shown that the distance distribution of all nodes from the maximally connected node of the network consists of two regimes. The first is characterized by rapid growth in the number of nodes, and the second decays exponentially. We also show analytically that the nodes degree distribution at each layer is a power law with an exponential cut-off. We then show that scale free networks with $\lambda < 3$ are robust to random breakdown, but vulnerable to intentional attack. We also describe the behavior of the network near the phase transition and show that the critical exponents are influenced by the scale free nature of the network. We show that the critical exponent for the infinite cluster size behaves as $\beta = 1/|\lambda - 3|$, and the exponent for the finite clusters size distribution behaves as $\tau = \frac{2\lambda-3}{\lambda-2}$, for $2 < \lambda < 4$. For $\lambda > 4$ the exponents are $\beta = 1$ and $\tau = 2.5$ as in normal infinite dimensional percolation. It is also shown that for all $\lambda > 3$ the exponent for the correlation length is $\nu = 1$ and formulas for the fractal dimensions are obtained. The size of the largest cluster at the transition point, known to scale as $N^{2/3}$ in regular random graphs, is shown to scale as $N^{(\lambda-2)/(\lambda-1)}$ for $3 < \lambda < 4$ and as $N^{2/3}$ for $\lambda > 4$.

1 Introduction

Much attention has been focused recently on the topic of complex network behavior [1–5]. Most of the interest has been on scale-free networks, which are believed to represent many phenomena in nature. Scale-free degree distributions have been observed in the Internet [6], World Wide Web (WWW) [7], metabolic networks [8] and many others. For recent reviews see [9–13]. In this paper we review the topics of structure [14] and percolation of such networks [1–4]. Understanding network structure can help devise better networks topologies. It may also help design more efficient algorithms for routing and searching in communications networks by taking advantage of the network structure. Percolation is especially important in forecasting and preventing network malfunctions in the Internet, as well as other realistic networks, and may also be important in the understanding of the stability of biological and chemical processes [15].

Percolation theory has been studied for some decades by physicists and mathematicians. In general it deals with the dilution of a fraction p (alternatively, the

T. Kalisky, R. Cohen, D. ben-Avraham, and S. Havlin, Tomography and Stability of Complex Networks, Lect. Notes Phys. **650**, 3–34 (2004)
http://www.springerlink.com/

occupation with a density $q = 1 - p$) of the sites or bonds in a graph [16,17]. It is known that for many graphs a finite threshold p_c exists, such that for dilution of $p < p_c$ a spanning cluster (*i.e.* a cluster of size proportional to that of the entire network) exists. While for $p > p_c$ the graph is fragmented into small clusters. When a spanning cluster exists, its size relative to the graph is denoted $P_\infty(p)$. Near the transition point $P_\infty \sim (p_c - p)^\beta$, where β (as well as other "critical exponents" such as ν, τ and σ) is universal – that is, depends only on the dimension and large scale properties of the graph and not on the local structure. At the transition point the clusters are fractals, while above and below that point the clusters are fractals up to length scale $\xi(p)$ (the correlation length) and have the dimension of the graph above ξ. Near criticality, $\xi \sim |p_c - p|^{-\nu}$.

The number of clusters of size s near criticality also follows a scaling form:

$$n_s \sim s^{-\tau} e^{-s/s^*} . \tag{1}$$

At $p = p_c$, the exponential cutoff $s^* \sim |p - p_c|^{-\sigma}$ diverges and the tail of the distribution behaves as a power law.

The structure of this paper is as follows: In Sect. 2 we discuss general results applicable to generalized random graphs with an arbitrary degree distribution. In Sect. 3 we discuss networks having a scale-free degree distribution, which will be the main concern of this paper. In Sect. 4 we discuss the tomography of scale-free networks, that is, their partition into layers surrounding the maximally connected node at different distances. Section 5 presents the model of random breakdown in scale-free networks and analytical and numerical results for this kind of failure. Section 6 offers a similar approach for an intentional attack on the most highly connected nodes. Section 7 presents an analytical derivation of the critical exponents for the percolation transition on scale-free networks, and finally Sect. 8 presents conclusions and prospects.

2 General Results

2.1 Condition for a Spanning Cluster

For a graph having degree distribution $P(k)$ to have a spanning cluster, a site which is reached by following a link from the giant cluster must have at least one other link in average to allow the cluster to exist[4]. For this to happen the average degree of a site must be at least 2 (one incoming and one outgoing link) given that the site i is connected to j:

$$\langle k_i | i \leftrightarrow j \rangle = \sum_{k_i} k_i P(k_i | i \leftrightarrow j) = 2. \tag{2}$$

---

[4] If we dilute the graph up to near p_c, the remaining structure resembles a tree, or a branching process. One can show that a branching process with an average branching factor that is less than 1 will die out with probability 1 after a finite number of steps [18].

Using Bayes rule we get

$$P(k_i|i \leftrightarrow j) = P(k_i, i \leftrightarrow j)/P(i \leftrightarrow j) = P(i \leftrightarrow j|k_i)P(k_i)/P(i \leftrightarrow j), \qquad (3)$$

where $P(k_i, i \leftrightarrow j)$ is the *joint* probability that node i has degree k_i and that it is connected to node j. For randomly connected networks (neglecting loops) $P(i \leftrightarrow j) = \langle k \rangle/(N-1)$ and $P(i \leftrightarrow j|k_i) = k_i/(N-1)$, where N is the total number of nodes in the network. Using the above criteria (2) reduces to [19,2]:

$$\kappa \equiv \frac{\langle k^2 \rangle}{\langle k \rangle} = 2, \qquad (4)$$

at the critical point. A spanning cluster exists for graphs with $\kappa > 2$, while graphs with $\kappa < 2$ contain only small clusters whose size is not proportional to that of the entire network. This criterion was derived earlier by Molloy and Reed [19] using somewhat different arguments.

The neglecting of loops can be justified below the threshold since the probability for a bond to form a loop in an s-node cluster is proportional to $(s/N)^2$ (i.e., proportional to the probability of choosing two sites in that cluster). Calculating the fraction of loops P_{loop} in the system yields:

$$P_{loop} \propto \sum_i \frac{s_i^2}{N^2} < \sum_i \frac{s_i S}{N^2} = \frac{S}{N}, \qquad (5)$$

where the sum is over all clusters in the system and s_i is the size of the ith cluster. Therefore, the fraction of loops in the system is less than or proportional to S/N, where S is the size of the largest cluster. Below the critical threshold there is no spanning cluster in the system and therefore the fraction of loops is negligible. Hence, until $\kappa = 2$ loops can be neglected. At the threshold the structure of the spanning cluster is almost a tree. Above the threshold loops can no longer be neglected, but since this only happens when a spanning cluster exists the criterion in (4) is valid as a criterion for finding the critical point. A derivation of the exact conditions under which (4) is valid can be found in [19].

2.2 Critical Threshold for Percolation

The above reasoning can be applied to the problem of percolation on a generalized random network. If we randomly remove a fraction p of the sites (or bonds), the degree distribution of the remaining sites will change. For instance, sites with initial degree k_0 will have, after the random removal of nodes, a different number a connections, depending on the number of removed neighbors. The new number of connections will be binomially distributed. If we begin with a distribution of degrees $P_0(k_0)$, the new distribution of degrees of the network will be:

$$P(k) = \sum_{k_0=k}^{\infty} P_0(k_0) \binom{k_0}{k} (1-p)^k p^{k_0-k}. \qquad (6)$$

Calculating the first moment for this distribution, given $\langle k_0 \rangle$ and $\langle k_0^2 \rangle$ for the original distribution leads to:

$$\langle k \rangle = \sum_{k=0}^{\infty} P(k)k = (1-p)\langle k_0 \rangle. \tag{7}$$

In the same manner we can calculate the second moment:

$$\langle k^2 \rangle = \sum_{k=0}^{\infty} P(k)k^2 = (1-p)^2 \langle k_0^2 \rangle + p(1-p)\langle k_0 \rangle. \tag{8}$$

Both quantities can be substituted into (4) to find the criterion for criticality. This yields:

$$\kappa \equiv \frac{\langle k^2 \rangle}{\langle k \rangle} = \frac{(1-p)^2 \langle k_0^2 \rangle + p(1-p)\langle k_0 \rangle}{(1-p)\langle k_0 \rangle} = 2. \tag{9}$$

Reorganizing (9), one gets the critical threshold for percolation [2]:

$$1 - p_{\mathrm{c}} = \frac{1}{\kappa_0 - 1}, \tag{10}$$

where $\kappa_0 \equiv \langle k_0^2 \rangle / \langle k_0 \rangle$ is calculated using the original distribution, before the removal of sites.

Equations (4) and (10) are valid for a wide range of generalized random graphs and distributions. For example for a Cayley tree – a graph with a fixed degree z and no loops – the criterion from (10) can be used. This yields the critical concentration $q_{\mathrm{c}} = 1 - p_{\mathrm{c}} = 1/(z-1)$, which is well known [16,17]. Another example is a random Erdös-Rényi (ER) graph. In those graphs edges are distributed randomly and the resulting degree distribution is Poissonian [20]. Applying the criterion from (4) to a Poisson distribution yields:

$$\kappa \equiv \frac{\langle k^2 \rangle}{\langle k \rangle} = \frac{\langle k \rangle^2 + \langle k \rangle}{\langle k \rangle} = 2, \tag{11}$$

which reduces to $\langle k \rangle = 1$ as known for ER graphs [20].

2.3 Generating Functions

A general method for studying the size of the infinite cluster and the residual network for a graph with an arbitrary degree distribution was first developed by Molloy and Reed [21]. They suggested viewing the infinite cluster as being explored and used differential equations for the number of un-exposed links and unvisited sites to find the size of the infinite cluster and the degree distribution of the residual graph (the finite clusters).

An alternative and very powerful derivation was given by Newman, Strogatz and Watts [5]. They have used the generating functions method to study the

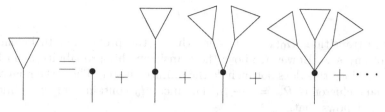

Fig. 1. An illustration of equations 14 and 15 for the probability to reach a branch of a given size by following a link. This is the sum of the probabilities to reach a vertex with zero outgoing links, of reaching a vertex with a single outgoing link connected to another such branch, of reaching a vertex with two outgoing links connected to two such branches etc. After Newman *et al.* [5].

size of the infinite cluster as well as other quantities (such as the diameter and cluster size distribution). They have also applied this method to other types of graphs (directed and bipartite). Here we closely follow their derivation in order to find the size of the infinite cluster and the critical exponents.

In [5] a generating function is built for the degree distribution:

$$G_0(x) = \sum_{k=0}^{\infty} P(k)x^k. \tag{12}$$

If we start from a randomly chosen site and follow each of its links to its nearest neighbors, the sites arrived will have a degree distribution $kP(k)/\langle k \rangle$ [19, 2,5,3]. The generating function describing the probability for k *outgoing* links (excluding the link we arrived along) will be:

$$G_1(x) = \frac{\sum kP(k)x^{k-1}}{\sum kP(k)} = \frac{d}{dx}G_0(x)/\langle k \rangle . \tag{13}$$

Let $H_1(x)$ be the generating function for the probability of reaching a branch of a given size by following a link[5]. If we denote the coefficients of $G_1(x)$ by q_k (i.e. the probability for k outgoing links from a site reached by following a random link), then $H_1(x)$ must satisfy the self-consistent equation (see Fig. 1):

$$H_1(x) = xq_0 + xq_1 H_1(x) + xq_2[H_1(x)]^2 + \cdots . \tag{14}$$

Which can be written as:

$$H_1(x) = xG_1(H_1(x)) . \tag{15}$$

If we start from a random site, we have one such branch at the end of each neighboring link. Since $G_0(x)$ is the generating function for the degree of the site, the generating function for the probability of a site to belong to an n-site cluster is:

---

[5] We assume that the finite clusters have almost no loops and are therefore tree-like structures.

$$H_0(x) = xG_0(H_1(x)) \,. \tag{16}$$

Below the transition, $H_0(1) = 1$, since this is the probability to belong to a cluster of any size. However, above the transition this probability is no longer normalized since this does not include the infinite cluster. Then, the relative size of the giant cluster is $P_\infty = 1 - H_0(1)$, since H_0 contains only the finite-size clusters. It follows that:

$$P_\infty = 1 - \sum_{k=0}^{\infty} P(k)u^k \,, \tag{17}$$

where $u \equiv H_1(1)$ is the smallest positive root of:

$$\langle k \rangle u = \sum_{k=0}^{\infty} kP(k)u^{k-1} \,. \tag{18}$$

This equation can be solved numerically and the solution can be substituted into (17) to calculate the size of the infinite cluster in a graph with a given degree distribution.

3 Scale-Free Networks

3.1 Description

Our main concern in this paper will be with the behavior of scale-free networks. Scale-Free networks are networks whose degree distribution (i.e. fraction of sites with k connections) behaves as:

$$P(k) \propto k^{-\lambda}, \qquad m \leq k \leq K, \tag{19}$$

where λ is the exponent, m is the lower cutoff, and K is the upper cutoff. There are no sites with degree below m and above K. For finite networks the upper cutoff $K(N)$ arises naturally since the fraction of high-degree sites decays with k. An estimate of this cutoff can be found by the assumption that the tail of the distribution above K is of the order of one site [2]:

$$\sum_{k=K}^{\infty} P(k) \sim \int_K^{\infty} P(k)dk = \frac{1}{N}. \tag{20}$$

The estimate obtained this way gives:

$$K \approx mN^{1/(\lambda-1)}. \tag{21}$$

This estimate allows the derivation of finite size effects in the network and allows calculations of moments of the distribution in (19), that would otherwise diverge. Newman *et al.* [5] use an exponential cutoff rather than a sharp one, but the

effect on the results is minor. Another difference between the distribution used
by us and by Newman *et al.* [5] is given in Sect. 3.2.

The importance of scale-free networks lies in the fact that this distribution
occurs in many natural and man-made networks [6,5]. An example of a scale-free
network is the physical Internet structure, that is the router to router (and end-
units) connectivity. This structure was studied by Faloutsos *et al.*[6]. They have
found that the inter-router network can be well approximated by a non-directed
scale-free network with $\lambda \approx 2.5$. The size of the Internet today is about 10^7 sites,
making it a fairly large network.

A few results about the structure of scale-free networks have also been derived
by Aiello *et al.* [22]. The size of the infinite cluster was calculated, and it was
found that for $\lambda \leq 2$ the infinite cluster is of the order of the size of the entire
graph (i.e. $P_\infty = 1 - o(1)$, where $o(1)$ is a function of the network size, N,
decaying to 0 as $N \to \infty$). For $\lambda > \lambda_c = 3.478...$ there is no infinite cluster at all
(since we use a somewhat different distribution, see below (23) and (24), we get
$\lambda_c \approx 4$). For $\lambda < \lambda_c$ the second largest cluster is of order lnN. For lower cutoff
$m \geq 2$ a spanning cluster exists for every λ.

The average distance between sites is also different in scale free networks from
its value for normal random graphs. While for ER graphs the average distance
between sites behaves as $d \sim \ln N$ [20], for scale free graphs with $2 < \lambda < 3$ the
distance behaves as $d \sim \ln \ln N$ [23,24], and for $\lambda = 3$ as $d \sim \ln N/ \ln \ln N$ [25].
The reason for this short distance is the small core, containing most high degree
sites, which has a very small diameter. For $\lambda > 3$ the normal behavior $d \sim \ln N$
is recovered.

3.2 Simulation

To simulate general networks we have used the following algorithm:

1. For each site choose a degree from the required distribution.
2. Create a list where each site is repeated as many times as the chosen degree.
3. Randomly choose pairs from the list and connect the chosen sites (by adding
   each node to the list of neighbors of the other site). Remove the chosen pair
   from the list (by replacing them with the last two entries in the list).

Double and self-edges are ignored, and if the number of entries in the list is
odd (overall odd number of connections) one entry can also be discarded. This
has a minor influence on the degree distribution if the network is large. The
justification of using this algorithm for simulation and analysis as an algorithm
generating the probability space of generalized random graphs can be found in
[19] by Molloy and Reed[6].

To generate degrees from a scale-free distribution we generate a random
number, u, between 0 and 1 from a uniform distribution, and then generate a
new number k using the formula:

---

[6] Actually the model that came to be known as the Molloy-Reed Model was first
introduced by Bollobás [26].

$$k = \frac{m}{u^{1/(\lambda-1)}}. \tag{22}$$

This generates a random *real* number greater than m, with a distribution of $P(k) \propto k^{-\lambda}$. To prevent the appearance of overflows, u can be chosen such that it is large enough to yield numbers smaller than $N-1$ (since this is the maximum degree per site). If an upper cutoff is required a new random number is drawn whenever k is too large. After k is chosen, the closest integer is taken as the degree of the site. The resulting degree distribution is thus:

$$P(k') = \int_{k'-1/2}^{k'+1/2} ck^{-\lambda}dk, \tag{23}$$

where c is the normalization factor, except for the lower cutoff whose probability is given by:

$$P(m) = \int_{m}^{m+1/2} ck^{-\lambda}dk. \tag{24}$$

This is in contrast to the probability chosen by [22] and [5,3], who use a discrete scale-free distribution. The main difference is in the fraction of sites of degree m (usually $m = 1$). Due to (24) the fraction of low degree sites in our derivation is lower and therefore the networks generated this way are more robust than the ones generated using the discrete distribution. However, the behavior of both distributions in the tail is approximately the same, and therefore the qualitative behavior of all phenomena influenced by the scale-free nature of the distribution should be the same.

The complexity of the algorithm depends upon the number of links in the network and therefore is of order $O(N\langle k\rangle)$, which is fast enough to execute for large graphs. The space needed is again of order $O(N\langle k\rangle)$. Those limits allow the creation of networks of about 10^7 sites.

An alternative method for the study of such networks is a variation of the Leath algorithm. In this method a table is created with number of sites of each degree. The Molloy-Reed construction [19] is then used to build the clusters.

3.3 Maximum Degree

In (21) we suggest that the upper cutoff of a scale free network scales as $K = N^{1/(\lambda-1)}$. However, for the spatially embedded graphs [27], we find that no graph with $\lambda < 3$ can be embedded in a lattice without sacrificing the natural cutoff. That is, the cutoff is limited to $k \approx \sqrt{N}$. This holds true for every d. Thus, we expect this to hold true even in the $d \to \infty$ case. Similar results are indeed obtained for mean field (*i.e.* non-embedded) graphs [28]. On the other hand, Warren *et al.* [29] find the natural cutoff even for graphs embedded in $d = 2$ lattices.

These differences may be explained by the fact that the cutoff depends on the ensemble from which the graphs are chosen. If the ensemble is defined as all graphs with the exact given scale free degree distribution and no self loops (loops connecting a site to itself) and no double edges (two edges or more connecting the same set of sites) then the upper cutoff can not be larger than $\sqrt{N}$, while if multigraphs are allowed (i.e. including self loops and double edges) the natural upper cutoff is achieved. The following two facts support this intuition:

Almost all graphs with the natural cutoff are multigraphs: Consider a graph with a site having degree $K \gg \sqrt{N}$. Since for $\lambda > 2$ the total number of links is of order N the number of self loops of the HUB is proportional to $K^2/cN \gg 1$. Similar results apply for double edges between two such sites.

Removing the double edges and self loops does not affect the behavior of the tail of the distribution: Suppose a site has degree $K \gg \sqrt{N}$. The number of degree 1 sites is of order N. Therefore, the number of connections from this site to sites of degree 1 is proportional to its degree K. Since edges leading to degree 1 sites are neither self loops nor double edges, the upper cutoff is at least proportional to the natural cutoff. In reality, the deviation from the exact degree distribution is quite small.

From the above one can conclude that starting from a scale free degree distribution a multigraph can be constructed, with the given degree sequence. Converting it into a simple graph will change the degree distribution, while keeping the power law form of the tail. However, if one is forced to keep the exact degree sequence while prohibiting self-loops and double edges, the upper cutoff may change.

4 Tomography of Scale Free Networks

In this section we study the network structure by describing the statistical behavior of layers surrounding the maximal connected node. First, we describe the process of generating the network, and define our terminology. Then, we analyze the degree distribution at each layer surrounding the maximally connected node. The results presented here are based on [14]. Similar analysis was done by [30].

4.1 Description

We base our construction on the Molloy-Reed model [21], also described in Sect. 3. The construction process tries to gradually expose the network, following the method introduced in [23,31], and is forcing a hierarchy on the Molloy-Reed model, thus enabling us to define layers in the graph.

We start by setting the number of nodes in the network, N. We then choose the nodes degrees according to the scale-free distribution function $P(k) = ck^{-\lambda}$, where $c \approx (\lambda - 1)m^{\lambda-1}$ is the normalizing constant and k is in the range $[m, K]$, for some chosen minimal degree m and the natural cutoff $K = mN^{1/(\lambda-1)}$ of the distribution [2,32].

At this stage each node in the network has a given number of outgoing links, which we term *open connections*, according to its chosen degree. Let us define V as the set of N chosen nodes, C as the set of unconnected outgoing links from the nodes in V, and E as the set of edges in the graph. Using these definitions, the set of links in E is empty at this point, while the set of outgoing open links in C contains all unconnected outgoing links in the graph. In the Molloy-Reed construction described in Sect. 3, the links in C are randomly matched, such that at the end of the process, C is empty, and E contains all the matched links $< u, v >$, $u, v \in V$.

Instead, here we proceed as follows: we start from the maximal degree node, which has a degree K, and connect it randomly to K available open connections, thus removing these open connections from C (see Fig. 2(a)). We have now exposed the first *layer* (or *shell*) of nodes, indexed as $l = 1$. We now continue to fill out the second layer $l = 2$ in the same way: We connect all open connections emerging from nodes in layer 1 to randomly chosen open connections. These open connections may be chosen from nodes of layer 1 (thus creating a loop) or from other links in C. We continue until all open connections emerging from layer 1 have been connected, thus filling layer $l = 2$ (see Fig. 2(b)). Generally, to form layer $l + 1$ from an arbitrary layer l, we randomly connect all open connections emerging from l to either other open connections emerging from l or chosen from the other links in C (see Fig. 2(c)). Note, that when we have formed layer $l + 1$, layer l has no more open connections. The process continues until the set of open connections, C, is empty.

4.2 Theory

We proceed now to evaluate the probability for nodes with degree k to reside outside the first l layers, denoted by $P_l(k)$.

The number of open connections outside layer l, is given by:

$$T_l = N \sum_k k P_l(k) \tag{25}$$

Thus, we can define the probability that a detached node with degree k will be connected to an open connection emerging from layer l by $\frac{k}{\chi_l + T_l}$, where χ_l is the number of open connections emerging from layer l (see Fig. 2(b)).

Therefore, the conditional probability for a node with degree k to be also outside layer $l + 1$, given that it is outside layer l, is the probability that it does not connect to *any* of the χ_l open connection emerging from layer l, that is:

$$P(k, l + 1|l) = \left[1 - \frac{k}{\chi_l + T_l} \right]^{\chi_l} \approx \exp\left(-\frac{k}{1 + \frac{T_l}{\chi_l}} \right), \tag{26}$$

for large enough values of χ_l.

Thus, the probability that a node of degree k will be outside layer $l + 1$ is:

$$P_{l+1}(k) = P_l(k) P(k, l+1|l) = P_l(k) \exp\left(-\frac{k}{1 + \frac{T_l}{\chi_l}} \right) \tag{27}$$

(a)

(b)

(c)

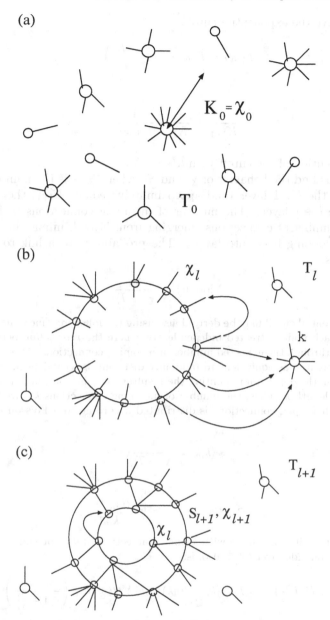

Fig. 2. Illustration of the exposure process. The large circles denote exposed layers of the giant component, while the small circles denote individual sites. The sites outside the circles have not been reached yet. (a) We begin with the highest degree node and fill out layer 1. (b) In the exposure of layer $l + 1$ any open connection emerging from layer l may connect to any open node (T_l connections) or loop back into layer l (χ_l connections). (c) The number of connections emerging from layer $l + 1$ is the difference between T_l and T_{l+1} after reducing the incoming connections S_{l+1} from layer l.

Thus we derive the exponential cutoff:

$$P_l(k) = P(k) \exp\left(-\frac{k}{K_l}\right) , \qquad (28)$$

where:

$$\frac{1}{K_{l+1}} = \frac{1}{K_l} + \frac{1}{1 + \frac{T_l}{\chi_l}} , \qquad (29)$$

gives the evolution of the cutoff with l.[7]

Now let us find the behavior of χ_l and S_l, where S_{l+1} is the number of links incoming to the $l+1$ layer (and approximately[8] equals N_{l+1}, the number of nodes in the $l+1$ layer). The number of incoming connections to layer $l+1$ equals the number of connections emerging from layer l, minus the number of connections looping back into layer l. The probability for a link to loop back into layer l is:

$$P(\text{loop}|l) = \frac{\chi_l}{\chi_l + T_l} \qquad (34)$$

---

[7] The exponential cutoff may be derived also using the following "mean field" approximation: Each node is treated independently, where the *interaction* between nodes is inserted through the expected number of incoming connections. At each node, the process is treated as equivalent to randomly distributing χ_l independent points on a line of length $\chi_l + T_l$ and counting the resultant number of points inside a *small* interval of length k. Thus, the number of incoming connections k_{in} from layer l to a node with k open connections is distributed according to a Poisson distribution with:

$$\langle k_{in} \rangle = \frac{k}{\chi_l + T_l} \chi_l , \qquad (30)$$

and:

$$P_{l+1}(k_{in}|k) = e^{-\langle k_{in} \rangle} \frac{\langle k_{in} \rangle^{k_{in}}}{k_{in}!} . \qquad (31)$$

The probability for a node with k open connections *not* to be connected to layer l, i.e. to be outside layer $l+1$ also, is:

$$P(k, l+1|l) = P_{l+1}(k_{in} = 0|k) = e^{-\langle k_{in} \rangle} = \exp\left(-\frac{k}{1 + \frac{T_l}{\chi_l}}\right) \qquad (32)$$

Thus the total probability to find a node of degree k outside layer $l+1$ is:

$$P_{l+1}(k) = P_l(k) P(k, l+1|l) = P_l(k) \exp\left(-\frac{k}{1 + \frac{T_l}{\chi_l}}\right) , \qquad (33)$$

and one obtains the exponential cutoff.

[8] This holds true assuming that almost no site in layer $l+1$ is reached by two connections from layer l. This is justified in the case where $m = 1$, and also for the first layers in case of $m > 1$.

and therefore:

$$S_{l+1} = \chi_l \left(1 - \frac{\chi_l}{\chi_l + T_l} \right).$$ (35)

The number of connections emerging from all the nodes in layer $l + 1$ is $T_l - T_{l+1}$. This is the sum of the number of incoming connections from layer l into layer $l + 1$, which is equal to S_{l+1}, and the number of outgoing connections χ_{l+1}. Therefore:

$$\chi_{l+1} = T_l - T_{l+1} - S_{l+1}$$ (36)

At this point we have the following relations: $T_{l+1}(K_{l+1})$ Equations (25) and (28), $S_{l+1}(\chi_l, T_l)$ (35), $K_{l+1}(K_l, \chi_l, T_l)$ (29), and $\chi_{l+1}(T_l, T_{l+1}, S_{l+1})$ (36). These relations may be solved numerically[9]. Note that approximate analytical results for the limit $N \to \infty$ can be found in [23,31,24][10].

4.3 Simulation

Figure 3 shows results from simulations (symbols) for the number of nodes on layer l, which can be seen to be in agreement with the analytical curves of S_l (lines). We can see that starting from a given layer $l = L$ the number of nodes decays exponentially. We believe that the layer index L is related to the radius of the graph [23,31]. It can be seen that S_l is a good approximation for the number of nodes at layer l. This is true in cases when only a small fraction of sites in each layer l have more than one incoming connection. An example for this case is when $m = 1$ so that most of the sites in the network have only one connection. Figure 4 shows results for $P_l(k)$ with similar agreement. Note the exponential cutoff which becomes stronger with l (i.e. K_l is a monotonically decreasing function of l).

It is important to note that the simulation results give the probability distribution for the giant percolation cluster, while the analytical reconstruction gives the probability distribution for the whole graph. This may explain the difference in the probability distributions for lower degrees: many low degree nodes are not connected to the giant percolation cluster and therefore the probability distribution derived from the simulation is smaller for low k.

Similar behavior was found in real Internet maps and multicast trees [14]. Deviations from theory may be attributed to correlations in node degrees [33, 30] and hierarchical structures [34], which were observed in the Internet. In

---

[9] We begin with $K_0 = K$ (the natural cutoff of the network), $\chi_0 = K_0 = K$, and $P_0(k) = ck^{-\lambda}$.

[10] An approximate analytical expression for the upper cutoff was found to be [23]:

$$K_l \sim A^{\frac{(\lambda-2)^{l-1}-1}{3-\lambda}} N^{\frac{(\lambda-2)^l}{\lambda-1}}.$$ (37)

where $A = \langle k \rangle m^{\lambda-2}/(3-\lambda) = \frac{(\lambda-1)m}{(\lambda-2)(3-\lambda)}$.

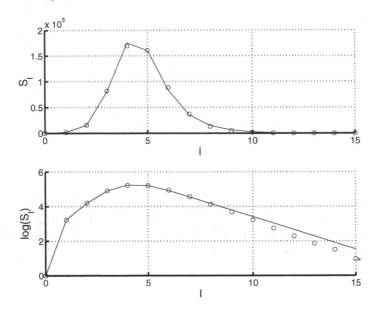

Fig. 3. Approximate number of nodes (S_l) vs. layer index l for a network with $N = 10^6$ nodes, $\lambda = 2.85$, and $m = 1$. Symbols represent simulation results while solid lines are a numerical solution for the derived recursive relations. Bottom: from the semi-log plot we see that there is an exponential decay of S_l for layers $l > L$ starting from a given layer L which we believe is related to the radius of the graph.

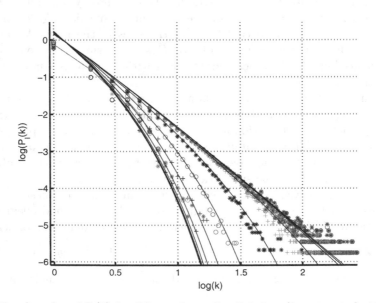

Fig. 4. Log-log plot of $P_l(k)$ for different layers $l = 0, 1, 2, ...$ (from top to bottom), for a network with $N = 10^6$ nodes, $\lambda = 2.85$, and $m = 1$. Symbols represent simulation results while solid lines are a numerical solution for the derived recursive relations.

another study, Brunet *et al.* [30] analyzed the layer structure for the Albert-Barabási model [1] and its randomized variations [35]. They found that in the Albert-Barabási model there are more nodes in the first layers due to dissortative mixing [33].

4.4 Bounds and Implications

The layer structure of the network has implications on several important topics. Since messages in a communication network travel between neighboring nodes the inter-node distances are important in understanding network performance and message routing. Another important subject is searching for nodes in a network. In [36], an efficient method for searching via the network HUBs is presented. This method is based on going up the degree sequence, from each site to a higher degree site, until the highest degree site is reached. Then, the search continues down the degree sequence to lower and lower degree sites. This method allows for a search much more efficient than a random one, as shown in [36]. However, as we show below, no search strategy based on local information can search a finite fraction of the network in less than $O(N)$ steps (with possible logarithmic corrections).

Some limits on the efficiency of such techniques can be obtained by using bounds on the structure of scale free networks as in [23]. These bounds follow from the scale free degree sequence of the network, and are independent of the exact model for network generation. Therefore, they apply to **every** network with a scale-free degree sequence.

If we consider the fraction, A, of the most most highly connected sites, their fraction of neighbors (relative to the network size), $n_1(A)$, can not exceed $\int_f^K ck^{1-\lambda}dk \approx f^{2-\lambda}m^{\lambda-1}(\lambda - 1)/(\lambda - 2)$, where f satisfies $N\int_f^K ck^{-\lambda}dk = A$. Hence, $f = mA^{1/(\lambda-1)}$, and $n_1(A) \leq mA^{(\lambda-2)/(\lambda-1)}(\lambda-1)/(\lambda-2)$. Similarly, the number of second neighbors of these sites can not exceed $n_2(A) \leq n_1(n_1(A)) \leq \left(m(\lambda - 1)/(\lambda - 2)\right)^{(2\lambda-3)/(\lambda-1)}A^{(\lambda-2)^2/(\lambda-1)^2}$. One can continue for the third nearest neighbors and so on.

Since the number of lth nearest neighbors of a group of AN sites behaves as $NA^{\left(\frac{\lambda-2}{\lambda-1}\right)^l}$ it follows that for $A \sim N^{-\epsilon}$ for any ϵ no finite number of layers can contain $O(N)$ sites in the limit $N \to \infty$. One can therefore conclude that the average distance between sites in a scale-free networks **can not be a constant**, and also that **no searching of $O(N)$ sites using less than $O(N)$ steps is possible**. The second conclusion is true, of course, only if no information other than lth nearest neighbors is allowed, and does not apply to methods such as the one suggested in [37], which can search in less than $O(N)$ step owing to the knowledge of some distance metric.

5 Random Breakdown

5.1 Description

Albert *et al.*[1] suggested to model the Internet as a scale-free network (in their original model $\lambda \approx 3$). They suggested a scenario in which nodes in the network fail randomly (due to random error or an external cause like power-failures etc.). To model this scenario, they suggested random removal of sites from the network, after which they calculated the size of the largest remaining cluster. They have compared the results of applying this process to a scale-free network to the same model on a random Erdös-Rényi (ER) graph. They found numerically that scale-free networks with $\lambda = 3$ are much more resilient to this kind of failure than ER graphs. In a previous paper [2] we have studied this problem analytically and numerically for the Molloy-Reed model [21][11]. The results are summarized below.

5.2 Theory

In the calculation of the threshold for random breakdown, the key parameter, according to (10), is the ratio second- to first-moment, κ_0, which we compute by approximating the distribution (19) to a continuum. (This approximation becomes exact for $1 \ll m \ll K$, and it preserves the essential features of the transition even for small m. Furthermore, for our continuous distribution this approximation is fairly accurate for all values of the cutoff):

$$\kappa_0 = \left(\frac{2-\lambda}{3-\lambda}\right) \frac{K^{3-\lambda} - m^{3-\lambda}}{K^{2-\lambda} - m^{2-\lambda}}. \tag{38}$$

When $K \gg m$, this may be approximated as:

$$\kappa_0 \to \left|\frac{2-\lambda}{3-\lambda}\right| \times \begin{cases} m, & \text{if } \lambda > 3; \\ m^{\lambda-2}K^{3-\lambda}, & \text{if } 2 < \lambda < 3; \\ K, & \text{if } 1 < \lambda < 2. \end{cases} \tag{39}$$

We see that for $\lambda > 3$ the ratio κ_0 is finite and there is a percolation transition at $1 - p_c \approx \left(\frac{\lambda-2}{\lambda-3}m - 1\right)^{-1}$: for $p > p_c$ the spanning cluster is fragmented and the network is destroyed. However, for $\lambda < 3$ the ratio κ_0 diverges with K and so $p_c \to 1$ when $K \to \infty$ (or $N \to \infty$). The percolation transition does not take place: a spanning cluster exists for arbitrarily large fractions of breakdown, $p < 1$. In *finite* systems a transition is always observed, though for $\lambda < 3$ the transition threshold is exceedingly high. For the case of the Internet ($\lambda \approx 5/2$), we have $\kappa_0 \approx K^{1/2} \approx N^{1/3}$. Considering the enormous size of the Internet, $N > 10^6$, one needs to remove over 99% of the nodes before the spanning cluster collapses. For $\lambda > 4$ calculation of κ shows that it is lower than 2 even before

---

[11] See Sect. 3.

the breakdown occurs. For $\lambda > 4$ and $m = 1$ the network will consist of only finite clusters and no spanning cluster exists (This is reminiscent of the result for $\lambda > 3.478...$ found in [22], where the different threshold stems from the different distribution taken). For $m \geq 2$, a spanning cluster exists for every λ.

The size of the spanning cluster can also be measured using the methods suggested in [5]. The distribution (6) can be substituted into (12) and the calculation of the other generating functions can be done using this distribution, giving the size of the spanning cluster relative to the undisturbed network. An alternative method [3] is to build a new generating function, taking the fraction of removed sites into account:

$$G_0(x) = \sum_{k=0}^{\infty} P(k)q(k)x^k, \tag{40}$$

where $q(k) = 1 - p(k)$ is the probability that a site of degree k is not removed. This equation replaces (12). For random breakdown $q(k) = 1 - p$ is independent of k. The size of the infinite cluster is then given by:

$$P_\infty = G_0(1) - G_0(u), \tag{41}$$

where u is the smallest positive solution of:

$$u = 1 - G_1(1) + G_1(u). \tag{42}$$

5.3 Simulation

To simulate the process of random breakdown of a fraction p of the sites, pN sites are chosen at random. Those sites are removed and all bonds connected to them are also removed. The links are followed to remove all connections to those sites from the other end of the link.

After the removal of the sites, the size of the spanning cluster (if exists) and the finite clusters is measured. The measurement is performed using breadth first search (BFS) from each site which hasn't been marked as probed. This method uncovers the graph cluster by cluster, and can also be used to find the distance (i.e. the shortest path) between a site on the cluster and all the other sites on that cluster. Numerical results for the largest cluster in random breakdown can be seen in Fig. 5. Numerical and analytical results of p_c, for different values of λ and m, are shown in Fig. 6.

6 Intentional Attack

6.1 Description

Another model suggested in [1] is that of intentional attack on the most highly connected nodes in the network. In this model an attacker (e.g. computer hackers

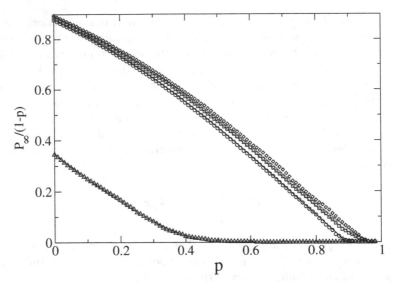

Fig. 5. Percolation transition for networks with power-law degree distribution. Plotted is the fraction of nodes (relative to the number of remaining nodes) that remain in the spanning cluster after breakdown of a fraction p of all nodes, $P_\infty(p)/(1-p)$, as a function of p, for $\lambda = 3.5 (\triangle)$ and $\lambda = 2.5$ (other symbols), as obtained from computer simulations of up to $N = 10^6$. In the former case, it can be seen that for $p > p_c \approx 0.5$ the spanning cluster disintegrates and the network becomes fragmented. However, for $\lambda = 2.5$ (the case of the Internet), the spanning cluster persists up to nearly 100% breakdown. The different curves for $K = 100$ ($\circ$), 400 ($\square$), and 2000 ($\diamond$) illustrate the finite size-effect: the transition exists only for finite networks, while the critical threshold p_c approaches 100% as the networks grow in size.

trying to cause damage to the network) leashes by some means an intentional attack on the most highly connected sites in the network, causing breakdown of those sites. An attack such as this will expectedly cause more extensive damage than a random attack, and as will be shown below, can cause even networks resilient to random breakdown to collapse [3,4].

6.2 Theory

Consider now intentional attack, or sabotage [1], whereby a fraction p of the sites with the highest degree is removed. (The links emanating from the sites are removed as well.) This has the following effect: (a) the cutoff degree K reduces to some new value, $\tilde{K} < K$, and (b) the degree distribution of the remaining sites is no longer the original distribution, but is changed, because of the removal of many of their links. The upper cutoff K before the attack may be estimated from (21). Similarly, the new cutoff $\tilde{K}$, after the attack, can be estimated from

$$\sum_{k=\tilde{K}}^{K} P(k) = \sum_{k=\tilde{K}}^{\infty} P(k) - \frac{1}{N} = p \,. \tag{43}$$

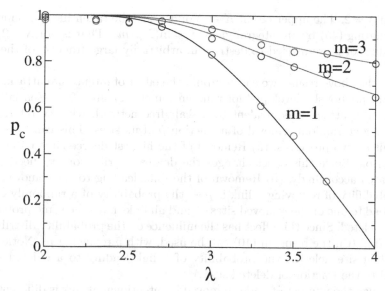

Fig. 6. The Percolation threshold p_c for different values of λ and m, for random breakdown of nodes. Circles represent simulation results for $N = 5 \cdot 10^5$ nodes, while solid lines give analytical results. The deviation from $p_c = 1$ for $\lambda \le 3$ is due to finite size effects (after K. Erez [38]).

If the size of the system is large, $N \gg 1/p$, the original cutoff K may be safely ignored. We can then obtain $\tilde{K}$ approximately by replacing the sum with an integral:

$$\tilde{K} = m p^{1/(1-\lambda)} \ . \tag{44}$$

We estimate the impact of the attack on the distribution of the remaining sites as follows. The removal of a fraction p of the sites with the highest degree results in a random removal of links from the remaining sites — links that had connected the removed sites with the remaining sites. The probability $\tilde{p}$ of a link leading to a deleted site equals the ratio of the number of links belonging to deleted sites to the total number of links:

$$\tilde{p} = \sum_{k=\tilde{K}}^{K} \frac{k P(k)}{\langle k_0 \rangle}, \tag{45}$$

where $\langle k_0 \rangle$ is the initial average degree. With the usual continuous approximation, and neglecting K, this yields

$$\tilde{p} = \left(\frac{\tilde{K}}{m} \right)^{2-\lambda} = p^{(2-\lambda)/(1-\lambda)} \ , \tag{46}$$

for $\lambda > 2$. For $\lambda = 2$, $\tilde{p} \to 1$, since just a few nodes of very high degree control the entire connectedness of the system. Indeed, consider a finite system of N

sites and $\lambda = 2$. The upper cutoff $K \approx N$ must then be taken into account, and approximating (45) by an integral yields $\tilde{p} = \ln(Np/m)$. That is, for $\lambda = 2$, very small values of p are needed to destroy an arbitrarily large fraction of the links as $N \to \infty$.

With the above results we can compute the effect of intentional attack, using the theory previously developed for random removal of sites [2]. Essentially, the network after attack is equivalent to a scale-free network with cutoff $\tilde{K}$, that has undergone random removal of a fraction $\tilde{p}$ of its sites. This can be seen as the result of two processes: (a) Removal of the highest degree sites reduces the upper cutoff. Since this effect changes the degree distribution, κ_0 needs to be recalculated accordingly. (b) Removal of the links leading to the removed sites. The probability of removing a link is $\tilde{p}$ — the probability of a randomly chosen link to lead to one of the removed sites — and all links have the same probability of being deleted. Since this effect has the influence on the probability distribution described in (6), the result in (10) can be used, with $\tilde{p}$ replacing p. (Notice that for random site deletion the probability of a link leading to a deleted site is identical to the fraction of deleted sites.)

Although the number of nodes removed in intentional attack is different than in the random breakdown model, this affects the size of the spanning cluster (see below) but not the critical point. This is because the transition point is defined as the point where the spanning cluster becomes a finite fraction of the whole network. A finite fraction of the remaining nodes is also a finite fraction of the original network, so the difference has no effect on p_c.

We therefore use (10) and (38), but with $\tilde{p} = (\tilde{K}/m)^{2-\lambda}$ and $\tilde{K}$ replacing p_c and K. This yields the equation:

$$(\tilde{K}/m)^{2-\lambda} - 2 = \frac{2-\lambda}{3-\lambda} m[(\tilde{K}/m)^{3-\lambda} - 1] , \qquad (47)$$

which can be solved numerically to obtain $\tilde{K}(m,\lambda)$, and then $p_c(m,\lambda)$ can be retrieved from (44). In Fig. 7 we plot p_c — the critical fraction of sites needed to be removed in the targeted attack strategy to disrupt the network — computed in this fashion, and compared to results from numerical simulations. A phase transition exists (at a finite and small p_c) for all $\lambda > 2$. The decline in p_c for large λ is explained from the fact that as λ increases the spanning cluster becomes smaller in size, even before attack. (Furthermore, for $m < 2$ the original network is disconnected for some large enough λ.) The decline in p_c as $\lambda \to 2$ results from the critically high degree of just a few sites: their removal disrupts the whole network. This was already argued in [1]. We note that for infinite systems $p_c \to 0$ as $\lambda \to 2$. The critical fraction p_c is rather sensitive to the lower degree cutoff m. As shown in Fig. 7, for larger m the networks are more robust, though they still undergo a transition at a finite p_c.

To calculate $P_\infty(p)$ one can use the generating function method in conjunction with $p(k) = \Theta(\tilde{K} - k)$ [3] (to be more exact, $p(k) = \Theta(\tilde{K} - k) + A\delta_{k,\tilde{K}}$, where A is chosen such that $\sum_{k=0}^{\infty} p(k) = 1 - p$). An alternative method is to use the generating functions in conjunction with the new distribution after the attack.

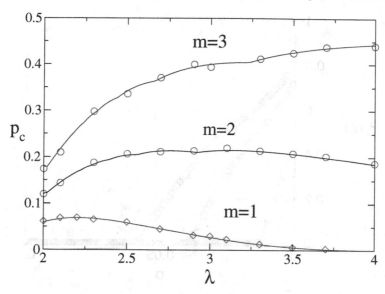

Fig. 7. Critical probability p_c as a function of λ, for networks of size $N = 5 \cdot 10^5$ under intentional attack (symbols) . Solid lines represent the analytical solution, obtained from (43–47), after K. Erez [38].

6.3 Simulation

To simulate the intentional attack the same method as in random removal of sites is applied. The only difference is that the removed sites are not selected randomly, but chosen as the highest degree nodes. Sorting the sites would take $O(N \ln N)$ operations, however since no site has degree greater than N an array of lists of sites for each degree can be produced, and the sites can be removed starting from the highest degree downwards. This only requires $O(N)$ operations, and therefore the entire execution of the program takes only order of $O(N)$ operations. Analytical and numerical result are shown in Fig. 8.

A somewhat different result would be achieved if the list is updated every time a site is deleted, since this influences the degree of the other sites. However, simulations show that this change has a small influence on the results, and the above analysis (in the previous section) is based on the static picture. Therefore, this is the method used to obtain the results given here.

7 Critical Exponents

7.1 Introduction

In the study of percolation [17,16] and many other critical phenomena it is very common to study the behavior of the system near the critical point. Many parameters of the system (such as the order parameter – in our case P_∞) display

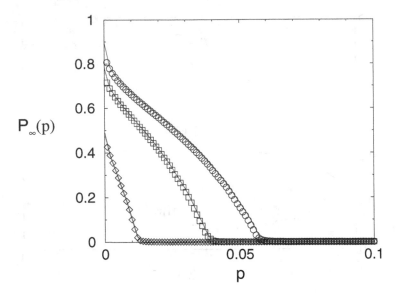

Fig. 8. Fraction of sites belonging to the spanning cluster, P_∞, as a function of the fraction of removed sites, p, for networks with $\lambda = 2.5$ (circles), $\lambda = 2.8$ (squares), and $\lambda = 3.3$ (diamonds). Lines represent the analytical result from (17) and (18). Both the simulation and analysis are for system size $N = 500,000$.

a power-law behavior near the critical point. That is, many properties of the system behave as $(p - p_c)^\omega$ near the transition point with a critical exponent ω. At the transition point itself, some properties also follow a power law in another parameter.

It is well known [16,17] that many of the critical exponents are universal. That is, the value of the exponent is independent of the exact microscopic setup, and depends only on the large scale topological properties of the problem (*e.g.* the embedding dimension). In many systems there is an upper critical dimension above which the behavior of the critical exponents is independent of the dimension and all obtain their "mean-field" values. In the case of percolation the critical dimension is $d_c = 6$ [16,17].

In the following we calculate the critical exponents for percolation in scale free networks and show that those networks, despite their mean-field nature, present a new universality class [39,31] (in fact, a new set of universality classes).

7.2 Infinite Cluster Size

In [3,5] the generating functions $G_0(x)$ and $G_1(x)$ are built for various distributions [(12) and (13)]. Let $H_1(x)$ be the generating function for the probability of reaching a branch of a given size by following a link. After a dilution of a fraction p of the sites (the remaining concentration is $q = 1 - p$), $H_1(x)$ satisfies the self-consistent equation

$$H_1(x) = 1 - q + qxG_1(H_1(x)) . \tag{48}$$

Since $G_0(x)$ is the generating function for the degree of a site, the generating function for the probability of a site to belong to an n-site cluster is

$$H_0(x) = 1 - q + qxG_0(H_1(x)) . \tag{49}$$

$H_0(1)$ is the probability that a site belongs to a cluster of any *finite* size. Thus, below the percolation transition $H_0(1) = 1$, while above the transition there is a finite probability that a site belongs to the infinite spanning cluster: $P_\infty = 1 - H_0(1)$. It follows that

$$P_\infty(q) = q \left(1 - \sum_{k=0}^{\infty} P(k)u^k \right) , \tag{50}$$

where $u \equiv H_1(1)$ is the smallest positive root of

$$u = 1 - q + \frac{q}{\langle k \rangle} \sum_{k=0}^{\infty} kP(k)u^{k-1} . \tag{51}$$

This equation can be solved numerically and the solution may be substituted into (50), yielding the size of the spanning cluster in a network of arbitrary degree distribution, at dilution q [3].

We now compute the order parameter critical exponent β. Near criticality the probability of belonging to the spanning cluster behaves as $P_\infty \sim (q - q_c)^\beta$. For infinite-dimensional systems (such as a Cayley tree) it is known that $\beta = 1$ [17, 16,40]. This regular mean-field result is not always valid, however, for scale-free networks. Equation (50) has no special behavior at $q = q_c$; the singular behavior comes from u. Also, at criticality $P_\infty = 0$ and (50) imply that $u = 1$. We therefore examine (51) for $u = 1 - \epsilon$ and $q = q_c + \delta$:

$$1 - \epsilon = 1 - q_c - \delta + \frac{(q_c + \delta)}{\langle k \rangle} \sum_{k=0}^{\infty} kP(k)(1 - \epsilon)^{k-1}. \tag{52}$$

The sum in (52) has the asymptotic form

$$\sum_{k=0}^{\infty} kP(k)(1 - \epsilon)^{k-1} \sim$$

$$\langle k \rangle - \langle k(k-1) \rangle \epsilon + \frac{1}{2} \langle k(k-1)(k-2) \rangle \epsilon^2 + \cdots + c\Gamma(2 - \lambda)\epsilon^{\lambda-2} , \tag{53}$$

where the highest-order analytic term is $O(\epsilon^n)$, $n = \lfloor \lambda - 2 \rfloor$. Using this in (52), with $q_c = 1/(\kappa - 1) = \langle k \rangle / \langle k(k-1) \rangle$, we get

$$\frac{\langle k(k-1) \rangle^2}{\langle k \rangle} \delta = \frac{1}{2} \langle k(k-1)(k-2) \rangle \epsilon + \cdots + c\Gamma(2 - \lambda)\epsilon^{\lambda-3} . \tag{54}$$

The divergence of δ as $\lambda < 3$ confirms the vanishing threshold of the phase transition in that regime. Thus, in the case $\lambda > 3$, keeping only the dominant term as $\epsilon \to 0$, (54) implies

$$
\epsilon \sim \begin{cases} \left(\frac{\langle k(k-1) \rangle^2}{c\langle k \rangle \Gamma(2-\lambda)} \right)^{\frac{1}{\lambda-3}} \delta^{\frac{1}{\lambda-3}} & 3 < \lambda < 4, \\ \frac{2\langle k(k-1)\rangle^2}{\langle k \rangle \langle k(k-1)(k-2)\rangle} \delta & \lambda > 4. \end{cases} \tag{55}
$$

Returning to P_∞, (50), we see that the singular contribution in ϵ is dominant only for the range of $\lambda < 2$. For $\lambda > 3$, we find $P_\infty \sim q_c \langle k \rangle \epsilon \sim (q - q_c)^\beta$. Comparing this to (55) we finally obtain

$$
\beta = \begin{cases} \frac{1}{\lambda-3} & 3 < \lambda < 4, \\ 1 & \lambda > 4. \end{cases} \tag{56}
$$

We see that the order parameter exponent β attains its regular mean-field value only for $\lambda > 4$. Moreover, since $\beta > 1$ for $\lambda < 4$ the percolation transition is smooth and is higher than 2nd-order: for $3 + \frac{1}{n-1} < \lambda < 3 + \frac{1}{n-2}$ the transition is of the nth-order (since all derivatives up to the $n - 1$th are continuous). The result (56) has been reported before in [4], and also found independently in a different but related model of virus spreading [41,42]. The existence of an infinite-order phase transition at $\lambda = 3$ for growing networks of the Albert-Barabási model, has been reported elsewhere [43,44]. These examples suggest that the critical exponents are universal and not model-dependent but depend only on λ.

For networks with $\lambda < 3$ the transition still exists, though at a vanishing threshold, $q_c = 0$. The sum in (52) becomes:

$$
\sum_{k=0}^{\infty} kP(k)u^{k-1} \sim \langle k \rangle + c\Gamma(2-\lambda)\epsilon^{\lambda-2} . \tag{57}
$$

Using this in conjunction with (51), and remembering that here $q_c = 0$ and therefore $q = \delta$, leads to

$$
\epsilon = \left(\frac{-c\Gamma(2-\lambda)}{\langle k \rangle} \right)^{\frac{1}{3-\lambda}} \delta^{\frac{1}{3-\lambda}} , \tag{58}
$$

which implies

$$
\beta = \frac{1}{3 - \lambda} , \qquad 2 < \lambda < 3 . \tag{59}
$$

In other words, the transition in $2 < \lambda < 3$ is a mirror image of the transition in $3 < \lambda < 4$. An important difference is that $q_c = 0$ is not λ-dependent in $2 < \lambda < 3$, and the amplitude of P_∞ diverges as $\lambda \to 2$ (but remains finite as $\lambda \to 4$).

7.3 Finite Cluster Size Distribution

Next, we determine the exponents τ and σ, which determine n_s, the number of clusters of size s.

In [5] it was shown that for a random graph of arbitrary degree distribution the finite clusters follow the usual scaling form:

$$n_s \sim s^{-\tau} e^{-s/s^*} . \tag{60}$$

At criticality $s^* \sim |q - q_c|^{-\sigma}$ diverges and the tail of the distribution behaves as a power law. We now derive the exponent τ. The probability that a site belongs to an s-cluster is $p_s = s n_s \sim s^{1-\tau}$, and is generated by H_0:

$$H_0(x) = \sum p_s x^s . \tag{61}$$

The singular behavior of $H_0(x)$ stems from $H_1(x)$, as can be seen from (49). $H_1(x)$ itself can be expanded from (48), by using the asymptotic form (53) of G_1. We let $x = 1 - \epsilon$, as before, but analyze at the critical point, $q = q_c$. With the notation $\phi(\epsilon) = 1 - H_1(1 - \epsilon)$, we finally get (note that at criticality $H_1(1) = 1$):

$$-\phi = -q_c + (1 - \epsilon) q_c \left[1 - \frac{\phi}{q_c} + \frac{\langle k(k-1)(k-2) \rangle}{2 \langle k \rangle} \phi^2 + \cdots + c \frac{\Gamma(2 - \lambda)}{\langle k \rangle} \phi^{\lambda - 2} \right] . \tag{62}$$

From this relation we extract the singular behavior of H_0: $\phi \sim \epsilon^y$. Then, using Tauberian theorems [45] it follows that $p_s \sim s^{-1-y}$, hence $\tau = 2 + y$.

For $\lambda > 4$ the term proportional to $\phi^{\lambda - 2}$ in (62) may be neglected. The linear term $\epsilon \phi$ may be neglected as well, due to the factor ϵ. This leads to $\phi \sim \epsilon^{1/2}$ and to the usual mean-field result

$$\tau = \frac{5}{2} , \qquad \lambda > 4 . \tag{63}$$

For $\lambda < 4$, the terms proportional to $\epsilon \phi$, ϕ^2 may be neglected, leading to $\phi \sim \epsilon^{1/(\lambda-2)}$ and [39]

$$\tau = 2 + \frac{1}{\lambda - 2} = \frac{2\lambda - 3}{\lambda - 2} , \qquad 2 < \lambda < 4 . \tag{64}$$

Note that for $2 < \lambda < 3$ the percolation threshold is strictly $q_c = 0$. In that case we analyze at $q = \delta$ small but fixed, taking the limit $\delta \to 0$ at the very end. For the case $2 < \lambda < 3$, τ in (64) represents the singularity of the distribution of branch sizes. For the distribution of cluster sizes in this range one has to consider the singularity of x in (49) leading to $\tau = 3$ for this range.

For growing networks of the Albert-Barabási model with $\lambda = 3$, it has been shown that $s n_s \propto (s \ln s)^{-2}$ [44]. This is consistent with $\tau = 3$ plus a logarithmic correction. Related results for scale free *trees* have been presented in [46].

At the transition point the largest cluster, S can be obtained from the finite cluster distribution by taking the integral over the tail of the distribution to be equal $1/N$. This results in

$$S \propto N^{\tau-1} = N^{(\lambda-2)/(\lambda-1)}. \tag{65}$$

For $\lambda = 4$ this reduces to the known $N^{2/3}$, termed by Erdös the "double jump", due to the transition of the largest cluster from order $\ln N$ for $q < q_c$, to $N^{2/3}$ at $q = q_c$, to order N at $q > q_c$ [20]. For $\lambda \to 3$, $S \propto N^{1/2}$. It is not yet clear whether the results have a meaningful interpretation for $\lambda < 3$.

7.4 Finite Cluster Size Cutoff

The critical exponent σ, for the cutoff cluster size, can also be derived. Finite-size scaling arguments predict [17] that

$$q_c(\infty) - q_c(N) \sim N^{-\frac{1}{d\nu}} = N^{-\frac{\sigma}{\tau-1}} , \tag{66}$$

where N is the number of sites in the network, ν is the correlation length critical exponent: $\xi \sim (q - q_c)^{-\nu}$, and d is the dimensionality of the embedding space. Using a continuous approximation of the distribution (19) one obtains [2]

$$\kappa \approx \left(\frac{2-\lambda}{3-\lambda}\right) \frac{K^{3-\lambda} - m^{3-\lambda}}{K^{2-\lambda} - m^{2-\lambda}} , \tag{67}$$

where $K \sim N^{1/(\lambda-1)}$ is the largest site degree of the network. For $3 < \lambda < 4$, this and (10) yield

$$q_c(\infty) - q_c(N) \sim \Delta\kappa \sim K^{3-\lambda} \sim N^{\frac{3-\lambda}{\lambda-1}} , \tag{68}$$

which in conjunction with (66) leads to

$$\sigma = \frac{\lambda-3}{\lambda-2} , \qquad 3 < \lambda < 4 . \tag{69}$$

For $\lambda > 4$ we recover the regular mean-field result $\sigma = 1/2$. Note that (66), (56), (64) are consistent with the known scaling relation: $\sigma\beta = \tau - 2$ [17,16,40]. For $2 < \lambda < 3$, $q_c(\infty) = 0$ and $q_c(N) \sim K^{\lambda-3} \sim N^{(\lambda-3)/(\lambda-1)}$ and therefore

$$\sigma = \frac{3-\lambda}{\lambda-2} , \qquad 2 < \lambda < 3 , \tag{70}$$

again consistent with the scaling relation $\sigma\beta = \tau - 2$ (cf (59)).

7.5 Fractal Dimension

It is well known that on a random network in the well connected regime, the average distance between sites is of order $\log_k N$ [20,47,5]. Even smaller distances have been shown to hold for scale-free networks [23,24]. However, the diluted case is essentially the same as infinite-dimensional percolation. In this case, there is no notion of geometrical distance (since the graph is not embedded in an Euclidean space), but only of a distance along the graph (which is the shortest distance

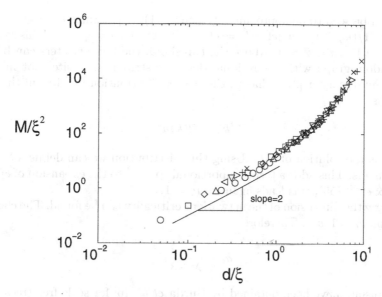

Fig. 9. Scaled mass (number of sites) M as a function of scaled distance d on the spanning cluster, for Erdos-Renyi graphs above criticality ($p < p_c$, $\langle k \rangle > 1$). Different symbols correspond to different values of $\langle k \rangle$ (1.01 − 1.06). The correlation length is $\xi = |p - p_c|^{-1}$. Note that for $d/\xi < 1$, the slope is 2, corresponding to the behavior in the critical regime, while for $d/\xi > 1$, M grows exponentially with d, corresponding to the well connected regime.

along bonds). It is known from infinite-dimensional percolation theory that the chemical fractal dimension at criticality is $d_l = 2$ [16]. Therefore the average (chemical) distance d between pairs of sites on the spanning cluster at criticality behaves as (Fig. 9)

$$d \sim \sqrt{M},\qquad(71)$$

where M is the number of sites in the spanning cluster. This is analogous to percolation in finite dimensions, where in length-scales smaller than the correlation length the cluster is a fractal with dimension d_l and above the correlation length the cluster is homogeneous and has the dimension of the embedding space. In our infinite-dimensional case, the crossover between these two behaviors occurs around the correlation length $\xi_l \approx |p_c - p|^{-\nu_l}$.

Next, we calculate ν_l for scale free networks with $3 < \lambda < 4$. Below the transition all clusters are finite and almost all finite clusters are trees. The correlation length can be defined using the formula [16]:

$$\xi_l^2 = \frac{\sum l^2 g(l)}{\sum g(l)}.\qquad(72)$$

Where $g(l)$, the correlation function, is the mean number of sites on the same cluster at distance l from an arbitrary occupied site. The number of sites in the

l shell can be seen to be approximately $\langle k \rangle (\kappa - 1)^{l-1}$ [5]. Since $\kappa - 1 = (\kappa_0 - 1)q$ and $q_c = 1/(\kappa_0 - 1)$ we get $g(l) = c(1 - \delta)^l$, where $\delta = q - q_c$. This leads to $\xi_l \sim (q - q_c)^{-1}$, *i.e.* $\nu_l = 1$. Above the threshold, the finite clusters can be seen as a random graph with the residual degree distribution of sites not included in the infinite cluster [21]. That is, the degree distribution for sites in the finite clusters is

$$P_r(k) = P(k)u^k, \tag{73}$$

where u is the solution of (51). Using this distribution we can define κ_r for the finite clusters. This adds a term proportional to $\epsilon^{\lambda-3}$ to the expansion of ξ_l. But, since $\delta \propto \epsilon^{\lambda-3}$ (55), this leads again to $\nu_l = 1$.

Using ν the dimension of the network at criticality can be found. The chemical dimension $d_l = 1/\sigma\nu_l$. Therefore,

$$d_l = \frac{\lambda - 2}{\lambda - 3}. \tag{74}$$

Similar results have been obtained by Burda *et al.*[46] for scale-free trees. Since every path when embedded in a space above the critical dimension can be seen as a random walk it is known that $\nu = \nu_l/2$ [16]. Therefore, the fractal dimension is,

$$d_f = \frac{1}{\nu\sigma} = 2\frac{\lambda - 2}{\lambda - 3}. \tag{75}$$

The dimension of the embedding space, which is the upper critical dimension, is:

$$d_c = \frac{1}{\nu\sigma(\tau - 1)} = 2\frac{\lambda - 1}{\lambda - 3}. \tag{76}$$

Those dimensions reduce to the known 2, 4, and 6, respectively, for $\lambda = 4$.

A direct method for calculating the chemical dimension is also possible. Denoting the generating function of the number of sites on the lth layer of some branch, as $N_l(x)$, we get

$$N_{l+1}(x) = G_1(N_l(x)) . \tag{77}$$

We are interested in the behavior of the average number of sites at a chemical distance l for those branches the have at least l layers. Since we expand exactly at criticality, the average branching factor is exactly 1, and therefore $N_l(1) = 1$ for any l. Therefore, A_l, the average number of sites for surviving branches is

$$A_l = \frac{1}{1 - N_l(0)} , \tag{78}$$

since $N_l(0)$ is the probability of the branching process to die out before the lth layer. At criticality the branching process will die out with probability $N_l(0) \to 1$

as $l \to \infty$, and therefore for large l we can take $N_l(0) = 1 - \epsilon_l$. Expanding G_1 at criticality one obtains ((52) and (53), with $\delta = 0$)

$$G_1(1 - \epsilon) = 1 - \epsilon + \frac{c\Gamma(2 - \lambda)}{\langle k^2 \rangle - \langle k \rangle} \epsilon^{\lambda - 2} + \ldots . \tag{79}$$

Substituting $N_l(0) = 1 - \epsilon_l$ into (77) one obtains

$$1 - \epsilon_{l+1} = 1 - \epsilon_l + \frac{c\Gamma(2 - \lambda)}{\langle k^2 \rangle - \langle k \rangle} \epsilon_l^{\lambda - 2} + \ldots . \tag{80}$$

Guessing a solution of the form $\epsilon_l \approx Bl^{-d}$ we get

$$B(l + 1)^{-d} \approx B(l^{-d} - dl^{-d-1}) = Bl^{-d} - \frac{c\Gamma(2 - \lambda)}{\langle k^2 \rangle - \langle k \rangle}(Bl^{-d})^{\lambda - 2} . \tag{81}$$

implying that $d = 1/(\lambda - 3)$, $\epsilon_l \sim l^{-d}$, and $N_l(0) \sim 1 - Dl^{-d}$ for some constant D [12]. The average number of sites for surviving branches at layer l is thus:

$$A_l = \frac{1}{\epsilon_l} \sim l^d . \tag{82}$$

Noting that the mass of the branch is the sum of the layers up to the lth one, we get $d_l = d + 1 = (\lambda - 2)/(\lambda - 3)$, similar to the results obtained in (74). The scaling relation therefore also proves to be correct in this case.

8 Conclusions

We have studied the structural properties and stability of scale-free graphs. We define a "layer" in a network as the set of nodes at a given distance from a chosen node. We have found that the degree distribution of the nodes of a scale free network at each layer obeys a power law with an exponential cutoff. We derived equations for this exponential cutoff and compared them with simulations. We also model the behavior of the number of nodes at each layer, and explain the observed exponential decay in the outer layers of the network. Our findings may have dual importance. First, they can help in devising better network algorithms that take advantage of the network structure [48]. Second, our analytical findings suggest a simple local test for the validity of the power law model as an exact model of the Internet [14]. Further parameters of the network such as betweenness centrality [49,50] should also be taken into account when efficient algorithms are considered.

---

[12] The exponent $d = 1/(\lambda - 3)$ is known in directed percolation as the *survivability* exponent. $\epsilon_l \sim l^{-d}$ gives the probability for a percolation cluster to survive up to chemical layer l at criticality. It may be proved by scaling relations that $d = \frac{\beta}{\nu}$, which is also the case here.

Furthermore, we have shown that many percolation properties of those graphs are different from generalized random graphs and from the regular mean-field percolation, due to their diverging moment. We have shown that scale free graphs with $2 < \lambda < 3$ are robust to random breakdown of almost 100% of the nodes, making this a favorable design for unmanaged networks, like the Internet, where nodes can be disconnected or fail unexpectedly. However, those networks are vulnerable to intentional attack on the most important nodes. The lack of percolation threshold also makes those networks sensitive to virus propagation [41,51], while their attack vulnerability makes them amenable to targeted immunization of the highest degree nodes [52,53]. An efficient strategy for the immunization of complex networks without any global information on the network topology has been introduced in [54,31].

We have also studied the behavior of these networks near the percolation transition. We have shown that scale free networks with $\lambda > 4$ near the percolation transition behave similar to infinite dimensional percolation theory, with $M \propto l^2$, where there is a crossover to the exponential behavior, $M \propto k^l$, far from the threshold. Thus, networks near the critical point become sparser, and communication becomes inefficient, as packets have to travel a long distance (many routers) on the way to their destination.

We have also shown that the critical exponents behave differently in scale-free networks than in regular networks both in the regime $3 < \lambda < 4$ where a transition occurs in a finite p_c, as well as in the regime $2 < \lambda < 3$ where $p_c \to 0$. In particular $\beta > 1$ in both those regimes, making the transition of higher order rather than of second order.

Acknowledgments

Support from the Israel Science Foundation and the NSF are gratefully acknowledged (DbA). The authors wish to thank Danny Dolev, Osnat Mokryn, Yuval Shavitt and Keren Erez for useful discussions.

References

1. R. Albert, H. Jeong, and A. L. Barabási, *Nature*, **406**, 6794, 378 (2000).
2. R. Cohen, K. Erez, D. ben-Avraham, and S. Havlin, *Phys. Rev. Lett.* **85**, 4626 (2000).
3. D. S. Callaway, M. E. J. Newman, S. H. Strogatz, and D. J. Watts, *Phys. Rev. Lett.* **85**, 5468 (2000)..
4. R. Cohen, K. Erez, D. ben-Avraham, and S. Havlin, *Phys. Rev. Lett.* **86**, 3682 (2001).
5. M. E. J. Newman, S. H. Strogatz, and D. J. Watts *Phys. Rev. E*, **64**, 026118 (2001).
6. M. Faloutsos, P. Faloutsos, and C. Faloutsos, *ACM SIGCOMM '99 Comput. Commun.* Rev. 29, 251 (1999).
7. A. Broder, R. Kumar, F. Maghoul, P. Raghavan, S. Rajagopalan, R. Stata, A. Tomkins, and J. Wiener, *Computer Networks* **33**, 309 (2000).

8. H. Jeong, B. Tombor, R. Albert, Z. N. Oltvai and A.-L. Barabási, *Nature*, **407**, 651 (2000).
9. R. Albert and A.-L. Barabasi, *Rev. of Mod. Phys.* **74**, 47 (2002).
10. S. N. Dorogovtsev, and J. F. F. Mendes, *Adv. in Phys.*, 51 (4), (2002).
11. M. E. J. Newman, *SIAM Review* **45**, 167 (2003).
12. J. F. F. Mendes, S. N. Dorogovtsev and A. F. Ioffe, *Evolution of Networks: From Biological Nets to the Internet and WWW*, (Oxford University Press, 2003).
13. R. Pastor-Satorras and A. Vespignani, *Evolution and Structure of the Internet : A Statistical Physics Approach*, (Cambridge University Press, 2003).
14. R. Cohen, D. Dolev, S. Havlin, T. Kalisky, O. Mokryn and Y. Shavitt, *cond-mat/0305582*.
15. H. Jeong, S. Mason, A.-L. Barabási and Z. N. Oltvai *Nature*, **411**, 41 (2001).
16. A. Bunde, and S. Havlin (editors), *Fractals and Disordered System* (Springer, New York, 1996).
17. D. Stauffer and A. Aharony, *Introduction to Percolation Theory*, 2nd edition (Taylor and Francis, London, 1991).
18. W. Feller, *An Introduction to Probability Theory and Its Applications* (John Wiley & Sons).
19. M. Molloy and B. Reed, *Random Structures and Algorithms* **6**, 161 (1995).
20. B. Bollobás, *Random Graphs.* pp. 123-136 (Academic Press, London, 1985).
21. M. Molloy and B. Reed, *Combinatorics, Probability and Computing* **7**, 295 (1998).
22. W. Aiello, F. Chung and L. Lu, *Proc. 32nd ACM Symp. Theor. Comp.*, (2000).
23. R. Cohen and S. Havlin, *Phys. Rev. Lett*, **90**, 058701 (2003).
24. S.N. Dorogovtsev, J. F. F. Mendes, and A.N. Samukhin, *Nuclear Physics B* **653** (3), 307 (2003).
25. B. Bollobas and O. Riordan, in *Handbook of Graphs and Networks*, Eds. S. Bornholdt and H. G. Schuster, (Wiley-VCH, 2002).
26. B. Bollobás, *Europ. J. Combinatorics* **1**, 311-316 (1980).
27. A. F. Rozenfeld, R. Cohen, D. ben-Avraham and S. Havlin, *Phys. Rev. Lett.* **89**, 218701 (2002).
28. Z. Burda and A. Krzywicki, *Phys. Rev. E* **67**, 046118 (2003).
29. C. P. Warren, L. M. Sander and I. M. Sokolov, *Phys. Rev. E* **66**, 56105 (2002).
30. R. Xulvi-Brunet, W. Pietsch and I.M. Sokolov, *Phys. Rev. E* **68**, 036119 (2003).
31. R. Cohen, S. Havlin and D. ben-Avraham, Chap. 4 in *Handbook of Graphs and Networks*, Eds. S. Bornholdt and H. G. Schuster, (Wiley-VCH, 2002).
32. S. N. Dorogovtsev, J. F. F. Mendes, and A. N. Samukhin, *Phys. Rev. E* **63**, 062101 (2001).
33. M. E. J. Newman, *Phys. Rev. Lett* **89**, 208701 (2002).
34. A. Vazquez and R. Pastor-Satorras and A. Vespignani *cond-mat/0206084* (2002).
35. S. Maslov and K. Sneppen, *Science* **296**, 910 (2002).
36. L. A. Adamic, R. M. Lukose, A. R. Punyani and B. A. Huberman, *Phys. Rev. E* **64**, 046135 (2001).
37. D. J. Watts, P. S. Dudds and M. E. J. Newman, *Science* **296**, 1302 (2002).
38. K. Erez, MSc. thesis (2001), Bar-Ilan University.
39. R. Cohen, D. ben-Avraham and S. Havlin, *Phys. Rev. E* **66**, 036113 (2002).
40. D. ben-Avraham and S. Havlin, *Diffusion and Reactions in Fractals and Disordered Systems* (Cambridge University Press, 2000).
41. R. Pastor-Sattoras and A. Vespignani, *Phys. Rev. Lett.* **86**, 3200 (2001).
42. Y. Moreno, R. Pastor-Satorras, and A. Vespignani *Eur. Phys. J. B* **26**, 521 (2002).
43. D. S. Callaway, J. E. Hopcroft, J. M. Kleinberg, M. E. J. Newman, and S. H. Strogatz, *Phys. Rev. E*, **64**, 041902 (2001).

44. S. N. Dorogovtsev, J. F. F. Mendes, and A. N. Samukhin, *Phys. Rev. E* **64**, 066110 (2001).
45. G. H. Weiss, *Aspects and Applications of the Random Walk* (North-Holland, Amsterdam, 1994).
46. Z. Burda, J. D. Curreira, and A. Krzywicki, *Phys. Rev. E* **64**, 046118 (2001).
47. F. Chung and L. Lu, *Adv. Appl. Math.*, **26**, 257, (2001).
48. D. Dolev, O. Mokryn and Y. Shavitt, *IEEE INFOCOM'03* (2003).
49. K.-I. Goh, B. Kahng and D. Kim, *Phys. Rev. Lett.* **87**, 278701 (2001).
50. K.-I. Goh, E. Oh, H. Jeong, B. Kahng and D. Kim *Proc. Natl. Acad. Sci. USA* **99**, 12583 (2002).
51. R. Pastor-Sattoras and A. Vespignani, *Phys. Rev. E.* **63**, 066117 (2001).
52. R. Pastor-Satorras, and A. Vespignani *Phys. Rev. E* **65**, 036104 (2002).
53. Z. Dezso, A.-L. Barabasi, *Phys. Rev. E* **65**, 055103 (R) (2002).
54. R. Cohen, D. ben-Avraham and S. Havlin, *Phys. Rev. Lett.* **91**, 247901 (2003).

Spectral Analysis of Random Networks

Sergei N. Dorogovtsev[1,2], Alexander V. Goltsev[1,2], José F.F. Mendes[2], and Alexander N. Samukhin[1,2]

[1] Departamento de Física, Universidade de Aveiro, Campus Universitário de Santiago, 3810-193 Aveiro, Portugal
[2] A.F. Ioffe Physico-Technical Institute, 194021 St. Petersburg, Russia

Abstract. We review a general approach that describes the spectra of eigenvalues for random graphs with a local tree-like structure. The exact equations to the spectra of networks with a local tree-like structure, are presented. The tail of the density of eigenvalues $\rho(\lambda)$ at large $|\lambda|$ is related to the behavior of the vertex degree distribution for large value of degree. In particular, as $P(k) \sim k^{-\gamma}$, $\rho(\lambda) \sim |\lambda|^{1-2\gamma}$. Under an effective medium approximation we propose a simple approximation, calculate spectra of various graphs analytically. We also analyse the spectra of various complex networks and discuss the role of vertices of low degree. We show that spectra of locally tree-like random graphs gives a good description of the spectral properties of real-life networks like the Internet.

1 Introduction

Complex networks, like the Internet or WWW, biological networks of interacting proteins, social networks, among others (see [1–7]), became in the recent past examples of intense research in Physics. Many properties of these networks have been studied after the pioneer paper of Watts and Strogatz [8]. More recently, a growing interest appears, in particular, in networks with a scale-free structure. They appear in many different contests, like in the examples given before, but more interestingly is the fact that most of them present the same non trivial statistical properties.

A network is described by nodes and links between them according some degree distribution. The structure of networks may be completely described by the associated adjacency matrices. The adjacency matrices of undirected graphs are symmetric matrices with matrix elements, equal to number of edges between the given vertices. The eigenvalues of an adjacency matrix are related to many basic topological invariants of networks such as, for example, the diameter of a network [9,10]. Recently, in order to characterize networks, it was proposed to study spectra of eigenvalues of the adjacency matrices as a fingerprint of the networks [11–18]. The rich information about the topological structure and diffusion processes can be extracted from the spectral analysis of the networks. Studies of spectral properties of the complex networks may also have a general theoretical interest. The random matrix theory has been successfully used to model statistical properties of complex classical and quantum systems such as complex nucleus, disordered conductors, chaotic quantum systems (see, for example, reviews [19]), the glassy relaxation [20] and so on. As the adjacency

S.N. Dorogovtsev, A.V. Goltsev, J.F.F. Mendes, and A.N. Samukhin, Spectral Analysis of Random Networks, Lect. Notes Phys. **650**, 35–50 (2004)
http://www.springerlink.com/

matrices are random, in the limit $N \longrightarrow \infty$ (N is the total number of vertices), the density of eigenvalues could be expected to converge to the semicircular distribution in accordance with the Wigner theorem [21]. However, Rodgers and Bray have demonstrated that the density of eigenvalues of a sparse random matrix deviates from the Wigner semicircular distribution and has a tail at large eigenvalues [22], see also [23]. Recent numerical calculations of the spectral properties of small-world and scale-free networks [13–15], and the spectral analyses of the Internet [11,12,16,17] have also revealed that the Wigner theorem does not hold. The spectra of the Internet [11,12] and scale-free networks [14,15] demonstrate an unusual power-law tail in the region of large eigenvalues. At the present time there is a fundamental lack in understanding of these anomalies. In order to carry out a complete spectral analysis of real networks it is necessary to take into account all features of these complex systems described by a degree distribution, degree correlations, the statistics of loops, etc.

This approach is valid for any network which has a *local tree-like structure*. In particular, these are uncorrelated random graphs with a given degree distribution [24,25], and their straightforward generalizations [26] allowing pair correlations of the nearest neighbors. These graph ensembles have one common property: almost every finite connected subgraph of the infinite graph is a tree. The tree is a graph, which has no loops. A random Bethe lattice is an infinite random tree-like graph. All vertices on a Bethe lattice are statistically equivalent. These features (the absence of loops and the statistical equivalence of vertices) are decisive for our approach. The advantage of Bethe lattices is that they frequently allow analytical solutions for a number of problems: random walks, spectral problems, etc.

Real-life networks, however, often contain numerous loops. In particular, this is reflected in a strong "clustering", which means that the (relative) number of loops of length 3 do not vanish even in very large networks. Nevertheless, we believe, that the study of graphs with a local tree-like structure may serve as a starting point in the description of more complex network architectures.

In this review we present the exact equations which determine the spectra of infinite random uncorrelated random tree-like graphs. For this, we use a method of random walks. We propose a method of an approximate solution of the equations. We shall show that the spectra of adjacency matrices of random tree-like graphs gave a tail at large eigenvalues. In the case of a scale-free degree distribution, the density of eigenvalues has a power-law behavior. A more detailed analyse of the spectra of correlated and uncorrelated graphs can be seen in [27].

2 Random Walk on a Tree

Lets call the probability to transit from node j to node i is P_{ij}. If node j has degree k_j and if the probability to transit from node j to each of his nearest neighbors is equal then, $P_{ij} \equiv \text{prob}\,(j \to i) = a_{ij}/k_j$, where a_{ij} is the adjacency matrix. P_{ij} obeys the following normalization condition: $\sum_{i=1}^{k_j} P_{ij} = 1$. Lets introduce the probability $g_i(t)$ to find a walker at node i if at time $t = 0$ the walker was at node 0 that is $g_0(t = 0) = 1$. The temporal evolution of this quantity obeys a master equation of the type,

$$\frac{dg_i(t)}{dt} = \sum_m P_{im} g_m(t) - g_i(t) \sum_m P_{mi} = -\sum_m (\delta_{im} - P_{im}) g_m(t). \qquad (1)$$

Lets define the Laplacian in the following way: $\mathcal{L}_{im} \equiv \delta_{im} - P_{im}$. So, one way to obtain information about the diffusion process in the network is through the spectrum of the Laplacian. As we will show the spectral analysis can be very useful in many aspects giving important information not only about the diffusion process but also about the structural properties of the net.

3 General Theory

Let $\hat{A} = (a_{vw})$ be the $N \times N$ symmetric adjacency matrix of an N-vertex graph G, $a_{vw}^2 = a_{vw}$, $a_{vv} = 0$. The degree k_v of a vertex v is defined as,

$$k_v = \sum_w a_{vw} . \qquad (2)$$

Every graph is characterized by a degree distribution $P(k)$,

$$P(k) = \left\langle \frac{1}{N} \sum_{v=1}^{N} \delta (k_v - k) \right\rangle . \qquad (3)$$

Graph ensembles with a given *uncorrelated* vertex degree distribution may be realized e.g. if we consider all possible graphs with $N(k)$ vertices of degree k, $k = 1, 2, \ldots$, assuming $N(k)/N \to P(k)$ in the thermodynamic limit $[N \to \infty, N(k) \to \infty]$.

The spectrum of $\hat{A}$ may be calculated by using the method of random walks on a tree-like graph G and generating functions [28]. We define a generating function,

$$R(z) = \frac{1}{N} \sum_{v=1}^{N} \sum_{n=0}^{\infty} \rho_v(n) z^n , \qquad |z| < 1 \qquad (4)$$

where $\rho_v(n)$ is the number of walks of length n from v to v, where v is any vertex of G,

$$\rho_v(n) = (\hat{A}^n)_{v,v} . \qquad (5)$$

In a tree-like graph the number of steps n is an even number. In order to return to v we must go back along all of the edges we have gone.

Let $q_v(n)$ be the number of walks of length n starting at v and ending at v for the first time. We define

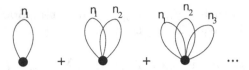

Fig. 1. Schematic representation of first contributions to $\rho_v(n)$ in (7).

$$Q_v(z) = \sum_{n=0}^{\infty} q_v(n) z^n \,. \tag{6}$$

One can note that $\rho_v(n)$ is given by (Fig. 1)

$$\rho_v(n) = q_n + \sum_{n_1,n_2} q_{n_1} q_{n_2} \delta_{n_1+n_2,n} + \sum_{n_1,n_2,n_3} q_{n_1} q_{n_2} q_{n_3} \delta_{n_1+n_2+n_3,n} + \cdots \tag{7}$$

Here the first term is the number of walks of length n starting at v and ending at v for the first time, the second term is the number of walks of length n starting at v and ending at v with one intermediate visit of v, the third term is the number of walks with two intermediate visit of v and so on.

Using this relationship gives

$$R(z) = \frac{1}{N} \sum_{v=1}^{N} \frac{1}{1 - Q_v(z)} \,. \tag{8}$$

Let $d(w,v) = m \geq 1$ be the distance from w to v and $t_{w,v}^{(m)}(n)$ be the number of paths of length n starting at w and ending at v for the first time. We define

$$T_{wv}^{(m)}(z) = \sum_{n=0}^{\infty} t_{w,v}^{(m)}(n) z^n \,. \tag{9}$$

One can prove

$$Q_v(z) \quad = z \sum_{w} T_{wv}^{(1)}(z) \,, \tag{10}$$

$$T_{wv}^{(m)}(z) = T_{wg_1}^{(1)}(z) T_{g_1 g_2}^{(1)}(z) \cdots T_{g_{m-1}v}^{(1)}(z) \,, \tag{11}$$

where $w \longrightarrow g_1 \longrightarrow g_2 \longrightarrow \cdots g_{m-1} \longrightarrow v$ is the shortest path from w to v. There is an important relationship:

$$T_{wv}^{(1)}(z) = z + z \sum_{g} T_{gv}^{(2)}(z) = z + z \sum_{g} T_{gw}^{(1)}(z) T_{wv}^{(1)}(z) \,. \tag{12}$$

In this sum the vertex g is the nearest neighbor of w and a second neighbor of the vertex v. Solving the recurrence equation (12), we can find $T_{wv}^{(1)}(z)$ and $Q_v(z)$.

Lets define $\widetilde{T}_{wv}^{(1)}(z) \equiv T_{wv}^{(1)}(z^{-1})$, then equation (12) may be written in the form,

$$\widetilde{T}_{wv}^{(1)}(z) = \frac{1}{z - \sum_g \widetilde{T}_{gw}^{(1)}(z)} \,. \tag{13}$$

If we can solve this equation one can find $Q_v(z)$, from which we get $R(z)$. Let us define $B(z) \equiv z^{-1}R(z^{-1})$. Then the density of the eigenvalues λ of a random graph is determined as follows:

$$\rho(\lambda) = -\operatorname{Im}\langle B(\lambda + i\varepsilon))\rangle / \pi \,, \tag{14}$$

where ε is positive and tends to zero. Note that the equations (6)–(13) are valid for both uncorrelated and correlated tree-like graphs.

In the case of a $k-$regular connected graph we have $\widetilde{T}_{wv}^{(1)}(z) \equiv T(z)$ and $Q_v(z) \equiv Q(z)$. In this case (13) gives becomes, $zT(z) - (k-1)T^2(z) = 1$. Solving this equation, we get the well known result:

$$\rho(\lambda) = \frac{k}{2\pi} \frac{\sqrt{4(k-1) - \lambda^2}}{k^2 - \lambda^2} \,. \tag{15}$$

This is a continuous spectrum of extended eigenstates with eigenvalues $|\lambda| < 2\sqrt{k-1}$.

4 Spectra of Uncorrelated Graphs

In the case of uncorrelated random tree-like graphs, $k_w - 1$ random parameters $\widetilde{T}_{gw}^{(1)}(z)$ on the right-hand side of (13) are equivalent and statistically independent. They are also independent on the degree k_w. We define the distribution function of $\widetilde{T}_{wv}^{(1)}(z)$ at $z = \lambda + i\varepsilon$ in the Fourier representation as:

$$F_\lambda(x) = \left\langle \exp\left[-ix\widetilde{T}_{wv}^{(1)}(\lambda + i\varepsilon)\right] \right\rangle , \tag{16}$$

where the brackets $\langle ... \rangle$ denote the averaging over the ensemble of random uncorrelated graphs associated with a degree distribution $P(k)$. The statistical independence of the $k-1$ random parameters $\widetilde{T}_{gw}^{(1)}(\lambda + i\varepsilon) \equiv T_i$, $i = 1, 2, ..., k-1$, $k \equiv k_w$, on the right hand side of (13) allows us to use the following identity:

$$F_\lambda(x) \equiv \langle \exp(-ixT) \rangle = \left\langle \exp\left(-\frac{ix}{\lambda + i\varepsilon - \sum_{i=1}^{k-1} T_i} \right) \right\rangle$$

$$= 1 - \sqrt{x} \int_0^\infty \frac{dy}{\sqrt{y}} J_1(2\sqrt{xy}) \left\langle \exp\left(iy[\lambda + i\varepsilon - \sum_{i=1}^{k-1} T_i] \right) \right\rangle$$

$$= 1 - \sqrt{x} \int_0^\infty \frac{dy}{\sqrt{y}} J_1(2\sqrt{xy}) e^{iy(\lambda+i\varepsilon)} \sum_k \frac{kP(k)}{\langle k \rangle} \langle \exp(-iyT) \rangle^{k-1} \tag{17}$$

where $J_1(x)$ is the Bessel function and $\langle k \rangle = \sum_k kP(k)$. Thus, we get the exact self-consistent equation for $F_\lambda(x)$:

$$F_\lambda(x) = 1 - \sqrt{x} \int_0^\infty \frac{dy}{\sqrt{y}} J_1(2\sqrt{xy}) e^{iy\lambda} \Phi_1(F_\lambda(y)), \tag{18}$$

where $\Phi_1(x) \equiv \sum_{k=1}^\infty k P(k) x^{k-1} / \langle k \rangle$. Solving (18) give the distribution of T, and so we can obtain Q, from which we get R. Equations (8), (10), and (14) gives

$$\rho(\lambda) = -\frac{1}{\pi} \operatorname{Im} \left\langle \frac{1}{\lambda - \sum_{i=1}^k T_i} \right\rangle = \frac{1}{\pi} \operatorname{Re} \int_0^\infty dy e^{iy\lambda} \Phi(F_\lambda(y)), \tag{19}$$

where $\Phi(x) \equiv \sum_{k=1}^\infty P(k) x^k$. From (18), we find the n-th moment of the distribution function $\Psi_\lambda(T)$, (16):

$$M_n \equiv \langle T^n \rangle = \frac{1}{(n-1)! \, i^n} \int_0^\infty dy \, y^{n-1} e^{iy\lambda} \Phi_1(F_\lambda(y)). \tag{20}$$

5 Effective Medium Approximation

In the absence of a technique to solve (18) exactly we will use an approximate solution. Neglecting fluctuations of T around a mean value $T(\lambda) \equiv \langle T \rangle$, a self-consistent equation for the function $T(\lambda)$ may be obtained if we insert

$$F_\lambda(x) \approx e^{-ixT(\lambda)} \tag{21}$$

into the right-hand side of (20) for $n = 1$. We get

$$T(\lambda) = \frac{1}{\langle k \rangle} \sum_k \frac{k P(k)}{\lambda + i\varepsilon - (k-1)T(\lambda)}. \tag{22}$$

This approach is known by "effective medium" (EM) approximation. At real λ, $T(\lambda)$ is a complex function, which is to be understood as an analytic continuation from the upper half-plane of λ, $T(\lambda) \equiv T(\lambda + i\varepsilon)$. Therefore, $\operatorname{Im} T(\lambda + i\varepsilon) < 0$. In the framework of the EM approach, the density $\rho(\lambda)$, (19), takes an approximate form

$$\rho(\lambda) = -\frac{1}{\pi} \sum_k \frac{k P(k) \operatorname{Im} T(\lambda)}{(\lambda - k \operatorname{Re} T(\lambda))^2 + k^2 (\operatorname{Im} T(\lambda))^2}. \tag{23}$$

6 Tail Behavior and Finite-Size Effects

Equation (22) may be solved analytically at $|\lambda| \gg 1$. We look for a solution in the region $\operatorname{Im} T(\lambda) \ll \operatorname{Re} T(\lambda) \ll 1$. It is convenient to use a continuum approximation in (22). The real and imaginary parts of this equation take a form

$$\operatorname{Re} T(\lambda) = \frac{1}{2\lambda \langle k \rangle} \int_{k_0}^{k_{cut}} \frac{dk\, kP(k)}{\left(1 - (k-1)R_1\right)^2 + \left((k-1)I_1\right)^2},$$

$$1 = \frac{1}{\lambda^2 \langle k \rangle} \int_{k_0}^{k_{cut}} \frac{dk\, k(k-1)P(k)}{\left(1 - (k-1)R_1\right)^2 + \left((k-1)I_1\right)^2}, \tag{24}$$

where $R_1 = \operatorname{Re}(T(\lambda)/\lambda)$, $I_1 = \operatorname{Im}(T(\lambda)/\lambda)$ and k_0 and k_{cut} are the smallest and largest degrees, respectively. A region $k_0 \leq k \ll k_\lambda$ gives a regular contribution into the integrals (24) while a region $k \sim k_\lambda \gg 1$ gives a singular contribution. Here $k_\lambda \equiv \lambda/\operatorname{Re} T(\lambda) + 1$. As a result we obtain

$$\operatorname{Re} \qquad T(\lambda) \cong \frac{1}{2\lambda} + \frac{\pi k_\lambda P(k_\lambda)}{2 \langle k \rangle \operatorname{Im} T(\lambda)}, \tag{25}$$

$$1 \cong \frac{1}{\lambda^2 \langle k \rangle} \int_{k_0}^{k_\lambda} dk\, k(k-1)P(k) + \frac{\pi \lambda k_\lambda P(k_\lambda)}{\langle k \rangle \operatorname{Im} T(\lambda)}. \tag{26}$$

If $P(k)$ decreases faster than k^{-2} at $k \gg 1$, i.e. $\langle k \rangle$ is finite, then in the leading order of $1/\lambda$ we find

$$T(\lambda) \cong \lambda^{-1} - i\pi |\lambda|\, k_\lambda P(k_\lambda)/ \langle k \rangle. \tag{27}$$

Within the same approach one can find from (23) that the density $\rho(\lambda)$ also has two additive contributions

$$\rho(\lambda) \cong -\frac{\langle k \rangle \operatorname{Im} T(\lambda)}{\pi \lambda^2} + \frac{k_\lambda P(k_\lambda)}{|\lambda|}. \tag{28}$$

Inserting (27) gives the density

$$\rho(\lambda) \cong 2 \frac{k_\lambda}{|\lambda|} P(k_\lambda). \tag{29}$$

Here $k_\lambda = \lambda/\operatorname{Re} T(\lambda) + 1 = \lambda^2 + O(1)$. The asymptotic expression (29) is our main result. The right-hand side of this expression originates from two equal, additive contributions: the contribution from the real part of $T(\lambda)$ and the one from the imaginary part of $T(\lambda)$. One can show that the asymptotic behavior of the real part, $\operatorname{Re} T(\lambda) = \lambda^{-1} + O(\lambda^{-3})$, in the leading order of $1/\lambda$ is universal and is valid even for graphs with finite loops. Contrastingly, the asymptotics of $\operatorname{Im} T(\lambda)$ in the leading order of $1/\lambda$ and the corresponding contribution to the right-hand side of (28) depend on details of the structure of a network.

The analysis of (22) shows that the main contribution to an eigenstate with a large eigenvalue λ is given by vertices with a large degree $k \propto k_\lambda \gg 1$. As we will show below, in the limit $\lambda \gg 1$, the result (29) is asymptotically exact. The relationship between largest eigenvalues and highest degrees, $\lambda^2 + O(1) = k$, for a wide class of graphs was obtained in two mathematical papers, [29,30].

The contribution of highly connected vertices may be compared with a simple spectrum of "stars", which are graphs with a vertex of degree k, connected to

k dead ends. The spectrum of this star-like graph consists of two eigenvalues $\lambda = \pm\sqrt{k}$ and a $(k-1)$-degenerate zero eigenvalue. Note that asymptotically, in the limit of large λ, (29) gives $\rho(\lambda) \cong 2|\lambda|P(\lambda^2)$ if $P(k)$ decreases slower than an exponent function at large k, that is, if higher moments of the degree distribution diverge.

A classical random graph [31,32] has the Poisson degree distribution $P(k) = e^{-\langle k \rangle} \langle k \rangle^k / k!$. The tail of $\rho(\lambda)$ is given by (29) with $k_\lambda \approx \lambda^2 + a \gg 1$ where a is of $O(1)$: $\rho(\lambda) \sim \lambda^{-2(\lambda^2+a)} \exp[(1 + \ln\langle k \rangle)\lambda^2]$. This equation agrees with the previous results [22,23] obtained by different analytical methods.

In the case of a "scale-free" graph with $P(k) \approx P_0 k^{-\gamma}$ at large k and $|\lambda| \gg 1$, we get an asymptotically exact power-law behavior: $\rho(\lambda) \approx 2|\lambda| P(\lambda^2) = 2P_0 |\lambda|^{-\delta}$, where the eigenvalue exponent $\delta = 2\gamma - 1$.

For finite $N \gg 1$, there is a finite-size cutoff of the degree distribution $k_{cut} \propto k_0 N^{1/(\gamma-1)}$ [33]. The cutoff determines the upper boundary of eigenvalues: $\lambda < k_{cut}^{1/2}$. This result agrees with an estimation of the largest eigenvalue of sparse random graphs obtained in [34].

7 Spectrum of a Transition Matrix

We will consider now the case of random walks on a graph with the transition probability $1/k_v$ of moving from a vertex v to any one of its neighbors. The transition matrix $\widehat{P}$ then satisfies,

$$P_{w,v} = a_{w,v}/k_v . \tag{30}$$

For each vertex v, $P_{w,v}$ obeys the normalization condition: $\sum_w P_{w,v} = 1$. Also, $\widehat{P}$ is related with the Laplacian of the graph by,

$$L_{v,w} = \begin{cases} 1 & if \quad v = u \\ -a_{v,w}/\sqrt{k_v k_w} & otherwise \end{cases}, \tag{31}$$

as follows: $\widehat{P} = \widehat{D}^{1/2}(1 - \widehat{L})\widehat{D}^{-1/2}$ where $D_{v,w} = \delta_{v,w}/k_v$. Therefore, if we know the density $\rho(\lambda)$ of eigenvalues of $\widehat{P}$, we can find the density of eigenvalues of the Laplacian: $\rho_L(\lambda) = \rho(1 - \lambda)$. We denote the eigenvalues of the matrix $\widehat{P}$ by $\lambda_1 \geq \lambda_2 \geq ... \geq \lambda_N$. The eigenfunction $f_v = k_v$ corresponds to the largest eigenvalue $\lambda_1 = 1$.

In order to calculate the spectrum of $\widehat{P}$ we use the same method of random walks described in Sect. 2. The probability of one step is given by (30). We define the generating function $Q_v(z)$ and $\widetilde{T}_{wv}^{(1)}(z^{-1}) \equiv T_{wv}^{(1)}(z)$ and obtain an exact equation which is similar to (13):

$$\widetilde{T}_{wv}^{(1)}(z) = \frac{1}{k_w z - \sum_g \widetilde{T}_{gw}^{(1)}(z)} , \tag{32}$$

where $g \sim w$ but $g \neq v$. At $z = \lambda + i\varepsilon$, we get exact equations for the function $F_\lambda(x) = \left\langle \exp(-ix\widetilde{T}_{wv}^{(1)}(\lambda + i\varepsilon)) \right\rangle$ and the density of the eigenvalues $\rho(\lambda)$:

$$F_\lambda(x) = 1 - \sqrt{x} \int_0^\infty \frac{dy}{\sqrt{y}} J_1(2\sqrt{xy}) e^{iy\lambda} \Phi_1(e^{i\lambda y} F_\lambda(y)) , \tag{33}$$

$$\rho(\lambda) = \frac{1}{\pi} \operatorname{Re} \sum_k P(k) k \int_0^\infty dy\, e^{ik\lambda y} F_\lambda^k(y) . \tag{34}$$

The function $F_\lambda(x) = e^{-ix}$ is an exact solution of (33). This solution corresponds to the eigenvalue $\lambda_1 = 1$ and gives the delta-peak $\delta(\lambda - 1)$ in the density $\rho(\lambda)$. In order to find the spectrum at $\lambda \leq \lambda_2$ we use, as before, the EM approach. We assume $F_\lambda(x) \approx e^{-ixT(\lambda)}$ and get an equation for a complex function $T(\lambda)$:

$$T(\lambda) = \frac{1}{\langle k \rangle} \sum_k \frac{kP(k)}{k\lambda + i\epsilon - (k-1)T(\lambda)} . \tag{35}$$

$\rho(\lambda)$ is given by

$$\rho(\lambda) = -\frac{1}{\pi} \operatorname{Im} \frac{1}{\lambda - T(\lambda)} . \tag{36}$$

The spectrum of the transition matrix of a k-regular tree can be obtained easily using (35) and (36). The second eigenvalue is given by $\lambda_2 = 2\sqrt{k-1}/k$.

8 Spectra of Different Topological Graphs

Let us compare available spectra of classical random graphs, scale-free networks [14,15], empirical spectra of the Internet [11,12,17], and of the random tree-like graphs.

We start by discuss the spectra of the adjacency matrices. The spectra were calculated using the EM approach from (22) and (23) for different degree distributions $P(k)$. The results can be seen in Figs. 2 and 3.

Classical random graphs — Classical random graphs have the Poisson degree distribution. The density of eigenvalues of the associated adjacency matrix has been obtained numerically in [14]. In Fig. 2 we display results of the numerical calculations and from results obtained within the EM approach. We found a good agreement in the whole range of eigenvalues. There are only some small differences in the region of small eigenvalues which may be explained by an inaccuracy of the EM approach in this range. In this region, the density $\rho(\lambda)$ has an elevated central part that differs noticeably from the semicircular distribution. The spectrum also has a tiny tail given by (29) which can hardly be seen in Fig. 2, see for detail Sect. 5 and [22,23]

Scale-free networks — Spectra of scale-free graphs with the degree distribution $P(k) = P_0 k^{-\gamma}$ differ strongly from the semi-circular law [14,15]. The

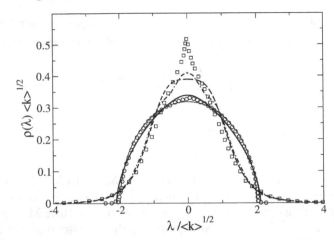

Fig. 2. Density of eigenvalues for the adjacency matrices of two different networks: (i) Classical random graph (the Erdős-Rényi model) with the average degree $\langle k \rangle = 10$: the effective medium (EM) approach (the solid line) and numerical calculations for the graphs of 20 000 vertices [14] (the open circles). (ii) Scale-free random tree-like graph with $\gamma = 3$ and the smallest degree $k_0 = 5$: the EM approach (the dashed line), the improved EM approach, see the text (the dashed-dotted line). The results of the simulations of the Barabási-Albert model of 7000 vertices [14] (the open squares). The semi-circular law is shown by the thin solid line.

Barabási-Albert model has a tree-like structure, the exponent $\gamma = 3$ of the degree distribution, and negligibly weak correlations between degrees of the nearest neighbors [4]. Therefore, one can assume that the spectrum of a random tree-like graph can mimic well the spectrum of the model. In Fig. 2 we compare the spectrum of the random tree-like graph with $\gamma = 3$ and the spectrum of the Barabási-Albert model obtained from simulations [14]. The density of states $\rho(\lambda)$ has a triangular-like form and demonstrates a power-law tail. There is only a noticeable deviation of the EM results from the results of simulations [14] at small eigenvalues λ. In order to improve the EM results, we used, as an ansatz, the distribution function $F_\lambda(x) = [1 + a(\lambda)x^2]e^{-ixT(\lambda)}$ instead of the function $F_\lambda(x) = e^{-ixT(\lambda)}$. In this case, there are two unknown complex functions $a(\lambda)$ and $T(\lambda)$ which were determined self-consistently from (18).

Power-law tail — The power-law behavior of the density of eigenvalues $\rho(\lambda) \propto \lambda^{-\delta}$ is an important feature of the spectrum of scale-free networks. The simulations [14] of the Barabási-Albert model having the degree exponent $\gamma = 3$ revealed a power-law tail of the spectrum, with the eigenvalue exponent $\delta \approx 5$. Our prediction $\delta = 2\gamma - 1 = 5$ is in agreement with the result of these simulations.

The study of the topology of the Internet at the Autonomous System (AS) level revealed a power-law behavior of eigenvalues of the associated adjacency matrix [11,12]. The degree distribution of the network has the exponent $\gamma \approx 2.1$[12]. The eigenvalues λ_i of the Internet graph are proportional to the power of the rank i of an eigenvalues (starting with the largest eigenvalue): $\lambda_i \propto i^\epsilon$ with some exponent ϵ. This leads to $\rho(\lambda) \propto \sum_i \delta(\lambda - \lambda_i) \approx \int \delta(\lambda - i^\epsilon)di \propto \lambda^{-1+1/\epsilon}$. The *Multi* dataset analyzed in [12] gave $\epsilon \approx -0.447$ and, hence, the eigenvalue exponent $1 - 1/\epsilon \approx 3.2$. The *Oregon* dataset [12] gave $\epsilon \approx -0.477$, $1 - 1/\epsilon \approx 3.1$. Our results with $\gamma = 2.1$ substituted, give the eigenvalue exponent $\delta = 2\gamma - 1 \approx 3.2$ in agreement with the results obtained from empirical data for this network. There are the following reasons for the agreement between the theory for tree-like graphs and the data for the Internet. At first, although the average clustering coefficient of the Internet at AS level is about 0.2, the local clustering coefficient rapidly decreases with increasing the degree of a vertex [36]. In other words, the closest neighborhood of vertices with large numbers of connections is "tree-like". Recall that vertices with large numbers of connections determine the large-eigenvalue asymptotics of the spectrum. So, we believe that our results for the asymptotics of the spectra of tree-like networks is also valid for the Internet and other networks with similar structure of connections. Secondly, the Internet is characterized by strong correlations between degrees of neighboring vertices [35]. However, as we have shown in Sect. 6, such short-range degree correlations do not affect the power-law behavior of eigenvalues.

The study of the Internet topology [12] also revealed a correspondence between the large eigenvalues λ_i and the degree k_i: $k_i = \lambda_i^2$. This result is in agreement with our theoretical prediction that it is the highly connected vertices with a degree about $k_\lambda \approx \lambda^2$ that produce the power-law tail $\rho(\lambda) \propto \lambda^{-\delta}$.

Pseudo-fractal network — The calculations of the eigenvalues spectrum of the adjacency matrix of a pseudofractal graph with $\gamma = 2.585\ldots$ [37] have revealed a power-law behavior with $\delta \approx 4.6$. The effective medium approximation gives lower value $\delta = 2\gamma - 1 \approx 4.2$. The origin of the difference is not clear. One should note that the pseudofractal is a deterministically growing graph with a very large clustering coefficient $C = 4/5$ and, what is especially important, with long-range correlations between degrees of vertices.

Weakly connected nodes — Let us study the influence of weakly connected vertices with degrees $1 \leq k \leq 5$ on the spectra of random tree-like graphs with the degree distribution $P(k) = P_0 k^{-\gamma}$. In Fig. 3a and 3b we represent the evolution of the spectrum of the network with $\gamma = 5$, when the smallest degree k_0 decreases from 5 to 1. The spectra were calculated in the framework of the EM approximation. Similar results are obtained at different γ. For $k_0 \leq 4$, two peaks at non-zero eigenvalues emerge in the density of states $\rho(\lambda)$. In order to understand an origin of the peaks one can note that for this degree distribution the average degree $\langle k \rangle$ is close to k_0. For example, at $k_0 = 3$ we have $\langle k \rangle = 3.49$. Therefore, in this network, the probability to find a vertex having three links is larger than the probability to find a vertex with a degree $k \geq 4$. There are large parts of the network which have a local $k = 3$–regular structure. In Fig. 3a we show a density of eigenvalues of an infinite $k = 3$–regular Bethe lattice [see (15) at $k = 3$]. At small eigenvalues, the density of the regular tree fits the density of the random network. At large λ, the density of eigenvalues demonstrates a power-law behavior with the exponent $\delta = 2\gamma - 1$.

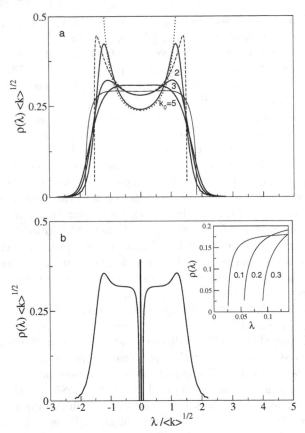

Fig. 3. Evolution of spectra of a random tree-like graph with the scale-free degree distribution for $\gamma = 5$ and the smallest degree $k_0 = 1, 2, 3$ and 5. The panel a shows the spectra of the graphs with $k_0 = 2, 3$ and 5. The dotted line corresponds to the density of eigenvalues of an infinite chain. The dashed and thin solid lines present the spectrum of the $k = 3$ and 6 regular Bethe lattices. The panel b shows the spectrum of a random uncorrelated graph having dead-end vertices with the probability $P(1) = 0.3$. The insert shows the behavior of the density of eigenvalues $\rho(\lambda)$ near the dip at $P(1) = 0.1, 0.2$, and 0.3. The central peak is produced by localized states.

In the case $k_0 = 2$ we have $\langle k \rangle = 2.23$. This network contains long chains which connect vertices with degrees $k \geq 3$. In Fig. 3a we display the density of eigenvalues of an infinite chain which is given by (15) at $k = 2$ [note that it presents a divergence at $\lambda = \pm 2$]. At small eigenvalues this density of eigenvalues fits well the density of eigenvalues of the random network.

Thus, it is the vertices with small degrees that are responsible for the formation of density $\rho(\lambda)$ of networks at small eigenvalues.

Dead-end vertices — Let us investigate the effect of dead-end vertices on the spectra of random tree-like graphs with different degree distributions. Figure 3 b shows a spectrum of a scale-free network with $\gamma = 5$ and the probability of dead-end vertices $P(1) = 0.3$. The EM approximation is used. The spectrum has a flat part and two peaks at moderate eigenvalues. As we have shown above, this (intermediate) part of the spectrum is formed mainly by the vertices with degree $k = 2$ and 3. The emergency of a dip at zero is a new feature of the spectrum. In fact, there is a gap in the spectrum obtained in the in the framework of the EM approach. The width of the gap increases with increasing $P(1)$. One can see this in the insert in Fig. 3b. The dead-end vertices also produce a delta peak at $\lambda = 0$. The central peak corresponds to localized eigenstates.

Note that the appearance of the central peak and a dip is a general phenomena in random networks with dead-end vertices. We also observed this effect in the classical random graphs. Spectral analysis of the Internet topology on the AS level revealed a central peak with a high multiplicity [17]. Thus the conjecture that localized and extended states are separated in energy may well hold in complex networks. A similar spectra was observed in many random systems, for example, in a binary alloy [38]. In order to estimate the height of the delta peak it is necessary to take into account all localized states. Unfortunately, so far this is an unsolved analytical problem [17].

Finite-size effects — In the present paper we studied the spectral properties of infinite random tree-like graphs. Numerical studies of large but finite random trees demonstrate that the spectrum of a finite tree consists, speaking in general terms, of a continuous component and an infinity of delta peaks. The components correspond to extended and localized states, respectively [18]. There is a hole around each delta peak in the spectrum. A finite regular tree has a spectral distribution function which looks like a singular Cantor function [39]. These results demonstrate that finite size effects in spectra may be very strong. In particular, the finite size of a network determines the largest eigenvalue in its spectrum. As was estimated in Sect. 5, the largest eigenvalue of the adjacency matrix associated with a scale-free graph is of the order of $k_{cut}^{1/2} = k_0^{1/2} N^{1/2(\gamma-1)}$.

Spectrum of the transition matrix — In Fig. 4 we represent a spectrum of the transition matrix $\widehat{P}$ defined by (30) for a tree-like graph with the scale-free degree distribution $P(k) \propto k^{-\gamma}$ at large degrees $k \geq 5$. The spectrum was calculated from (35) and (36) with the degree exponent $\gamma = 2.1$ and the probabilities $P(1), P(2), P(3)$ and $P(4)$ taken from empirical degree distribution of the Internet at the AS level [36].

The spectrum lies in the range $|\lambda| \leq \lambda_2 < 1$. In Fig. 4 we compare our results with the spectrum of the transition matrix $\widehat{P}$ of the Internet obtained in [16,17]. Unfortunately, the data [16,17] are too scattered to make a detailed comparison with our results. Nevertheless, one can see that the spectrum of $\widehat{P}$ of the tree-like graph reproduces satisfactory the general peculiarities of the real spectrum. Namely, the spectra have a wide dip at zero eigenvalue and a central delta-peak [17]. The multiplicity of the zero eigenvalue have been estimated in [17]. For a detailed comparison between the spectra, correlations in the Internet must also be taken into account.

In order to reveal an effect of dead-end vertices we calculated spectra of $\widehat{P}$ on a random tree-like graph with the Poisson and the scale-free degree distributions

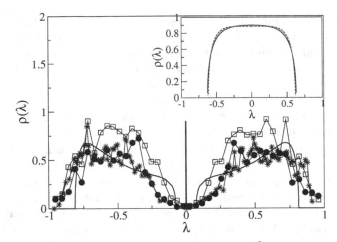

Fig. 4. Density of eigenvalues of the transition matrix $\hat{P}$ defined by (30). (i) The Internet data (the solid circles) and the results of simulations of a random scale-free networks (the open squares) from [16]. (ii) The spectrum of the Internet extracted from [17] (the stars). (iii) Our calculations (the solid line) with the degree distribution $P(k) = Ak^{-2.1}$ for $k \geq 4 P(2) = 0.4 \quad P(3) = 0.12$. These parameters are taken from [36]. The height of the central peak was estimated from [17]. The insert shows the spectra of the transition matrix of a random tree-like graph with excluded dead-end vertices: (i) a random tree-like graph with a scale-free degree distribution, $\gamma = 3$, the smallest degree $k_0 = 5$, and $\langle k \rangle = 9.06$ (the dashed line). (ii) A classical random graph with $\langle k \rangle = 9.06$ (the dotted line). (iii) A $k-$regular Bethe lattice with $k = 9$ (the solid line).

$\gamma = 3$ in the case when dead-end vertices are excluded, that is $P(1) = 0$, and $\langle k \rangle = 9.06$. These spectra are displayed in the insert in Fig. 4. In the whole range of eigenvalues these spectra are very close to the spectrum of a $k-$ regular Bethe lattice with the degree $k = 9$. These calculations confirm the fact that it is the dead-end vertices that produce the dip in the spectrum of the Internet.

9 Conclusions

The spectra of the adjacency and transition matrices of random uncorrelated tree-like complex networks was studied. We have derived exact equations which describe the spectrum of random tree-like graphs, and proposed a simple approximate solution in the framework of the effective medium approach. Our study confirms that spectra of scale-free networks as well as the spectra of classical random graphs do not satisfy the Wigner law.

We have demonstrated that the appearance of a tail of the density of the eigenvalues of sparse random matrices is a general phenomenon. The spectra of classical random graphs (the Erdős-Rényi model) have a rapidly decreasing tail. Scale-free networks demonstrate a power-law behavior of the density of eigenvalues $\rho(\lambda) \propto |\lambda|^{-\delta}$. We have found a simple relationship between the degree exponent γ and the eigenvalue exponent δ: $\delta = 2\gamma-1$. We have shown that correlations between degrees of neighboring vertices do not affect the power-law behavior of eigenvalues. Comparison with the available results of the simulations of the Barabási-Albert model and the analysis of the Internet at the Autonomous System level shows that this relationship is valid for these networks. We found that large eigenvalues $\lambda \gg 1$ are produced by highly connected vertices with a degree $k \approx \lambda^2$.

Many real-life scale-free networks demonstrate short-range correlations between vertices [40,41] and a decrease of a local clustering coefficient with increasing degree of a vertex. Therefore, the relationship $\delta = 2\gamma - 1$ between the degree-distribution exponent γ and the eigenvalue exponent δ may also be valid for these networks. We can conclude that the power-law behavior $\rho(\lambda) \propto \lambda^{-\delta}$ is a general property of real scale-free networks.

Weakly connected vertices form the spectrum at small eigenvalues. Dead-end vertices play a very special role. They produce localized eigenstates with $\lambda = 0$ (the central peak). They also produce a dip in the spectrum around the central peak. In conclusion, we believe that our general results for the spectra of tree-like random graphs are also valid for many real-life networks with a tree-like local structure and short-range degree correlations.

Acknowledgements

S.N.D, A.N.S., and J.F.F.M. were partially supported by the project POCTI/ 99/FIS/33141. A.G. acknowledges the support of the NATO program OUT-REACH.

References

1. S.N. Dorogovtsev and J.F.F. Mendes, *Evolution of Networks: From Biological Nets to the Internet and WWW* (Oxford, University Press, 2003).
2. A.-L. Barabási and R. Albert, Science **286**, 509 (1999).
3. S.H. Strogatz, Nature **401**, 268 (2001).
4. R. Albert and A.-L. Barabási, Rev. Mod. Phys. **74**, 47 (2002).
5. S.N. Dorogovtsev and J.F.F. Mendes, Adv. Phys. **51**, 4 (2002).
6. M.E.J. Newman, SIAM Review **45**, 167 (2003).
7. D.J. Watts, *Small Worlds: The Dynamics of Networks between Order and Randomness* (Princeton University Press, Princeton, NJ, 1999).
8. Watts and Strogatz, Nature **393**, 440 (1998).
9. D. Cvetković, M. Domb, and H. Sachs, *Spectra of Graphs: Theory and Applications* (Johann Ambrosius Barth, Heidelberg, 1995); D. Cvetković, P. Rowlinson, and S. Simić, *Eigenspaces of graphs* (Cambridge University Press, Cambridge, 1997).
10. F.R.K. Chung, *Spectral Graph Theory* (American Mathematical Society, Providence, Rhode Island, 1997).
11. M. Faloutsos, P. Faloutsos, C. Faloutsos, Comput. Commun. Rev., **29**, 251 (1999).
12. G. Siganos, M. Faloutsos, P. Faloutsos, C. Faloutsos, IEEE-ACM T. Network., to appear

13. R. Monasson, Eur. Phys. J. B **12**, 555 (1999).
14. I.J. Farkas, I. Derényi, A.-L. Barabási, and T. Vicsek, Phys. Rev. E **64**, 026704 (2001); I. Farkas, I. Derenyi, H. Jeong, Z. Neda, Z.N. Oltvai, E. Ravasz, A. Schubert, A.-L. Barabási, and T. Vicsek, Physica A **314**, 25 (2002).
15. K.-I. Goh, B. Kahng and D. Kim, Phys. Rev. E **64**, 051903 (2001).
16. K.A. Eriksen, I. Simonsen, S. Maslov and K. Sneppen, cond-mat/0212001.
17. D. Vukadinović, P. Huang, and T. Erlebach, Lect. Notes Comput. Sc., **2346**, 83 (2002).
18. O. Golinelli, cond-mat/0301437.
19. Th. Guhr, A. Müller-Groeling, and H.A. Weidenmüller, Phys. Rep. **299**, 189 (1998); A.D. Mirlin, Phys. Rep. **326**, 259 (2000).
20. A.J. Bray and G.J. Rodgers, Phys. Rev. B **38**, 11461 (1988).
21. E.P. Wigner, Ann. Math. **62**, 548 (1955); **65**, 203 (1957); **67**, 325 (1958).
22. G.J. Rodgers and A.J. Bray, Phys. Rev. B **37**, 3557 (1988).
23. G. Semerjian and L.F. Cugliandolo, J. Phys. A **35**, 4837 (2002).
24. A. Bekessy, P. Bekessy, and J. Komlos, Stud. Sci. Math. Hungar. **7**, 343 (1972); E.A. Bender and E.R. Canfield, J. Combinatorial Theory A **24**, 296 (1978); B. Bollobás, Eur. J. Comb. **1**, 311 (1980); N.C. Wormald, J. Combinatorial Theory B **31**, 156,168 (1981); M. Molloy and B. Reed, Random Structures and Algorithms **6**, 161 (1995).
25. S.N. Dorogovtsev, J.F.F. Mendes and A.N. Samukhin, Nucl. Phys. B (2003), cond-mat/0204111.
26. M.E.J. Newman, Phys. Rev. Lett. **89**, 208701 (2002); J. Berg and M. Lässig, Phys. Rev. Lett. **89**, 228701 (2002); S.N. Dorogovtsev, J.F.F. Mendes, and A.N. Samukhin, cond-mat/0206467.
27. S.N. Dorogovtsev, A. V. Goltsev, J.F.F. Mendes and A.N. Samukhin, Phys. Rev. E **68**, 046109 (2003).
28. S. Redner, *A Guide to First-Passage Processes*, (Cambridge University Press, Cambridge, 2001).
29. M. Mihail and C. Papadimitriou, Lect. Notes Comput. Sci. **254**, 2483 (2002).
30. F. Chung, L. Lu, and V. Vu, Ann. Combinatorics, **7**, 21 (2003).
31. R. Solomonoff and A. Rapoport, Bull. Math. Biophys. **13**, 107 (1951).
32. P. Erdős and A. Rényi, Publications Mathematicae **6**, 290 (1959); Publ. Math. Inst. Hung. Acad. Sci. **5**, 17 (1960).
33. One should note that the position of the cut-off may depend on subtleties of the ensemble of random graphs. In particular, Z. Burda and A. Krzywicki, cond-mat/0207020, showed that the exclusion of multiple connections in a network may diminish k_{cut}.
34. M. Krivelevich and B. Sudakov, Combinatorics, Probability and Computing, **12**, 61 (2003).
35. R. Pastor-Satorras, A. Vázquez, and A. Vespignani, Phys. Rev. Lett. **87**, 258701 (2001).
36. A. Vázquez, R. Pastor-Satorras and A. Vespignani, Phys. Rev. E **65**, 066130 (2002).
37. S.N. Dorogovtsev, A.V. Goltsev, and J.F.F. Mendes, Phys. Rev. E **65**, 066122 (2002).
38. S. Kirkpatrick and T.P. Eggarter, Phys. Rev. B **6**, 3598 (1972).
39. L. He, X. Liu, and G. Strang, Stud. Appl. Math. **110**, 123 (2003).
40. E. Ravasz, A.L. Somera, D.A. Mongru, Z.N. Oltvai, and A.-L. Barabási, Science **297**, 1551 (2002).
41. A. Vázquez, cond-mat/0211528.

A Tractable Complex Network Model Based on the Stochastic Mean-Field Model of Distance

David J. Aldous

Department of Statistics, 367 Evans Hall, University of California, Berkeley CA 94720 USA

Abstract. Much recent research activity has been devoted to empirical study and theoretical models of complex networks (random graphs) possessing three qualitative features: power-law degree distributions, local clustering, and slowly-growing diameter. We point out a new (in this context) platform for such models – the stochastic mean-field model of distances – and within this platform study a simple two-parameter proportional attachment (or copying) model. The model is mathematically natural, permits a wide variety of explicit calculations, has the desired three qualitative features, and fits the complete range of degree scaling exponents and clustering parameters; in these respects it compares favorably with existing models.

1 Introduction

The topic of *complex networks*, more precisely the design and theoretical analysis of stochastic models of large graphs which differ from the classical Erdős - Rényi model, has attracted intense recent attention, surveyed from a statistical physics viewpoint in [1–3] and from a rigorous mathematical viewpoint in [4].

Let us frame one aspect of this topic, by analogy. In freshman statistics we learn that bivariate data (e.g. heights and weights of n individuals) can be summarized by 5 *summary statistics*: average height, standard deviation of height, average weight, standard deviation of weight, correlation coefficient. And there is a 5-parameter probability model, the bivariate Normal, which (in several precise senses) exactly corresponds to these particular summary statistics. In the context of real-world graphs (where we will always regard the number n of vertices as large), one could analogously seek a crude statistical description by reporting a set of summary statistics. An evident choice is

- ∂ = average vertex-degree

and recently popular extra choices include

- an exponent γ characterizing power-law tail behavior of degree distribution
- a "clustering coefficient" κ measuring relative density of triangles
- the average distance $\bar{\ell}$ between vertex-pairs.

These choices reflect and seek to quantify three qualitative features claimed to hold in many interesting graphs (from WWW links to human social networks): power-law degree distribution, local clustering of edges, and diameter growing as $O(\log n)$. So from the viewpoint of classical mathematical statistics, it would

D.J. Aldous, A Tractable Complex Network Model Based on the Stochastic Mean-Field Model of Distance, Lect. Notes Phys. **650**, 51–87 (2004)
http://www.springerlink.com/

be natural to seek a several-parameter stochastic model of random graphs whose parameters could be readily identified with summary statistics of the kind above. In more detail, we propose three desiderata[1] for a satisfactory model, beyond possessing the three qualitative features mentioned above:

- *mathematical tractability*: one can find reasonably explicit formulas for a variety of quantities of interest
- *fitting flexibility*: by varying model parameters one can vary summary statistics (like the 4 listed above) broadly through their possible ranges
- *naturalness*: the qualitative properties emerge from some simple underlying mathematical structure rather than being forced by fiat.

Unfortunately no satisfactory such models are known. The statistical physics literature surveyed in [1,2] starts with a few elementary model-construction ideas (such as the *proportional attachment* and *small worlds* models mentioned in Sect. 6) and then explores numerous variations. Our purpose in this paper is to introduce a new class of model we call *metric copying*,[2] and to study a particular two-parameter model (*mean-field simple copying*, MFSC) within this class. The description and analysis of the MFSC model involve somewhat more sophisticated mathematical visualization than has been used in previous complex networks literature. So let us first address the first two desiderata by listing results for the model (Sect. 2), and only later (Sect. 3) describe the model. Section 4 derives most of the formulas in Sect. 2, and Sect. 5 exhibits further calculations. A briefer account of the model, aimed at mathematicians, appears in [5].

1.1 Some Notation

$P(\cdot)$ denotes probability, $E(\cdot)$ denotes expectation, and var $(\cdot)$ denotes variance. We assume familiarity with elementary probability notions of random variables and their distributions. We write $\mathrm{Geo}(p), \mathrm{Bin}(m,p), \mathrm{Exp}(\mu), \mathrm{Poi}(\eta)$ for the geometric, binomial, exponential and Poisson distributions in their usual parametrizations, reviewed below. We employ a "blackboard shorthand" of also writing $\mathrm{Geo}(p)$ etc for a random variable with that distribution. Thus the elementary reproductive property of the binomial distribution could be written as

$$\mathrm{Bin}(m_1 + m_2, p) \overset{d}{=} \mathrm{Bin}(m_1, p) + \mathrm{Bin}(m_2, p)$$

where the random variables on the right are independent, and where $\overset{d}{=}$ means equality in distribution. The point of this notation is that, analogous to "composition of functions" in which we interpret $\exp((x-1)^2)$ as the composition of the two functions $\exp(x)$ and $(x-1)^2$, we can "compose" (statisticians say

---

[1] From an applied viewpoint, one could regard "fitting empirical data" as the single criterion; we are of course taking a theoretical viewpoint

[2] We use mathematical terminology: a *metric* is a distance function. Confusingly, some engineers use "metric" to mean "summary statistic"

"mix") distributions. For instance (cf. (8) below), given a random variable Λ with values in $(0,1)$ we can write $\mathrm{Geo}(\Lambda)$ for a random variable whose conditional distribution given $\Lambda = p$ is the $\mathrm{Geo}(p)$ distribution.

Review of elementary distributions.

$$
\begin{aligned}
P(\mathrm{Geo}(p) = i) &= (1-p)^{i-1}p, \ i = 1, 2, \ldots \\
E\,\mathrm{Geo}(p) &= p^{-1}
\end{aligned}
$$

$$
P(\mathrm{Bin}(m,p) = i) = \binom{m}{i}p^{i}(1-p)^{m-i}, \ i = 0, 1, \ldots, m
$$

$$
\begin{aligned}
E\,\mathrm{Bin}(m,p) &= mp & (1) \\
\mathrm{var}\ \mathrm{Bin}(m,p) &= mp(1-p) & (2) \\
P(\mathrm{Poi}(\eta) = i) &= e^{-\eta}\eta^{i}/i!, \ i = 0, 1, 2, \ldots \\
E\,\mathrm{Poi}(\eta) &= \eta.
\end{aligned}
$$

The $\mathrm{Exp}(\mu)$ distribution has probability density function and expectation

$$
\begin{aligned}
f(x) &= \mu e^{-\mu x}, \ 0 < x < \infty \\
E\,\mathrm{Exp}(\mu) &= \mu^{-1}.
\end{aligned}
$$

A Poisson process of rate 1, say $(0 < \xi_1 < \xi_2 < \xi_3 < \ldots)$, is defined by the property

$$
\xi_1, \xi_2 - \xi_1, \xi_3 - \xi_2, \ldots \text{ are independent with } \mathrm{Exp}(1) \text{ distribution}
$$

and has the property

$$
P(\text{some } \xi_i \in [x, x+dx]) = 1 \cdot dx, \ 0 < x < \infty. \tag{3}
$$

1.2 Organization of Paper

Because the precise definition and *a priori* motivation of the model are lengthy to explain, we start by emphasizing the *a posteriori* motivation, the fact that the model permits many explicit calculations. In first reading the formulas in Sect. 2, focus on the left sides of equations, indicating what quantities can be calculated. The formulas on the right sides will be derived in Sect. 4.

2 Formulas

2.1 Key Methodology

Like other models involving vertices arriving and creating edges to existing vertices, the MFSC model defines a directed acyclic (no *directed* cycles) random graph $\mathcal{G}_n$ on n vertices. A key feature of the model is that there exists a well-defined limit infinite rooted graph $\mathcal{G}^*_\infty$ which represents the $n \to \infty$ limit of $\mathcal{G}_n$

rooted at a uniform random (we say "typical") vertex. So for "local" statistics of $\mathcal{G}_n$, one can give "exact formulas in the $n \to \infty$ limit" by doing calculations within the limit structure $\mathcal{G}_\infty^*$, and this methodology is how we will derive (Sect. 4) and interpret the formulas in Sects. 2.2–2.6 below.

Note that "rooting" is introduced merely as a convenient technical way to deal with infinite graphs. By analogy, one could study two-dimensional space without introducing the origin point (cf. Euclidean geometry) but for many purposes an origin and induced coordinate system are helpful.

The MFSC model has two parameters: α, λ. In explicit formulas, we distinguish between a *low clustering region* defined by parameter ranges

$$0 < \alpha < 1, \quad 0 < \lambda \leq 1/\alpha \quad \text{[low]} \tag{4}$$

and the complementary *high clustering region* defined by $\alpha\lambda > 1$; in the latter case it is convenient to reparametrize by using $\eta := \lambda^{-1}\log(\alpha\lambda)$ in place of α, and the parameter ranges are

$$0 < \eta < 1, \quad \eta + 1/\lambda < 1. \quad \text{[high]} \tag{5}$$

This distinction is purely notational; there is no intrinsic "non-analyticity" in the model's properties.

2.2 The Two Parameters Control Mean Degree and Clustering

(a). First consider D_{in} and D_{out}, the random in-degree and out-degree of a typical vertex. Then

$$ED_{\text{in}} = ED_{\text{out}}(= \partial, \text{ say}) = \begin{cases} \frac{\alpha}{1-\alpha} & \text{[low]} \\ \frac{\eta+1/\lambda}{1-\eta-1/\lambda} & \text{[high]}. \end{cases} \tag{6}$$

(b). Second, define a normalized *clustering coefficient* κ_{cluster} in words as

The proportion of directed 2-paths $v_1 \to v_2 \to v_3$ for which $v_1 \to v_3$ is also an edge.

(see (44) for a more precise definition and derivation of (7)). Then

$$\kappa_{\text{cluster}} = \begin{cases} \frac{\alpha(1-\alpha)\lambda}{2-\alpha^2\lambda} & \text{[low]} \\ \frac{(\eta+\frac{1}{\lambda})(1-\eta-\frac{1}{\lambda})}{(\eta+\frac{1}{\lambda})(1-\eta-\frac{1}{2\lambda})} & \text{[high]}. \end{cases} \tag{7}$$

By solving (6,7) we find (Sect. 4.6) that every pair of values of $\partial, \kappa_{\text{cluster}}$ in the complete range

$$0 < \partial < \infty, \quad 0 < \kappa_{\text{cluster}} < 1$$

occurs for a unique parameter pair (α, λ) or (η, λ). Moreover the two regions can be specified as

$$0 < \partial < \infty, \quad 0 < \kappa_{\text{cluster}} \leq \tfrac{1}{\partial+2} \text{ [low]}$$

$$0 < \partial < \infty, \quad \tfrac{1}{\partial+2} < \kappa_{\text{cluster}} < 1 \text{ [high]}$$

explaining our *low* and *high clustering* terminology. So the two model parameters α, λ have fairly direct interpretations in terms of mean degree and clustering; of course we could re-parametrize the model in terms of ∂ and κ_{cluster}, but the internal mathematical structure is more conveniently expressed using the given parameters.

2.3 Distributions of In- and Out-Degrees

(a). The distribution of D_{in} is specified as

$$1 + D_{\text{in}} \overset{d}{=} \text{Geo}(e^{-\beta T}) \text{ where } T \overset{d}{=} \text{Exp}(1) \tag{8}$$

and where

$$\beta = \begin{cases} \alpha & \text{[low]} \\ \eta + 1/\lambda & \text{[high].} \end{cases} \tag{9}$$

This works out explicitly as

$$P(D_{\text{in}} \geq d) = \frac{\Gamma(d+1)\Gamma(1/\beta)}{\beta \Gamma(d+1+\frac{1}{\beta})}, \quad d \geq 0 \tag{10}$$

$$P(D_{\text{in}} = d) = \frac{\Gamma(d+1)\Gamma(1/\beta)}{\beta^2 \Gamma(d+2+\frac{1}{\beta})}, \quad d \geq 0 \tag{11}$$

with asymptotics

$$P(D_{\text{in}} = d) \sim \beta^{-2} \Gamma(1/\beta) \, d^{-1-\frac{1}{\beta}}.$$

Formula (11) appears as a special case of recent results in two-parameter proportional attachment models [6–8], but in fact is a famous 80-year old calculation – see Sect. 4.2.

(b). The distribution of D_{out} is determined by the distributional equation

$$D \overset{d}{=} \begin{cases} \sum_{i=1}^{\infty} \text{Bin}(1 + D_i, \alpha \lambda e^{-\lambda \xi_i}) & \text{[low]} \\ \sum_{i=1}^{\text{Poi}(\eta)}(1 + D_i') + \sum_{i=1}^{\infty} \text{Bin}(1 + D_i, e^{-\lambda \xi_i}) & \text{[high]} \end{cases} \tag{12}$$

where $D, D_i, D_i', i \geq 1$ are i.i.d. random variables distributed as D_{out} and where $0 < \xi_1 < \xi_2 < \ldots$ are the points of a rate-1 Poisson point process on $(0, \infty)$.

We do not know how to extract a useful explicit formula from (12) but we can compute moments. For instance

$$\text{var } D_{\text{out}} = \begin{cases} \frac{\alpha(1-\alpha+\alpha^2\lambda/2)}{(1-\alpha)^2(1-\frac{1}{2}\alpha^2\lambda)} & \text{[low]} \\ \frac{(\eta+\frac{1}{2\lambda})(2-\eta-\frac{1}{\lambda})}{(1-\eta-\frac{1}{2\lambda})(1-\eta-\frac{1}{\lambda})^2} & \text{[high]} \end{cases} \tag{13}$$

In the case $\lambda = 1/\alpha$ and in the limits $\lambda \to 0$, $\lambda \to \infty$ we get explicit formulas for the distribution of D_{out} – see Sect. 4.3 – which show in particular that the tail of D_{out} has geometric rather than power-law decay.
(c).

$$D_{\text{in}} \text{ and } D_{\text{out}} \text{ are independent.} \tag{14}$$

Because both D_{out} and D_{in} can take the value 0, we see that $P(D_{\text{in}}+D_{\text{out}} = 0) > 0$, implying that $\mathcal{G}_n$ will typically not be connected (see Sect. 2.7 for further comments).

2.4 Densities of Induced Subgraphs

One of the major advantages of the model is that, for a fixed "small" graph G, one can (in principle, and often in practice) calculate explicitly an "asymptotic density" $\text{dens}_\infty(G)$ interpreted as the limit

$$\lim_{n\to\infty} \frac{\text{number of copies of } G \text{ in } \mathcal{G}_n}{n} = \text{dens}_\infty(G).$$

Precise definitions are fussy, and are deferred to Sect. 4.4, which also records the explicit formulas we have found. Here let us point out the formula for triangles K_3:

$$\text{dens}_\infty(K_3) = \begin{cases} \frac{\alpha^3\lambda}{(1-\alpha)(2-\alpha^2\lambda)} & \text{[low]} \\ \frac{(\eta+\frac{1}{\lambda})(\eta+\frac{1}{2\lambda})}{(1-\eta-\frac{1}{\lambda})(1-\eta-\frac{1}{2\lambda})} & \text{[high]}. \end{cases} \tag{15}$$

The formula above is the key ingredient in the formula for κ_{cluster}. Recall the verbal description of κ_{cluster}:

> The proportion of directed 2-paths $v_1 \to v_2 \to v_3$ for which $v_1 \to v_3$ is also an edge.

It is intuitively clear (and formalized at (40) that the asymptotic density for occurrence of directed 2-paths $v_1 \to v_2 \to v_3$, if one does not look whether or not a third edge $v_1 \to v_2$ is present, equals ∂^2 (because of independence of in-degree and out-degree at v_2). So the verbal definition translates to

$$\kappa_{\text{cluster}} = \frac{\text{dens}_\infty(K_3)}{\partial^2}$$

and then (15) immediately gives formula (7) for κ_{cluster}.

2.5 Triangle Density as a Function of Degree

The parameter κ_{cluster} gives an overall measure of triangle density. A more de-tailed description is provided by statistics $C(k)$, $k \geq 2$ defined by

$$C(k) = \frac{E(\text{number of triangles containing a random degree-}k\text{ vertex})}{\binom{k}{2}}.$$

In principle the methods of this paper could be used to obtain an exact formula for $C(k)$, but we shall be content with outlining (Sect. 5.5) the tail property

$$C(k) \sim \frac{2\beta_2}{\beta - \beta_2} \times \frac{1}{k} \text{ as } k \to \infty. \tag{16}$$

See Sect. 6 for further comments.

2.6 Edge-Lengths

Our model has a "metric structure", in that there is a distance $d_{\text{metric}}(v, w)$ between any two vertices which does not involve the realization of edges in the random graph. So each edge (v, w) of the graph has a real-valued length $d_{\text{metric}}(v, w)$, and so a typical edge has a random length L. The probability density function for L is given by the formula

$$f(x) = \frac{1 - \alpha}{\alpha} \sum_{i=0}^{\infty} \frac{(i + 1)\Gamma(\alpha + 3)(-\lambda x)^i}{\Gamma(i + \alpha + 3)}, \quad 0 < x < \infty \quad [\text{low}]. \tag{17}$$

Mathematica gives an equivalent expression as a sum of incomplete hyperge-ometric functions. One can readily observe that $f(x) = \exp(-(\lambda \pm o(1))x)$ as $x \to \infty$. In the underlying metric space, the number of vertices within distance x of a typical vertex grows as e^x. So the tail behavior of $f(x)$ suggests

> the chance that a vertex has an edge to its k'th nearest neighbor should scale as $k^{-\lambda-1}$,

though we have not attempted detailed calculations to verify this suggestion. Note this property appears without being explicitly built into the model.

2.7 Other Local Statistics

There are further questions, concerning exact behavior in the $n \to \infty$ limit, which are in principle solvable in terms of the limit network $\mathcal{G}_\infty^*$, but where we have been unable to obtain usefully explicit answers. A major question concerns the *percolation probability*

$$p_{\text{perc}}(\alpha, \lambda) = P(\text{typical vertex is in infinite connected component of } \mathcal{G}_\infty^*) \tag{18}$$

By analogy with classical facts about the Erdős - Rényi model, we expect that above the *percolation threshold*, that is when $p_{\text{perc}}(\alpha, \lambda) > 0$, the random graph $\mathcal{G}_n$ will have a *giant component* whose size $C_n(\alpha, \lambda)$ satisfies

$$n^{-1}EC_n(\alpha, \lambda) \to p_{\text{perc}}(\alpha, \lambda).$$

Unfortunately we do not see how to write $p_{\text{perc}}(\alpha, \lambda)$ as a solution of any simple equation. By studying an easier-to-analyze *directed percolation* problem, it is not hard to show (Sect. 5.3)

$$\text{if } 2\beta - \beta_2 > 1 \text{ then } p_{\text{perc}}(\alpha, \lambda) > 0. \tag{19}$$

2.8 Average Distance

In any graph, write $d_{\text{graph}}(v, w)$ for the minimal number of edges in any path from v to w. The *diameter* Δ and the *average vertex-vertex distance* Λ are defined by

$$\Delta = \max_{v,w} d_{\text{graph}}(v, w), \ \Lambda = \text{ave}_{v,w} d_{\text{graph}}(v, w).$$

In the context of a simple proportional attachment model it is known [9] that

$$E\Delta_n, \ E\Lambda_n = \frac{(1 + o(1)) \log n}{\log \log n} \text{ as } n \to \infty.$$

It is natural to conjecture, but hard to prove, the same result for our model (above the percolation threshold and restricted to the giant component). On the other hand it seems likely that standard techniques of abstract mathematical probability would be enough to show the weaker bound $E\Lambda_n = O(\log n)$ as $n \to \infty$. Such questions *cannot* in principle be answered completely using $\mathcal{G}_\infty^*$.

2.9 Summary of Advantages and Disadvantages of the Model

The previous sections convey some advantages of the model:

- it has the three qualitative features desired in a complex network model (power-law degree distribution, clustering, small diameter);
- it fits the complete possible range of mean degree (or scaling exponent) and clustering parameters;
- it permits a broad range of explicit calculations.

So to be fair let us list some disadvantages from a modeling viewpoint.

- $\mathcal{G}_n$ is not connected (for large n); cf. Sect. 2.7;
- there is no power law for distribution of out-degree;
- in-degree and out-degree are independent;
- the scaling exponent for in-degree is determined by the mean degree; one might prefer a model where these could be specified separately;
- in the $n \to \infty$ limit not every finite graph is possible as an induced subgraph (Sect. 4.5).

3 The Model

3.1 Metric Copying Models

Let us briefly outline a general modeling framework, *metric copying models*. Each vertex v is a point in a metric space; that is, there is some real-valued distance $d(v, w)$ between any two vertices v, w. Given some rule for the positions of successive vertices $1, 2, \ldots$, and given a function $p : [0, \infty) \to [0, 1]$, we can construct random directed graphs $\mathcal{G}_n$ inductively on n as follows. When vertex n arrives, then

(i) for each directed edge (i, j) of $\mathcal{G}_{n-1}$, a "copied" edge (n, j) is created with probability $p(d(n, i))$;

(ii) for each vertex i ($1 \leq i < n$), a new edge (n, i) is created with probability $p(d(n, i))$;

(iii) the events above are independent, except that repeat edges are censored.

Imagine $p(\cdot)$ to be rapidly decreasing. A moment's thought shows how this model resembles proportional attachment models. An existing vertex v with in-degree d has $d + 1$ opportunities to acquire an in-edge, due to the next arriving vertex being close to v or close to one of the d vertices with edges to v.

In principle one could study such models based on random points in d-dimensional space, but within such settings it is notoriously hard to do explicit calculations (see e.g. [10] for different models of random graphs based on d-dimensional random points), and the choice of d is arbitrary. We will avoid both problems by using a well known (in other contexts) model which is loosely interpretable as "random points in infinite-dimensional space". Note that in d-dimensional space, the number of points within distance r of a typical point grows as r^d; what will make our model "infinite-dimensional" is that this number grows as e^r.

For later use recall that a *pre-metric* $\bar{d}(i, j)$ is symmetric and strictly positive for $j \neq i$. A pre-metric can be used to specify a metric $d(v, w)$ as the minimum, over paths $v = i_0, i_1, \ldots, i_k = w$, of $\bar{d}(i_0, i_1) + \bar{d}(i_1, i_2) + \ldots + \bar{d}(i_{k-1}, i_k)$.

3.2 A d-Dimensional Analogy

As a final preliminary, the following analogy may be helpful. In d-dimensional space R^d, take a cube $[-n^{1/d}/2, n^{1/d}/2]^d$ of volume n, and put n uniform random points in that cube. This structure has a $n \to \infty$ limit, the *Poisson point process* in R^d with mean intensity 1 point per unit volume. Moreover the limit process, which is a spatial point process on all of R^d, can be represented as the distribution, at any fixed time, of a time-evolving process of points on all of R^d, where the evolution rules are

(i) points move away from the origin as deterministic motion with exponential rate $1/d$; a point at position x at time t will be at position $xe^{(t'-t)/d}$ at times $t' > t$.

(ii) New points arrive throughout R^d as a rate-1 space-time Poisson process; that is, the chance of a point arriving in a cube of volume dx during a time interval dt equals $1 \cdot dx \, dt$.

Thus if one takes a volume-1 region of space at time t_0, this space expands to become volume e^t at time $t_0 + t$, and the arrival rate per unit time within this expanding volume is e^t at time $t_0 + t$.

In the limit process, one may regard the "present time" as time 0, and regard the process as having evolved[3] over time $-\infty < t \leq 0$. Particles at the present time have ages which are independent $\mathrm{Exp}(1)$ random variables independent of present positions; from the present configuration of positions and ages one can deterministically reconstruct the past evolution of the process.

3.3 The Stochastic Mean-Field Model of Distance

Our model of an underlying metric space is specified by three rules.

(i) Point n arrives at time $t_n = \log n$.

(ii) At the arrival time t_n, define the pre-distances $(\bar{D}(n, j; t_n), 1 \leq j \leq n - 1)$ from n to the earlier-arriving points to be independent random variables with exponential, mean n, distribution.

(iii) Distances grow exponentially with time; $\bar{D}(n, j; t) = e^{t - t_n} \bar{D}(n, j; t_n)$ for $t > t_n$.

So at time t there are $n = \lfloor e^t \rfloor$ points, and the $\binom{n}{2}$ pre-distances $\bar{D}(i, j; t)$, $1 \leq i < j \leq n)$ are independent random variables with exponential, mean e^t, distribution. These particular pre-distances are an instance of a pre-metric, and this pre-metric specifies a metric $D(i, j; t)$, $1 \leq i < j \leq n)$. Write $(\mathcal{D}_t, 0 \leq t < \infty)$ for this process of arriving points and distances.

Here is a key feature of this construction. At time t pick a uniform random point V_t as a "root". Then there is a $t \to \infty$ limit (in distribution) structure, which is a metric space on a countable infinite number of points, one being distinguished as the root. The limit structure, called the PWIT, is described below. The meaning of "limit" is that, for arbitrary fixed $r < \infty$, the configuration of points in $\mathcal{D}_t$ within distance r of V_t converges in distribution to the configuration of points of the PWIT within distance r of the root (this is *local weak convergence* of random networks [11]).

3.4 The PWIT

The PWIT is defined by a construction, illustrated in Fig. 1[4]. Start with a single root vertex $\emptyset$. This root vertex is then given an *infinite* number of *near neighbors*, and the edges from the root to the near neighbors are assigned lengths according to a realization of a Poisson process $(\xi_i^0 : 1 \leq i < \infty)$ of rate 1 on $(0, \infty)$. Now, recursively, each vertex v arising as a near neighbor of a previous

---

[3] This model is reminiscent of the steady-state theory of the Universe advocated by Fred Hoyle in the 1950s

[4] Our figures are illustrations of the definitions, rather than honest Monte Carlo simulations

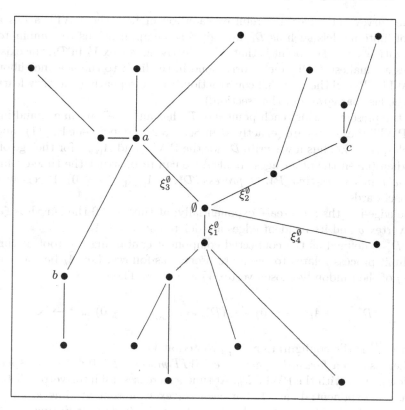

Fig. 1. The PWIT. Illustration of the vertices of the PWIT within a window of radius 3 centered on the root $\emptyset$. Lines indicate the near neighbor relationship, but are drawn only when both end-vertices are within the window. Thus the four near neighbors of $\emptyset$ shown are at distances $0 < \xi_1^\emptyset < \xi_2^\emptyset < \xi_3^\emptyset < \xi_4^\emptyset < 3$ from $\emptyset$, while there are an infinite number of near neighbors of $\emptyset$ at distances greater than 3. Orientation of lines in pictures is arbitrary. Labels a, b, c are included for later comparisons.

vertex is given an infinite number of near neighbors, and the edges to these near neighbors of v are again assigned lengths according to an independent realization of a Poisson process $(\xi_i^v : 1 \le i < \infty)$ of rate 1. This procedure is then continued *ad infinitum*. The resulting rooted infinite tree is a well defined random object, called the *Poisson weighted infinite tree* (PWIT).

The *distance* $D(v, w)$ between two vertices of the PWIT is just the sum of edge-lengths along the path from v to w. Though we have drawn a tree in Fig. 1, the lines merely indicate the near neighbor relationships; it is better to think of the edges as absent while retaining the distances $D(v, w)$. In this way we may regard the vertices of the PWIT as an infinite-dimensional analog of the d-dimensional Poisson point process in Sect. 3.2. Formula (29) later provides one formalization of "infinite-dimensional".

The survey [11] gives a careful explanation of how the PWIT arises as a limit of finite models such as $\mathcal{D}_t$, and gives some applications to combinatorial optimization[5]. The key point is that, for an arriving vertex V_t in $\mathcal{D}_t$, the existing vertices at smallest $\bar{D}$-distances correspond in the limit to the near neighbors in the PWIT. (Recall the Sect. 3.3 construction; we are repeating the "key feature" from the last paragraph of that section.)

In the present setting, each point v of $\mathcal{D}_t$ has an "age" at time t, and in the limit PWIT these ages are (exactly as in Sect. 3.2) independent Exp(1) random variables, A_v say. Thus if we write $\mathcal{D}_0^*$ for the PWIT and A_{root} for the age of the root, then (given the other ages A_v also) we can reconstruct the time-evolution of a *backwards space-time PWIT process* $(\mathcal{D}_s^*, -A_{\text{root}} \leq s \leq 0)$. Precisely, as s runs backwards

(a) the edge-lengths ξ decrease exponentially; at time $s < 0$ the length is ξe^s;

(b) a vertex v and its incident edges are deleted at $s = -A_v$.

Then $\mathcal{D}_s^*$ is defined as the connected component containing the root at time s. This limit process relates to the finite process as follows. Let $\tilde{A}_t$ be the age (at time t) of the randomly-chosen vertex V_t at time t. Then

$$(\mathcal{D}_{t+s}^{*(t)}, -\tilde{A}_t \leq s \leq 0) \stackrel{d}{\to} (\mathcal{D}_s^*, -\tilde{A}_{\text{root}} \leq s \leq 0) \text{ as } t \to \infty \qquad (20)$$

where $\mathcal{D}_{t+s}^{*(t)}$ is the configuration $\mathcal{D}_{t+s}$ rooted at V_t.

There is also a *forwards space-time PWIT process* $(\mathcal{D}_s^*, 0 \leq s < \infty)$ specified as follows. Start with the PWIT $\mathcal{D}_0^*$. At time s increases, all inter-vertex distances increase at exponential rate 1. For each vertex v present at time s, and each $0 < r < \infty$, there is (as explained below) chance $1 \cdot dr\, ds$ that during $[s, s+ds]$ a new vertex v' will appear at distance $\in [r, r+dr]$ from v as a near neighbor of v. Along with this vertex (which has current age 0) is an independent copy of the PWIT rooted at v', whose other vertex-ages are independent Exp(1). The relation between the finite-t and the limit process is analogous to (20):

$$(\mathcal{D}_{t+s}^{*(t)}, 0 \leq s < \infty) \stackrel{d}{\to} (\mathcal{D}_s^*, 0 \leq s < \infty) \text{ as } t \to \infty. \qquad (21)$$

Here is the calculation leading to the coefficient "1" in

there is chance $1 \cdot dr\, ds$ that during $[s, s+ds]$ a new vertex v'

will appear at distance $\in [r, r+dr]$ from v as a near neighbor of v .   (22)

In the process $(\mathcal{D}_t)$, during time $[t+s, t+s+ds]$ about $e^{t+s} ds$ vertices arrive; for each existing vertex, the chance an arriving vertex is within $\bar{D}$-distance $[r, r+dr]$ equals $e^{-(t+s)} \exp(-re^{-(t+s)}) dr$. So the chance in (22) equals

$$e^{t+s} ds \times e^{-(t+s)} \exp(-re^{-(t+s)}) dr \approx 1 \times dr\, ds.$$

---

[5] See also [12] for novel scaling exponents arising in the study of the mean-field traveling salesman problem.

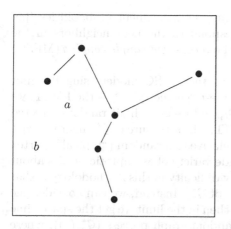 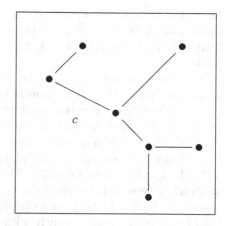

Fig. 2. The space-time PWIT process. Regarding Fig. 1 as showing the PWIT at a time t_+, Fig. 2 shows the space-time PWIT at an earlier time t_- at which only three vertices a, b, c of the vertices in the Fig. 1 window have arrived. Figure 2 shows smaller windows centered on a and on c. The other vertices in Fig. 2, and the near neighbor relation shown by lines, are still present at time t_+, but are not visible in Fig. 1 because the expansion of distances has placed them outside the Fig. 1 window.

Recursive Self-Similarity. Implicit in the model is the fact that the "geometry" of the space seen by a newly-arriving particle v^* is statistically the same as the geometry seen by a typical existing particle. This is the familiar PASTA (Poisson arrivals see time averages) property in queuing theory. In particular, at the arrival time of v^* the *geometric components* containing the different near neighbor vertices $v_1, v_2, \ldots$ are independent copies of the PWIT. This *recursive self-similarity* property of the PWIT process is fundamental to its analytic tractability.

Figure 2 and its legend may be helpful.

3.5 The MFSC Model

The process $(\mathcal{D}_{t_n}, n = 1, 2, 3, \ldots)$ of arrivals and inter-point distances described in Sect. 3.3 defines an "underlying geometry"; we now define the random graph process $(\mathcal{G}_n, n = 1, 2, 3, \ldots)$ which is the subject of this paper. Fix two parameters $0 < \alpha < \infty$ and $0 < \lambda < \infty$. Write

$$p(x) = \min(1, \alpha \lambda e^{-\lambda x}), \quad 0 \le x < \infty. \tag{23}$$

We now implement a version of the "metric copying" idea from Sect. 3.1. $\mathcal{G}_1$ consists of vertex 1 and no edges. When vertex n arrives at time $t_n = \log n$, then
(i) for each directed edge (i, j) of $\mathcal{G}_{n-1}$, a "copied" edge (n, j) is created with probability $p(\bar{D}(n, i; t_n))$;
(ii) for each vertex i $(1 \le i < n)$, a new edge (n, i) is created with probability $p(\bar{D}(n, i; t_n))$;
(iii) the events above are independent, except that repeat edges are censored.

Note that we use $\bar{D}$ instead of D in determining attachment probabilities (because $\bar{D}$-near vertices at finite time correspond to the near neighbors in the limit PWIT). We call $(\mathcal{G}_n, n = 1, 2, 3, \ldots)$ the *mean-field simple copying* (MFSC) model.

Our focus in this paper is the study of the MFSC model using its limit structure. Just as the $t \to \infty$ limit of the time-t "geometry" $\mathcal{D}_t$ is the PWIT, we can consider $\mathcal{G}_t$ as a structure built over $\mathcal{D}_t$, and we get a limit random directed graph $\mathcal{G}_\infty^*$ as a structure built over the PWIT. The structure of $\mathcal{G}_\infty^*$ near the root is exactly the $t \to \infty$ limit structure of $\mathcal{G}_t$ relative to a random ("typical") vertex V_t, and so we can obtain (in principle) a wide variety of asymptotic results about $\mathcal{G}_t$ by doing calculations upon $\mathcal{G}_\infty^*$. The only difficulty in this methodology is that we don't have a useful explicit description of $\mathcal{G}_\infty^*$. Instead, we can consider the space-time limits (20,21) jointly with $(\mathcal{G}_t)$; then in the limit we get the space-time PWIT processes existing jointly with a random graph process $(\mathcal{G}_\infty^*(s))$, where now $\mathcal{G}_\infty^*(0) = \mathcal{G}_\infty^*$. The process $(\mathcal{G}_\infty^*(s))$ evolves with s and the space-time PWIT process by the rules implied by (i)-(iii) above.

Precisely, the evolution rules (illustrated by Figs. 3–6) are

when a new vertex v^* arrives in the forwards space-time PWIT process at time s, it has near neighbors $(v_1, v_2, \ldots)$ at distances $(\xi_1, \xi_2, \ldots)$, and $\mathcal{G}_\infty^*(s)$ has put a random graph structure on the geometric component containing each v_i. For each i and each directed edge (v_i, w), a new edge (v^*, v_i) or (v^*, w) is created with probability $p(D(v^*, v_i))$, independently for different possible edges.

Recursive Self-Similarity. What makes this process tractable is that the recursive self-similarity property of the PWIT extends to the random graph process; each v_i defines a geometric component and a random graph on that component,

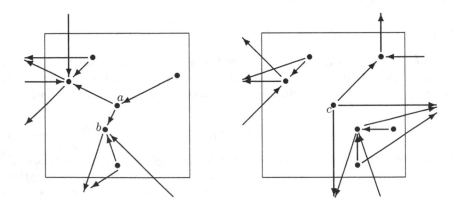

Fig. 3. The graph process $\mathcal{G}_\infty^*(t_-)$ on the realization of the space-time PWIT at time t_- in Fig. 2. For the graph process we show all edges with either end-vertex within the window. The following figures show the evolution of $\mathcal{G}_\infty^*(t)$ over $t_- < t \leq t_+$.

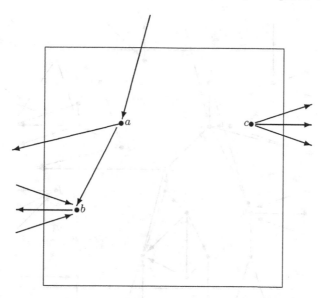

Fig. 4. Figures 4–6 build up the graph $\mathcal{G}^*_\infty(t_+)$ on the time-t_+ PWIT in Fig. 1. Figure 4 here shows only the edges that were present at time t_-, that is the edges shown in Fig. 3. Some edges crossing outside the window have been redrawn at different angles for later convenience.

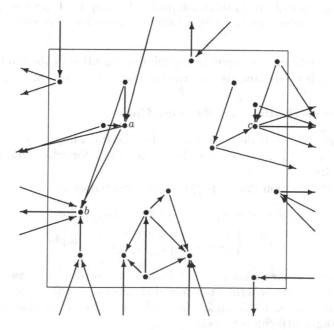

Fig. 5. Suppose in Fig. 1 that the center vertex $\emptyset$ arrives at time t_+; Fig. 5 illustrates the graph just before that arrival. Since the time-t_- configuration in Fig. 3, more vertices have arrived and formed edges, and distances have expanded.

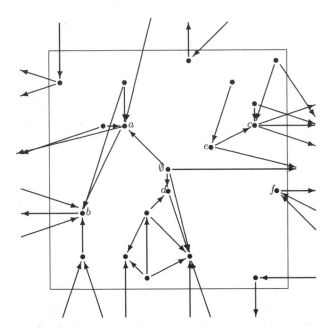

Fig. 6. Vertex $\emptyset$ arrives at time t_+ with near neighbors $d, e, a, f, \ldots$. Out-edges from $\emptyset$ appear according to the metric copying scheme of Sect. 3.1 with $p(x)$ given by (23). So for each near neighbor v and each existing edge (v, w), an edge $(\emptyset, v)$ or $(\emptyset, w)$ appears with probability $p(D(\emptyset, v))$. In this realization, edges appeared to the near neighbors d and a, and two other edges (one from d and one from e) were copied.

and these are independent copies of the joint distribution of the PWIT and $\mathcal{G}^*_\infty$. This property is used extensively in the calculations in Sect. 4.

3.6 Reparametrization and Extreme Cases

Although the MFSC model makes sense for the full range $(0 < \alpha, \lambda < \infty)$ of parameters, we will only consider the ranges (4,5) for which the limit mean degree is finite.

Note that definition (23) of $p(x)$ can be rewritten as

$$p(x) = \alpha \lambda e^{-\lambda x}, \qquad 0 < x < \infty \qquad \text{[low]} \tag{24}$$

$$p(x) = \begin{cases} 1 & 0 < x \leq \eta \\ e^{-\lambda(x-\eta)} & \eta < x < \infty \end{cases} \qquad \text{[high]} \tag{25}$$

where the reparametrization $\eta := \lambda^{-1} \log(\alpha\lambda)$ when $\alpha\lambda > 1$ is used in (25). In the two extremes of clustering, our model simplifies in different ways. For fixed α, when n is large and α is small, the model resembles the following variant of the proportional attachment model:

An arriving vertex has a Poi(α) number of out-edges, whose end-vertices are chosen with probabilities proportional to 1+ in-degree.

At the other extreme, for fixed η our model makes sense with $\lambda = \infty$, interpreting (25) to mean $p(x) = 0$, $x > \eta$. In this case the model becomes, for large n,

> An arriving vertex v^* chooses at random a $\text{Poi}(\eta)$ number of neighbors v_j, and creates edges (v^*, v_j), and also copies each existing edge (v_j, w) to a new edge (v^*, w).

Clearly in the former limit we have $\kappa_{\text{cluster}} = 0$ and in the latter limit we have $\kappa_{\text{cluster}} = 1$.

4 Calculations

In Sect. 4 we derive the formulas stated in Sects. 2.2–2.6. As described in Sect. 3.5, our methodology is to regard D_{in} and D_{out} as the (random) in-degree and out-degree of the root in $\mathcal{G}_\infty^*$, and to study this using the time-dynamics of $\mathcal{G}_\infty^*(s)$ derived from the space-time PWIT process and the evolution rules of the graph process.

4.1 Two Helpful Calculations

We will make frequent use of the next lemma.

Lemma 1. *For integers $u \geq 1$ write*

$$Z^{(u)} = \sum_{i=1}^{\infty} p^u(\xi_i) = \sum_{i=1}^{\infty} \left[\min(1, \alpha\lambda e^{-\lambda\xi_i}) \right]^u .$$

Then

$$\beta_u := EZ^{(u)} = \begin{cases} u^{-1}\alpha^u\lambda^{u-1} & \text{[low]} \\ \eta + \frac{1}{u\lambda} & \text{[high]} \end{cases} \tag{26}$$

$$\text{var } Z^{(1)} = \begin{cases} \frac{1}{2}\alpha^2\lambda & \text{[low]} \\ \eta + \frac{1}{2\lambda} & \text{[high]}. \end{cases} \tag{27}$$

In particular, $\beta := \beta_1 < 1$ for all parameter values.

Proof. We will do the low clustering density case – the high density case is similar. By (3) the chance that some ξ_i falls into an interval $[x, x + dx]$ is dx, so

$$\beta_u = \int_0^\infty (\alpha\lambda e^{-\lambda x})^u \, dx = \alpha^u\lambda^u \int_0^\infty e^{-u\lambda x} \, dx = \alpha^u\lambda^u/(u\lambda).$$

Moreover, there is a general formula for variance of a sum over a Poisson (rate 1) process (ξ_i):

$$\text{var} \left(\sum_i w(\xi_i) \right) = \int_0^\infty w^2(x) \, dx$$

and applying this formula for $Z^{(1)}$ gives

$$\text{var } Z^{(1)} = \int_0^\infty (\alpha\lambda e^{-\lambda x})^2 \, dx = \alpha^2\lambda^2/(2\lambda).$$

We next recall a classical result. Fix $0 < \theta < \infty$. Set $N(0) = 1$ and let $(N(t), t \geq 0)$ be the *Yule process* of rate θ, that is the Markov process which changes only by $+1$ steps and for which

$$P(N(t + dt) = n + 1 | N(t) = n) = \theta n \, dt.$$

A textbook result (e.g. [13] sec. 5.3) says

$$N(t) \stackrel{d}{=} \text{Geo}(e^{-\theta t}). \tag{28}$$

Note that in the PWIT, if $N(r)$ is the number of vertices within distance r from the root (counting the root itself), then the process $(N(r), r \geq 0)$ is a Yule process of rate 1, because for a vertex v at distance $r' < r$, the chance of v having a near neighbor at distance $\in [r - r', r - r' + dr]$ equals $1 \cdot dr$. So in particular,

$$EN(r) = E\text{Geo}(e^{-r}) = e^r. \tag{29}$$

4.2 Distribution of In-Degree

We start by giving the derivation of

$$1 + D_{\text{in}} \stackrel{d}{=} \text{Geo}(e^{-\beta T}) \text{ where } T \stackrel{d}{=} \text{Exp}(1) \tag{8}$$

for $\beta = EZ^{(1)}$. In the forwards space-time PWIT process, let $N(t)$ be $1+$ the in-degree of the root, when the root has age t. Thus $N(t)$ counts the set of vertices v for which $v \to$ root is an edge, or $v =$ root. When a new vertex v' arrives with some v in this set as a near neighbor, at distance r, there is chance $p(r)$ for the root's in-degree to increase by 1, and so from the dynamics (22) of the forwards space-time PWIT process we see that $N(t)$ is the Yule process of rate

$$\beta = \int_0^\infty p(r) \, dr.$$

Use formula (28) and the fact that the age of the root of the PWIT has $\text{Exp}(1)$ distribution to obtain (8).

We can quickly use (8) to calculate ED_{in}.

$$\begin{aligned}
1 + ED_{\text{in}} &= E(E(\text{Geo}(e^{-\beta T})|T)) \\
&= Ee^{\beta T} \text{ because } E\text{Geo}(p) = p^{-1} \\
&= \int_0^\infty e^{\beta t} e^{-t} \, dt = \frac{1}{1 - \beta}
\end{aligned}$$

giving $ED_{\text{in}} = \frac{\beta}{1-\beta}$ as at (6). We now calculate the distribution of D_{in} in the same way. Because $P(\text{Geo}(p) \geq i + 1) = (1 - p)^i$, $i \geq 0$ we have

$$P(D_{\text{in}} \geq i) = EP(D_{\text{in}} \geq i|T)$$
$$= EP(1 + D_{\text{in}} \geq i + 1|T)$$
$$= E\left(1 - e^{-\beta T}\right)^i$$
$$= \int_0^\infty (1 - e^{-\beta t})^i e^{-t}\, dt$$
$$= \tfrac{1}{\beta} \int_0^1 (1 - s)^i s^{\frac{1}{\beta} - 1}\, ds \text{ setting } s = e^{-\beta t}$$
$$= \frac{1}{\beta} \frac{\Gamma(i+1)\Gamma(\frac{1}{\beta})}{\Gamma(i+1+\frac{1}{\beta})} \text{ using the Beta integral formula.}$$

This is (10), and (11) follows.

Historical note. Yule [14] introduced what we now call the Yule process in 1924 in the context of a model for evolution of new species. It is interesting that his central mathematical results are the Geometric distribution (28) [his (5)] and the calculation starting from our (8) [representing, for Yule, a distribution of numbers of species in a typical genus] of the explicit distribution (11) [his (12)]. After 80 years we have slicker notation but the argument is the same! Moreover Yule's motivation was to find a simple model yielding a power-law distribution for number of species per genus, just as the motivation for the recent literature on proportional attachment models was to find a simple model yielding power-law degree distributions.

4.3 Distribution of Out-Degree

We will first derive (12). Because the out-edges are formed on arrival, we may suppose the root of the PWIT has just arrived. Consider a near neighbor v' at distance r. For each out-edge of v', and for v' itself, there is chance $p(r)$ that a corresponding out-edge is created at the root, giving a total number $\text{Bin}(1 + D(v'), p(r))$ of out-edges, where $D(v')$ is the out-degree of v'. The recursive self-similarity property (end of Sect. 3.5) implies that the $(D(v') : v'$ near neighbor of root) are i.i.d. random variables distributed as D_{out}, and independent of their distances (ξ_i) from the root. Rewriting $(D(v'))$ as $(D^{(i)})$ in increasing order of distance from root,

$$D_{\text{out}} = \sum_{i=1}^\infty \text{Bin}(1 + D^{(i)}, p(\xi_i)) \tag{30}$$

which becomes (12).

We now turn to the issue of using (30) to get information about the distribution of D_{out}. Because a directed edge contributes equally to total in-degree and to total out-degree, we know a priori that ED_{out} must equal ED_{in}, but let us first check that we can indeed use (12) to show $ED_{\text{out}} = \beta/(1 - \beta)$. Because (1) $E\text{Bin}(n, p) = np$ we see

$$E\text{Bin}(1 + D_i, p(\xi_i)) = (1 + ED_{\text{out}})\, Ep(\xi_i).$$

So (30) gives

$$ED_{\text{out}} = (1 + ED_{\text{out}}) \cdot EZ^{(1)} = (1 + ED_{\text{out}})\beta$$

giving $ED_{\text{out}} = \beta/(1 - \beta)$.

Variance. The calculation of the variance var D_{out} provides a textbook illustration of the utility of the general *conditional variance formula*

$$\text{var } X = E\text{var } (X|Y) + \text{var } E(X|Y).$$

We give the details in the low density case; the high density case is similar. In the defining equation (12) write D for D_{out} and write $\mathbf{D}$ and $\varXi$ for the random sequences (D_i) and (ξ_i). Because (2) var $\text{Bin}(n, p) = np(1 - p)$ we have

$$\text{var } (D|\mathbf{D}, \varXi) = \sum_i (1 + D_i)\alpha\lambda e^{-\lambda\xi_i}(1 - \alpha\lambda e^{-\lambda\xi_i}).$$

Recursive self-similarity, as used above, implies independence of the i.i.d. sequence $(D_i, i \geq 1)$ and the Poisson process $(\xi_i, i \geq 1)$. So

$$E\text{var } (D|\mathbf{D}, \varXi) = (1 + ED)(EZ^{(1)} - EZ^{(2)})$$
$$= \frac{\alpha(1 - \frac{\alpha\lambda}{2})}{1 - \alpha} \text{ using (6) and (26).} \tag{31}$$

Next consider the conditional expectation

$$E(D|\mathbf{D}, \varXi) = \sum_i (1 + D_i)\alpha\lambda e^{-\lambda\xi_i} = W, \quad \text{say.}$$

We will calculate var W by using the conditional variance formula. Because

$$\text{var } (W|\varXi) = \sum_i (\text{var } D) \cdot \alpha^2\lambda^2 e^{-2\lambda\xi_i}$$

we have

$$E\text{var } (W|\varXi) = (\text{var } D) \cdot EZ^{(2)} = (\text{var } D) \cdot \alpha^2\lambda/2. \tag{32}$$

And since $E(W|\varXi) = (1 + ED)Z^{(1)} = \frac{1}{1-\alpha}Z^{(1)}$ we have

$$\text{var } E(W|\varXi) = \frac{1}{(1 - \alpha)^2}\text{var } Z^{(1)} = \frac{\alpha^2\lambda}{2(1 - \alpha)^2}. \tag{33}$$

Using the conditional variance formula twice

$$\text{var } D = E\text{var } (D|\mathbf{D}, \varXi) + \text{var } W$$
$$= E\text{var } (D|\mathbf{D}, \varXi) + E\text{var } (W|\varXi) + \text{var } E(W|\varXi)$$
$$= \frac{\alpha(1 - \frac{\alpha\lambda}{2})}{1 - \alpha} + (\text{var } D) \cdot \alpha^2\lambda/2 + \frac{\alpha^2\lambda}{2(1 - \alpha)^2}.$$

Solving gives the equation (13) for var D_{out}.

Special cases. **(a).** Fix α. Because $p(x) \le \alpha\lambda$, in the $\lambda \to 0$ limit we can apply the Poisson limit of Binomials result to the defining equation (12) to obtain (cf. Sect. 3.6)

$$(\text{in } \lambda \to 0 \text{ limit}) \qquad D_{\text{out}} \overset{d}{=} \text{Poi}(\alpha). \tag{34}$$

(b). Fix η. In the $\lambda \to \infty$ limit we can use the limit process of Sect. 3.6 to show that $1 + D_{\text{out}}$ has the distribution of the total population size in a Galton-Watson branching process[6] with $\text{Poi}(\eta)$ offspring distribution. This is (see e.g. [15]) the *Borel-Tanner(η)* distribution

$$(\text{in } \lambda \to \infty \text{ limit}) \qquad P(1 + D_{\text{out}} = d) = \frac{(\eta d)^{d-1} e^{-\eta d}}{d!}, \quad d \ge 1. \tag{35}$$

(c). In the case $\alpha\lambda = 1$ it turns out (an argument is sketched in Sect. 5.4)

$$1 + D_{\text{out}} \overset{d}{=} \text{Geo}(1 - \alpha). \tag{36}$$

Independence of in-degree and out-degree. This independence, noted at (14), follows from the fact that in the forwards space-time PWIT process the out-degree of the root is determined at the arrival time of the root vertex; the subsequent evolution of the process of in-edges is clearly independent of the state of the graph immediately after arrival.

4.4 Densities of Induced Subgraphs

Here we give details of the definition and interpretation of "density of induced subgraphs" mentioned in Sect. 2.4, and list explicit formulas.

Let G and $\mathcal{G}$ be finite directed acyclic graphs; think of G as small and $\mathcal{G}$ as large. Define "density of G as an induced subgraph of $\mathcal{G}$" by

$$\text{dens}(G|\mathcal{G}) = \frac{\#\{V \subset \mathcal{G} : V \text{ isomorphic to } G\}}{\#\{\text{ vertices of } \mathcal{G}\}}$$

where $\#$ denotes cardinality ("number of") and V denotes a vertex-subset of $\mathcal{G}$ with its induced subgraph. See Fig. 7, where there are 3 such vertex-subsets $\{a, b, e\}$, $\{b, c, e\}$, $\{c, d, e\}$ and so where $\text{dens}(G|\mathcal{G}) = 3/5$.

We want to study $n \to \infty$ limits of $\text{dens}(G|\mathcal{G}_n)$ in our MFSC random graph model $(\mathcal{G}_n)$, for fixed G. To use our methodology we must first rephrase the definition of $\text{dens}(G|\mathcal{G})$ in terms of the rooted graph $\mathcal{G}^*$ obtained by giving $\mathcal{G}$ a uniform random root. For such $\mathcal{G}^*$, and for an arbitrarily-rooted directed graph G^*, define a random variable

---

[6] A population process starting with one individual in generation 0, individuals having i.i.d. random numbers of offspring in successive generations

Fig. 7. Two rootings G_1^*, G_2^* of a graph G.

$\mathbf{X}(G^*|\mathcal{G}^*)$ = number of vertex-subsets V of $\mathcal{G}^*$ including the root such that $\mathcal{G}^*$ restricted to V is isomorphic to G^* via a root-preserving isomorphism.

The randomness arises only from choice of root of $\mathcal{G}^*$; notation $\mathbf{X}$ has no special significance except to distinguish this from simpler random variables.

It is an easy fact that

$$\text{dens}(G|\mathcal{G}) = \frac{E\mathbf{X}(G^*|\mathcal{G}^*)}{\iota(G^*)} \tag{37}$$

where G^* is G with an arbitrary choice of root, and where $\iota(G^*) \geq 1$ is the number of different root-choices which would give a rooted graph isomorphic to this particular choice. Rather than write a formal proof, let us just illustrate identity (37) using Fig. 7. For the choice of root giving G_1^*, the number of isomorphic vertex-subsets V of $\mathcal{G}^*$ equals 2 (resp. 1) if the root of $\mathcal{G}^*$ chances to be b (resp. d), and so $E\mathbf{X}(G_1^*|\mathcal{G}^*) = 3/5$. For the choice of root giving G_2^*, the number of isomorphic vertex-subsets V of $\mathcal{G}^*$ equals 1 (resp. 2, 3) if the root of $\mathcal{G}^*$ chances to be a (resp. c, e), and so $E\mathbf{X}(G_2^*|\mathcal{G}^*) = 6/5$. Since $\iota(G_1^*) = 1$ while $\iota(G_2^*) = 2$, we have checked identity (37) in this example.

Note that in Fig. 7, the induced subgraph on $\{a, b, c\}$ is not isomorphic to G because of the extra edge $a \to c$. Obviously we can make parallel definitions allowing extra edges (pedantically: replace "isomorphism" by "vertex-bijection and edge-surjection") and we write $\overline{\text{dens}}\,(G|\mathcal{G})$ and $\bar{\mathbf{X}}(G^*|\mathcal{G}^*)$ in this setting. For instance, in Fig. 7 we have $\overline{\text{dens}}\,(G|\mathcal{G}) = 4/5$ and $E\bar{\mathbf{X}}(G_1^*|\mathcal{G}^*) = 4/5$.

The point of all this is that the definition of $\mathbf{X}(G^*|\mathcal{G}^*)$ makes sense when $\mathcal{G}^*$ is a rooted *infinite* graph. The key methodology in our analysis of the MFSC model $(\mathcal{G}_n)$ is that the randomly-rooted $\mathcal{G}_n^*$ converge locally to a limit random infinite rooted graph $\mathcal{G}_\infty^*$, implying via (37) that

$$\text{dens}(G|\mathcal{G}_n) \to \frac{E\mathbf{X}(G^*|\mathcal{G}_\infty^*)}{\iota(G^*)} := \text{dens}_\infty(G), \text{ say.} \tag{38}$$

In parallel.

$$\overline{\text{dens}}\,(G|\mathcal{G}_n) \to \frac{E\bar{\mathbf{X}}(G^*|\mathcal{G}_\infty^*)}{\iota(G^*)} := \overline{\text{dens}}_\infty\,(G), \text{ say.} \tag{39}$$

In Sects. 4.5 and 5.2 we calculate dens(G) for several cases of G; let us record the formulas below. Note that limit densities may be infinite, in which cases we will point out the conditions on parameters needed for finiteness.

(a). For a directed path π_r with $r \geq 1$ edges,

$$\overline{\mathrm{dens}}_\infty(\pi_r) = \partial^r. \tag{40}$$

(b). For the complete directed acyclic graph K_r on $r \geq 2$ vertices, (that is, vertices $\{1, 2, \ldots, r\}$ and edges $i \to j$ for $1 \leq i < j \leq r$),

$$\mathrm{dens}_\infty(K_r) = \prod_{u=1}^{r-1} \frac{\beta_u}{1 - \beta_u} \tag{41}$$

where $\beta_1 = \beta$ and for general $u \geq 1$

$$\beta_u := \begin{cases} u^{-1}\alpha^u\lambda^{u-1} & \text{[low]} \\ \eta + \frac{1}{u\lambda} & \text{[high]}. \end{cases} \tag{42}$$

In particular, for the case of *triangles* K_3 we have explicitly

$$\mathrm{dens}_\infty(K_3) = \begin{cases} \frac{\alpha^3\lambda}{(1-\alpha)(2-\alpha^2\lambda)} & \text{[low]} \\ \frac{(\eta+\frac{1}{\lambda})(\eta+\frac{1}{2\lambda})}{(1-\eta-\frac{1}{\lambda})(1-\eta-\frac{1}{2\lambda})} & \text{[high]}. \end{cases} \tag{43}$$

As already mentioned in Sect. 2.4, the formula above is the key ingredient in the formula for $\kappa_{\mathrm{cluster}}$. Recall its verbal description

> The proportion of directed 2-paths $v_1 \to v_2 \to v_3$ for which $v_1 \to v_3$ is also an edge.

This becomes

$$\kappa_{\mathrm{cluster}} = \frac{\mathrm{dens}_\infty(K_3)}{\overline{\mathrm{dens}}_\infty(\pi_2)} \tag{44}$$

and then (40,44) immediately give the formula (7).

(c). For a directed path π_r with $r \geq 1$ edges,

$$\mathrm{dens}_\infty(\pi_r) = \delta \left(\frac{\beta_1 - \beta_2}{(1 - \beta_1)(1 - \beta_2)} \right)^{r-1}. \tag{45}$$

(d). For the complete bipartite directed graph $K_{2,2}$, for $\beta_2 < \frac{1}{2}$ (which always holds in the low density case)

$$\overline{\mathrm{dens}}_\infty(K_{2,2}) = \frac{\partial\beta_2(\beta_2 + \frac{1}{2}\partial\beta)}{(1 - 2\beta_2)(1 - \beta_2)}. \tag{46}$$

(e). In principle one can calculate $\mathrm{dens}_\infty(G^*)$ for any G^*, but in practice it is not clear to what extent useful explicit formulas can be found – see Sect. 4.5 for further discussion, and for the observation that certain graphs G^* have $\mathrm{dens}_\infty(G^*) = 0$, "asymptotically negligible density".

4.5 Densities of Induced Subgraphs: Easy Explicit Formulas

Recall the basic result (38) on subgraph density:

$$\text{dens}(G|\mathcal{G}_n) \to \frac{E\mathbf{X}(G^*|\mathcal{G}_\infty^*)}{\iota(G^*)} := \text{dens}_\infty(G)$$

where G^* is an arbitrary rooting of G. In calculating the right side, to simplify notation we write

$$\chi(G^*) = E\mathbf{X}(G^*|\mathcal{G}_\infty^*)$$

and similarly for $\bar{\chi}(\cdot)$ and $\bar{\mathbf{X}}(\cdot)$ and $\overline{\text{dens}}\,(G)$.

First consider π_r, the directed path with r edges, rooted at the last-arriving vertex, which we will call the *head*. Clearly $\chi(\pi_1) = ED_{\text{out}} = \partial$. Let us write out the (rather obvious) inductive argument for calculating $\bar{\chi}(\pi_r)$. Whether or not the root vertex of $\mathcal{G}_\infty^*$ is the head of a r-path is determined at its arrival time. Consider a near neighbor v_i of the root, at distance ξ_i. The expected number of $r - 1$-paths headed by v_i equals $\bar{\chi}(\pi_{r-1})$. So the expected number of r-paths of the form root $\to v_i \to \ldots$ equals $\bar{\chi}(\pi_{r-1}) \times P((\text{root}, v_i)$ is edge of $\mathcal{G}_\infty^*)$. Summing over i gives

$$\bar{\chi}(\pi_r) = \bar{\chi}(\pi_{r-1}) \times ED_{\text{out}} = \bar{\chi}(\pi_{r-1}) \times \partial$$

and so $\bar{\chi}(\pi_r) = \partial^r$ by induction. This is formula (40).

The result for the complete directed graph K_r on r vertices is similar. For $r = 2$ we have $\chi(K_2) = \chi(\pi_1) = \partial$ and so to establish formula (41) by induction it is enough to show

$$\chi(K_{r+1}) = \chi(K_r) \times \frac{\beta_r}{1 - \beta_r}. \tag{47}$$

In the forwards space-time PWIT process, consider a vertex-set S_r isomorphic to K_r, headed by its latest-arriving vertex v_*. At time t after the arrival of v_*, let $N_t = 1+$ the number of K_{r+1}-subgraphs of the forwards space-time PWIT process which are of the form $\{v\} \cup S_r$ for some v; regard the "+1" as counting S_r itself. Then N_t is a Yule process of rate

$$E \sum_i p^r(\xi_i) = \beta_r \tag{48}$$

because for each vertex v counted in N_t, a new vertex v' arriving with near neighbor v at distance x has chance $[p(x)]^r$ to create the r edges needed to make $\{v'\} \cup S_r$ be a K_{r+1} subgraph. Moreover these are the only ways in which a new K_{r+1} of the form $\{v'\} \cup S_r$ can be formed. By the Yule formula (29) $N_t \overset{d}{=} \text{Geo}(\exp(-t\beta_r))$. Now regard K_{r+1} as rooted by its second-latest arriving vertex. In $\mathcal{G}_\infty^*$ the root has age $T \overset{d}{=} \text{Exp}(1)$. At its arrival time the root headed some random number of K_r's, with mean $\chi(K_r)$, so by considering the mean number of K_{r+1}'s at the present time

$$\chi(K_{r+1}) = \chi(K_r) \times (EN_T - 1)$$
$$= \chi(K_r) \times (E \exp(T \,\beta_r) - 1)$$
$$= \chi(K_r) \times \left(\frac{1}{1 - \beta_r} - 1\right)$$

giving (47).

Other subgraphs. The derivations of formulas (45,46) dealing with paths and $K_{2,2}$ are relegated to Sects. 5.1 and 5.2.

For the graph $\text{out} - \text{star}_r$ consisting of r out-edges at a root, it is clear that

$$\bar{\chi}(\text{out} - \text{star}_r) = E\binom{D_{\text{out}}}{r}$$

and similarly

$$\bar{\chi}(\text{in} - \text{star}_r) = E\binom{D_{\text{in}}}{r},$$

and these can in principle be evaluated using (12,8).

Some subgraphs have density zero.

It is easy to see that the graph G^* above (where no "vertical" edge is present) has $\chi(G^*) = 0$.

4.6 Reparametrization

Writing α, λ in terms of $\partial, \kappa = \kappa_{\text{cluster}}$ by solving (6,7) gives the formulas

$$\left.\begin{array}{l} \alpha = \frac{\partial}{\partial+1} \\ \lambda = \frac{2(1+\frac{1}{\partial})^2}{1+\frac{1}{\partial\kappa}} \end{array}\right\} 0 < \kappa \leq \frac{1}{\partial + 2} \tag{49}$$

$$\left.\begin{array}{l} \eta = \frac{\partial((\partial+2)\kappa-1)}{(\partial+1)(1+\partial\kappa)} \\ \lambda = \frac{(\partial+1)(1+\partial\kappa)}{2\partial(1-\kappa)} \end{array}\right\} \frac{1}{\partial + 2} < \kappa < 1. \tag{50}$$

4.7 Edge-Lengths

The previous calculations have not made very extensive use of the time-dynamics of the forwards space-time PWIT process, and in particular have not used the fact that edge-lengths grow exponentially at rate 1. To derive the formula (17) for edge length density we do need to exploit such time-dynamics. We consider only the low-density case; the high density case is more complicated because the distribution in (ii) below is no longer exponential.

Consider the lengths of the in-edges at a particular vertex v_0. Following a tradition in mathematical probability, we visualize an in-edge of length ℓ as a "particle" at position ℓ on a line; we also put a particle at position 0 to represent the vertex v_0 itself. If we start time τ with $\tau = 0$ at the arrival time of v_0, then the evolution of the "particle process" can be specified as follows.

(i) There is a particle at position 0 at all times $\tau \geq 0$.

(ii) For each particle (at position x at time τ, say), at stochastic rate α per unit time a new particle appears at position $x + \mathrm{Exp}(\lambda)$.

(iii) particle positions increase deterministically at exponential(1) rate: a particle at x at time τ will be at $xe^{\tau_0 - \tau}$ at time $\tau_0 > \tau$.

Rule (ii) derives from (22): for an existing edge (v', v_0), a new vertex arriving at distance r from near neighbor v' creates an edge to v_0 with probability $p(r)$, so the rate at which each existing edge is copied equals $\int_0^\infty p(x)\, dx = \alpha$; moreover conditional on copying, the distance r has $\mathrm{Exp}(\lambda)$ distribution, and so the length of the new edge equals the length of the old edge $+\mathrm{Exp}(\lambda)$.

To analyze this particle process of edge lengths, define

$$G(\tau, x) = E(\text{number of edges of length} > x \text{ at time } \tau)$$

so that

$$g(\tau, x) = -\frac{d}{dx}G(\tau, x) = \text{mean edge-length density at time } \tau.$$

We shall study

$$f(x)\, dx = E(\text{number of in-edges at a typical vertex with length} \in [x, x + dx]).$$

Because the age of a typical vertex has $\mathrm{Exp}(1)$ distribution, $f(x)$ can be written as

$$f(x) = \int_0^\infty g(\tau, x)e^{-\tau}\, d\tau.$$

The verbal description of the particle process leads to the equation

$$\frac{d}{d\tau}G(\tau, x) = xg(\tau, x) + \alpha \int_0^x g(\tau, y)e^{-\lambda(x-y)}\, dy + \alpha e^{-\lambda x}.$$

Here the first term on the right expresses the deterministic exponential growth, the second term expresses birth of particles to parents not at 0 (copying of

existing edges) and the third expresses births to the 0-particle (new edge to v_0).
Multiply the terms of the equation by $e^{-\tau}$ and integrate out τ; noting

$$\int_0^\infty \frac{d}{d\tau} G(\tau, x) e^{-\tau} \, d\tau = \int_0^\infty G(\tau, x) e^{-\tau} \, d\tau = F(x), \text{say,}$$

we obtain

$$F(x) = xf(x) + \alpha \int_0^x f(y) e^{-\lambda(x-y)} \, dy + \alpha e^{-\lambda x}. \tag{51}$$

Differentiate: $-f = (xf)' + \alpha f - \lambda \alpha \int_0^x f(y) e^{-\lambda(x-y)} \, dy - \lambda \alpha e^{-\lambda x}$.
Rewrite with the integral term on the left, and then substitute the integral term
by the expression implied in (51):

$$\lambda(-xf - \alpha e^{-\lambda x} + F) = f + (xf)' + \alpha f - \lambda \alpha e^{-\lambda x}.$$

Differentiate: $\lambda(-(xf)' + \lambda \alpha e^{-\lambda x} - f) = f' + (xf)'' + \alpha f' + \lambda^2 \alpha e^{-\lambda x}$.
Tidy: $(xf)'' + \lambda(xf)' + (1 + \alpha)f' + \lambda f = 0$.
Look for a series solution $f(x) = \sum_{n=0}^\infty a_n x^n$. Equating coefficients of x^n:

$$(n + 2)(n + 1)a_{n+1} + \lambda(n + 1)a_n + (1 + \alpha)(n + 1)a_{n+1} + \lambda a_n = 0.$$

That is,

$$\frac{a_{n+1}}{a_n} = \frac{-\lambda(n + 2)}{(n + 1)(n + 3 + \alpha)}$$

and so

$$a_n = \frac{(-\lambda)^n (n + 1) \Gamma(3 + \alpha)}{\Gamma(n + 3 + \alpha)} a_0.$$

One can directly check that $f(0+) = 1$, identifying $a_0 = 1$. Because the mean
in-degree is $\alpha/(1 - \alpha)$, the probability density function of a typical edge-length
must be $\frac{1-\alpha}{\alpha} f(x)$, establishing (17).

5 Further Calculations

5.1 Yule Arguments for Subgraph Density

The next Lemma abstracts the Yule process arguments used in Sect. 4.5. Recall
the reformulation there of limit subgraph density in terms of $\chi(G)$ and $\bar{\chi}(G)$.

Lemma 2. *Let G_0 be a rooted directed acyclic graph such that each vertex is a
descendant of the root (Fig. 8). Let q be the out-degree of the root. Let G_1 be a
directed acyclic graph obtained from G_0 by adding an extra vertex w and edges
(w, root) and c further edges from w to some children of the root (so $0 \leq c \leq q$).
Then*

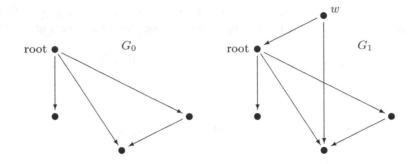

Fig. 8. Illustration of Lemma 2.

$$\bar{\chi}(G_1)/\bar{\chi}(G_0) = \frac{\beta_{c+1}}{1 - \beta_{c+1}}. \tag{52}$$

$$\chi(G_1)/\chi(G_0) = \sum_{j=0}^{q-c} (-1)^j \binom{q-c}{j} \frac{\beta_{c+1+j}}{1 - \beta_{c+1+j}}. \tag{53}$$

$$\tag{54}$$

In particular, if $q = c$ then

$$\chi(G_1)/\chi(G_0) = \frac{\beta_{c+1}}{1 - \beta_{c+1}}. \tag{55}$$

As a quick application let us derive formula (45) for $\chi(\pi_r)$ for the directed path π_r on r edges. Applying (53) with $q = 1, c = 0$,

$$\chi(\pi_{r+1})/\chi(\pi_r) = \frac{\beta_1}{1 - \beta_1} - \frac{\beta_2}{1 - \beta_2} = \frac{\beta_1 - \beta_2}{(1 - \beta_1)(1 - \beta_2)}.$$

Because $\chi(\pi_1) = \partial$ and $\iota(\pi_r) = 1$ we obtain formula (45).

Proof of Lemma 2. We will do the harder case (53). Consider a copy of G_0 (i.e. an isomorphic subgraph) at the root of the PWIT. In the space-time PWIT process, let $M(t)$ be the number of copies of G_1 which contain the given copy of G_0, at time t after the arrival of the root. Since the age T of the root has Exp(1) distribution,

$$\chi(G_1)/\chi(G_0) = EM(T).$$

Write S for the set of children of the root in G_0 and write A for a subset of S.

Consider the process of arriving vertices v which form an edge to the root. Such a v has a near neighbor v', where either $v' = $ root or (v', root) is already an edge. Writing $A(v') \subseteq S$ for the set of children of the root to which v' creates an edge, then $A(v) \subseteq A(v')$. We can now write

$$M(t) = \#\{v : (v, \text{root}) \text{ is an edge}, A(v) = A_1\}$$

where A_1 is the set of children of the root of G_1 to which w has an edge. In representing $M(t)$ as above, we are using the hypothesis "each vertex is a descendant of the root" to ensure that, in a subgraph of the space-time PWIT isomorphic to G_1, the last-arriving vertex must be w.

Consider a sequence root $= v_{(0)}, v_{(1)}, \ldots, v_{(i)}$ of arriving vertices such that each vertex $v_{(k)}$ arrives at distance x_k from its near neighbor $v_{(k-1)}$. The chance that each $v_{(k)}$ makes an edge to the root and to each child in A_1 equals $\prod_{k=1}^{i} p^{c+1}(x_k)$. The chance that furthermore no other child in S acquires an edge to $v_{(i)}$ equals $(1 - \prod_{k=1}^{i} p(x_k))^{q-c}$. By considering the times $0 < t_1 < t_2 < \ldots < t_i < t$ of arrivals of $v_{(i)}$,

$$EM(t) =$$

$$\sum_{i=1}^{\infty} \int_{0<t_1<\ldots<} \int_{t_i<t} dt_1 \ldots dt_i \int_0^{\infty} \cdots \int_0^{\infty} dx_1 \ldots dx_i \, \mathbf{p}^{c+1}(\mathbf{x})(1-\mathbf{p}(\mathbf{x}))^{q-c}$$

where $\mathbf{p}(\mathbf{x}) = \prod_{k=1}^{i} p(x_i)$. Because $\mathbf{p}^{c+1}(\mathbf{x})(1 - \mathbf{p}(\mathbf{x}))^{q-c} = \sum_{j=0}^{q-c}(-1)^j \binom{q-c}{j}$ $\mathbf{p}^{c+1+j}(\mathbf{x})$ and $\int_0^{\infty} p^{c+1+j}(x_k)\, dx_k = \beta_{c+1+j}$,

$$EM(t) = \sum_{i=1}^{\infty} \frac{t^i}{i!} \sum_{j=0}^{q-c}(-1)^j \binom{q-c}{j} \beta_{c+1+j}^i$$

$$= \sum_{j=0}^{q-c}(-1)^j \binom{q-c}{j}(\exp(\beta_{c+1+j}t) - 1).$$

Calculating $EM(T) = \int_0^{\infty} e^{-t} EM(t)\, dt$ establishes (53).

5.2 Subgraph Density of $K_{2,2}$

We have not pursued general methods for induced subgraph density beyond Lemma 2, but the argument that follows for the particular case of $K_{2,2}$, based on splitting into two cases, could clearly be applied somewhat more widely.

We first quote

Lemma 3.

$$E\binom{\mathrm{Geo}(p)}{2} = p^{-2} - p^{-1} \tag{56}$$

$$E\binom{\mathrm{Geo}(p) - 1}{2} = p^{-2} - 2p^{-1} + 1. \tag{57}$$

Next consider the graph G^* on the left of Fig. 9.

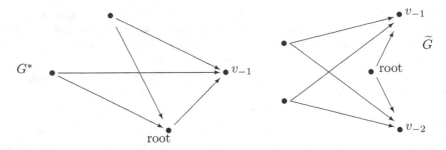

Fig. 9. Graphs related to $K_{2,2}$.

We will show

$$\bar{\chi}(G^*) = \frac{2\beta_2^2 \partial}{(1 - 2\beta_2)(1 - \beta_2)} \tag{58}$$

where $\beta_2 = EZ^{(2)}$.

We start by repeating the argument in the $r = 2$ case of (47). In the forwards space-time PWIT process, consider the newly-arrived root and an edge (root, v_{-1}). At time t after the arrival of the root, let $N_t = 1+$ the number of vertices v such that (v, root) and (v, v_{-1}) are both edges of the graph process; regard the "+1" as counting the root itself. Then N_t is a Yule process of rate $\beta_2 = \int_0^\infty p^2(x)\, dx$. Thus at time t there are $\binom{N_t - 1}{2}$ graphs of the desired form containing the edge (root, v_{-1}). Because the age T of the root has $\text{Exp}(1)$ distribution, we see

$$\bar{\chi}(G^*) = \partial E\binom{N_T - 1}{2}$$

where $\partial = ED_{\text{out}}$ is the expected number of edges of the form (root, v_{-1}). Using (28) and (57),

$$\bar{\chi}(G^*) = \partial E\left(e^{2\beta_2 T} - 2e^{\beta_2 T} + 1\right)$$
$$= \partial\left(\tfrac{1}{1 - 2\beta_2} - \tfrac{2}{1 - \beta_2} + 1\right)$$

leading to (58).

Next, in $\mathcal{G}_\infty^*$ consider

Q := number of unordered pairs (v_{-1}, v_{-2}) such that (root, v_{-1}) and (root, v_{-2}) are edges, and v_{-1} and v_{-2} were in different geometric components at the arrival time of the root.

By considering distances r_1, r_2 from the root to the near neighbors of the geometric components containing v_1, v_2,

$$EQ = \tfrac{1}{2}\int\int p(r_1)p(r_2)\, dr_1 dr_2 \times \partial^2 = \beta^2 \partial^2/2.$$

Now consider in $\mathcal{G}_\infty^*$ configurations $\widetilde{G}$ as on the right of Fig. 9, where there is no edge between v_{-1} and v_{-2}, and where the root is the first-arriving vertex to have edges to both v_{-1} and v_{-2}; these requirements are equivalent to saying that at the arrival time of the root, v_{-1} and v_{-2} were in different geometric components. Reuse a now-familiar argument. At time t after the arrival of the root, let $N_t = 1+$ the number of vertices $v \neq$ root such that (v, v_{-2}) and (v, v_{-1}) are both edges of the graph process; regard the "+1" as counting the root itself. Then N_t is a Yule process of rate $\beta_2 = \int_0^\infty p^2(x)\, dx$. Thus at time t the number of possible unordered pairs $\{v_1, v_2\}$ which give the configuration in the figure, *where we allow one of $\{v_1, v_2\}$ to be the root*, equals $\binom{N_t}{2}$. Because the age T of the root has Exp(1) distribution, we see

$$\bar{\chi}(\widetilde{G}) = E\binom{N_T}{2} \times EQ$$

where $\bar{\chi}(\widetilde{G})$ is the density of graphs as on the right of Fig. 9, perhaps with extra edges, but subject to the requirement that the root is the first-arriving vertex to have edges to both v_{-1} and v_{-2}. Using (28) and (57),

$$E\binom{N_T}{2} = E\left(e^{2\beta_2 T} - e^{\beta_2 T}\right) = \frac{1}{1-2\beta_2} - \frac{1}{1-\beta_2} = \frac{\beta_2}{(1-2\beta_2)(1-\beta_2)}.$$

One can now write

$$\overline{\text{dens}}\,(K_{2,2}) = \tfrac{1}{2}\bar{\chi}(K_{2,2}) = \bar{\chi}(G^*) + \bar{\chi}(\widetilde{G})$$

because a 4-vertex graph in $\mathcal{G}_\infty^*$ containing $K_{2,2}$ is either of the form G^* or is the restriction of a graph of the form $\widetilde{G}$, in which the extra root is specified by the requirement stated above (the factor $1/2$ reflects the fact $\iota(K_{2,2}) = 2$). Combining the formulas above gives

$$\tfrac{1}{2}\bar{\chi}(K_{2,2}) = \frac{\beta_2^2 \partial}{(1-2\beta_2)(1-\beta_2)} + \beta^2 \partial^2/2 \times \frac{\beta_2}{(1-2\beta_2)(1-\beta_2)}$$

which simplifies to (46).

5.3 Directed Percolation

Here we record some calculations without detailed explanation. In the context of the space-time PWIT and the evolving random graph process $\mathcal{G}_\infty^*(s)$, we can seek to grow a "core" graph $\mathcal{C}(s)$ inside $\mathcal{G}_\infty^*(s)$ via a greedy rule:

a newly-arriving vertex is included in $\mathcal{C}(s)$ if it creates an edge to some vertex already in $\mathcal{C}(s)$, in which case all such edges are included in $\mathcal{C}(s)$.

If this construction works, we expect the process $(\mathcal{C}(s))$ to have a a stationary distribution $\mathcal{C}(0)$, say, where $\mathcal{C}(0) \subset \mathcal{G}_\infty^*$. Consider

$$q = P(\text{root} \in \mathcal{C}(0))$$
$$Y = \text{out-degree of root in } \mathcal{C}(0), \text{ given root} \in \mathcal{C}(0).$$

Consider the relation

$$\tilde{Y} = \sum_{i=1}^{\infty} \text{Ber}_i(q)\text{Bin}_i(1 + Y_i, p(\xi_i))$$

where we write $\text{Ber}(p)$ for a Bernoulli(p) r.v. (taking value 1 with probability p and value 0 otherwise). Using the recursive structure of the limit random graph process, we see that q and Y solve the equations (for unknown $0 < q < 1$ and an unknown distribution Y on $\{1, 2, 3, \ldots\}$)

$$Y \overset{d}{=} \text{dist}(\tilde{Y}|\tilde{Y} \geq 1); \quad q = P(\tilde{Y} \geq 1). \tag{59}$$

Define $p_{\text{dir-perc}}(\alpha, \lambda)$ to be the solution q if it exists, and to be 0 otherwise. The interpretation of this quantity in terms of the finite random graph process $(\mathcal{G}_n, n \geq 1)$ is that

$$n^{-1}E\mathbf{T} \to p_{\text{dir-perc}}(\alpha, \lambda)$$

where $\mathbf{T}$ is the maximal size of a tree in $\mathcal{G}_n$ directed toward some root. So in particular, for $p_{\text{perc}}(\alpha, \lambda)$ defined at (18),

$$p_{\text{dir-perc}}(\alpha, \lambda) \leq p_{\text{perc}}(\alpha, \lambda).$$

Equation (59) in principle determines $p_{\text{dir-perc}}(\alpha, \lambda)$, but to get an explicit bound we reuse an underlying idea. Because $1 + Y_i \geq 2$,

$$\tilde{Y} \geq \sum_{i=1}^{\infty} \text{Ber}_i(q)\text{Bin}_i(2, p(\xi_i)) = Y^*, \text{ say.}$$

If the equation

$$q = P(Y^* \geq 1) \tag{60}$$

has a solution $q > 0$ then one can argue $p_{\text{dir-perc}}(\alpha, \lambda) \geq q$. But (60) is an explicit equation

$$1 - q = \exp\left(-\int (2p(x) - p^2(x))q \, dx\right) = \exp(-(2\beta - \beta_2)q).$$

If $2\beta - \beta_2 > 1$ there is a solution $q > 0$, establishing (19).

5.4 Out-Degree in the Case $\alpha\lambda = 1$

The special property of this case is that $p(x) = e^{-\lambda x}$. On the PWIT consider

$$Y = \sum_{v \neq \text{root}} \text{Ber}(e^{-\lambda d(v, \text{root})}).$$

This satisfies the same recursion (in the special case) as does D_{out}. But there is another way to study Y, which we sketch briefly. Either the root of the PWIT has no children within a small distance δ; or it does have a child, and the distances to the other descendants of the root and of this child are independent copies of the PWIT distances. Because the effect on Y of increasing distances by δ is to censor each Bernoulli success with probability $\lambda\delta$, we see that Y is the stationary distribution of the continuous-time Markov chain on states $\{0, 1, 2, \dots\}$ with dynamics

$$
\begin{aligned}
y &\to y - 1 &&: \text{ rate } \lambda y \\
y &\to y + \hat{Y} + 1 &&: \text{ rate } 1
\end{aligned}
$$

where $\hat{Y}$ is an independent copy of Y. One can now check algebraically that

$$P(Y = y) = (1 - \tfrac{1}{\lambda})(\tfrac{1}{\lambda})^y, \quad y \geq 0$$

solves the balance equations for this chain. That is, $1 + D_{\text{out}}$ has $\text{Geo}(1 - \tfrac{1}{\lambda}) = \text{Geo}(1 - \alpha)$ distribution, as asserted in (36).

Remark. Antar Bandyopadhyay (personal communication) has given a purely analytic verification of (36).

5.5 Triangle Density of a Function of Degree

Here we outline an argument for (16). Because D_{in} has power-law tail and D_{out} has geometric tail, when $D = D_{\text{in}} + D_{\text{out}}$ is large, say k, then $D_{\text{out}} = O(1)$ and $D_{\text{in}} = k - O(1)$. It is then not hard to argue that the large-k behavior for $C(k)$ will be the same as for

$$C^*(k) = \frac{E(\text{number of triangles with in-vertex } v_0 \mid v_0 \text{ has in-degree } k)}{\binom{k}{2}}$$

where the *in-vertex* of a triangle is the vertex with two in-edges.

Recall from Sect. 4.2 that

$$N(t) = \text{number of in-edges at a typical vertex } v_0 \text{ at time } t \text{ after its arrival}$$

is the Yule process of rate β. Write $v_1, v_2, v_3, \dots$ for the successive arriving vertices which create edges to v_0, and for $i \geq 2$ write

$$M_i = \text{number of edges from } v_i \text{ to } \{v_{i-1}, v_{i-2}, \dots, v_1\}.$$

After v_k arrives there are $M_2 + M_3 + \dots + M_k$ triangles with in-vertex v_0. If we can show

$$EM_k \to b \text{ as } k \to \infty$$

then we will have

$$C(k) \approx \frac{kb}{\binom{k}{2}} \sim \frac{2b}{k}. \tag{61}$$

Here we are sliding over the fact that $1 + D_{\text{in}}$ is the Yule process evaluated at an independent $\text{Exp}(1)$ time T; conditioning this to take a value k does not affect the properties used in the argument below.

Suppose vertices $v_1, \ldots, v_{k-1}$ have arrived and consider what edges will be created when v_k arrives. The dynamics (22) of the space-time PWIT say

the rate of arrival of new vertices with some one of $v_0, v_1, \ldots, v_{k-1}$ as near neighbor and at distance $\in [x, x+dx]$ from that near neighbor equals $k\,dx$. The index I of that near neighbor v_I is uniform on $\{0, 1, \ldots, k-1\}$.

Such an arriving vertex creates an edge to v_0 with probability $p(x)$. So conditional on that event (meaning the arriving vertex is v_k), the distance $\hat{\xi}$ from the near neighbor v_I and the index $I = I_k$ of that near neighbor satisfy
(i) $\hat{\xi}$ has probability density function $\hat{p}(x) = \frac{p(x)}{\int p(u)du}$;
(ii) I is uniform on $\{0, 1, \ldots, k - 1\}$.
Because v_k will copy each of the M_I out-edges from v_I with probability $p(\xi)$ each, and create an edge to V_I with the same probability, we obtain the recursion

$$M_k \overset{d}{=} \text{Bin}(1 + M_I, p(\hat{\xi}))$$

where $M_1 = 0$ and where we interpret the right side as 0 when $I = 0$. So the limit $\lim_k EM_k = b$ solves $b = (1 + b)Ep(\hat{\xi})$ and so $b = \frac{Ep(\hat{\xi})}{1 - Ep(\hat{\xi})}$. Finally,

$$Ep(\hat{\xi}) = \int p(x)\hat{p}(x)\,dx = \beta_2/\beta.$$

So $b = \frac{\beta_2}{\beta - \beta_2}$ and (16) follows from (61).

6 Comparison with Other Models

Recent complex networks models fall into two categories. In the *small worlds* models popularized by Strogatz and Watts, vertices are points in d-dimensional space, which automatically provides a metric distance between vertices, and the model uses some rule to create a random graph with short-range and long-range edges. In purely graph-theoretical models, such as the basic *proportional attachment* model popularized by Albert and Barabási[7] the vertices have no "intrinsic structure" other than that provided by the graph; we visualize this as

---

[7] but really just a minor variation of Yule's idea: see Lecture 4 of [16]

saying that each pair of vertices is metric distance 1 apart. In a metric copying model we visualize vertices as points in some abstract metric space, representing (in the case of web pages, say) the difference between the content of the pages, or (for people) some notion of "social distance" based on location, education, profession, interests etc of the individuals. In detail the mean-field model of distance model is used for mathematical tractability rather than any claimed realism. But the exponential growth of number of vertices with metric distance is intermediate between, and surely in many contexts more plausible than, the alternatives implicit in the two standard categories of model above.

Within graph-theoretic models, the idea of distance preferences in attachment has been explored (see [17] and citations therein). But the general idea of combining proportional attachment with metric geometry has scarcely been explored[8], and the specific use of the mean-field model is novel.

As a technical note, the mean-field model is a zero-parameter[9] model of distance. Our full network model has the two parameters (α, λ); in contrast a typical small-worlds network model has four parameters (dimension, number of short-range links, constant and exponent for probability of long-range edges).

As another technical note, the property (cf. (16)) $C(k) \sim c/k$ has been proposed [19] as a criterion for identifying networks which are "hierarchical" in some sense. But in our finite-n model (recall Sect. 3.5) each vertex has qualitatively the same behavior, rather than different vertices being a priori assigned different hierarchical roles. So our model is non-hierarchical, and we are inclined to regard the criterion as ineffective[10].

The specific model studied in this paper is intended as a "general purpose" model rather than being tuned to some particular subclass of real-world networks. Having as one ingredient the now-familiar proportional attachment feature, one could look at the many existing variant models in the literature and explore them within our platform. In other words, there are many ways to add a third parameter intended to express some presumed real-world feature or some theoretical desideratum. For instance

- One can impose connectivity by requiring that a new vertex always links to its nearest neighbor.
- one can add rules allowing a new vertex to immediately acquire in-edges, or for edges to randomly appear between existing edges. Such rules can be designed (as in e.g. [4] Sect. 11) to produce power law distributions for in-degree.

---

[8] [18] gives a simulation study of an explicitly power-law model, as well as interesting empirical study of a notion of *lexical distance* between web pages

[9] Zero *dimensionless* parameters, to be pedantic

[10] One could alternatively regard it as indicating some subtle emergent hierarchical structure; cf. [20]

6.1 Concluding Remarks

In this paper we have focused on

- describing the model and its conceptual background (section 3)
- listing explicit formulas (sections 2.2 – 2.6) and exhibiting the calculations which lead to these formulas (sections 4 and 5).

We are postponing to a later paper consideration of

- technical issues in the relation between the finite-n model and its infinite limit $\mathcal{G}_\infty^*$
- the open problems indicated in sections 2.7 – 2.8, whose study requires the "bounding" techniques of theoretical mathematical probability rather than explicit calculations.

Acknowledgement

I thank an anonymous referee for detailed constructive comments.

References

1. R. Albert and A.-L. Barabási. Statistical mechanics of complex networks. *Rev. Mod. Phys.*, 74:47–97, 2002.
2. S.N. Dorogovtsev and J.F.F. Mendes. Evolution of networks. *Adv. Phys.*, 51:1079–1187, 2002.
3. M.E.J. Newman. The structure and function of complex networks. *SIAM Review*, 45:167–256, 2003.
4. B. Bollobás and O. Riordan. Mathematical results on scale-free random graphs. In S. Bornholdt and H.G. Schuster, editors, *Handbook of Graphs and Networks*, pages 1–34. Wiley, 2002.
5. D.J. Aldous. A stochastic complex network model. *Electron. Res. Announc. Amer. Math. Soc.*, 9:152–161, 2003.
6. P.G. Buckley and D. Osthus. Popularity based random graph models leading to a scale-free degree sequence. *Discrete Mathematics*, to appear, 2004.
7. S.N. Dorogovtsev, J.F.F. Mendes, and A.N. Samukhin. Structure of growing networks with preferential linking. *Phys. Rev. Lett.*, 85:4633–4636, 2000.
8. P.L. Krapivsky, G.J. Rodgers, and S. Redner. Degree distribution of growing networks. *Phys. Rev. Lett*, 86:5401–5404, 2001.
9. B. Bollobás and O. Riordan. The diameter of a scale-free random graph. To appear in Combinatorica, 2004.
10. M. Penrose. *Random Geometric Graphs*. Oxford Univ. Press, 2003.
11. D.J. Aldous and J.M. Steele. The objective method: Probabilistic combinatorial optimization and local weak convergence. In H. Kesten, editor, *Probability on Discrete Structures*, volume 110 of *Encyclopaedia of Mathematical Sciences*, pages 1–72. Springer-Verlag, 2003.
12. David Aldous and Allon G. Percus. Scaling and universality in continuous length combinatorial optimization. *Proc. Natl. Acad. Sci. USA*, 100:11211–11215, 2003.

13. S. Ross. *Stochastic Processes*. Wiley, 1983.
14. G.U. Yule. A mathematical theory of evolution, based on the conclusions of Dr J. C. Willis. *Philos. Trans. Roy. Soc. London Ser. B*, 213:21–87, 1924.
15. D.J. Aldous and J. Pitman. Tree-valued Markov chains derived from Galton-Watson processes. *Ann. Inst. H. Poincarè Probab. Statist.*, 34:637–686, 1998.
16. David Aldous. Notes on STAT 206. http://www.stat.berkeley.edu/users/aldous/Networks/index.html, 2003.
17. J. Jost and M.P. Joy. Evolving networks with distance preferences. *Physical Review E*, 66:036126, 2002.
18. F. Menczer. Growing and navigating the small world web by local content. *Proc. Natl. Acad. Sci. USA*, 99:14014–14019, 2002.
19. E. Ravasz and A.-L. Barabási. Hierarchical organization in complex networks. *Physical Review E*, 67:026112, 2003.
20. H. Tangmunarunkit, R. Govindan, S. Jamin, S. Shenker, and W. Willinger. Network topology generators: Degree-based vs. structural. In *SIGCOMM 02*. ACM, 2002. http://topology.eecs.umich.edu/archive/sigcommo2.pdf.

The Small World Phenomenon
in Hybrid Power Law Graphs

Fan Chung and Linyuan Lu

Department of Mathematics, University of California, San Diego, La Jolla, CA 92093
USA

Abstract. The small world phenomenon, that consistently occurs in numerous existing networks, refers to two similar but different properties — small average distance
and the clustering effect. We consider a hybrid graph model that incorporates both
properties by combining a global graph and a local graph. The global graph is modeled
by a random graph with a power law degree distribution, while the local graph has
specified local connectivity. We will prove that the hybrid graph has average distance
and diameter close to that of random graphs with the same degree distribution (under
certain mild conditions). We also give a simple decomposition algorithm which, for
any given (real) graph, identifies the global edges and extracts the local graph (which
is uniquely determined depending only on the local connectivity). We can then apply
our theoretical results for analyzing real graphs, provided the parameters of the hybrid
model can be appropriately chosen.

1 Introduction

In 1967, the psychologist Stanley Milgram [1] conducted a series of experiments
which led him to the well known concept captured by the phrase "Six degrees
of separation". Namely, any two strangers (on the planet) are *connected* by a
short chain of intermediate acquaintances of length at most six. Since then, it
has been observed that many realistic networks possess the so-called *small world
phenomenon*, with two distinguishing traits —*small distance* between any pair
of nodes, and the *clustering effect* that two nodes are more likely to be adjacent
if they share a neighbor.

There have been various approaches to model networks that have the small
world phenomenon. Progress has been made in analyzing the aspect of small
distances by using generalized random graph theory and properties of the power
law distribution. However, the clustering effect seems much harder to model.

In 1999, several research groups independently observed that numerous networks such as the Internet graphs, call graphs and social networks, etc. all have
a *power law* distribution [2–15]. Namely, the number of nodes of degree k is proportional to $k^{-\beta}$ for some positive exponent β. By using a random graph model
for a given degree distribution, it can be shown [16] in a rigorous way that,
for example, a random power law graph with exponent β, where $2 < \beta < 3$,
almost surely has average distance of order $\log \log n$ and has diameter of order
$\log n$. (Note that the average distance is the average of distances between pairs

F. Chung and L. Lu, The Small World Phenomenon in Hybrid Power Law Graphs, Lect. Notes Phys.
650, 89–104 (2004)
http://www.springerlink.com/

of nodes that are connected and the diameter is the maximum distance between such pairs of nodes.)

To model the clustering effect, most common approaches just add random edges to grid graphs or the like (see Watts and Strogatz [17,18]). Kleinberg [19] introduced the network model of a grid graph with additional random edges joining two nodes u, v with probability proportional to $[d(u,v)]^{-r}$ (where $d(.,.)$ represents the distance in the grid graph and r is a constant that determines the effectiveness of decentralized algorithms for the network). In Kleinberg's model and the model of Watts and Strogatz, the graphs have the same expected degree at every node and do not have a power law degree distribution. Fabrikant, Koutsoupias and Paradimitriou [20] proposed a model of having vertices in the Euclidean plane and adding edges by optimizing the trade-off between (Euclidean) distances and "centrality" in the network. Such grid-based models are quite restrictive and far from satisfactory for modeling webgraphs or biological networks, for example.

The difficulty in reconciling these two aspects – small distance and clustering effect– resembles philosophically the challenge in physical world concerning the "weak force" and the "strong force". There is no unified model embracing both the weak and strong forces in spite of intense efforts of many great scientists. Random power law graphs are good for modeling the aspect of small distance, but fail miserably for modeling the clustering effect. As a matter of fact, the related graph-theoretical parameters involving small distances and clustering seem to be of an entirely different scale. For example, the clustering effect is quite sensitive to average degree but this is not so for the small (average) distance. Examples of real graphs tell the same story. According to Henzinger [21] at Google, random graphs are good for modeling interdomain hyperlinks but not the local links.

In this paper, we consider a general hybrid graph model that has both aspects of the small world phenomenon. Roughly speaking, a hybrid graph is a union of a global graph (consisting of "long edges" providing small distances) and a local graph (consisting of "short edges" respecting local connections). (Detailed definitions will be given in Sect. 2.) By using several tools for dealing with random graphs with given expected degree sequences, we will prove that our hybrid graphs have the following properties:

1. Power law degree distribution for a given power β.
2. Small average distance at the same order as that of random graphs.
3. Small diameter at the same order as that of random graphs.
4. Locally highly connected.

We will show that the average distance/diameter is bounded above by $c \log n$ where c depends on the "second-order" average degree (which will be defined later). Consequently, this implies a polylog upper bound for analyzing many distributed algorithms if the network can be well approximated by this general family of hybrid power law graphs.

Motivated by the hybrid graph model, we will give a simple decomposition algorithm. For any real network, the decomposition algorithm identifies the local graph and the global graph. We can then use our theorems to deduce properties of the real network if the local graph satisfies local connectivity conditions and the global graph can be approximated by a random power law graph.

This paper is organized as follows. In Sect. 2, we give basic definitions for power law graphs and random graphs. In Sect. 3, we consider local graphs and give the decomposition algorithm. In Sect. 4, we propose the hybrid graph model by combining a local graph and a random power law graph. We also show that the local graph can be extracted from the hybrid graph with an error estimate of lower order. In Sect. 5, we describe several useful facts about random graphs with given expected degrees. In particular, we summarize some facts concerning the average distance and diameter of random power law graphs. In Sect. 6, we establish the desired upper bounds for average distance/diameter for hybrid power law graphs. Section 7 includes further discussions and a number of remarks.

2 Preliminaries

Before we consider the hybrid graphs, we will discuss random graphs with given expected degree sequences and power law degree distribution.

2.1 Random Graphs with Given Expected Degrees

We consider a general class of random graphs with given expected degree sequence $\mathbf{w} = (w_1, w_2, \ldots, w_n)$. The probability p_{ij} that there is an edge between vertex v_i and vertex v_j is $w_i w_j \rho$ for any index i and index j. Here we choose ρ to be $(\sum w_i)^{-1}$ and we assume that $\max_i w_i^2 < \sum_k w_k$ so that $p_{ij} \leq 1$ for all i and j. It is then easy to check that the vertex v_i has expected degree w_i. We remark that the assumption $\max_i w_i^2 < \sum_k w_k$ implies that the sequence w_i is graphical (in the sense that it satisfies the necessary and sufficient condition for a sequence to be realized by a graph [22]) except that we do not require the w_i's to be integers. We note that this model allows a non-zero probability for self-loops. The expected number of loops is quite small (of lower order) in comparison with the total number of edges. Consequently, loops have little effect on various graph properties such as average distance, clusterness, etc.

We denote a random graph with a given expected degree sequence $\mathbf{w}$ by $G(\mathbf{w})$. For example, the typical random graph $G(n, p)$ (see [23]) on n vertices and edge density p is just a random graph with expected degree sequence $(pn, pn, \ldots, pn)$. The random graph $G(\mathbf{w})$ is different from the random graphs with a prescribed exact degree sequence (which involve dependency and are hard to analyze). For example, in [24,25], Molloy and Reed obtained results on the sizes of connected components for random graphs with prescribed exact degree sequences which are required to satisfy certain "smoothing" conditions. Our model is also different from the evolution models generated by simple growth rules (such as preferential attachment schemes as in [3,5,26,9]).

2.2 Power Law Degree Distribution

If a graph strictly follows the power law, then the average degree as well as its connectivity (i.e., the distribution of connected components) will be completely determined by the exponent of the power law (see [3]). However, for most realistic graphs, the power law holds only for a certain range of degrees, namely, for the degrees which not too small and not too large. We will consider the following model with the consideration that most examples of massive graphs satisfying power law have exponent $\beta > 2$.

Model $M(n, \beta, d, m)$ where

- n is the number of vertices,
- $\beta > 2$ is the power of the power law,
- d is the expected average degree,
- m is the expected maximum degree (or an upper bound for the range of degrees that obey the power law) and $m^2 = o(nd)$.

We assume that the $i - i_0 + 1$-th vertex v_i has expected degree

$$w_i = ci^{-\frac{1}{\beta-1}}$$

for $i_0 \leq i < n + i_0$. Here c depends on the average degree d and i_0 depends on the maximum expected degree m. It is easy to compute that the number of vertices of expected degree between k and $k+1$ is of order $c'k^{-\beta}$ where $c' = c^{\beta-1}(\beta-1)$ as required by the power law. To determine c, we consider

$$\text{Vol}(G) = \sum_i w_i = \sum_{i=i_0}^{n} ci^{-\frac{1}{\beta-1}} \approx c\frac{\beta-1}{\beta-2}n^{1-\frac{1}{\beta-1}}$$

Here we assume $\beta > 2$. Since $nd \approx \text{Vol}(G)$, we have

$$c = \frac{\beta-2}{\beta-1}dn^{\frac{1}{\beta-1}} \tag{1}$$

$$i_0 = n\left(\frac{d(\beta-2)}{m(\beta-1)}\right)^{\beta-1} \tag{2}$$

Here (2) is deduced from the (cut-off) condition $w_{i_0} = m$ and (1).

Let $f(x) = \frac{\beta-2}{\beta-1}dx^{-\frac{1}{\beta-1}}$. The expected degrees (or weights) are just $f(\frac{i}{n})$, $i_0 \leq i \leq n$.

We will also consider an alternative model $M'(n, \beta, d, m)$, in which each vertex x is assigned a weight $f(y)$, where y is a real number chosen uniformly in the range of (i_0, n). It can be easily shown that two models are equivalent (for $i_0 \ll n$) in the sense that a property that holds for a random graph in M almost surely must hold for M' and vice versa.

2.3 The Volume and the Second-Order Average Degree

For a subset S of vertices, the k-th volume of S, denoted by $\text{Vol}_k(S)$, is the sum of the k-th power of weights of vertices in S.

$$\mathrm{Vol}_k(S) = \sum_{v_i \in S} w_i^k$$

The expected average degree is defined to be

$$\frac{\mathrm{Vol}_1(G)}{\mathrm{Vol}_0(G)} = \frac{1}{n} \sum_i w_i = d(1 + o(1)).$$

We write $\mathrm{Vol}(G) = \mathrm{Vol}_1(G)$. Of particular interest is the second-order average degree $\tilde{d}$ defined by

$$\tilde{d} = \frac{\mathrm{Vol}_2(G)}{\mathrm{Vol}_1(G)} = \frac{\sum_i w_i^2}{\sum_i w_i}.$$

For power law graphs with exponent β, we have [16]

$$\tilde{d} = \frac{\mathrm{Vol}_2(G)}{\mathrm{Vol}_1(G)} = \begin{cases} d\frac{(\beta-2)^2}{(\beta-1)(\beta-3)}(1 + o(1)) & \text{if } \beta > 3. \\ \frac{1}{4}d \ln m(1 + o(1)). & \text{if } \beta = 3. \\ d\frac{(\beta-2)^2}{(\beta-1)(3-\beta)}m^{3-\beta}(1 + o(1)) & \text{if } 2 < \beta < 3. \end{cases}$$

3 Local Graphs

Roughly speaking, a local graph is locally highly connected. To be precise, here we use two parameters to describe the local connectivity. For any fixed two integers $k \geq 2$ and $l \geq 2$, a graph L is called "locally (k, l)-connected" if for any edge uv, there are at least l edge-disjoint paths (i.e, no two paths share a common edge) with length at most k joining from u to v (including the edge uv). For example, the grid graph $C_n \square C_n$ is locally $(3, 3)$-connected as well as locally $(5, 9)$-connected.

By the definition, the union of two locally (k, l)-connected graphs is locally (k, l)-connected. The maximum locally (k, l)-connected subgraph H is the union of all locally (k, l)-connected subgraphs of G. Thus, for any graph G, the maximum locally (k, l)-connected subgraph is unique. We remark that a (k, l)-connected graph is not necessarily connected. For example, the disjoint union of two (k, l)-connected graphs is still (k, l)-connected.

Here is a simple greedy algorithm for finding the maximum locally (k, l)-connected subgraph.

Algorithm(k, l):
For each edge $e = uv$, check whether there are l edge-disjoint paths with length at most k connecting u and v in the current graph G. If not, delete the edge e from G. Then iterate the procedure until no edge can be removed.

Theorem 1. *For any graph G, Algorithm(k, l) finds the unique maximum locally (k, l)-connected subgraph regardless of the order of edges chosen.*

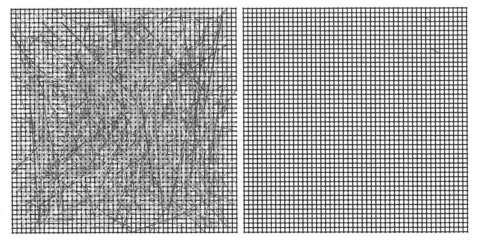

Fig. 1. A hybrid graph, which contains the grid graph $C_{50} \square C_{50}$ as the local graph, and 528 additional random edges.

Fig. 2. After removing all global edges (with $k = l = 3$), the local graph is almost perfectly recovered.

Proof. Let H' be a graph produced by the Algorithm(k, l) where the order that edges are removed is arbitrary. It is sufficient to show $H = H'$.

Let $G = H_0 \supset H_1 \supset H_2 \supset \cdots \supset H_r = H'$ be the sequence of the intermediate subgraphs produced by Algorithm(k, l). We will prove $H \subset H_i$ for all $i = 0, 1, 2, \dots, r$ by induction on i. It is trivial for $i = 0$ since $H \subset H_0 = G$. Now we assume $H \subset H_i$. For $i + 1$, let $e_{i+1} = uv$ be the edge being removed at the $(i+1)$-st stage. It is sufficient to show uv is not an edge of H. Otherwise, there are l edge-disjoint paths of H joining from u to v. Since $H \subset H_i$, these paths are also paths of H_i. According to the algorithm, it can not be removed, which is a contradiction. Thus, we have $H \subset H_{i+1}$ and $H \subset H_r = H'$.

In the other direction, since H' is locally (k, l)-connected, we have $H' \subset H$. H is the maximum subgraph with this property. The proof is complete. $\qquad \square$

The edges removed are considered the "global" edges. For certain classes of graphs, the local graph can be almost perfectly recovered as shown by Figs. 1 and 2.

Another example is the Collaboration Graph of the second kind with 237,426 vertices (as authors of *Math Review*) and about 226,194 edges (each of which is associated with a paper with *exactly* two coauthors), (see http://www.oakland.edu/~grossman/erdoshp.html for detailed explanations). The local graph L (with $k = l = 3$) has 1979 vertices and 4221 edges. Their degree distributions are showed by Figs. 3, 4 and 5.

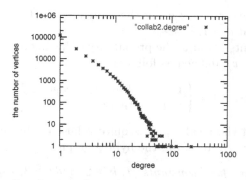

Fig. 3. The degree distribution of the Collaboration Graph of the second kind.

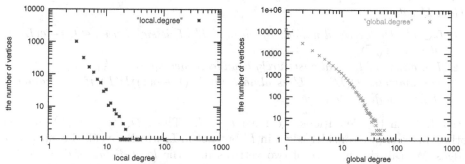

Fig. 4. The degree distribution of the local graph of the Collaboration Graph of the second kind (with $k = l = 3$) .

Fig. 5. The degree distribution of the global graph of the Collaboration Graph of the second kind.

4 The Hybrid Power Law Model

A hybrid graph consists of two parts – a global graph and a local graph. The edge set of the hybrid graph is a disjoint union of the edge set of the global graph and that of the local graph. The related parameters include:

β, the power law exponent,

d, the average degree,

m, the expected maximum degree (or an upper bound for the range of degrees that obey the power law), and

L, the local graph.

We remark that for a given network, all these parameters are straightforward to compute and estimate. Therefore it is quite easy to build a simulation for a network with given parameters.

The hybrid graph $H(n, \beta, d, m, L)$:
The local graph L is a locally (k, l)-connected graph with bounded degrees. The vertex v_i of H has weight w_i where $w_1, w_2, \ldots, w_n$ satisfy a power law

distribution with power $\beta > 2$ using the model $M'(n, \beta, d, m)$ in the Sect. 2.2. We also assume that $d \geq 1$.

For any two points u and v, the probability of having an edge between u and v is denoted by $p(u, v)$, defined as follows:

$$p(u, v) = \begin{cases} 1 & \text{if } uv \text{ is an edge of } L; \\ w_u w_v \rho & \text{otherwise.} \end{cases}$$

We will see that the local graph is quite robust in the sense that it can be almost completely recovered from the hybrid graph.

Theorem 2. *For any fixed constants M, $k \geq 2$, and $l \geq 3$, suppose L is a locally (k, l)-connected graph with degrees bounded by M. Let L' be the maximum locally (k, l)-connected subgraph in the hybrid graph $H(n, \beta, d, m, L)$ with $m = o(n^{\frac{1}{2} - \frac{1}{l}})$. Then L' satisfies*

1. *$L \subset L'$. The expected number of edges in $L' \setminus L$ is small, i.e., $e(L') - e(L) = O(m) = o(\sqrt{n})$.*
2. *The degree of L' is almost surely bounded above by $M + \lceil \frac{l}{2} \rceil - 1$.*
3. *The diameter $D(L')$ of L' is almost surely $(1 + o(1))D(L)$ if the diameter $D(L)$ is sufficiently large.*

Proof. From the definitions, we have $L \subset L'$. Thus, $D(L') \leq D(L)$. In the other direction, we consider edges in L' but not in L, which we call the *surviving edges*. We call the distance of two vertices in L the *local distance*, denoted by d_L. The neighborhood of a vertex in L is said to be a *local neighborhood*. A i-th local neighborhood of v consists of all vertices within local distance i from v. We will prove the following:

Claim. Almost surely all surviving edges uv have endpoints with local distance $d_L(u, v)$ at most k.

For any vertices u and v, if $d_L(u, v) > k$, any path of length at most k in L' from u to v must contain at least one surviving edge. Since this edge uv survives after the algorithm terminates, there exist at least l edge-disjoint edges in L' from the i-th local neighborhood of u to the j-th local neighborhood of v with some $i + j = k - 1$. Since the local degrees are bounded by M, the number of vertices in the i-th local neighborhood of u is at most

$$\sum_{s=0}^{i} M^s = \frac{M^{i+1}}{M - 1} \leq 2M^i.$$

Similarly, the number of vertices in the j-th local neighborhood of u is at most $2M^j$. There are at most $2M^i \times 2M^j = 4M^{k-1}$ pairs of such vertices. For each pair, the probability of being randomly chosen for the hybrid graph is less than $m^2 \rho$. Thus, the probability that uv survives is at most

$$\binom{4M^{k-1}}{l} (m^2 \rho)^l = o\left(\frac{1}{n^2}\right).$$

Therefore, almost surely, all surviving edges have endpoints with local distance at most k.

Now we consider the expected number of surviving edges, which almost surely have endpoints within local distance at most k. We choose a vertex u. There are at most $2M^k$ vertices with local distance at most k from u. The expected number of surviving edge uv with $d_L(u,v) \leq k$ is at most $\sum_u 2M^k w_u m\rho = 2M^k m$. For m large, the number of surviving random edges is well-concentrated on its expected value.

For any fixed u, we examine the number of surviving edges uv which are incident to u. Since almost surely v is within local distance k from u, there are at most $2M^k$ of possible v's. The probability that there are at least $\lceil \frac{l}{2} \rceil$ v's with uv surviving is at most

$$\binom{2M^k}{\lceil \frac{l}{2} \rceil}(m^2\rho)^{\lceil \frac{l}{2} \rceil} = o\left(\frac{1}{n}\right).$$

Thus, almost surely the degree of L' is at most $M + \lceil \frac{l}{2} \rceil - 1$.

Let $g(n)$ be a (very) slowly growing function of n, which approaches infinity as n increases. There are at most $4M^{2g(n)}$ pairs of vertices within local distance $g(n)$ from any given vertex u. The probability that l surviving edges are within local distance $g(n)$ from any given vertex u is at most

$$\binom{4M^{2g(n)}}{l}(m^2\rho)^l = o\left(\frac{1}{n^2}\right),$$

for some slowly growing function $g(n) = o(\log \frac{n^{\frac{1}{2}-\frac{1}{t}}}{m})$. Almost surely, for all vertex u, there are at most $l-1$ surviving edges with local distance at most k from u.

Let (u,v) be a pair of vertices with $d_L(u,v) = D(L)$. The distance between u and v in the hybrid graph can be reduced (from the local distance of u and v) by surviving edges. Each surviving edge can reduce the distance from u to v by at most $k-1$. The total number of surviving edges which can be used on the path from u to v is at most $\frac{l}{g(n)} = o(1)$. Hence $d_{L'}(u,v) \geq (1 - \frac{kl}{g(n)})d_L(u,v) = (1 - o(1))D(L)$. Thus, the diameter $D(L')$ is at least $(1-o(1))D(L)$. This completes the proof of Theorem 2   □

5 Several Facts Concerning Random Power Law Graphs

In this section we state several useful facts for random power law graphs $G(\mathbf{w})$ with given expected degree sequence $\mathbf{w}$. Proofs of these facts can be found in [16].

The expected degree sequence $\mathbf{w}$ for a graph G on n vertices in $G(\mathbf{w})$ is said to be *strongly sparse* if we have the following :
(i)   The second order average degree $\tilde{d}$ satisfies $0 < \log \tilde{d} \ll \log n$.
(ii)   For some constant $c > 0$, all but $o(n)$ vertices have expected degree w_i

satisfying $w_i \geq c$. The average expected degree $d = \sum_i w_i/n$ is strictly greater than 1, i.e., $d > 1 + \epsilon$ for some positive value ϵ independent of n.

The expected degree sequence $\mathbf{w}$ for a graph G on n vertices in $G(\mathbf{w})$ is said to be *admissible* if the following condition holds, in addition to the assumption that $\mathbf{w}$ is strongly sparse.

(iii)   There is a subset U satisfying:

$$\mathrm{Vol}_2(U) = (1 + o(1))\mathrm{Vol}_2(G) \gg \frac{\mathrm{Vol}_3(U)\log \tilde{d}\log\log n}{d\log n}.$$

The expected degree sequence $\mathbf{w}$ for a graph G on n vertices is said to be *specially admissible* if (i) is replaced by (i') and (iii) is replaced by (iii'):

(i') $\log \tilde{d} = O(\log d)$.

(iii') There is a subset U satisfying $\mathrm{Vol}_3(U) = O(\mathrm{Vol}_2(G))\frac{\tilde{d}}{\log d}$, and $\mathrm{Vol}_2(U) > d\mathrm{Vol}_2(G)/\tilde{d}$.

Fact 1. *For a random graph G with admissible expected degree sequence $(w_1, \dots, w_n)$, the average distance is almost surely $(1 + o(1))\frac{\log n}{\log \tilde{d}}$.*

Fact 2. *For a random graph G with a specially admissible degree sequence $(w_1, \dots, w_n)$, the diameter is almost surely $\Theta(\log n/\log \tilde{d})$.*

Fact 3. *For a power law random graph with exponent $\beta > 3$ and average degree d strictly greater than 1, almost surely the average distance is $(1+o(1))\frac{\log n}{\log \tilde{d}}$ and the diameter is $\Theta(\log n)$.*

Fact 4. *Suppose a power law random graph with exponent β has average degree d strictly greater than 1 and maximum degree m satisfying $\log m \gg \log n/\log\log n$. If $2 < \beta < 3$, almost surely the diameter is $\Theta(\log n)$ and the average distance is at most $(2 + o(1))\frac{\log\log n}{\log(1/(\beta-2))}$.*

For the case of $\beta = 3$, the power law random graph has diameter almost surely $\Theta(\log n)$ and has average distance $\Theta(\log n/\log\log n)$.

The proofs of the above facts use the following lemmas concerning the distances and neighborhood expansions in $G(\mathbf{w})$. These lemmas (as proved in [27]) are useful later for proving the main theorems in the next section.

Lemma 1. *In a random graph G in $G(\mathbf{w})$ with a given expected degree sequence $\mathbf{w} = (w_1, \dots, w_n)$, for any fixed pairs of vertices (u, v), the distance $d(u, v)$ between u and v is greater than $\left\lfloor \frac{\log \mathrm{Vol}(G) - c}{\log \tilde{d}} \right\rfloor$ with probability at least $1 - \frac{w_u w_v}{d(d-1)}e^{-c}$.*

Lemma 2. *In a random graph $G \in G(\mathbf{w})$, for any two subsets S and T of vertices, we have*

$$\mathrm{Vol}(\Gamma(S) \cap T) \geq (1 - 2\epsilon)\mathrm{Vol}(S)\frac{\mathrm{Vol}_2(T)}{\mathrm{Vol}(G)}$$

with probability at least $1 - e^{-c}$ where $\Gamma(S) = \{v : v \sim u \in S \text{ and } v \notin S\}$, provided $\mathrm{Vol}(S)$ satisfies

$$\frac{2c\mathrm{Vol}_3(T)\mathrm{Vol}(G)}{\epsilon^2\mathrm{Vol}_2^2(T)} \leq \mathrm{Vol}(S) \leq \frac{\epsilon\mathrm{Vol}_2(T)\mathrm{Vol}(G)}{\mathrm{Vol}_3(T)} \tag{3}$$

Lemma 3. *For any two disjoint subsets S and T with $\mathrm{Vol}(S)\mathrm{Vol}(T) > c\mathrm{Vol}(G)$, we have*

$$Pr(d(S,T) > 1) < e^{-c}$$

where $d(S,T)$ denotes the distance between S and T.

6 The Diameter of the Hybrid Model

Most local graphs have large diameters and large average distances. For example, the average distance of the grid graph on n vertices is $O(\sqrt{\frac{n}{\log n}})$. However, with additional "hyperlinks", (e.g., edges from the global random power law graph), the average distance of the hybrid graph can be significantly reduced.

In a hybrid graph H, let G denote its global power law graph as defined in Sect. 4. Let $\mathbf{w} = (w_1, w_2, \ldots, w_m)$ denote the degree sequence of G. We will say that vertex v_i has weight w_i and we recall that for a subset S of vertices, we have $\mathrm{Vol}(S) = \sum_{v_i \in S} w_i$ and $\mathrm{Vol}(G) = \sum w_i$. Also for $k \geq 1$, we have $\mathrm{Vol}_k(S) = \sum_{v_i \in S} w_i^k$. In particular, the second order average degree $\tilde{d}$ is just $\mathrm{Vol}_2(G)/\mathrm{Vol}(G)$. The following are immediate consequences of Fact 3 and Fact 4.

Theorem 3. *For a hybrid graph $H(n, \beta, d, m, L)$ with $\beta > 3$, almost surely, the average distance is $(1 + o(1))\frac{\log n}{\log \tilde{d}}$ and the diameter is $O(\log n)$.*

Theorem 4. *For a hybrid graph $H(n, \beta, d, m, L)$ with $2 < \beta < 3$, almost surely, the average distance is $O(\log \log n)$ and the diameter is $O(\log n)$.*

For a hybrid graph $H(n, \beta, d, m, L)$ with $\beta = 3$, almost surely, the average distance is $O(\log n/\log \log n)$ and the diameter is $O(\log n)$.

For the range of $2 < \beta < 3$, the power law graphs include many real networks. We can further reduce the diameter if additional conditions are satisfied. A local graph L is said to have isoperimetric dimension δ if for every vertex v in L and every integer $k < (\log \log n)^{1/\delta}$, there are at least k^δ vertices in L of distance k from v. For example, the grid graph in the plane has isoperimetric dimension 2. The d-dimensional grid graph has isoperimetric dimension d.

Theorem 5. *In a hybrid graph $H(n, \beta, d, m, L)$ with $2 < \beta < 3$, suppose that the local graph has isoperimetric dimension δ with $\delta \geq \log \log n/(\log \log \log n)$. Then almost surely, the diameter is $O(\log \log n)$.*

The main idea of the proof of Theorem 5 is to use the "octopus" structure of the random powerlaw graph with exponent β between 2 and 3. The proof is quite similar to that in [16] except that here we have the additional help from the local graph. For the sake of completeness, we include the proof here.

Proof of Theorem 5.
First, we define the core of a power law graph with exponent β to be the set S_t of vertices of degree at least $t = n^{1/\log \log n}$.

Claim 1. The diameter of the core is almost surely $O(\log \log n)$. This follows from the fact that the core contains an Erdős-Renyi graph $G(n', p)$ with $n' = cnt^{1-\beta}$ and $p = t^2/\text{Vol}(G)$. From [23], this subgraph is almost surely connected. Using a result in [28], the diameter of this subgraph is at most $\frac{\log n'}{\log pn'} = (1 + o(1))\frac{\log n}{(3-\beta)\log t} = O(\log \log n)$.

Claim 2. Almost all vertices with degree at least $\log n$ are almost surely within distance $O(\log \log n)$ from the core. To see this, we start with a vertex u_0 with degree $k_0 \geq \log^C n$ for some constant $C = \frac{1.1}{(\beta-2)(3-\beta)}$. By applying Lemma 3, with probability at least $1 - n^{-3}$, u_0 is a neighbor of some u_1 with degree $k_1 \geq (k_0/\log^C n)^{1/(\beta-2)}$. We then repeat this process to find a path with vertices $u_0, u_1, \dots, u_s$, and the degree k_s of u_s satisfies $k_s \geq (k_0/\log^C n)^{1/(\beta-2)^s}$ with probability $1 - n^{-2}$. By choosing s to satisfy $\log k_s \geq \log n/\log \log n$, we are done.

Claim 3. Each vertex v is within distance $O(\log \log n)$ from a vertex of degree at least $\log^C n$.

Proof of Claim 3. The main tools are Lemma 2. Let S be i-th neighborhood of u, consisting of all vertices within distance i_0 from u where $i_0 = \log \log n$. Let $T = S(w_{\min}, a)$ denote the set of vertices with weights between $w_{\min}$ and $aw_{\min}$. Here a is some large value to be chosen later. We have

$$\text{Vol}(T) \approx nd(1 - a^{2-\beta}).$$

$$\text{Vol}_2(T) \approx nd^2 \left(1 - \frac{1}{\beta-1}\right)^2 \frac{\beta-1}{3-\beta} a^{3-\beta}$$

$$\text{Vol}_3(T) \approx nd^3 \left(1 - \frac{1}{\beta-1}\right)^3 \frac{\beta-1}{4-\beta} a^{4-\beta}$$

To apply Lemma 2, $\text{Vol}(\Gamma(S))$ must satisfy:

$$\text{Vol}(\Gamma(S)) \geq \frac{2c}{\epsilon^2} \frac{\text{Vol}_3(T)}{\text{Vol}_2^2(T)} \text{Vol}(G)$$

$$\approx \frac{2c}{\epsilon^2} \frac{(3-\beta)^2}{(\beta-2)(4-\beta)} a^{\beta-2}$$

and

$$\text{Vol}(\Gamma(S)) \leq \epsilon \frac{\text{Vol}_2(T)}{\text{Vol}_3(T)} \text{Vol}(G)$$

$$\approx \epsilon \frac{(\beta-2)(3-\beta)}{(\beta-1)(4-\beta)a} n.$$

Both the above equations are easy to satisfy by using the assumption on the local graph. Namely, we can select $a = 10$, $c = 3 \log n$ and for each vertex u,

$$\mathrm{Vol}(\Gamma_{i_0}(u)) \geq \frac{60}{\epsilon^2(\beta - 2)} \log n.$$

By Lemma 2, with probability at least $1 - e^{-c} = 1 - \frac{1}{n^3}$, the volume of $\Gamma_i(u)$ for $i > i_0$ will grow at a rate greater than

$$(1 - 2\epsilon)\frac{\mathrm{Vol}_2(T)}{\mathrm{Vol}(G)} \approx \frac{(1 - 2\epsilon)d(\beta - 2)^2}{2(\beta - 1)(3 - \beta)}a^{3-\beta},$$

if $\Gamma_i(u)$ has volume not too large ($< \sqrt{n}$). After at most $(1 + o(1))\frac{2\log\log n}{(3-\beta)\log a} = O(\log \log n)$ steps, the volume of the reachable vertices is at least $\log^2 n$. Lemma 3 then implies that with one additional step we can reach a vertex of weight $\log^C n$ with probability at least $1 - e^{-\log^2 n}$ The total number of steps is at most

$$i_0 + O(\log \log n) + 1 = O(\mathrm{loglog} n).$$

The total failure probability for u to reach a vertex of weight at least $\log^C n$ is at most

$$O(\log \log n)\frac{1}{n^3} + e^{-O(\log^2 n)} = o\left(\frac{1}{n^2}\right).$$

Thus, the total failure probability that some vertex u can not reach a vertex of weight at least $\log^C n$ is at most

$$o(1) + O(\log \log n)O\left(\frac{1}{n}\right) + ne^{-O(\log^2 n)} = o(1).$$

Claim 3 is proved.   □
This completes the proof of Theorem 5.   □
The proof of the above theorem implicitly implies the following results:

Theorem 6. *In a hybrid graph $H(n, \beta, d, m, L)$ with $2 < \beta < 3$, suppose that the local graph has isoperimetric dimension δ. Then almost surely, the diameter is $O((\log n)^{1/\delta})$.*

Theorem 7. *In a hybrid graph $H(n, \beta, d, m, L)$ with $2 < \beta < 3$, suppose that every vertex is within distance $\log \log n$ of some vertex of degree $\log n$. Then almost surely, the diameter is $O(\log \log n)$.*

7 Concluding Remarks

In this paper, we consider the hybrid model for further understanding the "landscape" of real networks. Here we mention a number of remarks concerning the flexibility and possible extensions of our model.

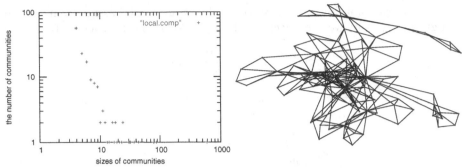

Fig. 6. The distribution of the sizes of local communities in the local graph (with $k = l = 3$) of the Collaboration Graph of the second kind.

Fig. 7. A "community" of size 87 in the Collaboration Graph of the second kind.

1. In our hybrid model, the global graph was chosen to be a random graph with given degree distribution satisfying a power law. If the global graph is to be taken to be an admissible graph or specially admissible graph (as defined in Sect. 5), similar results on average distance and diameter can be established by using methods in the the proofs of Theorem 5. There are several reasons for selecting the global graph to be a power law graph. Namely, many real networks have power law degree distribution. In addition, random power law graph G has the "scale-free" property [29] in the sense that if a fraction of vertices or edges are deleted from G, the remaining graph is still a power law graph with the same exponent (but with different average degree).

2. It is of interest to further analyze the local graphs for various classes of networks. In addition to local connectivity, are there other distinct properties that local graphs have? One such example is the isoperimetric dimension (as defined in Sect. 6) or its variations. Different types of networks (Internet graphs versus biological graphs, and so on) can have different kinds of local graphs. Are there good characterizations for different local graphs? In particular, are there special characterizations for local graphs for networks arising in epidemics and percolation?

3. The local graph in our hybrid model is a (k, l)-connected graph, with parameters k and l which can be chosen to suit the actual network under consideration. We note that (k, l)-connected graphs include the grid graphs, disjoint union of grid graphs, and grid graphs of higher dimensions, depending on the choice of k and l. In fact, by appropriately choosing several pairs of k and l, the algorithm given in Sect. 3 can result in a (k, l)-connected subgraph with a number of distinct connected components and thereby identify local "communities" within the (large) network. For example, for $k = l = 3$, the local graph of Collaboration graph of the second kind is the disjoint union of 149 (non-trivial) components as shown in Fig. 6. Each component can be viewed as a community (Fig. 7).

Acknowledgements

This research was supported in part by NSF Grants DMS 0100472 and ITR 0205061.

References

1. S. Milgram, The small world problem, *Psychology Today*, **2** (1967), 60-67.
2. L. A. Adamic and B. A. Huberman, Growth dynamics of the World Wide Web, *Nature*, **401**, September 9, 1999, pp. 131.
3. W. Aiello, F. Chung and L. Lu, A random graph model for massive graphs, *Proceedings of the Thirty-Second Annual ACM Symposium on Theory of Computing*, (2000) 171-180.
4. R. B. R. Azevedo and A. M. Leroi, A power law for cells, *Proc. Natl. Acad. Sci. USA*, vol. **98**, no. 10, (2001), 5699-5704.
5. A.-L. Barabási and R. Albert, Emergence of scaling in random networks, *Science* **286** (1999) 509-512.
6. A.-L. Barabási, R. Albert, and H. Jeong, Scale-free characteristics of random networks: the topology of the world wide web, *Physica* A 272 (1999), 173-187.
7. A. Broder, R. Kumar, F. Maghoul, P. Raghavan, S. Rajagopalan, R. Stata, A. Tompkins, and J. Wiener, "Graph Structure in the Web," *proceedings of the WWW9 Conference*, May, 2000, Amsterdam. Paper version appeared in *Computer Networks* **33**, (1-6), (2000), 309-321.
8. K. Calvert, M. Doar, and E. Zegura, Modeling Internet topology. *IEEE Communications Magazine*, **35(6)** (1997) 160-163.
9. C. Cooper and A. Frieze, A general model of web graphs, *Random Structures and Algorithms* Vol. **22**, (2003), 311-335.
10. M. Faloutsos, P. Faloutsos, and C. Faloutsos, On power-law relationships of the Internet topology, *Proceedings of the ACM SIGCOM Conference*, Cambridge, MA, 1999.
11. S. Jain and S. Krishna, A model for the emergence of cooperation, interdependence, and structure in evolving networks, *Proc. Natl. Acad. Sci. USA*, vol. **98**, no. 2, (2001), 543-547.
12. J. Kleinberg, S. R. Kumar, P. Raghavan, S. Rajagopalan and A. Tomkins, The web as a graph: Measurements, models and methods, *Proceedings of the International Conference on Combinatorics and Computing*, 1999.
13. S. R. Kumar, P. Raghavan, S. Rajagopalan and A. Tomkins, Extracting large-scale knowledge bases from the web, *Proceedings of the 25th VLDB Conference*, Edinburgh, Scotland, 1999.
14. M. E. J., Newman, The structure of scientific collaboration networks, *Proc. Natl. Acad. Sci. USA*, vol. **98**, no. 2, (2001), 404-409.
15. E. Zegura, K. Calvert, and M. Donahoo, A quantitative comparison of graph-based models for Internet topology. *IEEE/ACM Transactions on Networking*, **5** (6), (1997), 770-783.
16. F. Chung and L. Lu, Average distances in random graphs with given expected degree sequences, *Proceedings of National Academy of Science*, **99** (2002), 15879-15882.
17. D. J. Watts, *Small Worlds — The Dynamics of Networks between Order and Randomness*, Princeton University Press, New Jersey, 1999.

18. D. J. Watts and S. H. Strogats, Collective dynamics of 'small world' networks, *Nature* **393**, 440-442.
19. J. Kleinberg, The small-world phenomenon: An algorithmic perspective, *Proc. 32nd ACM Symposium on Theory of Computing*, 2000.
20. A. Fabrikant, E. Koutsoupias and C. H. Papadimitriou, Heuristically optimized trade-offs: a new paradigm for power laws in the Internet, *STOC* 2002.
21. M. Henzinger, private communication.
22. P. Erdős and T. Gallai, Gráfok előírt fokú pontokkal (Graphs with points of pre-scribed degrees, in Hungarian), *Mat. Lapok* **11** (1961), 264-274.
23. P. Erdős and A. Rényi, On random graphs. I, *Publ. Math. Debrecen* **6** (1959), 290-291.
24. M. Molloy and B. Reed, A critical point for random graphs with a given degree sequence. *Random Structures and Algorithms*, Vol. **6**, no. 2 and 3 (1995), 161-179.
25. M. Molloy and B. Reed, The size of the giant component of a random graph with a given degree sequence, *Combin. Probab. Comput.* **7**, no. 3 (1998), 295-305.
26. B. Bollobás, O. Riordan, J. Spencer and G. Tusnády, The Degree Sequence of a Scale-Free Random Graph Process, Random Structures and Algorithms, Vol. **18**, no. 3 (2001), 279-290.
27. F. Chung and L. Lu, Connected components in a random graph with given degree sequences, *Annals of Combinatorics*, **6** (2002), 125-145.
28. F. Chung and L. Lu, The diameter of random sparse graphs, *Advances in Applied Math.*, **26** (2001), 257-279.
29. W. Aiello, F. Chung and L. Lu, Random evolution in massive graphs, Extended abstract appeared in *The 42th Annual Symposium on Foundation of Computer Sciences*, October, 2001. Paper version appeared in *Handbook on Massive Data Sets*, (Eds. J. Abello, et. al.), Kluwer Academic Publishers (2002), 97-122.

Classes of the Shortest Pathway Structures in Scale Free Networks

Kwang-Il Goh, Eulsik Oh, Chul-Min Ghim, Byungnam Kahng, and
Doochul Kim

School of Physics, Seoul National University, Seoul 151-747, Korea

Abstract. We study a problem of data packet transport between a pair of vertices on scale-free network, and introduce load of a vertex as the accumulated sum of a fraction of data packets traveling along the shortest pathways between every pair of vertices. It is found that the load distributions for many real-world networks follow a power law with an exponent δ which is close to either 2.2(1) (class I) or 2.0 (class II), insensitive to different values of the degree exponent γ in the range, $2 < \gamma < 3$. The classification of scale-free networks into the two classes may stem from the characteristics of the shortest pathways structures. While the shortest pathways between a pair of vertices are multiply connected in the class I, they are almost singly connected in the class II. Such distinct topological features of the shortest pathways produce different behaviors in diverse problems such as the distribution of diameter change by the removal of a single vertex. Finally, we remark that since the two numerical values of the load exponent are too close, it still remains an open question if they are really robust. Analytic solution resolving this controversial issue is needed.

1 Introduction

Complex systems comprise of many elements such as individuals, substrates, and companies in social, biological, and economic systems, respectively, showing cooperative phenomena between elements through diverse interactions and adaptations to the pattern they create [1,2]. Such complex systems may be described in terms of graphs, consisting of vertices and edges, where vertices (edges) represent elements (their interactions) [3–6]. This approach was initiated by Erdős and Rényi (ER) [7]. In their model, the number of vertices N is fixed and edges are connected among vertices randomly. They found that the entire system can be connected with a small fraction of edges, being of order $\mathcal{O}(1/N)$. While such random graph theory is a pioneering work, it cannot describe realistic networks recently observed. In graph theory, one of interesting quantities is degree, defined as the number of edges connecting to a certain vertex. The degree distribution of the ER network follows a Poisson distribution.

Recently, there were findings that the degree distribution of the world-wide web follows a power law,

$$P_D(k) \sim k^{-\gamma}, \tag{1}$$

where k means degree and γ is the degree exponent. The network displaying a power-law degree distribution is called scale-free (SF) network. SF networks [8]

K.-I. Goh, E. Oh, C.-M. Ghim, B. Kahng, and D. Kim, Classes of the Shortest Pathway Structures
in Scale Free Networks, Lect. Notes Phys. **650**, 105–125 (2004)
http://www.springerlink.com/

are ubiquitous in real-world networks such as the world-wide web (WWW) [9–11], the Internet [12–14], the citation network [15] and the author collaboration network of scientific papers [16,17], and the metabolic networks in biological organisms [18]. To illustrate the mechanism of SF network formation, Barabási and Albert (BA) [19–21] introduced an evolving network model where the number of vertices N increases linearly with time rather than fixed, and a newly introduced vertex is connected to m already existing vertices with probability proportional linearly to the degree of the selected vertex, called preferential attachment (PA) rule. Then the degree exponent follows a power law with the exponent $\gamma = 3$. A generalized version assigns the probability proportional to $k + m(a - 1)$, $a(> 0)$ being a tunable parameter. Then the degree exponent is $\gamma = 2 + a$ [21].

Recently, we introduced another type of SF network model, called the static model, in which the number of vertices N is fixed from the beginning, and vertices are indexed by an integer i $(i = 1, \ldots, N)$ [22]. We assign the weight $p_i = i^{-\alpha}$ to each vertex, where α is a control parameter in $[0, 1)$. We select two different vertices (i, j) with probabilities equal to the normalized weights, $p_i / \sum_k p_k$ and $p_j / \sum_k p_k$, respectively, and add an edge between them unless one exists already. This process is repeated until mN edges are made in the system. Then the mean degree is $2m$. Since edges are connected to a vertex with frequency proportional to the weight of that vertex, the degree at that vertex is given as

$$\frac{k_i}{\sum_j k_j} \approx \frac{(1 - \alpha)}{N^{1-\alpha} i^\alpha}, \tag{2}$$

where $\sum_j k_j = 2mN$. Then it follows that the degree distribution follows the power law, (1), where γ is given by

$$\gamma = (1 + \alpha)/\alpha. \tag{3}$$

Thus, adjusting the parameter α in $[0,1)$, we can obtain various values of the exponent γ in the range, $2 < \gamma < \infty$.

Emergence of a power-law in the degree distribution in complex networks is a self-organized pattern in complex systems. From the viewpoint of the theory of equilibrium critical phenomena, the power-law behavior implies the absence of a characteristic scale, implying that the network is in a critical state. However, the degree exponent depends on the details of network structure, varying mostly in the range $2 < \gamma \leq 3$. Thus it would be interesting to find a physical quantity showing universal feature, insensitive to the details of network structure. In Sect. 2, we will introduce a quantity, called load, which has been used in the study of transport phenomena of data packet on networks. Interestingly, the load distribution displays a power law and the load exponent is likely to be insensitive to the details of SF network structures. Currently it seems that there exist two different classes displaying different load exponents for real world networks. Furthermore, in Sect. 3, we will introduce the load-load correlation function, which behaves differently from the degree-degree correlation function in social networks. In Sect. 4, we will consider the distribution of diameter changes when a single vertex is removed. This problem is another example that can distinguish the shortest pathways characteristics between a pair of vertices in SF networks.

2 Load or Betweenness Centrality

Let us suppose that a data packet is sent from a vertex i to j, $(i \to j)$. It is transmitted along the shortest pathway between them. If there exist more than one shortest pathways, the data packet would encounter one or more branching points. In this case, the data packet is presumed to take one of them with equal probability, and the data packet is substantially divided evenly by the number of branches at each branching point as it travels. The contribution to the load at vertex k due to the packet flow from i to j is denoted as $\ell_k^{(i \to j)}$ as shown in Fig. 1a. Then we define the load ℓ_k of a vertex k, $\ell_k = \sum_{i,j} \ell_k^{(i \to j)}$. Note that $\ell_k^{(i \to j)}$ may be different from $\ell_k^{(j \to i)}$ even for undirected networks (See Fig. 1a). Here, we do not take into account the time delay of data transfer at each vertex or edge, so that all data are delivered in a unit time, regardless of the distance between any two vertices. So the load is a static variable for a given number of nodes N. Since the packets are conserved, the total load contributed by one pair is simply related to the shortest pathway length d_{ij} between them, by $\sum_k \ell_k^{(i \to j)} = d_{ij} + 1$. Thus we have the sum rule for ℓ_k:

$$\sum_k \ell_k = \sum_{i \neq j}(d_{ij} + 1) = N(N-1)(D+1) \sim N^2 D , \qquad (4)$$

where D is the average of d_{ij} over all pairs, called the diameter.

The quantity, load is closely related to the one used in sociology called "betweenness centrality" (BC) to quantify how much power is centralized to people in social networks [23,24]. This is defined as follows. Let us consider a communication between two persons (i,j), represented by a pair of vertices (i,j). The communication is supposed to travel along the shortest pathway between the two. When there are more than one geodesic between them, one of them is taken with equal probability. The BC for a certain vertex is defined as the accumulated fraction of the total number of the shortest pathways passing on that vertex over all pairs. That is, the betweenness centrality at a vertex k is

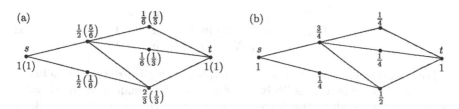

Fig. 1. The definitions of the load (a) and the BC (b). In (a), the load at each vertex due to a unit packet transfer from the vertex s to the vertex t, $\ell^{(s \to t)}$ is shown. In the parentheses, $\ell^{(t \to s)}$ is also shown for the packet transport from t to s for comparison. In (b), the BC at each vertex due to a packet transfer between s and t, $b^{(s,t)}$, is shown. The BC $b^{(s,t)}$ is equal to $b^{(t,s)}$.

$$b_k = \sum_{i \neq j} b_k^{(i,j)} = \sum_{i \neq j} \frac{C_k(i,j)}{C(i,j)}, \tag{5}$$

where $C(i,j)$ is the total number of geodesics connecting the vertices i and j, and $C_k(i,j)$ the number of those passing through the vertex k among them. The definition of BC is also illustrated in Fig. 1b. Slightly different in the definitions, the two quantities, load or BC, behave closely and their distributions are indistinguishable within our numerical resolution. Hence we shall not distinguish them throughout this paper, unless otherwise noted explicitly.

2.1 The Load Distribution

Once a SF network is generated artificially or adopted from the real world, we select an ordered pair of vertices (i,j) on the network, and identify the shortest pathway(s) between them and measure the load on each vertex along the shortest pathway using the modified version of the breath-first search algorithm introduced by Newman [24] and independently by Brandes [25]. We have measured load ℓ_i of each vertex i for SF networks with various γ. It is found numerically that the load distribution $P_L(\ell)$ follows the formula,

$$P_L(\ell) \sim \ell^{-\delta}. \tag{6}$$

When the index of the vertices are ordered according to the rank of the load, we have $\ell_1 \geq \cdots \geq \ell_N$. Then, the power-law behavior of the load distribution implies that

$$\frac{\ell_i}{\sum_j \ell_j} \sim \frac{1}{N^{1-\beta} i^{\beta}}. \tag{7}$$

with

$$\delta = 1 + 1/\beta. \tag{8}$$

The relation, (7), is valid in the region, $\ell_{\min} < \ell < \ell_{\max}$, where

$$\ell_{\min} \sim \ell_{\max}/N^{\beta} \sim \begin{cases} ND & \text{if } \beta < 1 \\ ND/\ln N & \text{if } \beta = 1 \\ N^{2-\beta} D & \text{if } \beta > 1. \end{cases} \tag{9}$$

Based on numerical measurements of load exponents for a variety of SF networks, we find that load exponent is likely to be robust, independent of the details of network structure such as the degree exponent γ in the range, $2 < \gamma < 3$, and mean degree, directionality of edge, etc. We find that SF networks are likely to be classified into two classes, say, class I and II. For the class I, the load exponent is $\delta \approx 2.2(1)$ and for the class II, it is $\delta \approx 2.0(1)$. However, see Sect. 2.4 for further discussions on the possibility of slight variation of δ on γ. The load exponent for the class II seems to be exactly $\delta = 2$ since it can be derived analytically for

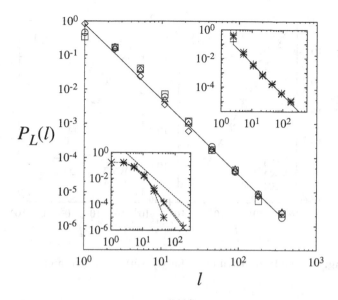

Fig. 2. Plot of the load distribution $P_L(\ell)$ versus ℓ for various $\gamma = 2.25$ (◇), 2.5 (△), 2.75 (○) and 3.0 (□) in double logarithmic scales. The data are obtained from the static model. The linear fit (solid line) has a slope -2.2. Simulations are performed for $N = 10^4$ and $m = 2$ and all data points are averaged over 10 configurations. Lower Inset: Same plot for $\gamma = 4$ (+), 5 (×), and ∞ (∗). The line having a slope -2.2 is drawn to compare the data with the case for $2 < \gamma \leq 3$. Upper Inset: Plot of $P_L(\ell)$ versus ℓ for different $m = 2$, 4 and 6, but for the same $\gamma = 2.5$.

simple models. We may show that such different universal behaviors in the load distribution originate from different generic topological features of networks.

For $\gamma > 3$, however, δ depends on γ in a way that it increases as γ increases. Eventually, the load distribution decays exponentially for $\gamma = \infty$ as shown in the lower inset of Fig. 2. Thus, the transport properties of the SF networks with $\gamma > 3$ are fundamentally different from those with $2 < \gamma < 3$. Note that (1) and (6) combined gives a scaling relation between load and degree of a certain vertex as

$$\ell \sim k^{(\gamma-1)/(\delta-1)}, \tag{10}$$

provided the ranks of vertices in load and degree do not vary significantly. See Fig. 9 and discussions in Sect. 3.

2.2 Real World and Artificial Networks Investigated

The networks that we find to belong to the class I with $\delta = 2.2(1)$ include:

(i) The co-authorship network in the field of the neuroscience, published in the period 1991-1998 [17], where vertices represent scientists and they are connected if they wrote a paper together.

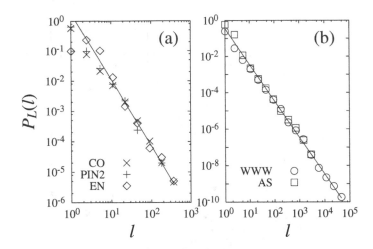

Fig. 3. The load distribution for (i), (iii), (iv), (xi), and (xiii).

(ii) The protein interaction network (PIN) of the yeast *Saccharomyces cerevisiae* compiled by Jeong *et al.* (PIN1) [26], where vertices represent proteins and the two proteins are connected if they interact.

(iii) The core PIN of the yeast *S. cerevisiae* by Ito *et al.* (PIN2) [27].

(iv) The metabolic networks for 5 species of eukaryotes and 32 species of bacteria in [18], where vertices represent substrates and they are connected if a reaction occurs between two substrates via enzymes. The reaction normally occurs in one direction, so that the network is directed.

(v) The Barabási-Albert (BA) model [19] when the number of incident edges of an incoming vertex $m \geq 2$ where $\gamma = 3$.

(vi) The geometric growth model by Huberman and Adamic [28].

(vii) The copying model [29] where $\gamma = 3$.

(viii) The undirected or the directed static model [22] where $\gamma = 3$ or $\gamma_{in} = \gamma_{out} = 3$.

(ix) The accelerated growth model proposed by Dorogovtsev *et al.* [30].

(x) The stochastic model for the PIN by Solé *et al.* [31].

For both (ii) and (x), the degree distribution is likely to follow a power-law but there needs an exponential cutoff to describe its tail behavior for finite system. However, it converges to a clean power-law for (x) as system size increases, but the converging rate is rather slow [32]. Despite this abnormal behavior in the degree distribution for finite system, the load distribution follows a pure power law with the exponent $\delta \approx 2.2(1)$ in (ii) and (x). The representative load distributions for real world networks (i), (iii), and (iv) are shown in Fig. 3a.

The networks that we find to belong to the class II with $\delta = 2.0$ include:

(xi) The Internet at the autonomous systems (ASes) level [33].

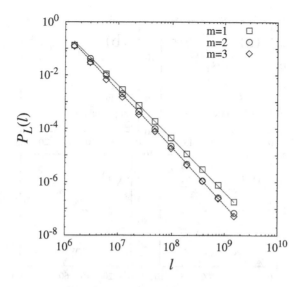

Fig. 4. The load distribution for the BA model for $\gamma = 3$ with $m = 1, 2$, and 3.

(xii) The metabolic networks for 6 species of archaea in [18].
(xiii) The WWW within www.nd.edu domain [9].
(xiv) The BA model with $m = 1$ for $\gamma = 3$ [19].
(xv) The deterministic model by Jung *et al.* [34].

In particular, the networks (xiv) and (xv) are of tree structure, where the edge load distribution can be solved analytically. The load distributions for real world networks (xi) and (xiii) are shown in Fig. 3b. Since the load exponents of each class are very close numerically, one may wonder if there exist really two different universal classes apart from error bar. To make this point clear, we plot the load distributions for the BA model with $m = 1, 2$ and 3 in Fig. 4, obtained from large system size, $N = 3 \times 10^5$. We can see clearly different behaviors between the two load distributions for the cases of $m = 1$ (class II) and of $m = 2$ and 3 (class I) in the BA model with $\gamma = 3$.

2.3 Topology of the Shortest Pathways

To understand the generic topological features of the networks in each class, we particularly focus on the topology of the shortest pathways between two vertices separated by a distance d. Along the shortest pathways, we count the total number of vertices $\mathcal{M}(d)$ lying on these roads, averaged over all pairs of vertices separated by the same distance d. Adopting from the fractal theory, $\mathcal{M}(d)$ is called the "mass-distance" relation. We find that it behaves in different ways for each class; For the class I, $\mathcal{M}(d)$ behaves nonlinearly (Figs. 5a–b), while for the class II, it is roughly linear (Fig. 5c–d). For the networks belonging to the class I such as the PIN2 (iii) and the metabolic network for eukaryotes (iv),

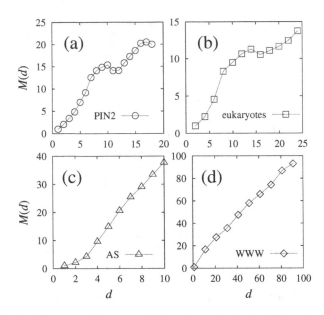

Fig. 5. The mass-distance relation for prototypical SF networks

$\mathcal{M}(d)$ exhibits a non-monotonic behavior (Fig. 5a–b), *viz.*, it exhibits a hump at $d_h \approx 10$ for (iii) or $d_h \approx 14$ for (iv). To understand why such a hump arises, we visualize the topology of the shortest pathways between a pair of vertices, taken from the metabolic network of a eukaryote organism, *Emericella nidulans* (*EN*), as a prototypical example for the class I. Fig. 6a shows such a graph with linear size 26 edges ($d = 26$), where an edge between a substrate and an enzyme is taken as the unit of length. From Fig. 6a, one can see that there exists a blob structure inside which vertices are multiply connected, while vertices outside are singly connected. What is characteristic for the class I is that the blob is localized in a small region.

For the class II, the mass depends on distance linearly, $\mathcal{M}(d) \sim Ad$ for large d (Fig. 6c–d). Despite the linear dependence, the shortest pathway topology for the case of $A > 1$ is more complicated than that of the simple tree structure where $A = 1$. Therefore, the SF networks in the class II are subdivided into two types, called the class IIa and IIb, respectively. For the class IIa, $A > 1$ and the topology of the shortest pathways includes multiply connected vertices (Figs. 6b and c), while for the class IIb, $A \sim 1$ and the shortest pathway is almost singly connected (Fig. 6d). Examples in real world networks in the class IIa are the Internet at the AS level ($A \sim 4.5$) and the metabolic network for archaea ($A \sim 2.0$), while that in the class IIb is the WWW ($A \sim 1.0$).

The WWW is an example belonging to the class IIb. For this network, the mass-distance relation exhibits $\mathcal{M}(d) \sim 1.0d$, suggesting that the topology of the shortest pathway is almost singly connected, which is confirmed in Fig. 6d.

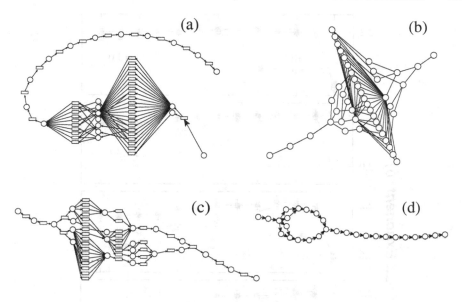

Fig. 6. Topology of the shortest pathways. (a) The metabolic network of *E. nidulans* (eukaryote) of length 26. (b) The Internet AS of length 10. (c) The metabolic network of *Methanococcus jannaschii* (Archaea) of length 20. (d) WWW of www.nd.edu of length 20. In a and c, circles denote substrates, and rectangles denote intermediate states.

2.4 Crossover in Load Distributions Between Tree and Loop Scale-Free Networks

The load exponents for various SF trees have been obtained analytically to be $\delta = 2.0$, independent of the degree exponent γ [35,36]. We investigate how the value of the exponent $\delta = 2.0$ changes as the number of loops increases. To this end, we modify the BA model in such a way that a new vertex attaches one or two edges to existing network $1 - p$ or p, respectively [37]. Then the mean number of edges emanating from a new vertex is given by $\langle m \rangle = 1 + p$. When $p = 0$, the network is a tree, and the load exponent is confirmed to be $\delta = 2.0$. In addition, we find that the load exponent increases to $\delta \simeq 2.2$ by increasing $\langle m \rangle$ to $\langle m \rangle \simeq 1.1$ at which the edges connecting different branches of the tree structure form sparse loops in a nontrivial manner. As shown in Fig. 7, the value $\delta \simeq 2.2$ at $\langle m \rangle \simeq 1.1$ turns out to be robust, largely independent of the degree exponent γ for $2 < \gamma < 3$. Such a universal behavior persists as long as $\langle m \rangle$ is smaller than a γ-dependent characteristic value, $\langle m \rangle_c$, beyond which δ depends on γ. Moreover, we find that the plateau region of $\delta \simeq 2.2$ is extended as the system size N increases as shown in Fig. 7. These data suggest that the universal behavior of δ may hold in some finite region of parameter space in the thermodynamic limit, for the sparse graph of BA model. Note that if one measures the load exponent δ at $\langle m \rangle = 2$ in finite size systems, the universal

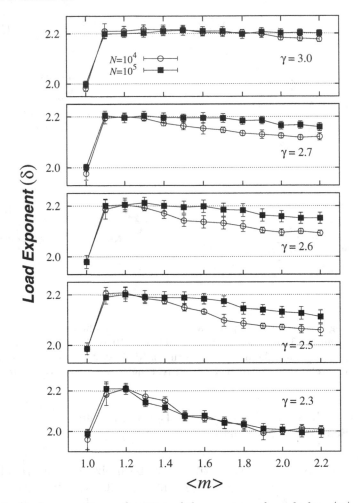

Fig. 7. The load exponent as a function of the mean number of edges $\langle m \rangle$ emanating from a new vertex for various degree exponents γ in the BA model and different system sizes, $N = 10^4$ (◯) and $N = 10^5$ (■).

features could not be observed as Barthélemy noticed recently [38] in the static model.

Barthélemy performed extensive numerical simulations of the load distribution of the static model and the BA model, finding that the robustness of the load exponent fails even for $2 < \gamma < 3$. Note that the definition of load he used is slightly different from ours. He ignored the contributions of $N-1$ packets leaving and another $N-1$ packets arriving at the vertex to load of a certain vertex, while we counted them. The difference can be neglected for vertices with large load ℓ in the limit of $N \to \infty$, however, in finite-size systems particularly those compatible with most real world networks comprising $N = 10^3 \sim 10^4$ vertices, this difference produces somewhat different value of δ. We also performed extensive

Table 1. Comparison of the load exponent for the static model with $m = 2$: δ_1 and δ_2 are the measured values using the definitions of [22] and [38], respectively. "log" and "cum" mean the measurement by log-binned and cumulative ways, respectively.

N	γ	$\delta_1(\log)$	$\delta_1(\text{cum})$	$\delta_2(\log)$	$\delta_2(\text{cum})$
1×10^4	3.0	2.30(1)	2.34(5)	2.27(3)	2.30(3)
1×10^4	2.8	2.23(3)	2.22(5)	2.17(1)	2.19(1)
1×10^4	2.5	2.15(4)	2.13(3)	2.07(1)	2.10(1)
1×10^4	2.2	2.05(5)	2.00(4)	1.96(2)	2.01(2)
1×10^5	2.5	2.13(3)	2.10(2)	2.08(1)	2.10(1)
5×10^5	2.5	2.11(2)	2.08(2)	2.08(1)	2.07(1)

numerical simulations on a larger scale $N = 5 \times 10^5$ than the size $N = 10^4$ shown in Fig. 2 for the static model with $\gamma \approx 2.5$ and $m = 2$ following both the original definition and the new one of [38], confirming that indeed, δ turns out to have weak γ dependency. Numerical values of δ measured newly are listed in Table 1. However, the behavior of δ as a function of γ with $m = 2$ for the static model is similar to the one shown in Fig. 7 for the BA model. Therefore, whether the universal behavior of the load exponent indeed breaks down or it still holds in some range of the model parameter space is still an open problem. Finally, notice that real-world networks contain nontrivial degree-degree correlations, while the static model does not have it. Thus it would be interesting to investigate how the load exponent depends on the degree-degree correlation, which is in progress. Conclusively, the problem of the universal behavior of the load exponent is not established yet, and related analytic work is needed to clarify this issue.

3 Load-Load Correlation

Recently, SF networks are grouped into three types according to the behavior of the degree-degree correlation coefficient [39,40], in a different aspect from the classification of the load exponent. They are the ones exhibiting the assortative, dissortative, and neutral mixing on their degree. For the network of the assortative (dissortative) mixing, called the assortative (dissortative) network, a vertex with large degree tends to connect to vertices with large (small) degree, while for the network of the neutral mixing, there is no such tendency. The assortative network can be found in social networks such as the coauthorship network, the actor network and so on, and the dissortative network in information networks such as the Internet and the www, and in biological networks such as protein interaction networks and neural networks. While such assortative and dissortative networks appear in real world, the neutral network, *i.e.*, the network of the neutral mixing on their degree, appears in *in silico* networks such as the BA model and the copying model [29] with $\gamma = 3$.

The degree-degree correlation [41,42] was investigated in terms of the correlation function between the remaining degrees of the two vertices on each side of an edge. Here the remaining degree means the degree of that vertex minus one [39]. First one defines the joint probability $e_D(j,k)$ that the two vertices on each side of a randomly chosen link have j and k remaining degrees, respectively. Then the normalized correlation coefficient is defined as

$$r_D = \frac{1}{\sigma_D(q)^2} \sum_{j,k} jk\{e_D(j,k) - q_D(j)q_D(k)\}, \tag{11}$$

where $q_D(k)$ is the normalized distribution of the remaining degree $q_D(k) = (k+1)P_D(k+1)/\sum_j jP_D(j)$, and $\sigma_D(q)^2 = \sum_k k^2 q_D(k) - [\sum_k k q_D(k)]^2$. Recently Newman called this quantity the degree assortativity coefficient [40]. For the assortative (dissortative) networks, r_D is positive (negative), and for the neutral networks, $r_D = 0$. On the other hand, the degree-degree correlation was also investigated in terms of the mean degree of neighbors of a vertex with degree k, denoted by $\langle k_{\text{nn}}\rangle(k)$ [13]. For the assortative (dissortative) networks, $\langle k_{\text{nn}}\rangle(k)$ increases (decreases) with increasing k, while the neutral networks, $\langle k_{\text{nn}}\rangle(k)$ is independent of k.

For the BA-type model, it was shown that the load is related to the degree via the relation, (10). Thus the vertices with larger degree are much more influential to others in communications. Due to this relation, one may think that the load-load correlation would behave similarly to the degree-degree correlation. Here we show that while for the dissortative and neutral network, the load-load correlation coefficients behave similarly to the degree-degree correlation coefficients, for the assortative network, the relation (10) is nontrivial, leading to that the load-load correlation is very weakly assortative, i.e., the mean load of neighbors of a certain vertex with load ℓ_i is almost independent of ℓ_i.

To study the load-load correlation, we introduce the load-load correlation coefficient, called the load assortativity coefficient, in analogy with (11):

$$r_L = \frac{1}{\sigma_L(q)^2} \sum_{\ell,m} \ell m\{e_L(\ell,m) - P_L(\ell)P_L(m)\}, \tag{12}$$

where $e_L(\ell,m)$ is the joint probability that the loads of the two vertices of a link are ℓ and m and $\sigma_L(q)^2 = \sum_\ell \ell^2 P_L(\ell) - [\sum_\ell \ell P_L(\ell)]^2$. Moreover, similarly to $\langle k_{\text{nn}}\rangle$, we define the mean load of neighbors of a vertex with load ℓ, denoted by $\langle \ell_{\text{nn}}\rangle(\ell)$, through which we can check if the load-load correlation is assortative or dissortative.

We first check the load-load correlation for the network of the Internet on the level of autonomous systems as of January 2000 [33] and the so-called non-degenerate configuration model with $\gamma = 3$ [42,44,45], which belong to the dissortative and the neutral network, respectively. For these networks, r_L is -0.16 (< 0) and 0.02, respectively, which is close to their r_D values of -0.18 and 0.01, respectively. Moreover, $\langle \ell_{\text{nn}}\rangle(\ell)$ behaves similarly to $\langle k_{\text{nn}}\rangle(k)$ as shown in Fig. 8a and b. However for the assortative networks, the coauthorship network for example, r_L is considerably smaller than r_D often by one order of magnitude and is close to zero. The comparison of r_L and r_D for various social networks are tabulated in Table 2. The mean load $\langle \ell_{\text{nn}}\rangle(\ell)$ of neighbors of a vertex with load

Table 2. Size N, mean degree $\langle k \rangle$, degree assortativity coefficient r_D, load assortativity coefficient r_L for a number of social networks.

Type	Name	N	$\langle k \rangle$	r_D	r_L	Ref.
Actor	Videomovie	29824	33.7	0.22	0.024	[43]
	TVminiseries	33980	73.0	0.38	0.033	[43]
	TVcablemovies	117655	55.5	0.14	0.035	[43]
	TVseries	79663	118.4	0.53	0.013	[43]
Coauthor	Neuroscience	205202	11.8	0.60	0.057	[17]
	Mathematics	78835	5.50	0.59	0.091	[17]
	cond-mat	16264	5.85	0.18	0.086	[16]
	arXiv.org	52909	9.27	0.36	0.057	[16]

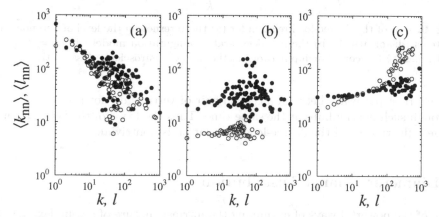

Fig. 8. Plot of $\langle k_{nn} \rangle (k)$ (○) and $\langle \ell_{nn} \rangle (\ell)$ (●) for (a) the Internet on the level of autonomous systems (dissortative), (b) the non-degenerate configuration model with $\gamma = 3$ (neutral), and (c) the coauthorship network in the field of neuroscience (assortative). All data are obtained from a single configuration.

ℓ increases with increasing ℓ, however, the increasing rate is very low compared with that of $\langle k_{nn} \rangle (k)$, i.e., it depends on ℓ very weakly (Fig. 8c). Such a behavior appears in other social networks too. Since load is regarded as a good measure of centrality, it implies that the mean influence of neighbors of a person is almost the same regardless of the influence of the centered person. So a person is surrounded by almost the same influential people on average no matter how influential the centered person may be, although a person who is acquainted with many people is likely to connect to people who also is acquainted with many others.

To understand the abnormal behavior of the load-load correlation in detail, we examine the degree-load relation. In Fig. 9, we compare the degree-load relation $\ell(k)$ for the three types. While the relation of (10) holds for the dissortative and the neutral networks, it breaks down for large k for the assortative networks. Rather the loads of large k vertices cover wide range of values. Since the vertices with large degree are located next to each other in the assortative network,

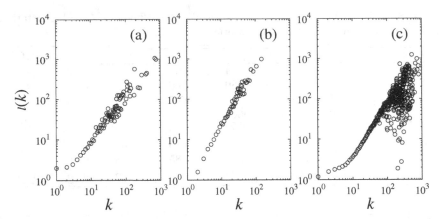

Fig. 9. Plot of the degree-load relation for (a) the Internet on the level of autonomous systems (dissortative), (b) the non-degenerate configuration model with $\gamma = 3$ (neutral), and (c) the coauthorship network in the field of neuroscience (assortative).

the shortest pathways between a certain pair of vertices do not necessarily pass through such nearby hubs at the same time. Thus the loads of the vertices with large k fluctuate and the degree-load correlation is nontrivial.

4 Diameter Change Distribution

One of the powerful ways of examining the intrinsic nature of a complex system is to observe how emerging patterns change by small perturbation applied to the system. In complex systems, such a change or response is so sensitive to the details of the perturbation that it is extremely diverse. In such a case, it is not adequate to *predict* how much the change would be definitely. Recently, Parisi has argued [46] that the prediction for the responses to small perturbations in complex systems can be made in a probabilistic way. He showed examples of protein structures in biological systems and spin glasses in physical systems. In case of proteins, subject to small external perturbations such as the change in pH or the substitution of a single amino acid, they would fold to a completely different 3D structure but with practically the same free energy. In case of the disordered magnetic systems, each spin responds to a slowly varying external field by changing its orientation, forming a series of bursts, known as Barkhausen noise [47]. The number of spins bursting depends on the disorder strength of the system, following a power-law distribution at a critical strength of disorder. The prediction of the number of spins burst in this case can only be probabilistic. All these examples aptly illustrate how the concept of probabilistic prediction may apply as a new paradigm in modern science.

   In this section we study how SF networks respond to small perturbations and check if the concept of probabilistic prediction can be applied. For this purpose, we investigate a simple problem of diameter change when a single vertex is removed from the system. Diameter, defined as the average distance between every pair of vertices in a network, is a simple yet fundamental quantity of SF networks to characterize the small-world nature, and can be thought of as a measure reflecting the efficiency of a network. Our main interest is how much the efficiency of a network would be affected by the removal of a single vertex. When a vertex is removed, each pair of remaining vertices whose shortest pathway had passed through the removed vertex should find detours, resulting in the rearrangement of shortest pathways over the network. Thus the diameter change occurs in a collective manner. From the extensive numerical calculations for a number of SF network models and real-world examples, we find that the diameter changes indeed are very diverse and crucially depend on the degree of the removed vertex. When a vertex with a few number of connections is removed, the diameter changes little. However, when a vertex with a large number of connections is removed, the diameter change is drastic, exhibiting a power-law distribution with an exponent ζ,

$$P_c(\Delta) \sim \Delta^{-\zeta} \tag{13}$$

for large Δ. Here Δ is the dimensionless relative diameter change defined as the diameter change caused by the removal of a certain vertex divided by the original diameter before the removal, and $P_c(\Delta)$ is its distribution. Moreover the exponent ζ turns out to be robust for various SF networks, insensitive to the degree exponent γ for $2 < \gamma \leq 3$.

To be specific, we consider an undirected SF network with finite number of vertices N and measure the diameter of the network. Note that we limit our interest to undirected networks only in this work. Next we remove a certain vertex i and measure the diameter D_i of the rest of the network. Measuring a dimensionless quantity, $\Delta_i = (D_i - D_0)/D_0$ for all i, where D_0 is the diameter of the original unperturbed network, we obtain the distribution of Δ for the network. Note that our case is different from the previous study of the robustness of SF networks [48,49] where vertices are removed successively. In our case, on the other hand, only a single vertex is removed each time. When a certain vertex is removed, the network may disintegrate into more than one cluster. In such cases, D_i is calculated only within the largest cluster. The diameter can be measured via a simple breadth-first search algorithm. To obtain the distribution of the diameter changes, we need the computation time of order $\mathcal{O}(N^3)$.

The diameter change Δ_i by the removal of a certain vertex i in such SF networks can be positive or negative, and the histogram of the diameter changes is highly centralized around $\Delta = 0$ (Fig. 10). However, it exhibits a fat tail for $\Delta > 0$ (the inset of Fig. 10). For the static model with $\gamma = 3$, for example, the case of small diameter changes in the range $|\Delta| < 2 \times 10^{-4}$ occurs with frequency as high as 96%. Thus the effect of a vertex removal usually is negligible as a whole, which is manifested by the exponentially bounded fluctuations of the diameter around its original value. We estimate the N-dependence of such small diameter changes in a mean-field-type approach. It is known that the diameter d_0 depends on the number of vertices as $D_0 \sim \ln N$ for random graph and $D_0 \sim \ln N / \ln \ln N$

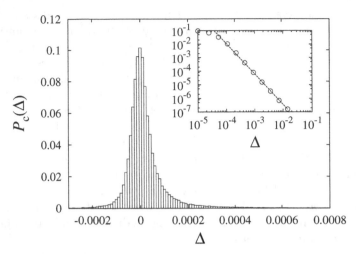

Fig. 10. Normalized histogram of the diameter changes for the static model with $\gamma = 3$ and $N = 10^4$, averaged over 10 configurations. Horizontal range is truncated for clearance, but runs up to 2×10^{-2}. Inset: Plot of $P_c(\Delta)$ in log-log scale for $\Delta > 0$. Dashed line is a fit line having a slope -2.2. Data points are logarithmically binned.

[52] for the BA model [19] with $\gamma = 3$. When a vertex is removed, the diameter may be reduced as $D \sim \ln(N - 1)$ or $D \sim \ln(N-1)/\ln\ln(N-1)$, both leading to $\Delta \approx -1/N \ln N$ for large N. Thus when $N = 10^4$, $\Delta \sim \mathcal{O}(10^{-5})$, which is comparable to numerical values of the central part in Fig. 10. On the other hand, substantial (about 4%) vertices have a serious impact on the system's efficiency and they indeed contribute to the positive tail of the histogram, showing the power-law behavior, (13). We find that such large diameter changes are mainly due to the removal of a vertex with large degree.

Let us investigate the power-law behavior for large Δ in details. The exponent ζ seems to be robust as $\zeta \approx 2.2(1)$ as long as $2 < \gamma \leq 3$ for the static model. Similar behaviors are found in other model networks (ii)–(vii) listed in Table 3. These include the SF networks showing nontrivial degree-degree correlations [39]. For $\gamma > 3$, on the other hand, as γ increases, the power-law behavior sets in only for larger values of Δ and the exponent ζ increases with γ. Eventually the diameter change distribution for the Erdős-Rényi random networks decays exponentially.

To see such universal behavior of ζ in real world, we consider a couple of real-world networks, the PIN and the Internet. For the PIN of the yeast *S. cerevisiae* [51], we also find a power law in the diameter change distribution with an exponent $\zeta \approx 2.3(1)$ (Fig. 11), consistent with the one obtained for various model networks, including the one proposed as its own *in silico* model (vii) [31]. For the Internet at the autonomous systems level [33], the diameter change distribution again follows a power law, however, with a different exponent $\zeta \approx 1.7(1)$ (Fig. 12). The smaller exponent ζ indicates that the effect of the removal of vertices contributing to the tail of the distribution is much more

Table 3. Summary of the results for various SF networks. Tabulated for each network are the degree exponent γ, the diameter change exponent ζ, and the betweenness centrality exponent η [35].

	System	γ	ζ	η	ref.
(i)	Static model	2.2–3.0	2.2(1)	2.2(1)	[22]
(ii)	Barabási-Albert model	2.2–3.0	2.2(1)	2.2(1)	[19]
(iii)	Copying model	2.2–3.0	2.2(1)	2.2(1)	[29]
(iv)	Fitness model	2.25	2.2(1)	2.2(1)	[50]
(v)	Accelerated-growth model	3.0(1)	2.2(1)	2.2(1)	[30]
(vi)	Huberman-Adamic model	3.0(1)	2.2(1)	2.2(1)	[28]
(vii)	PIN model	–	2.2(1)	2.2(1)	[31]
(viii)	PIN of the yeast	3.2(2)	2.3(1)	2.3(1)	[51]
(ix)	Internet at the autonomous systems level	2.1(1)	1.7(1)	2.0(1)	[33]
(x)	Adaptation model	2.1	1.7(1)	2.0(1)	[14]

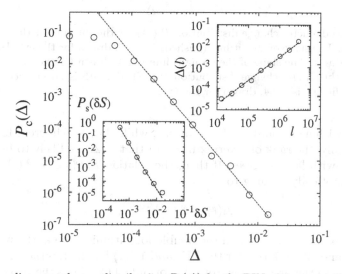

Fig. 11. The diameter change distribution $P_c(\Delta)$ for the PIN of the yeast *S. cerevisiae*. The slope of the fit line (dashed) is -2.3, drawn for the eye. Upper inset: Plot of $\Delta(\ell)$ vs. ℓ. The slope of the straight line is 1.1, drawn for the eye. Lower inset: The largest-cluster-size change distribution $P_s(\delta S)$. Here δS is normalized by N. The slope of the fit line is -3.0, drawn for the eye.

severe than the previous cases with $\zeta \approx 2.2$ [(i)–(viii) in Table 3]. To confirm the novel value of ζ for the Internet, we perform the same calculations for its *in silico* model, called the adaptation model [14], and indeed obtain $\zeta \approx 1.7$ for it, too. The two different behaviors of the diameter change distribution are rooted from distinct topological features of shortest pathways of each case.

Interestingly, the networks (i)–(viii) in Table 3 having the diameter change exponent $\zeta \approx 2.2$ belong to the class I, and the values of ζ and δ coincide with

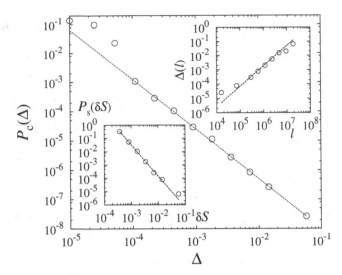

Fig. 12. The diameter change distribution $P_c(\Delta)$ for the Internet at the autonomous system level. The slope of the fit line (dashed) is -1.7, drawn for the eye. Upper inset: Plot of $\Delta(\ell)$ *vs.* ℓ. The slope of the straight line is 1.4, drawn for the eye. Lower inset: The largest-cluster-size change distribution $P_s(\delta S)$. Here δS is normalized by N. The slope of the fit line is -2.4, drawn for the eye.

each other within our numerical resolutions, while they are different for the class II. Empirically, the rank of a vertex in ℓ and that in Δ are likely to be the same for vertices with large degrees. If then, the relation $P_L(\ell)d\ell \sim P_c(\Delta)d\Delta$ would hold asymptotically, leading to

$$\Delta(\ell) \sim \ell^{(\delta-1)/(\zeta-1)}, \qquad (14)$$

for large ℓ. Indeed, the slopes in the double logarithmic scale in the upper insets of Figs. 11 and 12 are 1.1(1) for the PIN and 1.4(1) for the Internet, respectively, consistent with the predictions from the formula, (14). Thus the two classes, the classes I and II, are also categorized by the diameter change distribution and the distinction between them can be observed more clearly through it.

Our finding that the diameter change distribution is also classified into the classes I and II following those for load distribution may be rooted from the fact that both quantities, diameter and load, depend on universal features of the shortest pathways topology between a vertex pair in networks. When the sum rule [53], $\sum_k \ell_k \sim D$, is applied, one can see immediately that the diameter change distribution is the same as the total load change distribution. On the other hand, the networks belonging to the class II are more sparse and ramified than those in the class I, so that the Internet is more fragile by the removal of a single vertex than the PIN. We compare the distribution of the size change δS of the largest cluster for the PIN and the Internet by a single vertex removal. As shown in the lower insets of Figs. 11 and 12, the giant cluster in the Internet becomes much smaller than in the PIN. Thus the number of vertex pairs connected after the removal becomes much smaller in the Internet than in the PIN. Consequently, the difference of the exponent ζ between the two classes appears much larger than that of the exponent δ in the class II. However, it is not clear how the power-law behavior in $P_c(\Delta)$ arises and what determines its exponent.

5 Conclusions and Discussion

We have introduced a quantity called *load* to study a transport property of data packet on SF network, finding that the load distribution follows a power law with exponent δ. Interestingly, the load exponent δ seems to be robust, independent of the details of network structures. So far, we have found that there exist two different classes where $\delta \approx 2.2(1)$ or 2.0. The classification is mainly rooted from generic topological features of the shortest pathways between a pair of vertices. For the former, most shortest pathways are made of multiple pathways and have a blob structure, while for the latter, it is effectively a tree. Whether the load exponent δ is strictly universal in the sense of critical phenomena and renormalization group is now an open question. Initially, we conjectured it to be so based on real world network data and topological difference of the shortest pathways. Recently, however, it was observed [38] that the universality does not hold for the static model as γ approaches 2. On the other hand, we have studied the crossover behavior of the load exponent from a tree to looped structure as the number of loops increases in the BA model. The numerical result indicates two aspects: On one hand the universal behavior may still hold in the thermodynamic limit of some *in silico* model networks, however, on the other hand the universality of the load distribution may not hold for real world class I networks. Further studies are needed to resolve this controversial issue. The contents of this review article are mainly based on our works published in [22,35,37,53–55].

Acknowledgements

This work is supported by the KOSEF Grant No. R14-2002-059-01000-0 in the ABRL program. BK would like to thank the organizing committee for inviting him to the conference.

References

1. K. Ziemelis and L. Allen, Nature **410**, 241 (2001) and following review articles on complex systems.
2. R. Gallagher and T. Appenzeller, Science **284**, 87 (1999) and following viewpoint articles on complex systems.
3. S. H. Strogatz, Nature **410**, 268 (2001).
4. R. Albert and A.-L. Barabási, Rev. Mod. Phys. **74**, 47 (2002).

5. S. N. Dorogovtsev and J. F. F. Mendes, *Evolution of Networks: From Biological Nets to the Internet and WWW* (Oxford University Press, Oxford, 2003).
6. M. E. J. Newman, SIAM Review **45**, 167 (2003).
7. P. Erdős and A. Rényi, Publ. Math. Inst. Hung. Acad. Sci. Ser. A **5**, 17 (1960).
8. A.-L. Barabási, R. Albert, and H. Jeong, Physica A **272**, 173 (1999).
9. R. Albert, H. Jeong, and A.-L. Barabási, Nature **401**, 130 (1999).
10. B. A. Huberman and L. A. Adamic, Nature **401**, 131 (1999).
11. A. Broder, *et al.*, Computer Networks **33**, 309 (2000).
12. M. Faloutsos, P. Faloutsos, and C. Faloutsos, Comput. Commun. Rev. **29**, 251 (1999).
13. R. Pastor-Satorras, A. Vázquez, and A. Vespignani, Phys. Rev. Lett. **87**, 258701 (2001).
14. K.-I. Goh, B. Kahng, and D. Kim, Phys. Rev. Lett. **88**, 108701 (2002).
15. S. Redner, Eur. Phys. J. B **4**, 131 (1998).
16. M. E. J. Newman, Proc. Natl. Acad. Sci. USA **98**, 404 (2001).
17. A.-L. Barabási, H. Jeong, R. Ravasz, Z. Neda, T. Vicsek, and A. Schubert, Physica A **311**, 590-614 (2002).
18. H. Jeong, B. Tombor, R. Albert, Z. N. Oltvani, and A.-L. Barabási, Nature **407**, 651 (2000).
19. A.-L. Barabási and R. Albert, Science **286**, 509 (1999).
20. P. L. Krapivsky, S. Redner, and F. Leyvraz, Phys. Rev. Lett. **85**, 4629 (2000).
21. S. N. Dorogovtsev, J. F. F. Mendes, A. N. Samukhin, Phys. Rev. Lett. **85**, 4633 (2000).
22. K.-I. Goh, B. Kahng, and D. Kim, Phys. Rev. Lett. **87**, 278701 (2001).
23. L. C. Freeman, Sociometry **40,** 35 (1977).
24. M. E. J. Newman, Phys. Rev. E **64**, 016132 (2001).
25. U. Brandes, J. Math. Sociol. **25**, 163 (2001).
26. H. Jeong, S. P. Mason, A.-L. Barabási, and Z. N. Oltvai, Nature **411**, 41 (2001).
27. T. Ito, T. Chiba, R. Ozawa, M. Yoshida, M. Hattori, and Y. Sakaki, Proc. Natl. Acad. Sci. USA **98**, 4569 (2000).
28. B. A. Huberman and L. A. Adamic, e-print (cond-mat/9901071) (1999).
29. R. Kumar, P. Raghavan, S. Rajagopalan, D. Sivakumar, A. Tomkins, and E. Upfal, in *Proc. IEEE FOCS 2000* (IEEE Computer Society Press, Los Alamitos, CA, 2000), pp. 57-65.
30. S. N. Dorogovtsev and J. F. F. Mendes, Phys. Rev. E **63**, 025101(R) (2001).
31. R. Solé, R. Pastor-Satorras, E. Smith, and T. Kepler, Adv. Complex. Syst. **5**, 43 (2002).
32. J. Kim, P. L. Krapivsky, B. Kahng, and S. Redner, Phys. Rev. E **66**, 055101(R) (2002).
33. Meyer, D. (2001) *University of Oregon Route Views Archive Project* (http://archive.routeviews.org).
34. S. Jung, S. Kim, and B. Kahng, *Phys. Rev. E* **65**, 056101 (2002).
35. K.-I. Goh, E. Oh, H. Jeong, B. Kahng and D. Kim, Proc. Natl. Acad. Sci. USA **99**, 12583 (2002).
36. G. Szabó, M. Alava, and J. Kertész, Phys. Rev. E **66,** 026101 (2002).
37. K.-I. Goh, C.-M. Ghim, B. Kahng, and D. Kim, Phys. Rev. Lett. **91,** 189804 (2003).
38. M. Barthélemy, Phys. Rev. Lett. **91**, 189803 (2003).
39. M. E. J. Newman, Phys. Rev. Lett. **89**, 208701 (2002).
40. M. E. J. Newman, Phys. Rev. E **67**, 026126 (2003).
41. P. L. Krapivsky and S. Redner, Phys. Rev. E **63**, 066123 (2001).

42. S. N. Dorogovtsev, J. F. F. Mendes, and A. N. Samukhin, e-print (cond-mat/0206467) (2002).
43. http://www.imdb.com.
44. M. Molloy and B. Reed, Random Structures and Algorithms **6**, 161 (1995); Z. Burda and A. Krzywicki, Phys. Rev. E **67**, 046118 (2003).
45. Given a degree sequence $\mathcal{D} \equiv \{k_i\}$ ($i = 1, \ldots, N$), we generate a maximally random network whose degree sequence is $\mathcal{D}$, with a constraint that any self-loops and multiple edges are forbidden. By choosing $\mathcal{D} = \left\{ \left\lfloor (N/i)^{1/(\gamma-1)} \right\rfloor \right\}$, we get a SF network with degree exponent γ, where $\lfloor \cdot \rfloor$ is the floor function.
46. G. Parisi, Physica A **263**, 557 (1999).
47. J. P. Sethna, K. A. Dahmen, and C. R. Myers, Nature **410**, 242 (2001) and references therein.
48. R. Albert, H. Jeong, and A.-L. Barabási, Nature **406**, 378 (2000).
49. Z. Liu, Y.-C. Lai, and N. Ye, Phys. Rev. E **66**, 036112 (2002).
50. G. Bianconi and A.-L. Barabási, Europhys. Lett. **54**, 436 (2001).
51. K.-I. Goh, B. Kahng, and D. Kim, e-print (q.bio-MN/0312009).
52. R. Cohen and S. Havlin, Phys. Rev. Lett. **90**, 058701 (2003).
53. K.-I. Goh, B. Kahng, and D. Kim, Physica A **318**, 72 (2003).
54. K.-I. Goh, B. Kahng and D. Kim, Phys. Rev. E **67**, 017101 (2003).
55. J.-H. Kim, K.-I. Goh, B. Kahng, and D. Kim, Phys. Rev. Lett. **91**, 058701 (2003).

The Optimal Path
in an Erdős-Rényi Random Graph

Lidia A. Braunstein[1,2], Sergey V. Buldyrev[1], Sameet Sreenivasan[1], Reuven Cohen[3], Shlomo Havlin[1,3], and H. Eugene Stanley[1]

[1] Center for Polymer Studies and Department of Physics, Boston University, Boston, MA 02215, USA
[2] Departamento de Física, Facultad de Ciencias Exactas y Naturales, Univ. Nacional de Mar del Plata, Funes 3350, 7600 Mar del Plata, Argentina
[3] Minerva Center and Department of Physics, Bar-Ilan University, 52900 Ramat-Gan, Israel

Abstract. We study the optimal distance ℓ_{opt} in random networks in the presence of disorder implemented by assigning random weights to the links. The optimal distance between two nodes is the length of the path for which the sum of weights along the path ("cost") is a minimum. We study the case of strong disorder for which the distribution of weights is so broad that its sum along any path is dominated by the largest link weight in the path. We find that in Erdős-Rényi (ER) random graphs, ℓ_{opt} scales as $N^{1/3}$, where N is the number of nodes in the graph. Thus, ℓ_{opt} increases dramatically compared to the known small world result for the minimum distance ℓ_{min}, which scales as $\log N$. We also find the functional form for the probability distribution $P(\ell_{opt})$ of optimal paths. In addition we show how the problem of strong disorder on a random graph can be mapped onto a percolation problem on a Cayley tree and using this mapping, obtain the probability distribution of the maximal weight on the optimal path.

1 Introduction

Much attention has been focused on the topic of complex networks characterizing many biological, social, and communication systems [1–3]. The networks can be visualized by nodes representing individuals, organizations, or computers and by links between them representing their interactions. The classical model for random networks is the Erdős-Rényi (ER) random graph where two nodes are chosen randomly from the total N nodes in the system and are connected by a link [4]. An important quantity characterizing networks is the minimum distance ℓ_{min} between two nodes in the network. For the Erdős-Rényi random graph, ℓ_{min} scales as $\log N$, consistent with the "six degrees of separation" concept (e.g., if $N = 10^6$, $\ell \approx 6$).

Here we study a more realistic problem in which all links are not assumed to be equivalent. Hence we assign to each link a weight or "cost." For example, the cost could be the time required to transit the link, e.g., there are often many traffic routes from point A to point B with a set of delay times τ_i associated with each link along the path. The fastest (optimal) path is the one for which $\sum_i \tau_i$ is a minimum, and often the optimal path has more links than the shortest path.

L. Braunstein, S. Buldyrev, S. Sreenivasan, R. Cohen, S. Havlin, and H. Stanley, The Optimal Path in an Erdős-Rényi Random Graph, Lect. Notes Phys. **650**, 127–137 (2004)
http://www.springerlink.com/

If the distribution of weights is such that all the links have the same weight, the average length of the optimal path between any two nodes is the minimal length $\ell_{\min}$. In that case it is well known that $\ell_{\min} \sim \log N$ [5]

If the distribution is narrow, the average length of the optimal path ℓ_{opt}, in general, is greater than $\ell_{\min}$ but scales the same as $\ell_{\min}$ [6,7]. If the random distribution is broad, in the limit of infinite broadness, the disorder is called "strong" and only the largest weight in the path dominates the sum. The strong disorder limit is implemented by assigning to each link a potential barrier ϵ_i so that τ_i is the waiting time to cross this barrier. Thus $\tau_i = e^{\beta \epsilon_i}$, and the optimal path corresponds to the minimum $(\sum_i \tau_i)$ over all possible paths. When $\beta = 1/kT \to \infty$, only the largest τ_i dominates the sum. Thus $T \to 0$ (very low temperatures) corresponds to the strong disorder limit.

We focus here on the case of strong disorder. This is believed to be the case for many computer and traffic networks, since the slowest link in communication networks determines the connection speed. We study this problem both theoretically and numerically and find that for ER random graphs ℓ_{opt}, the average length of the optimal path, scales as $N^{1/3}$.

2 Theoretical Arguments

To obtain the optimal path in the strong disorder limit, we present the following theoretical argument. It has been shown [8,9] that the optimal path for $\beta \to \infty$ between two nodes A and B on the network can be obtained by the following algorithm:

1. Sort the links by descending weight.
2. If the removal of the highest weight link will not disconnect A from B – remove it.
3. Go back to step 2 until all links have been processed.

Since the link weights are random, so is the ordering. Therefore, in fact, one needs not even select link weights to begin with. This "bombing" algorithm can be replaced by simply removing randomly chosen links one at a time, where an link is not removed if its removal will cause the connectivity between A and B to be lost. The final path left is the optimal path between A and B in the limit $\beta \to \infty$.

Since randomly removing links is a percolation process, the optimal path must be on the percolation backbone connecting A and B. Since the network is not embedded in space but has an infinite dimensionality, we expect from percolation theory that at criticality loops are not relevant and the random graph can be approximated by a Cayley tree with a Poisson degree distribution. Thus, the shortest path must be the same as the optimal path. It is also known from percolation theory on the Cayley tree that at criticality the average mass S of the cluster with $\ell_{\min}$ branching generations scales as $\ell_{\min}^2$ [10]. Since the mass S of the giant component of the graph scales at criticality as $N^{2/3}$ [11], it follows that

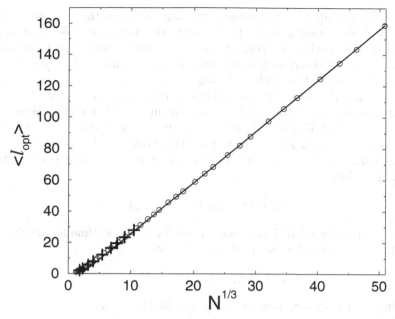

Fig. 1. The optimal path length $\langle \ell_{\text{opt}} \rangle$ averaged over 10^5 realizations of strong disorder as a function of $N^{1/3}$ computed by the two methods discussed in the text:(i) the results obtained using the "bombing" approach (∘) and (ii) the results obtained using the ultrametric approach (×).The straight line is the linear fitting of the results showing for large N the linear relation $\langle \ell_{\text{opt}} \rangle = 3.27 N^{1/3} - 7.11$. This result supports the theoretical value 1/3.

$$\ell_{\min} \sim \ell_{\text{opt}} \sim \sqrt{S} \sim N^{\nu_{\text{opt}}}, \tag{1}$$

where $\nu_{\text{opt}} = 1/3$ [12].

To test (1), we apply two numerical approaches (Fig. 1). The first approach is to find the optimal path (which minimizes the sum of weights) using the ultrametric approach described in [8]. The second approach is based on the "bombing" algorithm of [8].

3 Numerical Analysis

Next we describe in detail the two numerical methods for computing ℓ_{opt} between any two nodes in strong disorder. We can assume that the energy spectra ϵ_i is discrete. We can make β so large that, even for the closest values of energy spectra, the waiting times $\tau_i = \exp[\beta \epsilon_i]$ differ by at least a factor of 2. In this limit, the sum is dominated by the maximum value $\exp[\beta \epsilon_{\max}]$. When all the links on the paths have different weights, the optimal path is the one that has the smallest maximal link weight between all the paths. In general, as a consequence of the existence of loops, there are links in common between different paths. Such a link might provide the maximum ϵ_i of both paths. In this case we compare the second highest weight and take the path with the lower value and so forth until the optimal path is determined. This procedure is equivalent to comparing integers written in binary codes and hence indeed minimizes $\sum \tau_i$ for $\beta \to \infty$.

First, we describe the ultrametric algorithm [9]. We assign weights to all the links $\tau_i = \exp[\beta \epsilon_i]$, where $0 \leq \epsilon_i < 1$ is taken from a uniform distribution. Next, we start from one node (the origin) and visit all the other nodes connected to the origin using the Dijkstra algorithm [13]. If a node at distance ℓ_0 (from the origin) is being visited for the first time, this node will be assigned a list S_0 of weights τ_{0i}, $i = 1 \cdots \ell_0$ of the links by which we reach that node sorted in descending order,

$$S_0 = \{\tau_{01}, \tau_{02}, \tau_{03}, \ldots, \tau_{0\ell_0}\}, \qquad (2)$$

with $\tau_{0j} > \tau_{0j+1}$ for all j. If we reach a node for a second time by another path of length ℓ_1, we define for this path a new list S_1,

$$S_1 = \{\tau_{11}, \tau_{12}, \tau_{13}, \ldots, \tau_{1\ell_1}\}, \qquad (3)$$

and compare it with a S_0 previously defined for this node.

Different sequences can have weights in common because some paths have links in common, so it is not enough to identify the sequence by its maximum weight; in this case it must also be compared with the second maximum, the third maximum, etc. We define $S_p < S_q$ if there exists a value m, $1 \leq m \leq \min(\ell_p, \ell_q)$ such that

$$\begin{aligned} \tau_{pj} &= \tau_{qj} \quad &\text{for} \quad &1 \leq j < m \quad &\text{and} \\ \tau_{pj} &< \tau_{qj} \quad &\text{for} \quad &j = m, \end{aligned} \qquad (4)$$

or if $\ell_q > \ell_p$ and $\tau_{pj} = \tau_{qj}$ for all $j \leq \ell_p$.

If $S_1 < S_0$, we replace S_0 by S_1. The procedure continues until all paths have been explored and compared. At this point, $S_0 = S_{\text{opt}}$, where S_{opt} is the sequence of weights for the optimal path of length ℓ_{opt}. In reality it is highly inefficient to compare all possible paths. This is why we use the Dijkstra algorithm. The Dijkstra algorithm explores only a limited set of paths, guaranteeing that the optimal path belongs to this set. The algorithm is implemented as follows. At the beginning we assign to every node i except one that we choose as our "origin", a value $S_i = \{\infty\}$. The origin is assigned a value $S_0 = \{0\}$. The search for the optimal path follows a procedure akin to "burning" where the "fire" starts from our chosen origin. In the first step, we burn all the neighbors of the origin and replace the values assigned to them by the weight of the link that connects them to the origin. For example, if node i is a neighbor of the origin connected to it through a link which carries weight 10, then $S_i = \{10\}$ after the first step. At this point all the neighbors of the origin form what we call the "burning set" and the origin is deemed "extinguished". Now the algorithm proceeds as follows. That member of the burning set, which has the lowest value of S_i assigned to it, is deemed extinguished, and the same burning procedure starts with the node

i as the origin. The only difference is that now we burn a node j if and only if its weight sequence S_j is larger than $S_i \bigcup \tau_{ij}$, where τ_{ij} is the weight of the link connecting nodes i and j and $\bigcup$ denotes conjunction. If node j is already burning, we do not include it into the burning set again, but just replace its S_j with $S_i \bigcup \tau_{ij}$. This procedure guarantees that a node, once extinguished, can never again become part of the "burning set". Moreover the weight sequence S_i for an extinguished node yields weight of the optimal path connecting it with the origin. Once Dijkstra algorithm is completed, we have a minimal spanning tree [14] constructed on our graph. Dijkstra algorithm in the strong disorder limit is also equivalent to that of invasion percolation [15,16].

Using this method, we obtain systems of sizes up to 4000 nodes, typically 10^5 realizations of disorder. We compute $\langle \ell_{\text{opt}} \rangle$ by averaging the length of the optimal path for all the nodes of the configuration and over all realizations.

An alternative method of obtaining the optimal path in strong disorder is called the "bombing" algorithm [8]. We first choose a pair of nodes on the graph and begin removing links randomly, making sure that the connectivity between the two chosen nodes is not destroyed as each link is removed. The last path remaining is equivalent to the optimal path obtained by the ultrametric algorithm.

The bombing algorithm is slow, as one must test the connectivity after removal of each link. To improve the speed, we first find the minimal path in the graph and then select links in random order. We remove the selected link from the graph. If the removed link belongs to the minimal path, we check if the connectivity between the two nodes is still present and recompute the new minimal path. If the connectivity between the two nodes is destroyed, we restore the link.

The advantage of this procedure is that one has to test for connectivity only if the selected link appears to belong to the minimal path. Since checking the connectivity is the most time consuming part in the original "bombing" algorithm, we could reach systems of sizes up to 2^{17} nodes with 10^5 realizations of weight disorder.

Figure 1 demonstrates that both algorithms yield very similar results, supporting the theoretical result $\langle \ell_{\text{opt}} \rangle \sim N^{1/3}$. Indeed, numerical values of $\langle \ell_{\text{opt}} \rangle$ averaged over 10^5 realizations of disorder have a linear behavior as a function of $N^{1/3}$, confirming the theoretical value $\nu_{\text{opt}} = 1/3$.

We also study the probability distribution $P(\ell_{\text{opt}})$ of optimal path lengths on the network. The scaled curve for $P(\ell_{\text{opt}})$ for different network sizes is shown in Fig. 2 on a log-log plot. We find that there are two regimes in this distribution, the first one being a power law $P(\ell_{\text{opt}}) \sim (\ell_{\text{opt}})^\alpha$ which is evident from the figure, with $\alpha \approx 2$. The second regime can be well approximated by a stretched exponential $P(\ell_{\text{opt}}) \sim e^{-C\ell_{\text{opt}}^\theta}$ where C is a constant and θ is close to 2. This leads us to the conjecture that the distribution may have a Maxwellian functional form:

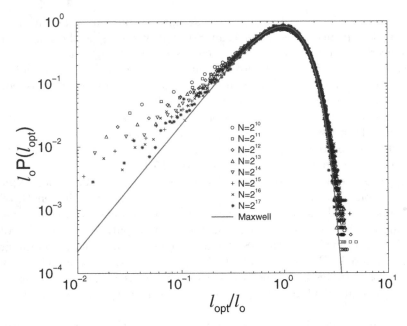

Fig. 2. Scaled probability distributions $P(\ell_{opt})$ of optimal path lengths for network sizes $N = 2^{10}, 2^{11}, \ldots, 2^{17}$. The curve represents a Maxwellian fit given by (5).

$$P(\ell_{opt}) = \frac{4\ell_{opt}^2 e^{-(\ell_{opt}/l_o)^2}}{\sqrt{\pi}l_o^3}, \qquad (5)$$

Where $l_o = \sqrt{\pi}\langle\ell_{opt}\rangle/2$ is the most probable value of ℓ_{opt}. The solid line in the figure is the plot of this function and as seen it agrees well with our numerical results, especially for large N.

Finally, we repeat our simulations for the case in which disorder weights are associated with the nodes of the graph, and obtain the same scaling laws as for the disordered links case.

It should be pointed out that the above results concerning ℓ_{opt} practically do not depend on the average degree $\langle k \rangle$ of the random graph for large $\langle k \rangle \gg 2$ and are the same even for the complete graph with $k = N - 1$. However, as we will see in the next section, the distribution of the maximal weight τ_{max} drastically depends on $\langle k \rangle$.

4 Probability Distribution of the Maximal Weight on the Optimal Path

Now, we address another aspect of the problem, which is the probability distribution of the maximal weight τ_{max} or equivalently the maximal random number ϵ_{max} along the optimal path in a strongly disordered random graph. As we mentioned earlier, the problem of the optimal path on a random graph in the strong disorder limit can be mapped onto a percolation problem on a Cayley tree with a degree distribution corresponding to the random graph and with a fraction p of its links conducting. In order to further develop this analogy, we will show that the distribution of the maximal random number ϵ_{max} along the optimal path can be expressed in terms of the order parameter $P_\infty(p)$ in the percolation problem on the Cayley tree, where $P_\infty(p)$ is the probability that randomly chosen node on the Cayley tree belongs to the infinite cluster. The motivation for the mapping on the Cayley tree comes from the following. Suppose A and B are two nodes in the random graph. Now, if we start our search for the optimal path between nodes A and B beginning at node A, then in the limit of the graph being of infinite size, the probability that we will visit a previously visited node after a finite number of steps tends to zero. Hence, we can assume that our search is equivalent to the search on a Cayley tree. If the original graph has a degree distribution p_k, the probability that we reach a node with a degree k by following a randomly chosen link on the graph, is equal to $kp_k/\langle k\rangle$, where $\langle k\rangle \equiv \sum_{k=0}^\infty kp_k$ is the average degree. This is because the probability of reaching a given node by following a randomly chosen link is proportional to the number of links or the degree k of that node. Also, if we arrive at a node with degree k, the total number of outgoing branches is $k - 1$. Therefore, from the point of view of the branching process [17–22] represented by the Cayley tree , the probability to arrive at a node with $k - 1$ outgoing branches (descendants) by following a randomly chosen link is $kp_k/\langle k\rangle$.

In the asymptotic limit, where the optimal path between the two points is very long, the probability distribution for the maximal weight link can be obtained from the following analysis. Let us assume that the probability of *not* reaching the nth generation of descendants starting from a given node of the Cayley tree whose links conduct with a probability p, is Q_n. Suppose we are at a node whose outgoing degree is 2. Then the probability that starting from this node, we will not reach the nth generation of its descendants is the sum of three terms:

1. The probability that both outgoing nodes are not conducting : $(1 - p)^2$
2. The probability that both outgoing links conduct, but the nodes reached by following them, do not have $n - 1$ generations of descendants : $p^2 Q_{n-1}^2$
3. The probability that one of the two outgoing links conduct but the node reached by following the conducting link does not have $n - 1$ generations of descendants : $2(1 - p)pQ_{n-1}$

Therefore, in this case

$$Q_n(p) = (1 - p)^2 + p^2 Q_{n-1}^2 + 2(1 - p)pQ_{n-1} \tag{6}$$

which on simplification becomes

$$Q_n(p) = ((1 - p) + pQ_{n-1})^2. \tag{7}$$

Following this argument for the case where a node has m outgoing links, the probability that starting from this node, we can not reach n generations, is

$$Q_n(p) = ((1 - p) + pQ_{n-1})^m. \tag{8}$$

Now in the case of a Cayley tree with a variable degree such as ours, we also have to incorporate a factor which accounts for the probability that the node under consideration has a given number of outgoing links. Thus for a node on the Cayley tree, the probability that it does not have descendants in the nth generation can be obtained by applying a recursion relation

$$Q_l(p) = \sum_{k=1}^{\infty} p_k k ((1 - p) + pQ_{l-1})^{k-1} / \langle k \rangle \tag{9}$$

for $l = 1, 2, ..., n$ and the initial condition $Q_0 = 0$, which indicates that a given node is always present in generation zero of its descendants.

A randomly chosen node A of a random graph has k outgoing links with probability p_k which differs from the corresponding probability on the Cayley tree, $(k + 1)p_{k+1}/\langle k \rangle$. Thus this node has a slightly different probability $\tilde{Q}_n$ of not having descendants in its nth generation:

$$\tilde{Q}_n(p) = \sum_{k=1}^{\infty} p_k ((1 - p) + pQ_{n-1})^k. \tag{10}$$

If we denote by $f_n(p)$, the probability that starting at a randomly chosen node we can reach, or survive up to, the nth generation, then

$$\tilde{f}_n = 1 - \tilde{Q}_n, \quad f_l = 1 - Q_l \tag{11}$$

for $0 \le l < n$ and hence,

$$\tilde{f}_n = 1 - \sum_{k=1}^{\infty} p_k (1 - pf_{n-1})^k \tag{12}$$

while for $1 \le l < n$

$$f_l = 1 - \sum_{k=1}^{\infty} p_k k (1 - pf_{l-1})^{k-1} / \langle k \rangle \tag{13}$$

and $f_0 = 1$.

If n goes to infinity, this formula converges exponentially to the probability $P_\infty(p) \equiv \lim_{l \to \infty} f_l(p)$ for a node to be connected to infinity for any p except for $p_c = \langle k \rangle / \sum_{k=1}^{\infty} p_k k(k - 1)$, where the convergence is a power law [17]. Analogously, we define the probability that node A on a random graph is connected to infinity as $\tilde{P}_\infty \equiv \lim_{n \to \infty} \tilde{f}_n$. In the asymptotic limit of the optimal path problem, we have a pair of nodes A and B separated by a very long path ℓ_{opt} on a giant component of a random graph. The probability $\Pi(p)$, that they will be connected at given p, provided that they are connected at $p = 1$, can in fact be approximated by the probability that both of them are connected to infinity and hence

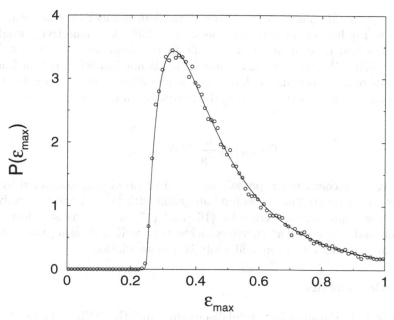

Fig. 3. The probability distribution of the maximal random number ϵ_{max} along the optimal path obtained using simulations on a random graph with $\langle k \rangle = 4$ ($\circ$) and using the analytical method on a Cayley tree with Poisson degree distribution and $\langle k \rangle = 4$ (line). The simulations involve 10^5 network realizations and are carried out on a network of 2^{16} nodes. The distribution of ϵ_{max} for small ℓ_{opt} does not obey (17), since it is derived in the limit of large ℓ_{opt}. Also the distribution of ϵ_{max} does not obey (17) for very large ℓ_{opt} when the effect of finite size N becomes evident. Thus, to achieve the best agreement with (17), we construct the histogram of ϵ_{max} only if ℓ_{opt} is in the range $40 < \ell_{opt} < 120$.

$$\Pi(p) = \left[\tilde{P}_\infty(p) / \tilde{P}_\infty(1) \right]^2. \tag{14}$$

Using (13) in the limit $l \to \infty$, for the Poisson degree distribution $p_k = x^k e^{-x}/k!$ with $x \equiv \langle k \rangle$, we conclude that $P_\infty(p)$ must satisfy the following transcendental equation

$$P_\infty(p) = 1 - e^{-\langle k \rangle p P_\infty(p)}, \tag{15}$$

which always has a trivial root $P_\infty = 0$. For $p > p_c = 1/\langle k \rangle$, (15) has a positive root, which gives us the required solution for $P_\infty(p) > 0$. For $p \leq p_c$, positive roots do not exist and $P_\infty(p) = 0$. Finally, using (12) and (14), we have

$$\Pi(p) = \left[1 - e^{-\langle k \rangle p P_\infty(p)} \right]^2 / \left[1 - e^{-\langle k \rangle P_\infty(1)} \right]^2, \tag{16}$$

where $P_\infty(p)$ and $P_\infty(1)$ are the solutions of (15).

In the bombing algorithm, the largest random number on the path, ϵ_{max}, is equal to the fraction of remaining bonds at which the connectivity would be lost for the first time, if we remove bonds in the descending order of ϵ. Thus the probability $P(\epsilon_{max} \leq p)$ that connectivity is not lost when only a fraction p of bonds remains is equal to $\Pi(p)$. The probability density of the maximum random number ϵ_{max} is thus equal to the derivative of this function with respect to p:

$$P(\epsilon_{max}) = \frac{d}{dp}\Pi(p)|_{p=\epsilon_{max}} \tag{17}$$

In Fig. 3 we compare the probability distribution of ϵ_{max} computed by simulations on a strongly disordered random graph with $\langle k \rangle = 4$ and the analytical Cayley tree approximation given by (16) and (17) with a Poisson degree distribution and $\langle k \rangle = 4$. The curves coincide very well, indicating the excellent agreement between the theoretical analysis and simulation.

Acknowledgements

We thank A.-L. Barabási for helpful discussions, and the ONR and Israel Science Foundation for financial support.

References

1. R. Albert and A.-L. Barabási, Rev. Mod. Phys. **74** (2002) 47–97.
2. J. F. F. Mendes, S. N. Dorogovtsev, and A. F. Ioffe, *Evolution of Networks: From Biological Nets to the Internet and the WWW* (Oxford University Press, Oxford, 2003).
3. R. Pastor-Satorras and A. Vespignani, *Evolution and Structure of the Internet: A Statistical Physics Approach* (Cambridge University Press, in press).
4. P. Erdős and A. Rényi, Publicationes Mathematicae **6** (1959) 290–297.
5. B. Bollobás, *Random Graphs* (Academic Press, London, 1985).
6. I. Smailer, J. Machta, and S. Redner, Phys. Rev. E **47** (1993) 262.
7. P. van der Hofstad, G. Hooghienstra, and P. van Mieghen, Prob. Eng. Inf. Sciences **15** (2001) 225.
8. M. Cieplak, A. Maritan, and J. R. Banavar, Phys. Rev. Lett. **72** (1994) 2320–2323; **76** (1996) 3754–3757.
9. L. A. Braunstein, S. V. Buldyrev, S. Havlin, and H. E. Stanley Phys. Rev. E **65** (2001) 056128.
10. A. Bunde and S. Havlin, eds., *Fractals and Disordered Systems* (Springer, New York, 1996).
11. P. Erdős and A. and Rényi, Publications of the Mathematical Institute of the Hungarian Academy of Sciences **5** (1960) 17–61.
12. L. A. Braunstein, S. V. Buldyrev, R. Cohen, S. Havlin, and H. E. Stanley Phys. Rev. Lett. **91** (2003) 168701.
13. T. H. Cormen et al., *Introduction to Algorithms* (MIT Press, Cambridge MA, 1990).

14. R. Dobrin and P. M. Duxbury, Phys. Rev. Lett. **86** (2001) 5076.
15. A.-L. Barabási, Phys. Rev. Lett. **76**, 3750 (1996);
16. N. Schwartz, A. Nazaryev, and S. Havlin, Phys. Rev. E **58**, 7642 (1998);
17. T. E. Harris, *The Theory of Branching Processes*, (Dover Publication Inc., New York, 1989).
18. R. Cohen, S. Havlin, and D. ben-Avraham, in *Handbook of Graphs and Networks*, edited by S. Bornholdt and H. G. Shuster (Willey-VCH, New York, 2002), Chapt. 4.
19. R. Cohen, K. Erez, D. ben-Avraham and S. Havlin, Phys. Rev. Lett. **85** (2000) 4626.
20. S. V. Buldyrev, S. Havlin, J. Kertész, R. Sadr, A. Shehter, and H. E. Stanley, Phys. Rev. E **52** (1995) 373–388.
21. A.-L. Barabási. S. V. Buldyrev, H. E. Stanley, and B. Suki, Phys. Rev. Lett. **76** (1996) 2192–2195.
22. B. Derrida and H. Spohn, J. Stat. Phys. **51**(1988) 817–840

Clustering in Complex Networks

Gábor Szabó[1], Mikko Alava[2], and János Kertész[1]

[1] Department of Theoretical Physics, Institute of Physics, Budapest University of Technology, 8 Budafoki út, H-1111 Hungary
[2] Laboratory of Physics, Helsinki University of Technology, P. O. Box 1100, FIN-02015 HUT, Finland

Abstract. Real, mostly socially rooted networks have the tendency to form the simplest morphological structures observable: triangles, or in other words mutually linked triples of nodes. This is captured by the concept of clustering. We review a few milestones in the quantitative description of this feature, together with selected models that reproduce certain attributes of clustering in various networks. Lastly, we present an example of problems faced when solving a simple relevance-based model by one of the common methods in widespread use.

1 Introduction

In this paper we review a few aspects of what can be considered to be one of the simplest ways of describing nontrivial network structures: clustering or transitivity. The term clustering has been borrowed from the social sciences, but its meaning has become somewhat narrowed down to refer to the prevalence of connected triangles in the network. It is in other words a common observed feature in many real networks that if a node is connected to either node of a connected pair of vertices, then it is likely that it is also connected to the other node of the pair. In large networks, it becomes a matter of quantitative analysis to define a measure of clustering, either locally or globally. The local clustering coefficient C_i of a node i has been introduced [1] to express the connectedness of the node's neighbors with each other:

$$C_i = \frac{\text{number of direct links between neighbors of } i}{\text{number of all such possible links}}. \tag{1}$$

When the degree of the node is given as k_i, then the denominator in (1) is yielded as $k_i(k_i - 1)/2$. An overall clustering coefficient C characteristic of the network can be taken to be the algebraic average of all C_i's. The clustering coefficient used by social scientists differs from the expression above, insofar it is defined only globally and basically relates the number of triangles to the number of paths of length two [2].

Indeed, the clustering coefficients measured on many types of networks are markedly larger than in random graphs, and most probably have a non-zero thermodynamic limit, in contrast to random graphs again. Thinking about networks in the human societies or networks based on mutual introduction rules, it is easy to conceive how or why dense clustering may emerge. On the other

G. Szabó, M. Alava, and J. Kertész, Clustering in Complex Networks, Lect. Notes Phys. **650**, 139–162 (2004)
`http://www.springerlink.com/`

hand, networks on the technological front may be expected to be different even qualitatively from the "spontaneous" networks in the above sense, for the reason that they often evolve through any kind of nonlocal optimization minimizing a cost function. In this sense the WWW, though, is in fact closer in nature to human contacts when shared interest dominates over utility or energy costs of man-made networks.

Many models are now able to give account for ever more observed details of networks, the clustering coefficient being one of them. Qualitative comparison is often made on the basis of the clustering coefficient dependence on the node degree, that is, the function that gives the expected value of the coefficient for nodes with the same degree. As the following sections illustrate, this function many times is simple enough to warrant further investigation.

As a factor of network structure, clustering influences the dynamical processes on the network (diffusion, random walk, search), so far as it is related to the the number of loops of the shortest size possible. Results that directly connect clustering and the processes are more scarce. A less-researched subject is the number of loops of higher order than 3, which is particularly appealing when one is to characterize the networks better or when the goal is to find the best matching model to a given network, for instance [3].

Large local or global clustering can be an indication of hidden modularity in the network, especially when the network itself is the representation of certain structures or functions as is the case for example in social or metabolic networks [4,5]. Modularity is inherently subject to a wide variety of interpretations, due to the many degrees of freedom that can go into designing the clustering algorithms that classify the nodes into modules [6]. The clustering coefficient inside modules can still be expected to be considerably higher than in intermodular regions.

The empirical data is usually analyzed in terms of the average clustering coefficient C and the degree-dependence of the same, $C(k)$. Network motifs [7, 8], as triangles, are subgraphs that are more frequent than one would expect in, say, the basic configuration model to be outlined below.

One of the interesting issues in looking at local structures, as triangle formation, is how they correlate with other properties. This holds information about the mechanism that drives the network growth. The "friends of friends become friends" one is a typical example—as we show in the last section about the mean-field theory this leads to correlations in clustering that can be detected by computing $C(k)$. For other possibilities, similar analysis reveals that the dynamics of the local neighborhood—as reflected in the average degree of the neighbors, $\langle k_{nn} \rangle$—should also be exhibited in the clustering of a node of degree k. This in turn directly implies that e.g. in assortative networks degree-degree correlations and the clustering are coupled [9].

2 Examples of Clustering

A most basic, clear-cut way to look at clustering is to consider the average clustering coefficient, C, and this is displayed in Table 1 for various examples. The clustering seen in non-social networks is of about the magnitude one would expect for a random graph model with parameters (number of nodes, average degree) similar to real networks. The configuration model [10,11] for instance produces similar numbers for graphs of such sizes. Notice the relatively high clustering in *directed* networks, from which the clustering has been computed neglecting the asymmetry. Ravasz *et al.* [12] and Vázquez [13] provide numerous examples in real-world networks for $C(k)$. Such plots imply, possibly, the presence of *power-law* dependence of the coefficient on the node degree k. In the section about mean-field behavior we shall see that there is a rather trivial mechanism, of the "friends of friends" type, that can create $1/k$-behavior [13–15]. A number of examples is shown in Fig. 1, supporting this observation. Other kinds of functional dependencies are less obvious. The analysis of growing networks allows in principle to understand various effective decays of $C(k)$, but also highlights the fact that there are large corrections to scaling. In particular, the structure of the Internet is an intriguing example, with a power-law decay in between $3/4$ and 1, as shown in Fig. 2. Note that the reason for a *decay* of the

Table 1. Average clustering for various networks. The table contains the type of the graph, its nature (directed or undirected), the number of vertices N, the average clustering coefficient C, and finally there is a citation to the source of the data. According to Newman [2].

network	type	N	C	Ref(s).
film actors	undirected	449 913	0.20	[1,16]
math coauthorship	undirected	253 339	0.15	[17,18]
physics coauthorship	undirected	52 909	0.45	[19,20]
biology coauthorship	undirected	1 520 251	0.088	[19,20]
email messages	directed	59 912	0.16	[21]
email address books	directed	16 881	0.17	[22]
student relationships	undirected	573	0.005	[23]
WWW nd.edu	directed	269 504	0.11	[24,25]
Roget's Thesaurus	directed	1 022	0.13	[26]
Internet	undirected	10 697	0.035	[27,28]
power grid	undirected	4 941	0.10	[1]
software packages	directed	1 439	0.070	[29]
software classes	directed	1 377	0.033	[30]
electronic circuits	undirected	24 097	0.010	[31]
peer-to-peer network	undirected	880	0.012	[32,33]
metabolic network	undirected	765	0.090	[34]
protein interactions	undirected	2 115	0.072	[35]
marine food web	directed	135	0.16	[36]
freshwater food web	directed	92	0.20	[37]
neural network	directed	307	0.18	[38,1]

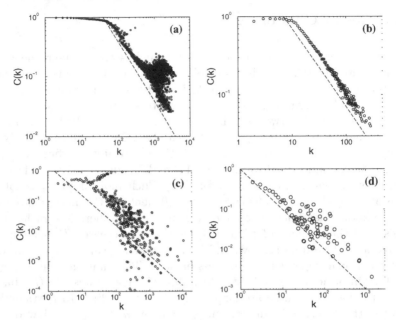

Fig. 1. The scaling of $C(k)$ for four networks. The dashed lines indicate inverse power laws in all cases. **(a)** The actor network based on the `imdb.com` database. Two actors are connected if they ever played together. **(b)** The semantic web of English words. Words are connected if the Merriam-Webster dictionary lists them together. **(c)** WWW data collected by Albert *et al.* [24] **(d)** The Internet at the Autonomous System level, where each node represents a domain. The figures are presented by Ravasz *et al.* [12].

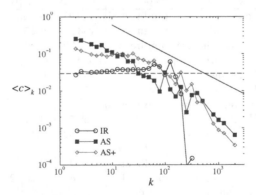

Fig. 2. Clustering coefficient as a function of the node degree for data collected on the Internet on both the Autonomous System level at two different times (AS and AS+) and the Internet Router level (IR). The solid line indicates the power law decay $C(k) \sim k^{-0.75}$. After Vázquez *et al.* [39].

distribution is clear in many cases. Effectively, hubs connect (in particular in graphs with a hierarchical structure) subgraphs or cliques to each other. These are of course not connected with others, so the largest-k nodes have a small $C(k)$.

Loops up to size 5 are much more frequent in the Internet than in random scale-free networks [3]. For triangles, the data implies that $C(k)$ for the Autonomous System level Internet indeed scales as $k^{-0.7}$, regardless of the time at which the data has been taken [39]. The same data compares also with the fitness model outlined below.

Thus, vertices of high degree k tend to be connected, on average, to others of low degree, and *vice versa*. Newman and collaborators [9,29] argue that only social networks tend to have a positive so-called Pearson correlation coefficient r that quantifies this. Small networks can accidentally have large clustering, though r is negative. The existence of modular structures may have geographic (the Internet [40]), functional (metabolic [5] or protein interaction networks [41]) bearings, or arise due to social network activities [6,42].

3 Models That Create Clustering

There are several possible recipes for models that result in local structures with the desired properties. On the level of computer simulations one can separate two classes of dynamics, depending on whether the graph is growing (nodes are added) or whether the edge population is allowed to equilibrate. Or, one can set-up the graph from the very beginning, from an ensemble like in the Erdős-Rényi graph example [43].

3.1 Static Graphs

In this respect, a perfect example is the *Newman and Park model* of networks based on groups [44]. The idea is simple: create local clustering inside groups, mimicking social networks. One takes N individuals and divides them into M groups ("community structures" [6]). Like in the Holme-Kim model of growth individuals have strong, probabilistic triangle formation inside groups. One can complicate this idea further, by e.g. labeling the groups [45]. This graph can be presented as a bipartite one, with connections between groups and individuals, only. The clustering properties are obviously something that one can tune, at will, by changing the local intra-group connection probability p. The outcome is that the average clustering coefficient is

$$C = pC_b, \tag{2}$$

C_b being the clustering coefficient of the simple one-mode projection of the bipartite graph [46]. Likewise one can determine $C(k) \sim 1/k$ again in analogy with the "friends of friends" principle. The factor C_b can be determined exactly in a number of cases.

An even simpler way is to modify Erdős-Rényi graphs by tuning the degree distribution to the wanted one: the *configuration model* [10,11]. Here [2,21]

$$C = \frac{1}{Nz_1} \left[\frac{z_2}{z_1}\right]^2 = \frac{z_1}{N} \left[\frac{\langle k^2 \rangle - \langle k \rangle}{\langle k \rangle^2}\right]^2, \tag{3}$$

where z_i denotes the coordination of the i'th layer (i.e. 1 means neighbors), and N is the size. This is the Poisson random graph value $C = z_1/N$ times an extra factor. Notice that $C \sim N^{-1}$, but for wide degree distributions the actual value can be significantly larger and thus noticeable. For power-law distributions, this results in [2]

$$C \sim N^{-\beta}, \qquad \beta = \frac{3\alpha - 7}{\alpha - 1}. \tag{4}$$

For truly SF networks ($P(k) \sim k^{-\alpha}$), with an α in the range between two and three, this leads to interesting behaviors. For small enough α values C even increases with N, while above $7/3$ the decay is slower than in random graphs. Note that this is in slight contrast with the fact that the configuration model is assumed to have no loop structure. Similar results can be derived for the projections of bipartite graphs [46]. Recently, Caldarelli *et al.* [47] and Söderberg [48] (see also [49]) have proposed different models of inhomogeneous random graphs that represents a natural generalization of the classical Erdős-Rényi random graph model [43].

For the film-actor collaborations [1,16], collaboration networks of mathematicians [17,50], company directors [51], and an email network [22], the configuration model-based argumentation results in an average clustering coefficient of 0.0098, 0.00015, 0.0035, and 0.017, respectively, much smaller than the actual values (see Table 1). The suggested reason is the presence of community structures or groups [2,12,6,42], but of course direct "friends of friends get to know each other" mechanisms would work as well (e.g. [52–56]).

A more complicated idea to create networks, with an influence on clustering, is to assign *"fitnesses"* to the vertices and make the linking to be dependent on the quenched values for any pair of nodes (i, j): f_i and f_j [57]. This is in fact just a generalization—for a static network—of the Erdős Rényi graphs (where $f \equiv p$). The properties of this kind of networks depend on the probability distribution $\rho(f)$ of the fitnesses, assuming that links are established simply by a probability linearly dependent on both f_i and f_j. A power-law ρ leads simply to a power-law distribution for the degrees. Even for other distributions, one may get SF-networks using threshold rules for the fitnesses in the linking process. Power-law clustering is not created by a $\rho(f)$ that is a power-law, but for an exponential one: with $N = 10^4$ and $\langle k \rangle = 10$, the average clustering coefficient is $C \simeq 0.1$, $\langle k_{nn} \rangle(k) \sim k^{-0.85}$ and $C(k) \sim k^{-1.6}$ has been demonstrated [47]. It remains to be seen how the threshold rule incorporated in this can be justified. The authors argue that protein networks may provide an example, since in the "two-hybrid" method the hybridizing of two proteins depends on the interaction time of the two proteins which if large enough allows transcription.

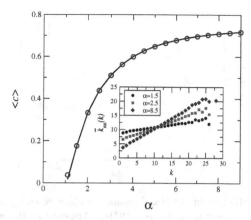

Fig. 3. Average clustering coefficient in the social network formation model as a function of the connecting tendency α and fixed average degree, $\langle k \rangle = 10$. The solid line is the analytical estimate, while the symbols are from simulations. The inset shows the average nearest neighbor degree, $\langle k_{nn} \rangle$. Originally presented by Boguñá et al. [58].

One of the important advances in the understanding of network properties based on the "rules of the game" is the analysis of the *hidden-variable models* by Boguñá et al., applicable in particular to models with static fitnesses [57]. One example is given by social networks. One can for instance define *distances*, by assuming the presence of some effective metric. Since social networks empirically exhibit large clustering, the metric should also lead to comparable behavior. This is naturally coupled to the assortative mixing exhibited by such networks, i.e., the function $\langle k_{nn} \rangle$ is not trivial. These are often in social networks coupled to the presence of community structures [6,44]. In the social network formation model of Boguñá et al. [58] the connections between individuals are established as a function of the social distance. Communities arise naturally, and the average clustering coefficient remains non-zero in the thermodynamic limit. The central idea is to use (in a d-dimensional space of distances) a connection probability that is approximately inversely proportional to the social distance to a power α. Social distance, in this case, is measured in a d-dimensional space of social features (geographical location, wealth, religion, etc.). $\alpha > 1$ measures the tendency of people to connect to similar people. Using the aforementioned hidden-variable theory [57] one can compute many properties of the model.

One-dimensional simulations with initial random conditions for the individuals clearly exhibit the fact that α is the essential control parameter (Fig. 3). The asymptotic value is $C = 3/4$ for $\alpha \to \infty$.

Finally, another approach akin to equilibrium statistical mechanics is to assume an *energy or a Hamiltonian* [59] for the structure. One attempt is to use an energy for an $i \leftrightarrow j$-link [60]

$$\epsilon_{ij} = \frac{\min\{k_i, k_j\}}{\max\{k_i, k_j\}} - 1, \tag{5}$$

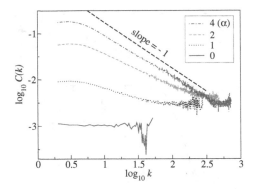

Fig. 4. Clustering coefficient in the Hamiltonian random rewiring model as a function of the degree, for four values of α, for $N = 8192$. Appears in the work by Baiesi and Manna [60].

and ϵ_{ij} is 0 if i and j are not connected. Equation (5) favors large differences between the respective degrees, together with the Hamiltonian being $H(\mathcal{G}) = \sum_{i<j} \epsilon_{ij}$ for the configuration $\mathcal{G}$. One may now do random rewiring (which is easy since one does not need to conserve properties such as the local degree). In particular, for Metropolis-like rewiring dynamics the outcome depends on the effective temperature α chosen.

Numerical data implies $C \sim N^{-\sigma(\alpha)}$ for $N \to \infty$, with σ ranging from ≈ 1 to ≈ 0.25 for $0 \leq \alpha \leq 6$, and $C(k) \sim k^{-\beta}$ is also obtained for various values of α. The exponent β reflects the fact that there is hierarchical clustering (like before, locally highly clustered subnetworks are connected via hubs). $\beta = 0$ for $\alpha = 0$, but for larger α a power-law $C(k)$ takes over, with β increasing with α towards $\beta = 1$ (Fig. 4).

The rule of an *underlying geometry* has natural effects on clustering. This may be behind the fact that the Internet exhibits such geometric properties. The existing models in which links are added to regular lattices to create SF degree distributions have rather large clustering coefficients. Manna *et al.* studied another version of the same theme [61], in which the cost (Euclidean length) is minimized starting from a N node Barabási-Albert (BA) network [62] on a two-dimensional plane, with link interchange. This is done by rewiring two disjoint links so that either ends of the links will be exchanged to point to the pruned vertex of the other pair, preserving the disjointness. A random rewiring version also exists. The ensuing clustering—note that the degree properties of the original BA model are maintained—shows interesting features. The average clustering coefficient does not show any dependence on N, while $C(k)$ becomes a power law with a logarithmic correction (Fig. 5). The behavior can be characterized empirically with

$$C(k) \sim \{k/[\ln(k)]^{1/2}\}^{-b}, \tag{6}$$

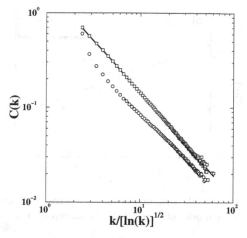

Fig. 5. $C(k)$ as a function of the degree k in the geometrically positioned nodes model of Manna *et al.* The exponent b referred to in the text is $b \approx 0.94$ and 1.1 for two different kinds of rewiring strategies, respectively. Provided by Manna *et al.* [61].

where $b \approx 0.94$ and 1.1 for the normal and random rewirings, respectively. The presence of the logarithmic correction is interesting, and note the similar discussion in Section 4.

3.2 Growing Networks

The easiest complication to the Barabási-Albert model is to allow for both the addition of new nodes, with preferential attachment, and the *addition of new edges* [63]. The edge addition can then be made such as to indirectly favor clustering over the BA-case. If the formation of new links is such that we require assortative mixing [9], one would take the nodes to which new links are being connected with a probability function that uses as the argument $|k_1 - k_2|$, i.e., the difference in the degress of the trial nodes. Such assortative networks however can produce *increasing* clustering with k, in stark contrast to usual data, which simply arises from a tendency of hubs to be interconnected. In social networks, this is often not the case, as for instance in preprint databases.

In more concrete terms, the idea is to combine vertex addition (with probability p) and link addition (with probability $1 - p$), incorporating into the latter a probability that takes into account the degrees of the nodes the link would connect. For instance, the probability to actually place a link being

$$P(k_1, k_2) \propto \frac{1}{|k_1 - k_2| + 1} \tag{7}$$

would result in assortative mixing, and

$$P(k_1, k_2) \propto |k_1 - k_2| \tag{8}$$

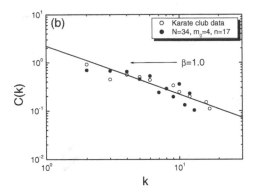

Fig. 6. Plot of the Karate Club data of Zachary [68], together with the clustering as given by the node coloring model of Kim *et al.* As shown by the same authors [67].

in disassortative mixing, in analogy to many non-social networks. The latter allows to solve the degree distribution, a power law with the exponent $\gamma(p) = 2 + \frac{p}{2-p}$. The usual clustering related quantities show again the effect of controlled triangle-formation. For $C(k)$ a power-law can be established, in analogy to the behavior of $\langle k_{nn}\rangle(k)$ again. For small enough values of p, β is in the range of 0.7...0.8, while for $p > 0.7$ the BA tree behavior is approached.

The *relevance variables driven model* of Caldarelli *et al.* [64] is another example of "hidden variables", which has the explicit motivation to attain as close a match to the Internet variables as possible. The model has the basic idea that a node added at time t with a relevance r_t connects only to nodes with higher relevances $r_\tau > r_t$, with a preferential attachment (PA) rule applied in that case with m links. One may proceed to solve the degree distribution using standard rate equation techniques. More interestingly, perhaps, the $\langle k_{nn}\rangle(k)$ and $C(k)$ turn out to have properties influenced by the "community" mechanism implied by the relevances. Both decay with k, and it is found that $\langle k_{nn}\rangle(k) \simeq k^{-0.57}$, close to the Internet value [65,66]. The $C(k)$ decays as $k^{-0.72}$, close again to the Internet Autonomous System data scaling.

A recent modification of the BA model is such that a color is assigned to each vertex, to denote an *association with a group*. The model of Kim *et al.* [67] starts as follows: first a BA graph is grown up to a size n. Then the vertices are split into two groups, by using the nodes with the two largest degrees and, for the others, the topological distances to these two for classification. Subsequently, more vertices are added with randomly chosen colors, and with preferential attachment linking only to those with the same color. Once the largest group grows to the threshold size n, it is split again.

Figure 6 demonstrates a typical $1/k$ -scaling for $C(k)$; note that $\langle k_{nn}\rangle(k) \sim k^{-0.5}$ for the data. The degree distribution is interestingly "narrower" than for the BA model. Notice that the scalings are dependent on the split-up parameter n in the model. A cross-over in $C(k)$ from constant to power law can be obtained by tuning the maximum number of links per new node towards n.

Another question is the role of an *underlying (two-dimensional) geography*: previously we met an example where the geometry was optimized while keeping the distance properties as an objective. A more general one is how the geometry and clustering combine if spatial distances are involved in general, e.g., as a growing network develops [56,69–72]. An easy example is the formation of social networks, with face-to-face contacts being obviously of importance. One attempt is to change the generalized PA rule $[P(new \to i) \sim k_i^\beta]$ to such that a factor ℓ^α is added, where ℓ is the distance between the two nodes, and α is a parameter. Notice that without any spatial structure, scale-free graphs are obtained only with $\beta = 1$ [73], but the incorporation of α may enlarge the parameter space such that SF features are obtained in a region of α-β phase space. Simply adding the spatial correlations to a BA network maintains the power-law characteristics if α is large enough [70,71].

Simulations in [74] with a one dimensional Euclidian structure imply a complicated phase diagram if the preservation of the SF properties are considered. Roughly, if $\alpha < 0.5$, β has to be increased from unity, and meanwhile the degree decay exponent γ decreases from the BA value of 3. As regards to clustering, the addition of spatial background is able to change the behavior from the quasi-trivial BA limit $C(N) \sim N^{-1}$. The behavior of the average clustering coefficient $C(N)$ on the SF phase boundary is interesting: clustering is enhanced when α decreases, the power-law exponent of $C(N)$ is reduced, and for very large negative α clustering even seems to increase (obviously, this should be reconciled with the fact that $C \leq 1$).

In real problems one usually cannot expect that the global rule of preferential attachment governs the growth mechanism. E.g., in biochemical applications, or autocatalytic reactions some local chemistry or, in biological systems, evolution determines the adjacency, which, however, may result in a network with similar properties [34,35,75–85].

Such local dynamics in protein network-related models (see e.g. [86]) typically contains the idea that a gene is duplicated but not completely, and thus proteins that the gene produces or controls are not exactly the same [87]. Thus the protein-protein interaction network evolves as a duplicated gene creates two vertices with the same interactions (addition of a new node). Now one can postulate mechanisms that slightly modify the copied interactions [81,88,89]. Note that this is very reminiscent of the World Wide Web growth, since one often copies (outgoing) links to new pages from existing ones. Such mechanisms give easily rise to an effective PA rule, which as usual may imply a power-law degree distribution with an exponent dependent on the relative rates of the microscopic processes. The clustering in such models has to arise so that the creation of edges gives also rise to real correlations between the degrees of neighboring vertices. Node-duplication of course reproduces already existing triangles. A constant probability for this induces automatically a $\sim 1/k$-behavior for $C(k)$, since the probability for a node to get increased clustering from duplication is linearly proportional to its degree. This scaling might be changed with any possible edge removal process. A variant is the duplication-divergence model [89], where vertices are first duplicated and the copies made to self-interact with a finite probability, and then out of each pair of edges coupled to the twin vertices one is removed with another probability. One particular twist is that the model has a highly non-trivial degree distribution that depends on the probability to remove edges, and may be called multi-fractal. Simulations of clustering [89] reveal that while the degree distributions are extremely non-trivial and follow only partly mean-field descriptions, the behavior of $C(k)$ does follow an approximate power-law with $C(k) \sim k^{-\beta}$, and $\beta \geq 1$. Increased loss of edges makes the deviation from the $1/k$-scaling more dramatic. The fast decay of $C(k)$ couples to a negative correlation between the degree of a vertex and its neighbors, in analogy to some empirical and modelling results [83,90].

Further model variants, still keeping away from global PA dynamics, have been devised by Vázquez [13]. The idea is as follows. Typically, it could be envisioned that a *random walk* is performed on a set of connected web pages. If one now assumes that there is a constant probability, per vertex visited, to be joined to the first node, this naturally creates an effective PA rule—nodes with k links have a linear probability, on the average, to be added. If one forgets about the difference of in- and out-degrees, the number of pairs of nearest neighbors connected follows a rate equation, which, after integration, yields the local clustering coefficient for nodes with sufficiently large degrees as

$$C(k) \approx \frac{2(1 + q_e)}{k}, \tag{9}$$

where the random walker with probability q_e follows an outgoing link, or else jumps to a random page. Simulations with a version of the random walk model, with a vertex addition mechanism similar to the Holme-Kim model [15], reproduce this behavior. The other possiblity to consider is to follow all the new edges emanating from the last chosen one, with a probability q_e. This recursive search dynamics gives rise to a phase diagram in terms of the control parameter.

Local dynamics can also be made to mimic the friends-of-friends characteristics of social graphs. Davidsen *et al.* have presented a model [91], in which a joint friend directly creates a link between two vertices (or individuals), in analogy with the Holme-Kim modification of the BA model (see also [54]). Vázquez moreover has presented a model in the language of social dynamics [13]: links are formed at random between nodes that do not have joint neighbors with a rate $\sim 1/N^2$ and links between such with a rate $\sim 1/N$. The equations of [13] are in fact, for clustering, an adaptation of the mean-field theory to this particular model, with the analytical outcome that $C(k) \sim k^{-1}$. Simulations of the model demonstrate that mean-field approximations do not necessarily apply here after all, and likewise the exponent of $C(k)$ does not scale inversely with k, but decays slower. This is coupled with the presence of an increase in $\langle k_{nn} \rangle (k)$ with k.

4 Rate-Equation Approach

4.1 Rate Equations for Growing Models

Growing models are particularly apt to be described by rate equations. When these kinds of models are considered, we can write down the rate equations for the clustering in a general form. We thus need to examine the rate of change averaged over many realizations,

$$\frac{\partial n_i}{\partial t} = R(k_i, \Pi) \sum_{n \in \Omega} R(k_n, \Pi), \tag{10}$$

where n_i is the average number of connected neighbors of the site i, and the local clustering coefficient $C_i = n_i/[k_i(k_i-1)/2]$. Here R is the rate at which i gets new links (or even loses them, if applied to processes with re-attachment or deletion of links). We allow the rate to depend on both the degrees of the node in question and a parameter set Π. This can be "annealed" or "quenched", depending on whether the parameters describe stochastic rules (as in the example below) or a fixed property of each node i. E.g., R can simply follow from the preferential attachment rule. Ω is the set of neighbors of node i and the sum accounts for the probability that a new node linked to i also links to one of the neighbors of i. This increases n_i and enhances clustering. In order to make (10) more concrete, we briefly discuss the *triad formation model* [15] as an example.

The complications in solving a rate equation like (10) arise from the correlations that are embedded between the degree of node i and the properties of its neighborhood. For the triad formation model, the rules consist of a BA model extended by a triad formation step. Initially, the network contains m_0 vertices and no edges, and in every time step a new vertex is added with m undirected edges. The m edges are then one-by-one subsequently linked to m different nodes in the network. One performs a preferential attachment step for the first edge as defined in the BA model. With probability p, the second and further edges are joined to a randomly chosen neighbor of the node selected in the previous PA step. Alternatively, with probability $1 - p$, a PA step is performed again.

In the limit when p approaches zero, one recovers the original BA model, and by setting p to a value between 0 and 1 the average clustering can be adjusted continuously and grows monotonically with an increasing p. The microscopic mechanisms that increase n_i are illustrated in Fig. 7 and are (I): the new node connects to node i in a PA step, which is potentially followed by several triad formation (TF) steps; (II): the new node connects to one of the neighbors of i in a PA step and then i conversely gets linked to the new node in one of the subsequent TF steps; (III): the new node connects to node i in a PA step and a neighbor of i is also selected for connection to the new node in another PA step.

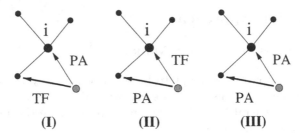

Fig. 7. Three different options to connect to node i with $m \geq 2$. In (I), a PA step is performed first linking to i and then a TF step creates a link between neighbors of i. In (II), the same happens, in a different order. (III) shows how two PA steps may contribute to n_i. Bold edges increase n_i.

Using the above for $R(k_i, p)$, the rate equation for n_i reads

$$
\frac{\partial n_i}{\partial t} = m_{PA} \frac{k_i}{2mt} m_{TF} + m_{PA} \sum_{n \in \Omega} \frac{k_n}{2mt} \frac{1}{k_n} m_{TF} +
$$

$$
+ \, m_{PA} \frac{k_i}{2mt} (m_{PA} - 1) \sum_{n \in \Omega} \frac{k_n}{2mt}. \tag{11}
$$

The three terms correspond to the three mechanism of Fig. 7. The solution [14] requires the calculation of the average degree of the neighbors, and yields a $C(k)$ function composed of a $C(k) \sim k^{-1}$ part for small degrees, and $C(k) \sim$ const. for large degrees. It is summarized in Fig. 8.

4.2 AB Model Definition

In the following we demonstrate a few of the implications on a model that is a simplification of the one proposed by Capocci *et al.* [92], and is called the AB model.

The AB model is an extension of the Barabási-Albert model, and it proceeds by assigning a type (fitness, relevance) to every new node, which does not change afterwards. It can be of type A with probability p, or type B with probability $1 - p$. New nodes can then link to a fixed number of m different nodes in the system, with the condition that B nodes are allowed to attach to B nodes only. A nodes are permitted to link to any type of other node. These rules result in bipartite graphs, as shown in Fig. 9.

4.3 Rate-Equation Approach for the AB Model

In this section we analyze how the rate-equation approach can be applied in attempting to calculate the clustering coefficient for nodes of the AB model. We first formulate the time evolution of the node degrees with respect to their type. The rate equations read

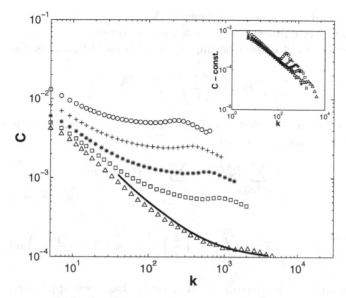

Fig. 8. Clustering coefficient as a function of the node degree for $m = 5$ and different sizes (10^4 for circles, 25119 for + symbols, 63096 for stars, 158489 for boxes, and 10^6 for triangles). The triad formation probability is uniformly $p = 0.01$. The bold line is the prediction given for the largest system, $C(k) \approx 0.04\,k^{-1} + 9.5 \cdot 10^{-5}$. The inset shows the data collapse of the power-law part of $C(k)$.

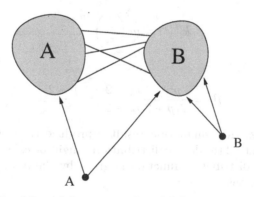

Fig. 9. The AB model as two nodes of different kinds are added.

$$\frac{\partial k_i^A(t)}{\partial t} = p\,m\,\frac{k_i^A(t)}{2mt} \tag{12}$$

for A nodes, since only new incoming A nodes are able to increase the degree of a given A node. On the other hand, B nodes satisfy

$$\frac{\partial k_i^B(t)}{\partial t} = p\,m\,\frac{k_i^B(t)}{2mt} + (1-p)\,m\,\frac{k_i^B(t)}{\sum_{j \in \{B\}} k_j^B(t)}, \tag{13}$$

where the second term accounts for new nodes of type B, and therefore the sum is to be taken over only existing B nodes in the system.

The solution for A nodes is straightforward from (12), and is given by .

$$k_i^A(t) = m \left(\frac{t}{t_i}\right)^{\beta_A}, \qquad \beta_A = \frac{p}{2} \tag{14}$$

in compliance with the initial condition that $k_i^A(t_i) = m$. This enables us to estimate for the sum in (13) using

$$\sum_{j \in \{A\}} k_j^A(t) + \sum_{j \in \{B\}} k_j^B(t) = 2mt \tag{15}$$

as

$$\sum_{j \in \{B\}} k_j^B(t) \approx 2mt - \int_1^t p\, m \left(\frac{t}{\tau}\right)^{\beta_A} d\tau \approx \left(2 - \frac{p}{1 - \beta_A}\right) mt, \tag{16}$$

where the sum over the degrees of all A nodes has been approximated by the integral. We furthermore assumed that the probability that a given node is of type A is uniformly p independently of the time of its introduction, as defined in the model.

The rate equation (13) for B nodes can now be solved by the same virtue as (12), yielding

$$k_i^B(t) = m \left(\frac{t}{t_i}\right)^{\beta_B} \tag{17}$$

and

$$\beta_B = \frac{p^2 + 2\beta_A - 2}{2(p + 2\beta_A - 2)} = \frac{p + 2}{4}. \tag{18}$$

Figure 10 shows that simulation result reproduce the exponents obtained above. The exponents of the degree distributions might be calculated accordingly, and a joint degree distribution function is given by the p-weighted sum of the individual distributions.

4.4 Clustering Coefficient

To calculate the clustering coefficient, one needs to consider the neighborhood the node in question is surrounded by. The coefficient for the given node changes only when a new node is linked to both to it and to one of its neighbors. In order to compute the linking probabilities, we need to know the expected values of the degrees for the node and its neighbors. This is not particularly difficult for models like for instance the Barabási-Albert one, and here we can take a similar approach. Observe that the average rate of increase in n_i is the probability that a new node links to both i and one of its neighbors:

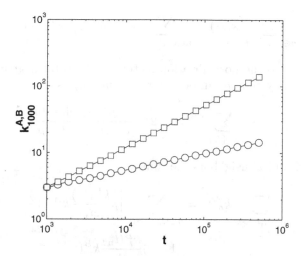

Fig. 10. Time evolution of degrees of the 1000th node of preset type A (circles) and B (squares), respectively. Systems of maximum size $5 \cdot 10^5$ and parameters $p = 0.5$, $m = 3$ were chosen. Power-law fits at the upper end of the scale result in exponents of 0.24 for the A node and 0.63 for the B node. Values other than $p = 0.5$ also give consistent results with (14) and (18).

$$\frac{\partial n_i}{\partial t} = P(\text{new} \to i)\, P(\text{new} \to \Omega_i), \qquad (19)$$

where the set Ω_i denotes the nodes belonging to the neighborhood of i. The expressions for both probabilities above depend on the details of the model, and in our case will be different for the two possible kinds of nodes.

To illustrate these, let us now give approximations for the total degree of neighboring nodes for both types of nodes. For A nodes, the sum of all degrees of neighbors that are *older* than the node itself is

$$\sum_{n \in \Omega_i, t_n < t_i} k_n(t) = \sum_{n \in \Omega_i, t_n < t_i} \left[p \frac{k_n^A(t_i)}{2t_i} k_n^A(t) + (1-p) \frac{k_n^B(t_i)}{2t_i} k_n^B(t) \right]. \qquad (20)$$

The first term in the sum stands for older neighboring A nodes, while the second term for older neighboring B nodes, since an A node can attach to either types. Both of the terms are composed of two probabilities indicating that the neighbor is of type A or B, respectively, that i connected to it when it was introduced, and a third multiplier to provide for the expected degree of the neighbor.

Likewise, for *newer* neighbors the sum becomes

$$\sum_{n \in \Omega_i, t_n > t_i} k_n(t) = \sum_{n \in \Omega_i, t_n > t_i} p \frac{k_i^A(t_n)}{2t_n} k_n^A(t), \qquad (21)$$

for only A nodes can get linked to the A node.

Very similarly, the corresponding equations can be written for the sum of the neighbors' degrees of a B node. For *earlier* B neighbors it gives

$$\sum_{n \in \Omega_i, t_n < t_i} k_n(t) = \sum_{n \in \Omega_i, t_n < t_i} (1-p) \frac{k_n^B(t_i)}{bt_i} k_n^B(t), \qquad (22)$$

where b is related to the normalization factor for the total degrees of B nodes at any time in the system. b can be inferred from (16) as

$$b = \frac{1}{mt} \sum_{j \in \{B\}} k_j^B(t) \approx 2 - \frac{p}{1-\beta_A} = \frac{4-4p}{2-p}. \qquad (23)$$

For *newer* A and B neighbors the sum is

$$\sum_{n \in \Omega_i, t_n > t_i} k_n(t) = \sum_{n \in \Omega_i, t_n > t_i} \left[p \frac{k_i^B(t_n)}{2t_n} k_n^A(t) + \right.$$
$$\left. (1-p) \frac{k_i^B(t_n)}{bt_i} k_n^B(t) \right]. \qquad (24)$$

For a node i of type A, the rate equation for the number of links between its neighbors, n_i^A, is

$$\frac{\partial n_i^A(t)}{\partial t} = p \cdot m \frac{k_i^A(t)}{2mt} \cdot (m-1) \sum_{n \in \Omega_i} \frac{k_n(t)}{2mt}, \qquad (25)$$

since only a new node of type A can increase it, precisely by linking to it and to one of its neighbors. The sum over the neighbors $\sum_{n \in \Omega_i} \frac{k_n(t)}{2mt}$ can be calculated by using (20–21).

The same for a B node is

$$\frac{\partial n_i^B(t)}{\partial t} = p \cdot m \frac{k_i^B(t)}{2mt} \cdot (m-1) \sum_{n \in \Omega_i} \frac{k_n(t)}{2mt} +$$
$$(1-p) \cdot m \frac{k_i^B(t)}{bmt} \cdot (m-1) \sum_{n \in \Omega_i, n=B} \frac{k_n(t)}{bmt}. \qquad (26)$$

The only difference here from (25) is that new B nodes have to be incorporated in the second term of the sum as well.

The clustering coefficient is then calculated as $C_i(t) \approx 2n_i(t)/k_i(t)^2$, where we have to integrate (25) or (26) in order to obtain the total change in n_i relative to $n_i(t_i)$. $n_i(t_i)$ is the initial value of the links between neighbors of i acquired when i is introduced and linked to m other nodes. To calculate this, one would need to consider any pairs of neighbors that i might get linked to, and also possibly higher-order neighbors so that for instance a pair of second neighbors could increase $n_i(t_i)$ by two even when i uses only three edges to link to nodes, as illustrated in Fig. 11. This factor is relatively significant, since nodes with large degrees are very likely to be clustered together, and thus new nodes often link among these.

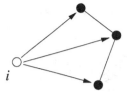

Fig. 11. Higher-order neighbors increase $n_i(t_i)$ in an accelerated manner.

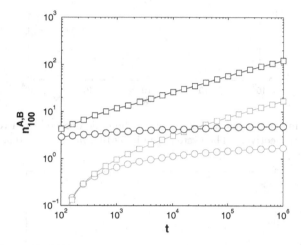

Fig. 12. Number of links between nearest neighbors at time t for the $i = 100$ node, of type A (circles) and B (squares), respectively, in darker color. The curves in light color represent the analytical solutions of (25) and (26). In both cases, the simulations were performed by setting the 100th node to the desired type, and measuring $n_i(t)$ for as long as 10^6 nodes were added. $p = 0.5$.

An approximation for $n_i(t)$ of both kinds of nodes can then be given by taking integrations instead of all the sums above, using (14) and (17) for k_n^A and k_n^B, respectively. Substituting the appropriate sums for the degrees of nearest neighbors into (25) and (26) and integrating both sides yields $n_i^A(t)$ and $n_i^B(t)$ at any given time, up to the constant initial values discussed above. This is illustrated in Fig. 12, where the neglected initial offset is apparent.

The model growth rules give rise to nontrivial scaling characteristics: it can be calculated that apart from various power-law terms, $n_i(t)$ will also contain logarithmic terms in t (Fig. 12). For reasons outlined above, it is inherently difficult to predict the initial value of n_i at the introduction time $t_i (\equiv i)$, just after the node connected to m other nodes, and thus we can calculate n_i in the thermodynamic limit only, where a constant offset becomes negligible.

The above has the consequence that the local clustering coefficient will also have logarithmic corrections to it, as shown in Fig. 13. Furthermore, if we consider the clustering coefficient of A nodes, we can see that contrary to most (but not all) measurement results, it increases with the degree for the given p

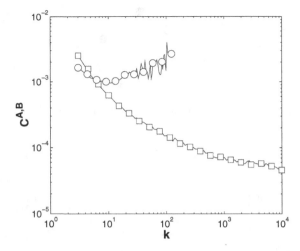

Fig. 13. Clustering coefficient as a function of the node degree, for A (circles) and B (squares) nodes, respectively. The system size is $5 \cdot 10^5$, $m = 3$, $p = 0.5$.

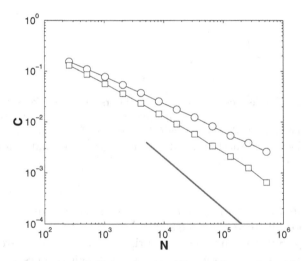

Fig. 14. Average clustering coefficient in the AB model as a function of the network size. Circles represent $p = 0.3$, squares $p = 0.7$. An approximate power-law exponent is -0.55 for the former, and -0.7 for the latter. The line is a guide to the eye, with an exponent of -1 for comparison.

value. In general, the p-weighted average of the two kinds of clustering coefficient functions needs to be taken to obtain the overall clustering coefficient, and thus $C(k)$ will be an even more intricate function. A perhaps general feature is the nonmonotonic behavior of the weighted average, however.

As far as the average clustering coefficient is concerned, it is shown for two p values in Fig. 14 by means of simulations. In both the $p \to 0$ and $p \to 1$ limits the original BA model is recovered, and so a $C(N) \sim N^{-1}$ thermodynamic behavior is expected. Apparently in the mixed cases, no true power law is obtained, as also in the original BA model a $(\log N)^2$ correction is in place.

5 Conclusions

The clustering aspect has been crucial in the revolutionary development of modern theory of complex networks. In the seminal paper by Watts and Strogatz [1] this property is in the focus. In fact, the Small World model could explain two out of three generally observable properties of many complex networks: it has a small average distance and a large clustering, but it fails to reproduce the scale-free character of the degree distribution. The model by Barabási and Albert [62] again reflects a pair of the properties, namely it is scale-free, also a small world, but the clustering is too small as compared to the observations. Meanwhile a series of simple models have been introduced, which obey all three properties and which are usually generalizations of the BA model [5,15,55,70,74].

Clustering is a natural way to characterize nontrivial correlations in a network. These are very important from the point of view of applications: Newman showed that significant high clustering—besides assortativeness—is characteristic for social networks. Bianconi and collaborators [3] showed that clustering can be considered as a first element in an infinite hierarchy of quantities, namely the number of h-loops.

It seems that clustering is closely related to the functionality of the network. If it is defined locally it could give insight into the modularity of certain networks [4], at least in such cases where the length scales are separated [93]. An aspect where this becomes obvious is *weighted clustering*, where the triangles are given weights, e.g., depending on the edges and/or vertices they contain. One may envision in weighted networks [94,95] scenarios in which either "weak" or "strong" triangles and clustering dominates.

One of the most important applications of complex networks is that of classification or taxonomy. The task is the following: given a similarity function, order entities into groups, possibly in a hierarchical manner (which is essentially the construction of a network). Several techniques can be applied to the problem [96], many of them leading to tree structures like the minimum spanning tree, or the average linkage cluster algorithms. The advantage of the latter methods is that they provide immediate hierarchies. The price to pay, however, for the enforced tree structure of the network is that important links forming loops are neglected. In non-hierarchical agglomerative classification, where gradually the links in descending strength order are inserted, the concept of local clustering seems to play a guiding role with respect to the information content of similarity matrices [97].

Considering these aspects, clustering is of crucial importance in analyzing complex networks. We have shown that besides simulation techniques approximate analytic tools like mean-field type rate equations can be very useful in treating this quantity. As often for complex networks mean-field methods work quite efficiently where usually the hand-waving argument of long range interactions is mentioned in support of this observation. At the moment, however, we do not have a clear view about the applicability of mean-field methods but we do see their limitations. It would be desirable to apply a Ginzburg-type criterion, or, even more ambitiously, to see how renormalization group ideas could be used for clustering properties.

Acknowledgements

This research was partially carried out within the framework of the Center for Applied Mathematics and Computational Physics of the Budapest University of Technology.

References

1. Watts DJ, Strogatz SH (1998) Nature 393:440–442
2. Newman MEJ (2003) Random graphs as models of networks. In: Bornholdt S, Schuster HG (eds) Handbook of Graphs and Networks, pp. 35–68. Wiley-VCH Berlin. Also preprint cond-mat/0303516
3. Bianconi G, Caldarelli G, Capocci A (2003) preprint cond-mat/0310339
4. Oltvai ZN, Barabási A-L (2002) Science 298:763–764
5. Ravasz E, Somera AL, Mongru DA, Oltvai ZN, Barabási A-L (2002) Science 297:1551–1555
6. Girvan M, Newman MEJ (2002) Proc Natl Acad Sci USA 99:7821–7826. Also preprint cond-mat/0308217, preprint cond-mat/0309508
7. Itzkovitz S, Milo R, Kashtan N, Ziv G, Alon U (2003) Phys Rev E 68:026127
8. Milo R, Shen-Orr S, Itzkovitz S, Kashtan N, Chklovskii D, Alon U (2002) Science 298:824–827
9. Newman MEJ (2002) Phys Rev Lett 89:208701
10. Molloy M, Reed B (1995) Random Structures and Algorithms 6:161–179
11. Molloy M, Reed B (1998) Combinatorics, Probability and Computing 7:295–305
12. Ravasz E, Barabási A-L (2003) Phys Rev E 67:026112
13. Vázquez A (2003) Phys Rev E 67:056104
14. Szabó G, Alava M, Kertész J (2003) Phys Rev E 67:056102
15. Holme P, Kim BJ (2002) Phys Rev E 65:026107
16. Amaral LAN, Scala A, Barthelemy M, Stanley HE (2000) Proc Natl Acad Sci USA 97, 11149–11152
17. Grossman JW, Ion PDF (1995) Congressus Numerantium 108:129–131
18. de Castro R, Grossman JW (1999) Mathematical Intelligencer 21:51–63
19. Newman MEJ (2001) Proc Natl Acad Sci USA 98:404–409
20. Newman MEJ (2001) Phys Rev E 64, 016131
21. Ebel H, Mielsch L-I, Bornholdt S (2002) Phys Rev E 66:035103
22. Newman MEJ, Forrest S, Balthrop J (2002) Phys Rev E 66:035101
23. Bearman PS, Moody J, Stovel K (2002) Preprint, Department of Sociology, Columbia University
24. Albert R, Jeong H, Barabási A-L (1999) Nature 401:130–131
25. Barabási A-L, Albert R, Jeong H (2000) Physica A 281:69–77
26. Knuth DE (1993) The Stanford GraphBase: A Platform for Combinatorial Computing. Addison-Wesley, Reading, MA

27. Faloutsos M, Faloutsos P, Faloutsos C (1999) Computer Communications Review 29:251–262
28. Chen Q, Chang H, Govindan R, Jamin S, Shenker SJ, Willinger W (2002) The origin of power laws in Internet topologies revisited. In: Proceedings of the 21st Annual Joint Conference of the IEEE Computer and Communications Societies, IEEE Computer Society
29. Newman MEJ (2003) Phys Rev E 67:026126
30. Valverde S, Cancho RF, Solé RV (2002) Europhys Lett 60:512–517
31. Ferrer i Cancho R, Janssen C, Solé RV (2001) Phys Rev E 64:046119
32. Adamic LA, Lukose RM, Puniyani AR, Huberman BA (2001) Phys Rev E 64:046135
33. Ripeanu M, Foster I, Iamnitchi A (2002) IEEE Internet Computing 6:50–57
34. Jeong H, Tombor B, Albert R, Oltvai ZN, Barabási A-L (2000) Nature 407:651–654
35. Jeong H, Mason SP, Barabási A-L, Oltvai ZN (2001) Nature 411:41–42
36. Huxham M, Beaney S, Raffaelli D (1996) Oikos 76:284–300
37. Martinez ND (1991) Ecological Monographs 61:367–392
38. White JG, Southgate E, Thompson JN, Brenner S (1986) Phil Trans R Soc London 314:1–340
39. Vázquez A, Pastor-Satorras R, Vespignani A (2002) preprint cond-mat/0206084
40. Eriksen KA, Simonsen I, Maslov S, Sneppen K (2003) Phys Rev Lett 90:148701
41. Rives AW, Galitski T (2003) Proc Natl Acad Sci USA 100:1128–1133
42. Guimerà R, Danon L, Díaz-Guilera A, Giralt F, Arenas A (2002) preprint cond-mat/0211498
43. Erdős P, Rényi A (1959) Publicationes Mathematicae, 6:290–297
44. Newman MEJ, Park J (2003) preprint cond-mat/0305612
45. Watts DJ, Dodds PS, Newman MEJ (2002) Science 296:1302–1305
46. Newman MEJ, Strogatz SH, Watts DJ (2001) Phys Rev E 64:026118
47. Caldarelli G, Capocci A, De Los Rios P, Muñoz MA (2002) Phys Rev Lett 89:258702
48. Söderberg B (2002) Phys Rev E 66:066121
49. Goh K-I, Kahng B, Kim D (2001) Phys Rev Lett 87:278701
50. Batagelj V, Mrvar A (2000) Social Networks 22:173–186
51. Davis GF, Yoo M, Baker WE (2001) The small world of the corporate elite. Preprint, University of Michigan Business School
52. Banks DL, Carley KM (1996) Journal of Mathematical Sociology 21:173–196
53. Watts DJ (1999) Am J Sociol 105:493–592
54. Jin EM, Girvan M, Newman MEJ (2001) Phys Rev E 64:046132
55. Klemm K, Eguíluz VM (2002) Phys Rev E 65:036123
56. Jost J, Joy MP (2002) Phys Rev E 66:036126
57. Boguña M, Pastor-Satorras R (2003) Phys Rev E 68:036112
58. Boguña M, Pastor-Satorras R, Díaz-Guilera A, Arenas A (2003) preprint cond-mat/0309263
59. Berg J, Lässig M (2002) Phys Rev Lett 89:228701
60. Baiesi M, Manna SS (2003) preprint cond-mat/0305054
61. S. S. Manna, Kabakçıoğlu A (2003) J Phys A 36:L279. Also preprint cond-mat/0302224
62. Barabási AL, Albert R (1999) Science 286:509–512
63. Caldarelli G, De Los Rios P, Pietronero L (2003) preprint cond-mat/0307610
64. Capocci A, Caldarelli G, De Los Rios P (2002) preprint cond-mat/0206336
65. Pastor-Satorras R, Vázquez A, Vespignani A (2001) Phys Rev Lett 87:258701

66. Goh K-I, Kahng B, Kim D (2002) Phys Rev Lett 88:108701
67. Kim D-H, Rodgers GJ, Kahng B, Kim D (2003) preprint cond-mat/0310233
68. Zachary WW (1977) J Anthropol Res 33:452
69. Yook SH, Jeong H, Barabási A-L, Tu Y (2001) Phys Rev Lett 86:5835
70. Manna SS, Sen P (2002) Phys Rev E 66:066114
71. Xulvi-Brunet R, Sokolov IM (2002) Phys Rev E 66:026118
72. Barthelemy M (2003) Europhys Lett 63:915–921
73. Krapivsky PL, Redner S (2001) Phys Rev E 63:066123
74. Sen P, Manna SS (2003) preprint cond-mat/0301617
75. Fell DA, Wagner A (2000) Nature Biotechnology 18:1121–1122
76. Wagner A, Fell D (2001) Proc R Soc London B 268:1803–1810
77. Solé RV, Pastor-Satorras R (2003) Complex networks in genomics and proteomics. In: Bornholdt S, Schuster HG (eds) Handbook of Graphs and Networks, pp. 145–167. Wiley-VCH, Berlin
78. Stelling J, Klamt S, Bettenbrock K, Schuster S, Gilles ED (2002) Nature 420:190–193
79. Kleinberg JM, Kumar SR, Raghavan P, Rajagopalan S, Tomkins A (1999) The Web as a graph: Measurements, models and methods. In: Proceedings of the International Conference on Combinatorics and Computing, no. 1627 in Lecture Notes in Computer Science, pp. 1–18. Springer, Berlin
80. Kumar R, Raghavan P, Rajagopalan S, Sivakumar D, Tomkins AS, Upfal E (2000) Stochastic models for the Web graph. In: Proceedings of the 42st Annual IEEE Symposium on the Foundations of Computer Science, pp. 57–65. Institute of Electrical and Electronics Engineers, New York
81. Solé RV, Pastor-Satorras R, Smith E, Kepler TB (2002) Advances in Complex Systems 5:43–54
82. Kim J, Krapivsky PL, Kahng B, Redner S (2002) Phys Rev E 66:055101
83. Berg J, Lässig M, Wagner A (2002) preprint cond-mat/0207711
84. Jain S, Krishna S (1998) Phys Rev Lett 81:5684–5687
85. Jain S, Krishna S (2001) Proc Natl Acad Sci USA 98:543–547
86. Dorogovtsev SN, Mendes JFF, Samukhin AN (2002) Europhys Lett 57:334–340
87. Ohono S (1970) Evolution by gene duplication. Springer-Verlag, Berlin
88. Wagner A (2001) Mol Biol Evol 18:1283
89. Vázquez A, Flammini A, Maritan A, Vespignani A (2003) ComplexUs 1:38. Also preprint cond-mat/0108043
90. Maslov S, Sneppen K (2002) Science 296:910–913
91. Davidsen J, Ebel H, Bornholdt S (2002) Phys Rev Lett 88:128701
92. Capocci A, Caldarelli G, De Los Rios P (2002) preprint cond-mat/0206336
93. Vicsek T, Kertész J (unpublished)
94. Barthelemy M, Barrat A, Pastor-Satorras R, Vespignani A, (2003) preprint cond-mat/0311501
95. Barrat A, Barthelemy M, Vespignani A, (2004) preprint cond-mat/0401057
96. Aldenderfer MS, Blashfeld RK (1984) Cluster Analysis, Sage Publications
97. Onnela J-P, Kaski K, Kertész J (2004) Eur Phys J B in press. Also preprint cond-mat/0312682

Equilibrium Statistical Mechanics
of Network Structures

Illés Farkas[1], Imre Derényi[2], Gergely Palla[1], and Tamás Vicsek[1,2]

[1] Biological Physics Research Group of HAS
[2] Department of Biological Physics, Eötvös University, Pázmány P. stny. 1A,
   1117 Budapest, Hungary

Abstract. In this article we give an in depth overview of the recent advances in the
field of equilibrium networks. After outlining this topic, we provide a novel way of
defining equilibrium graph (network) ensembles. We illustrate this concept on the classical random graph model and then survey a large variety of recently studied network
models. Next, we analyze the structural properties of the graphs in these ensembles
in terms of both local and global characteristics, such as degrees, degree-degree correlations, component sizes, and spectral properties. We conclude with topological phase
transitions and show examples for both continuous and discontinuous transitions.

1 Introduction

A very human way of interpreting our complex world is to try to identify subunits
in it and to map the interactions between these parts. In many systems, it is
possible to define subunits in such a way that the network of their interactions
provides a simple but still informative representation of the system. The field of
discrete mathematics dealing with networks is *graph theory*.

Research in graph theory was started by Leonhard Euler [1]. In the 1950s
another major step was taken by Erdős and Rényi: they introduced the notion
of classical random graphs [2–4]. By the late 1990s more and more actual maps
of large networks had become available and modeling efforts were directed towards the description of the newly recognized properties of these systems [5–9].
A network is constructed from many similar subunits (vertices) connected by
interactions (edges), similarly to the systems studied in statistical physics. Because of this analogy, the methods by which some of the central problems of
statistical physics are effectively handled, can be transferred to networks, e.g.,
to graph optimization and topological phase transitions.

In this article we will discuss the construction of network ensembles that fit
into the concept of equilibrium as it is used in statistical physics, with a focus
on structural transitions [10–14]. Note that even though structural transitions in
growing networks are non-equilibrium phenomena [15–17], some of the main features of the structures constructed by growth can be reproduced by non-growing
models (see, e.g., . [11,12,18]). Similarly, non-growing graphs are not necessarily
equilibrium systems (see, e.g., [8]). Closely related real-world phenomena and
mathematical models are the configurational transitions of branched polymers
[19], structural transitions of business networks during changes of the "business"

I. Farkas, I. Derényi, G. Palla, and T. Vicsek, Equilibrium Statistical Mechanics of Network Structures, Lect. Notes Phys. **650**, 163–187 (2004)
http://www.springerlink.com/   © Springer-Verlag Berlin Heidelberg 2004

climate [20,21], the transitions of collaboration networks [22], networks defined by the potential energy landscapes of small clusters of atoms [23], and potentials on tree graphs as introduced by Tusnády [24].

In the present review we intend to go beyond those that have been published previously [8,12,25], both concerning the scope and the depth of the analysis.

We will *focus on the structure of networks*, represented by graphs, and will not consider any dynamics *on* them. Thus, several widely studied models are beyond the scope of the present review: models using, e.g., spins on the vertices [12,26–31], disease spreading [32–34] agent-based models on networks [35], or weighted edges and traffic on a network [36–39].

Definition: Natural networks mostly arise from non-equilibrium processes, thus, the notion of equilibrium in the case of networks is essentially an abstraction (similarly to any system assumed to be in perfect equilibrium). We *define equilibrium network ensembles* as stationary ensembles of graphs generated by restructuring processes obeying *detailed balance* and ergodicity. During such a restructuring process, edges of the graph are removed and/or inserted.

This definition raises a few issues to be discussed. First of all, the characteristic timescale of rewiring one particular link varies from system to system. For example, the network of biochemical pathways [40] available to a cell can undergo structural changes within years to millions of years, in contrast to business interactions [20,22], which are restructured over time scales of days to years, while the characteristic times of technological networks may be even shorter. With a finite number of measurements during the available time window it is often difficult to decide whether a graph that has not been observed has a low probability or it is not allowed at all. Hence, the set of allowed graphs is often unclear. A simple way to by-pass this problem is to enable all graphs and tune further parameters of the model to reproduce the statistics of the observed typical ones.

Once the set of allowed graphs has been fixed, the next step in the statistical physics treatment of a network ensemble is to fix some of the thermodynamic variables, e.g., for the canonical ensemble one should fix the temperature and all extensive variables except for the entropy[3]. At this point, an *energy function* would be useful. Unfortunately, unlike in many physical systems, the energy of a graph cannot be derived from first principles. A possible approach for deriving an energy function is reverse engineering: one tries to reproduce the observed properties of real networks with a suitable choice of the energy in the model. Another possibility can be to explore the effects of a wide range of energy functions on the structures of networks. Alternatively, to suppress deviations from a prescribed target property, one can also introduce a cost function (energy). Having defined the energy, one can proceed towards a detailed analysis of the equilibrium system using the standard methods of statistical physics.

Often a complete analogy with statistical physics is unnecessary, and shortcuts can be made to simplify the above procedure. It is very common to define graph ensembles by assigning a statistical weight to each allowed graph, or to

---

[3] In the mathematics literature, the entropy of graphs has been analyzed in detail [41–43].

supply a set of master equations describing the dynamics of the system, and to find the stable fixed point of these equations. Of course, skipping, e.g., the definition of the energy will leave the temperature of the system undefined.

This article is organized as follows. In Sect. 2 we introduce the most important notions. Section 3 will concentrate on currently used graph models and the construction of equilibrium graph ensembles. Section 4 will discuss some of the specific properties of these sets of graphs. In Sect. 5 examples will be given for topological phase transitions of graphs and Sect. 6 contains a short summary.

2 Preliminaries

Except where stated otherwise, we will consider *undirected simple graphs*, i.e., non-degenerate graphs where any two vertices are connected by zero or one undirected edge, and no vertex is allowed to be connected to itself[4]. The number of edges connected to the ith vertex is called the degree, k_i, of that vertex. Two vertices are called neighbors, if they are connected by an edge. The degree sequence of a graph is the ordered list of its degrees, and the degree distribution gives the probability, p_k, for a randomly selected vertex to have degree k. The degree-degree correlation function, $p(k, k')$, gives the probability that one randomly selected end point of a randomly chosen edge will have the degree k and the other end point the degree k'.

The clustering coefficient of the ith vertex is the ratio between the number of edges, n_i, connecting its k_i neighbors and the number of all possible edges between these neighbors:

$$C_i = \frac{n_i}{k_i(k_i - 1)/2} .\tag{1}$$

The clustering coefficient of a graph is C_i averaged over all vertices. The shortest distance, $d_{i,j}$, is defined as the smallest number of edges that lead from vertex i to j. Finally, a set of vertices connected to each other by edges and isolated from the rest of the graph is called a *component* of the graph.

The two basic constituents of a simple graph are its vertices and edges, therefore it is essential whether a vertex (or edge) is distinguishable from the others. In this article, we will consider *labeled graphs*, i.e., in which both vertices and edges are distinguishable. A graph with distinguishable vertices can be represented by its adjacency matrix, $\mathbf{A}$. The element A_{ij} denotes the number of edges between vertices i and j if $i \neq j$, and twice the number of edges if $i = j$ (unit loops). For simple graphs, this matrix is symmetric, its diagonal entries are 0, and the off-diagonal entries are 0 or 1. Note, that the adjacency matrix is insensitive to whether the edges of the graph are distinguishable: swapping any two edges will result in the same $\mathbf{A}$.

---

[4] In a degenerate (or pseudo) graph multiple connections between two vertices and edges connecting a vertex to itself are allowed. Some additional extensions are to assign, e.g., weights and/or fitnesses to the edges and vertices.

Also, it is possible to define equivalence classes of labeled graphs using graph isomorphism: two labeled graphs are equivalent, if there exists a permutation of the vertices of the first graph transforming it into the second one. As a consequence, each equivalence class of labeled graphs can be represented by a single unlabeled graph (in which neither the edges nor the vertices are distinguishable). These equivalence classes will be referred to as *topologies*, i.e., two graphs are assumed to have the same topology, if they belong to the same equivalence class. This definition is the graph theoretical equivalent of the definition of topology for geometrical objects, where two objects have the same topology, if they can be transformed into each other through deformations without tearing and stitching.

The focus of this article is on graph restructuring processes. Denoting the transition rates between graphs a and b by $r_{a \to b}$, the time evolution of the probability of the graphs in the ensemble can be written as a set of master equations:

$$\frac{\partial P_a}{\partial t} = \sum_b \left(P_b r_{b \to a} - P_a r_{a \to b} \right), \tag{2}$$

where P_a is the probability of graph a.

If the dynamics defined in a system has a series of non-zero transition rates between any two graphs (ergodicity), and there exists a stationary distribution, P_a^{stat} fulfilling the conditions of detailed balance,

$$P_a^{\text{stat}} r_{a \to b} = P_b^{\text{stat}} r_{b \to a}, \tag{3}$$

then the system will always converge to this stationary distribution, which can thus be called *equilibrium distribution*.

In the reverse situation, when the equilibrium distribution is given, one can always create a dynamics that leads to this distribution. Such a dynamics must fulfill the conditions of detailed balance and ergodicity. Since the detailed balance condition (3) fixes only the ratio of the rates of the forward and backward transitions between each pair of graphs (a and b), the most general form of the transition rates can be written as

$$r_{a \to b} = \nu_{ab} P_b, \tag{4}$$

where all $\nu_{ab} = \nu_{ba}$ values are arbitrary factors (assuming that they do not violate ergodicity).

3 Graph Ensembles

Similarly to Dorogovtsev et. al [8,12] and Burda et. al [44], we will discuss graph ensembles in this section. According to statistical physics, for a rigorous analysis one needs to define the microcanonical, canonical, and grand canonical ensembles. However, even if some of the necessary variables, (e.g., the energy) are not defined, it is still possible to define similar graph ensembles.

In equilibrium network ensembles, the edges (links) represent particles and one graph corresponds to one state of the system. In this article we will keep the number of vertices constant, which is analogous to the constant volume constraint.

3.1 Ensembles with Energy

Energy is a key concept in *optimization problems*. Even if it is not possible to derive an energy for graphs from first principles, one can find analogies with well-established systems, and also phenomenological and heuristic arguments can lead to such energy functions [10–14], as described in the Introduction.

Microcanonical Ensemble

In statistical physics, the microcanonical ensemble is defined by assigning identical weights to each state of a system with a given energy, E, and a given number of particles; all other states have zero weight. Thus, the definition of a microcanonical ensemble is straightforward: assign the *same* weight,

$$P^{\mathrm{MC}} = n^{-1},$$ (5)

to each of the n graphs that has M edges and energy E, and *zero* weight to all other graphs.

Canonical Ensemble

The canonical ensemble is composed of graphs with a fixed number of edges, and each graph a has a weight

$$P_a^{\mathrm{C}} = \frac{e^{-E_a/T}}{Z^{\mathrm{C}}},$$ (6)

where T is the temperature, E_a is the energy of this graph, and

$$Z^{\mathrm{C}} = \sum_b e^{-E_b/T}$$ (7)

denotes the partition function. Network ensembles with a constant edge number and a cost function to minimize the deviations from a prescribed feature (e.g., a fixed total number of triangles), belong to this category.

Grand Canonical Ensemble

The grand canonical ensemble is characterized by a fixed temperature (T) and a fixed chemical potential (μ). The energy and the number of edges (particles) can vary in the system, and the probability of graph a is

$$P_a^{\mathrm{GC}} = \frac{e^{-(E_a - \mu M_a)/T}}{Z^{\mathrm{GC}}}, \tag{8}$$

where E_a and M_a denote the energy and edge number of graph a respectively, and

$$Z^{\mathrm{GC}} = \sum_b e^{-(E_b - \mu M_b)/T} \tag{9}$$

is the partition function.

3.2 Ensembles Without Energy

Microcanonical Ensemble

Numerous network models are defined through a static set of allowed graphs, and no restructuring processes are involved. Even if no energies and no probabilities are provided for these graphs, the microcanonical ensemble can still be defined by assigning equal weight to each allowed graph [8,45]. This is equivalent to assigning the same energy to each allowed graph (and a different energy to all the others).

Canonical Ensemble

If a graph model provides probabilities, $\{P_a\}$, for a set of graphs with an identical number of edges, then it can be considered as a canonical ensemble. One can easily construct an energy function from the probabilities using (6):

$$E_a = -T \log P_a + \log Z. \tag{10}$$

That is, the energy can be defined up to a factor, T, and an additive term, $\log Z$.

The Grand Canonical Ensemble

This ensemble is very similar to the canonical ensemble except that even the number of edges is allowed to vary. In this case an energy function can be constructed from (8):

$$E_a = -T \log P_a + \mu M_a + \log Z, \tag{11}$$

and a new, arbitrarily chosen parameter, μ, appears.

3.3 Basic Examples

The Classical Random Graph

We will discuss this classical example to *illustrate the concept* of equilibrium network ensembles. The classical random graph model is based on a fixed number

(N) of vertices. The model has two variants. The first one [2] is the $\mathcal{G}(N, M)$ model: M edges are placed randomly and independently between the vertices of the graph. The second variant [3] is the $\mathcal{G}(N, p)$ model: each pair of vertices in the graph is connected via an edge with a fixed probability, p. In both variants the degree distribution converges to a Poisson distribution in the $N \to \infty$ limit:

$$p_k \to \frac{\langle k \rangle^k e^{-\langle k \rangle}}{k!}, \qquad (12)$$

where $\langle k \rangle = 2M/N$ in $\mathcal{G}(N, M)$ and $\langle k \rangle = pN$ in $\mathcal{G}(N, p)$. Viewing the edges as particles, the constant edge number variant of the classical random graph model corresponds to the microcanonical ensemble, since each particular configuration is generated with the same probability. In the constant edge probability variant, only the expectation value of the number of "particles" is constant, and can be described by the grand canonical ensemble.

At this point one should also mention the notion of the *random graph process* [2,4], a possible method for generating a classical random graph. One starts with N vertices, and adds edges sequentially to the graph at independent random locations. In the beginning, there will be many small components in the graph, but after a certain number of inserted edges – given by the critical edge probability, p_c – a giant component[5] will appear. This transition is analogous to percolation phase transitions. The fraction of nodes belonging to the largest component in the $N \to \infty$ limit is [4]

$$G(\langle k \rangle) = 1 - \frac{1}{\langle k \rangle} \sum_{n=1}^{\infty} \frac{n^{n-1}}{n!} \left(\langle k \rangle e^{-\langle k \rangle} \right)^n. \qquad (13)$$

This analytical result and actual numerical data [46] showing the appearance of the giant component are compared in Fig. 1.

The Small-World Graph

Another well-known example for a graph ensemble is the small-world model introduced by Watts and Strogatz [5]. The construction of a small-world graph starts from a one-dimensional periodic array of N vertices. Each vertex is first connected to its k nearest neighbors, where k is an even positive number. Then, each edge is moved with a fixed probability, r, to a randomly selected new location. This construction leads to a canonical ensemble: the number of edges is constant and the probabilities of the individual graphs in the ensemble are different, because the number of rewired edges can vary.

Ensembles with a Fixed Degree Distribution

Many real-world graphs have a degree distribution that decays slowly, as a power law, as measured and described by Barabási, Albert and Jeong [6,48]. These

---

[5] Note that the growth rate of this component is sublinear: it grows as $\mathcal{O}(N^{2/3})$ [47].

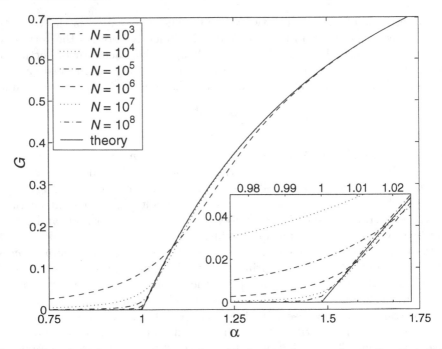

Fig. 1. Size of the largest component in a classical random graph as a function of the average degree, $\alpha = \langle k \rangle$, of a vertex. Note that for $N > 10^6$ the Monte Carlo data is almost indistinguishable from the theoretical result in (13). Error bars are not shown, because in all cases the error is smaller than the width of the lines. The inset shows the transition in the vicinity of the percolation threshold, $\alpha_c = 1$. Figure from [46].

graphs are often referred to as *scale-free*. On the other hand, the classical random graph's degree distribution has a quickly decaying $(1/k!)$ tail (see (12)). The degree distributions of graphs have become central to numerous analyses and various graph ensembles with fixed degree distributions have been developed [8, 12,44].

Given a network with the degree distribution p_k, there exist several rewiring algorithms that retain the degrees of all nodes at each rewiring step and generate an equilibrium ensemble of graphs. Two examples are the *link randomization* [49] and the *vertex randomization* [50] methods. In both methods, two edges are selected first, and then one of the end points of each edge is picked and swapped, ensuring that none of the degrees are changed. The two methods are explained in detail in Fig. 2. The resulting canonical ensembles will have the degree distribution p_k in common, but can have different equilibrium weights for the individual graphs. As pointed out by Xulvi-Brunet et. al [50], upon link randomization the degree-degree correlations are removed from a network, but vertex randomization builds up positive degree-degree correlations.

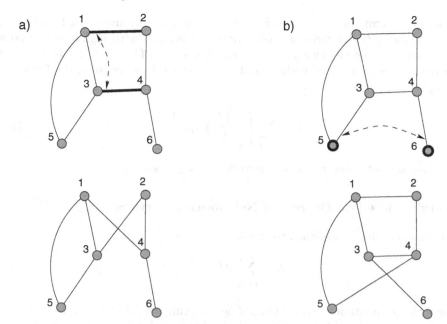

Fig. 2. Generating graph ensembles by randomization methods that leave the degree sequence of a graph unchanged. (a) Link randomization. First, one selects two edges of the graph randomly. These are indicated by heavier lines: edges $1-2$ and $3-4$. Then, one end point of each edge is selected randomly – the end points at vertices 1 and 3, respectively – and the selected end points of these two edges are swapped. (b) Vertex randomization. One starts with selecting two vertices at random (vertices 5 and 6 in the example). Next, one of the edges at each vertex is picked randomly and their end points at the selected vertices are swapped.

3.4 Examples for Graph Energies

Energies Based on Vertex Degrees

The most obvious units in a graph are the vertices themselves. Therefore, it is plausible to assign the energy to each vertex separately:

$$E = \sum_{i=1}^{N} f(k_i) \,. \tag{14}$$

Note that if the number of edges is constant, then the linear part of f is irrelevant (since its contribution is proportional to the number of edges in the graph), and simply renormalizes the chemical potential in case of the grand canonical ensemble. In the infinite temperature limit any f will produce the classical random graph ensemble. If f decreases faster than linear, e.g., quadratically,

$$E = -\sum_{i=1}^{N} k_i^2 \,, \tag{15}$$

then at low temperatures the typical graphs will have an uneven distribution of degrees among the vertices: a small number of vertices with high degrees and a large number of vertices with low degrees. In a model of Berg et. al [10], to avoid the occurrence of isolated vertices and vertices with large degrees, the following energy was proposed:

$$E = \sum_{i=1}^{N} \left[-\frac{k_i^2}{2} + \eta k_i^3 \right] , \tag{16}$$

and graphs containing vertices of zero degree were not allowed.

Energies Based on Degrees of Neighboring Vertices

Energies can also be assigned to edges,

$$E = \sum_{(i,j)} g(k_i, k_j) , \tag{17}$$

where the summation goes over pairs of neighboring vertices (i.e., over the edges). Energy functions of this type inherently lead to correlations between vertices, as demonstrated by Berg and Lässig [10] using

$$g(k_i, k_j) = \zeta \, \delta_{k_i,1} \delta_{k_j,1} , \tag{18}$$

see Fig. 3. Another example for this type of energy is

$$g(k_i, k_j) = \frac{\min(k_i, k_j)}{\max(k_i, k_j)} - 1 , \tag{19}$$

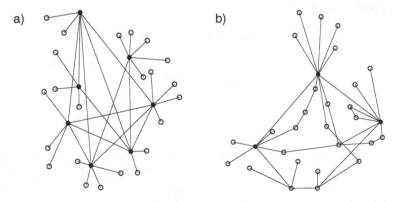

Fig. 3. Optimized networks generated by Berg and Lässig [10] using (a) energies with local correlations, see (17), and (b) energies based on global properties, see (23). In both cases, the temperature, T, was low. Notice that in both graphs disassortativity is present (see Sect. 4.1): vertices with high degrees (hubs, indicated by filled circles) are preferentially connected to vertices with low degrees (empty circles). Figure from [10].

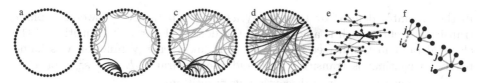

Fig. 4. (a)-(d) Snapshots of the simulation used by Baiesi and Manna [11] to generate an ensemble of scale-free equilibrium networks from a Hamiltonian dynamics, and (e)-(f) the Monte-Carlo rewiring method used during the simulation. The initial network is constructed from N links and N vertices: the vertices are connected as a ring. Later, $M-N$ ($M > N$) further edges are added to the network and the Monte-Carlo dynamics with the energy of (19) is used to decide whether a randomly selected edge, connecting vertices i and j_1, should be rewired to connect the previously unconnected pair of vertices, i and j_2. Blue (dark) edges meet at the vertex with the highest degree. Figure from [11].

which favors different degrees at the end points of an edge [11] (see Fig. 4 and Sect. 4.1).

To account for correlations over longer distances, a logical next step would be to add terms containing second neighbor interactions, e.g.,

$$E = -\frac{1}{6} \operatorname{Tr} \mathbf{A}^3 , \qquad (20)$$

which counts the number of triangles in the graph with a negative sign. If the number of edges can be written as $M = n(n-1)/2$ with an integer n ($n < N$), then at low temperatures this energy leads to a complete (fully connected) subgraph on n vertices, leaving the rest of the vertices ($N - n$) isolated.

Energies Based on Global Properties

The most apparent global properties of a network are the sizes[6] of its components, and especially, the size of the largest component, $s_{\max}$. A simple form of an energy containing component sizes is [13]

$$E = \sum_{i=1}^{n} f(s_i) , \qquad (21)$$

where n is the number of components in the graph and s_i is the size of the ith component.

The simplest form of the energy is proportional to the size of the largest component

$$E = -s_{\max} . \qquad (22)$$

---

[6] Component sizes are usually defined as the number of vertices in a component, however, in this article, because of the edge–particle analogy, s_i is the number of edges in the ith component.

In the ensemble defined by this energy as the temperature is lowered a phase transition occurs which is analogous to the density dependent transition of the classical random graph (see Sect. 3.3). This linear energy function was found to give a continuous transition [14], and the quadratic, $E = -s_{\max}^2$ or $E = -\sum_{i=1}^{N} s_i^2$ energies result in discontinuous transitions.

A possible goal of optimization can be to decrease the graph's diameter. This can be realized with, e.g., the energy [10]

$$E = \sum_{i,j} d_{i,j}, \qquad (23)$$

where the summation goes over all pairs of vertices. See also Fig. 3 for a typical network generated with this energy function.

3.5 Mapping the Graph onto a Lattice Gas

A simple, natural mapping of a graph with N vertices onto a lattice gas with $N(N-1)/2$ lattice sites[7] is shown in Fig. 5. One particle of this lattice gas corresponds to one edge of the original graph, and can be at any of the $N(N-1)/2$ lattice sites. Two lattice sites are neighbors, if the corresponding two edge locations (not necessarily occupied by edges) of the original graph have one end point in common. Note that this lattice strongly differs from the lattices generally used for lattice gases. Taking an arbitrary edge of the graph, there are $2(N-2)$ other possible edges sharing an end point with this edge: in the lattice gas, therefore, each site has $2(N-2)$ first neighbors. All the other $N(N-1)/2 - 2(N-2) - 1$ sites are second neighbors.

The quadratic single-vertex energy is analogous to the usual definition of the energy for a lattice gas with nearest neighbor attraction,

$$E = -\sum_{(\alpha,\beta)} n_\alpha n_\beta = -\sum_{i=1}^{N} \frac{k_i(k_i-1)}{2}, \qquad (24)$$

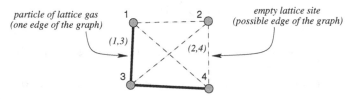

Fig. 5. Mapping a graph onto a lattice gas. One edge of the graph corresponds to one particle. There are $N(N-1)/2$ possible locations for an edge in a graph: these locations correspond to the sites of the lattice.

---

[7] This lattice is called the edge-dual graph of the complete (fully connected) graph of N vertices [4]. One lattice site corresponds to an edge in the complete graph.

a standard choice to describe the nucleation of vapors. Here n_α is the occupation number of lattice site α, which is 0 or 1, depending on whether the corresponding edge exists in the original graph. The summation in the first sum goes over all pairs of neighboring particles in the lattice gas, which corresponds to all pairs of edges sharing an end point in the original graph.

This analogy can be extended to an Ising model with a Kawasaki-type dynamics, where spins have $s_\alpha = 2n_\alpha - 1 = \pm 1$ values and from (24) the energy of the system is

$$E = - \sum_{(\alpha,\beta)} \frac{s_\alpha + 1}{2} \frac{s_\beta + 1}{2}, \tag{25}$$

which can be written as

$$E = -\frac{1}{4} \sum_{(\alpha,\beta)} s_\alpha s_\beta - \frac{1}{2} \sum_{\alpha=1}^{N(N-1)/2} s_\alpha - \frac{N(N-1)(N-2)}{8}. \tag{26}$$

Mapping the equilibrium graph ensemble with the $-\sum_i k_i^2$ energy onto a lattice gas shows that the only difference between this equilibrium graph ensemble and a lattice gas with the nearest neighbor attraction $E = -\sum_{(\alpha,\beta)} n_\alpha n_\beta$ on, e.g., a cubic lattice is the *underlying lattice*.

3.6 Ensembles of Degenerate Graphs

Degenerate graphs occur in almost all kinds of real-world networks, e.g., in food webs (cannibalism), biochemical interaction networks (autocatalytic or multiple reactions), technological networks (multiple connections between subunits), collaboration networks (repeated co-authorships), and also in field theoretic expansions of particle interactions in the form of Feynman graphs [51].

Ensembles of degenerate graphs [8,12,44] can be introduced similarly to the case of simple graphs. The *microcanonical ensemble* on the set of all labeled degenerate graphs can be defined by assigning the same weight to each graph with N vertices and M edges. The number of these elements can be given as follows. There are $N(N + 1)/2$ possible locations for an edge in a degenerate graph: one can pick two different vertices to be connected by an edge in $N(N - 1)/2$ different ways, and the number of locations for self-connections is N. Each of the M (distinguishable) edges can be placed into any of these possible $N(N+1)/2$ locations yielding

$$P^{MC} = \left(\frac{N(N + 1)}{2} \right)^{-M} \tag{27}$$

for the microcanonical probability distribution.

It is straightforward to define a *microcanonical ensemble on a subset* of labeled degenerate graphs. Since the degree distribution is a characteristic property of most real-world graphs, it can be used to select a subset: labeled degenerate graphs with N vertices and a *fixed degree distribution*, p_k, meaning that for each value of k there are exactly $N(k) = Np_k$ vertices in the graph with that degree. A given degree distribution is realized by many adjacency matrices, and each adjacency matrix is further realized by many labeled degenerate graphs (because the edges are distinguishable).

Since each graph has the same weight in the microcanonical ensemble, the probability of a given adjacency matrix, $\mathbf{A}$, is proportional to the number of different graphs, $\mathcal{N}(\mathbf{A})$, that realize this particular adjacency matrix[8]:

$$P^{\mathrm{MC}}(\mathbf{A}) \propto \mathcal{N}(\mathbf{A}) = M! \prod_{i=1}^{N} \frac{1}{(A_{ii}/2)!} \prod_{j<k=1}^{N} \frac{1}{A_{jk}!}. \tag{28}$$

Dorogovtsev et. al [12] have constructed canonical ensembles of degenerate graphs by equilibrium processes that keep the degree distribution and the number of edges fixed. At each step of such a process one end of a randomly chosen edge is moved to a new vertex, i, selected with a weight $w(k_i)$. Similarly, the removal of edges (with a rate λN) together with the insertion of new edges between vertices i and j (with a rate proportional to $w(k_i)w(k_j)$) lead to grand canonical ensembles.

4 Main Features of Equilibrium Graphs: Local and Global Properties

In this section, the characteristic features of equilibrium graph ensembles will be discussed. We will start with local properties and will proceed towards properties taking into account larger groups of vertices.

4.1 Local Correlations

Most networks obtained from experimental data contain significant correlations. Therefore, it is a natural requirement that the models describing them should also contain correlated quantities. The frequent occurrence of connections between vertices of similar properties such as, e.g., similar degrees, has been termed *assortativity*, and the higher probability of connections between vertices with different degrees was termed *disassortativity*. In social and biological networks, both assortativity and disassortativity have been observed [49,52].

One possible way of constructing a random graph with a given degree-degree correlation, $p(k, k')$, is the following [53]. First, the degree distribution, p_k, of such a graph has to be determined from

---

[8] The M edges of the graph can be permuted in $M!$ ways. There are $A_{jk}!$ equivalent permutations of the edges between vertices j and k, but they all represent the very same graph. Similarly, there are $(A_{ii}/2)!$ such equivalent permutations of the unit loops at vertex i.

$$\sum_{k'} p(k, k') = \frac{k p_k}{\langle k \rangle},$$ (29)

where $\langle k \rangle = \sum_k k p_k$ is a condition for self-consistence. Next, one needs to assign a random number, q_i, to each vertex i from the degree distribution, p_k. Finally, one should go through each pair of vertices, i and j, in the graph and put a link between them with probability

$$\frac{\langle k \rangle}{N} \frac{p(q_i, q_j)}{p_{q_i} p_{q_j}}.$$ (30)

A short technical comment here is that not all $p(k, k')$ functions can ensure that the degree-degree correlation of the networks constructed with this algorithm converges to $p(k, k')$ in the $N \to \infty$ limit. The necessary condition is that $p(k, k')$ should decay slower than $\exp(-\sqrt{k} - \sqrt{k'})$ [54].

An alternative approach could be to generate a canonical ensemble with a cost function (energy) to suppress deviations from the prescribed $p(k, k')$ (see Sect. 3.1).

4.2 Global Characteristics

Component Sizes

One of the often studied global properties of networks has been the size of the largest component. Whenever the number of vertices in this component, $s_{\max}$, is in the order of the total number of vertices, it is called the "giant component". In the classical random graph, the giant component appears at the *critical edge density*, $\langle k \rangle = 1$. (see [4] and Sect. 3.1). Below this density the largest component contains $\mathcal{O}(\log N)$ vertices and above this density it will start to grow linearly.

In a random graph with a *fixed degree distribution*, p_k, the condition for the giant component to exist is [55,56]

$$\sum_{k=3}^{N} k(k-2) p_k > p_1.$$ (31)

At the transition point, the component size distribution of a random graph with any fixed degree distribution is known to decay as a power law with the exponent $-3/2$ [55]. Near the transition, the component size distribution follows a power law with an exponential cutoff. This is in analogy with percolation phenomena, where the component sizes also have a power law distribution at the critical point. An analytic treatment of connected components in random graphs with fixed degree sequences is available in [57].

Spectral Properties

Work related to the spectral properties of random structures was launched by Wigner's semicircle law [58]. His result enabled the modeling of complex quantum mechanical systems lying far beyond the reach of exact methods and later it was found to have numerous applications in statistical and solid state physics as well [59,60]. As one particular extension of Wigner's work, Füredi and Komlós [61] proved that the spectral density of a classical random graph also converges to a semicircle. It is important to note that in the classical random graph the number of edges is $pN^2/2$ with p=const., i.e., it grows quadratically with the number of vertices. The general form of the semicircle law valid for the classical random graph [62] states that the spectral density of $\mathbf{A}/\sqrt{pN}$, apart from the largest eigenvalue, will converge to

$$\rho(\lambda) = \begin{cases} (2\pi)^{-1}\sqrt{4-\lambda^2}, & if |\lambda| < 2; \\ 0, & \text{otherwise.} \end{cases} \tag{32}$$

The largest eigenvalue is detached from the rest of the spectrum, and scales as $pN = \langle k \rangle$, while the second largest one is about $2\sqrt{pN} = 2\sqrt{\langle k \rangle}$ [63,64]. Note that reducing the density of edges may destroy the semicircular distribution. In the case of a *sparse*[9] classical random graph (i.e., with a fixed average degree) $\rho(\lambda)$ converges to a distribution rich in singularities [65,66].

The next class of networks to be analyzed is graph ensembles with a fixed power law degree distribution. For both real-world networks and graph models having a power law degree distribution, the overall shape of the spectral density differs from the semicircle and the largest eigenvalues follow a power law distribution [65,71,72] (see Fig. 6). Chung et. al [73] have found that a fixed power law degree distribution with the exponent γ can be analytically connected to a power law tail of the spectral density with the exponent α:

$$\alpha = 2\gamma - 1, \quad \text{if} \quad \gamma > 2.5. \tag{33}$$

The findings of related numerical and analytical studies [65,68,70] are in agreement with this result. Evidently, the large eigenvalues are caused by the large degrees in the graph. More precisely, it can be shown that the largest eigenvalues can be approximated by the square roots of the largest degrees [69]. We mention here that the spectral properties of a graph are closely related to random walks on the graph and to the electrical resistance of the graph as a network of resistors. For a concise review we refer the reader to [74].

5 Topological Phase Transitions
in Equilibrium Network Ensembles

As already mentioned, a widely studied phase transition in an equilibrium network ensemble is the occurrence of the giant component in the classical random

---

[9] Sparse graphs are a common version of graphs expressing the fact that there is a cost associated to each connection, therefore, the average degree is limited even when $N \to \infty$.

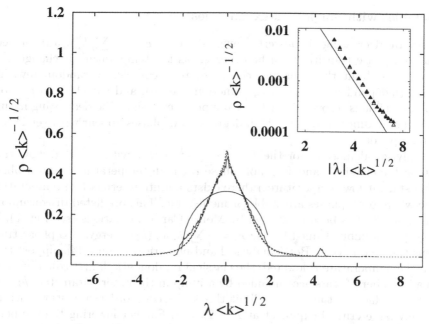

Fig. 6. Average spectral densities of scale-free graphs. (The average degree is $\langle k \rangle = 10$.) **Main panel:** Graphs with $N = 100$ (—), $N = 1000$ (- -), and $N = 7000$ (- - -) vertices and a degree distribution decaying as $p_k \propto k^{-\gamma}$ ($\gamma = 3$). A continuous line shows the semi-circular distribution for comparison. The central part of the scale-free graph's spectral density is spiked in contrast to the flat top of the semi-circle. Also, the scale-free graph's spectrum decays as a power law, while the semicircular distribution decays exponentially at its edges [67]. **Inset:** The upper and lower tails of $\rho(\lambda)$ (open and full triangles) for scale-free graphs with $N = 40,000$ vertices. Note that both axes are logarithmic and $\rho(\lambda)$ has a power law tail with the same decay rate at both ends of the spectrum. The line with the slope -5 (i.e., the exponent $\gamma = 5$) in this figure is a guide to the eye, and at the same time a numerical prediction also that was later confirmed by analytic results [68–70]. Figure from [65].

graph model as a function of the density of edges. For $\langle k \rangle < 1$, there is a similar transition in the ensemble with the $E = -s_{\max}$ energy as a function of the temperature (see later). An appropriate order parameter for such transitions is the normalized *size of the largest component*, $\Phi_s = s_{\max}/M$. In transitions where a condensation of edges onto one vertex (or a small number of vertices) occurs, the normalized *largest degree*, $\Phi_k = k_{\max}/M$, is the most appropriate order parameter. In general, such transitions where some global statistical property of the topology changes (measured by an order parameter), will be referred to as *topological phase transitions*.

Ensembles with Single-Vertex Energies

For several decreasing single-vertex energies (e.g., $E = -\sum_i k_i^2$), a dispersed-connected phase transition can be observed as the temperature is changed. In the $T \to \infty$ limit the dynamics converges to a completely random rewiring process (independent of the energy function chosen), and the classical random graph ensemble is recovered. At lower temperatures, since the decreasing nature of the energy function rewards high degrees, new phases appear with vertices of macroscopic degrees.

Analytic calculations for the $E = -\sum_i k_i^2$ energy with $\langle k \rangle < 1$ show that between the classical random graph phase (at high temperatures) and a phase with a star (at low temperatures) there exists a finite intermediate temperature range where both phases are stable or metastable. This predicted discontinuous transition [13], has been confirmed by Monte-Carlo simulations (see Fig. 7). In the graph ensemble defined by the $E = -\sum_i k_i \log(k_i)$ energy, two phase transitions can be observed. Both analytical and numerical results [14] support that when the temperature is lowered, the classical random graph first collapses onto a small number of stars accompanied by a jump in the order parameter, Φ_k (see Fig. 8a). In fact, in the $N \to \infty$ limit this is a second order transition with an infinitely large critical exponent at $T = T_c = 1$. Further lowering the temperature will lead to another transition: a compactification where all edges collapse onto the minimum possible number of vertices (see Fig. 8b). This transition is discontinuous with a hysteresis.

Note that for both single-vertex energy functions discussed here the total energy of the system in the different topological phases scales differently with N, which is connected to the singular changes in the average degree (see caption of Fig. 7).

Transient Ensembles vs. Growing Networks

Non-equilibrium processes, such as, growth, can produce a high variety of network ensembles. Some of these ensembles can also be constructed with the help of an equilibrium dynamics as a *transient ensemble*, i.e., as an intermediary and temporary ensemble between an initial set of graphs and the final, equilibrium ensemble. In the case of the $E = -\sum_i k_i \log(k_i)$ energy, during the process of relaxation from the classical random graph phase to the star-like phase near the critical temperature, $T_c = 1$ (see Fig. 8b for two typical graphs illustrating these two phases), the degree distribution of the graphs in the transient ensemble decays continuously, as a power law (see Fig. 8c). The qualitative description of this phenomenon is the following. During the transition there is a "pool" of edges attached to vertices of small degrees, and a small number of vertices with higher degrees serve as centers of condensation. The change of energy associated with moving edges within the "pool" is negligible, whereas the nucleation centers are accumulating edges at a rate proportional to their degrees[10]. This

---

[10] If an edge from the "pool" is moved to a vertex with a large degree (k), then the energy of the system changes by approximately $\Delta E = \partial E / \partial k = -\log k - 1$. In

a)

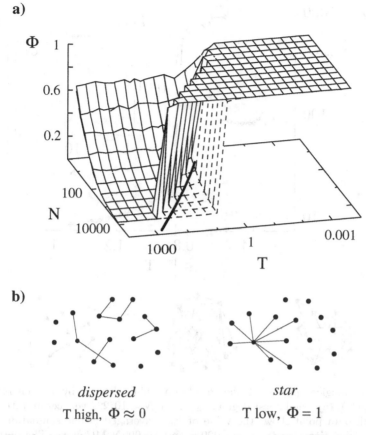

b)

dispersed

T high, $\Phi \approx 0$

star

T low, $\Phi = 1$

Fig. 7. Topological phase transition in the graph ensemble defined by the $E = -\sum_i k_i^2/2$ graph energy. (a) The order parameter $\Phi = \Phi_k = k_{max}/M$ as a function of the temperature and the system size ($\langle k \rangle = 0.5$). The simulations were started either from a star (corresponding to $T = 0$, solid line) or a classical random graph (corresponding to $T = \infty$, dashed line). Each data point represents a single run, and averaging was carried out between the simulation times of $t = 100N$ and $200N$ Monte-Carlo steps. The thick solid line shows the analytically calculated spinodal $T_1 = M/\log(N)$. This panel is from [14]. (b) Two typical graphs from the two phases of the graph ensemble. At low temperatures, edges are condensed onto one vertex ($\Phi = 1$), and the total energy of the system is *non-extensive*: it scales as N^2. At high temperatures, one has a dispersed classical random graph with $\Phi \approx 0$, and the total energy of the system scales as N.

mechanism, produced by an equilibrium dynamics, is analogous to the preferential attachment rule of growing (non-equilibrium) models of scale-free networks [6,48], which also lead to power law degree distributions.

---

the equilibrium dynamics the rate of this step will be $e^{-\Delta E/T} \propto k$ at the critical temperature, $T = T_c = 1$.

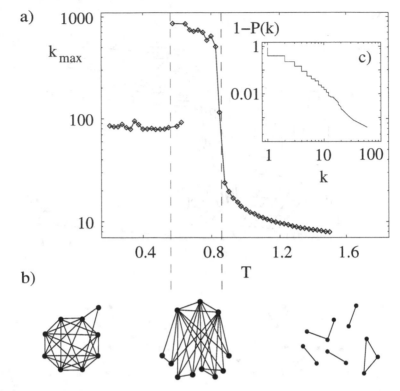

Fig. 8. Topological phases of the graph ensemble defined by the energy $E = -\sum_i k_i \log(k_i)$. (a) The largest degree, k_{max}, for $N = 10,224$ vertices and $M = 2,556$ edges. Each data point shows the value of k_{max} averaged in one simulation run between the simulation times of $t = 5,000N$ and $20,000N$ MC steps. The data points are connected to guide the eye. There is a sharp, continuous transition near $T = 0.85$ and a discontinuous transition (with a hysteresis) around $T = 0.5 - 0.6$. (b) The three different plateaus in (a) correspond to distinct topological phases: $k_{max} = \mathcal{O}(1)$ to the classical random graph, $k_{max} = \mathcal{O}(M)$ to the star phase (a small number of stars sharing most of their neighbors) and $k_{max} = \mathcal{O}(\sqrt{M})$ to the fully connected subgraph. (c) At $T = 0.84$ and $t = 600N$, one minus the cumulative degree distribution, i.e., $1 - P(k) = \int_0^k dk' p_{k'}$, follows a power law, thus, the degree distribution decays as a power law also. Figure from [14].

Ensembles with Neighboring Vertex Energies

Baiesi and Manna [11] have analyzed the canonical ensemble of connected graphs defined by the energy shown in (19). This energy favors degree dissasoratativity, i.e., a negative degree-degree correlation. As a function of temperature three phases have been identified in this ensemble: the classical random graph at $T \to \infty$, scale-free graphs at intermediate temperatures, and a phase with a small number of stars at low temperatures.

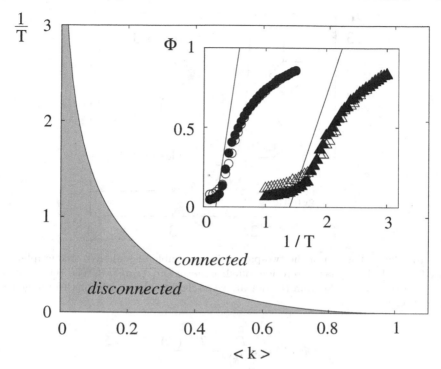

Fig. 9. Analytical phase diagram and Monte-Carlo simulation results for the graph ensemble defined by the $E = -s_{\max}$ energy. **Main panel:** The white and shaded areas correspond to the ordered phase (containing a giant component) and the disordered phase, respectively as given by (34). **Inset:** The order parameter $\Phi = \Phi_s = s_{\max}/M$ obtained from Monte-Carlo simulations as a function of the inverse temperature for $\langle k \rangle = 0.1$ (triangles) and $\langle k \rangle = 0.5$ (circles). Each data point shows averages taken for 10 runs between the simulation times of $t = 100N$ and $500N$ Monte-Carlo steps. The open and closed symbols represent $N = 500$ and $1,000$ vertices, respectively. The critical exponent, in agreement with the analytical approximations (solid lines, see [13]), was found to be 1. Figure from [14].

Ensembles with Component Energies

Similarly to the single-vertex case, a decreasing, component-size dependent energy can also lead to phase transitions. The simplest case which we analyze in this paragraph, is $E = -s_{\max}$. At low densities ($\langle k \rangle < 1$) one can observe a classical random graph at $T \to \infty$, whereas at low temperatures a giant component is present. It can be shown that the dividing line between the two topological phases is [14]

$$T_c(\langle k \rangle) = \frac{1}{\langle k \rangle - 1 - \log(\langle k \rangle)}, \tag{34}$$

which is also supported by numerical results (see Fig. 9). In the vicinity of the critical temperature the order parameter can be approximated as

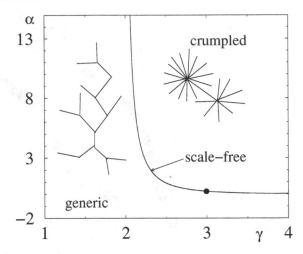

Fig. 10. Phase diagram of the two-parameter ensemble of scale-free tree graphs presented in [44]. Two phases were identified: *generic* and *crumpled*. The points of the dividing line to the right from the dot are scale-free graphs and belong to the generic phase. Figure from [44].

$$\Phi_s^*(T) = 2\,\frac{T^{-1} - T_c^{-1}(\langle k\rangle)}{\langle k\rangle^2 - 3\langle k\rangle + 2}\,,\tag{35}$$

indicating that the phase transition is *continuous* (see Fig. 9 for details).

Further Ensembles

For the ensembles of degenerate graphs introduced by Dorogovtsev et. al [12] (see Sect. 3.6), with $w(k) \propto k + 1 - \gamma$, a critical line, $\langle k \rangle = k_c(\gamma)$, was found. Below this line the degree distribution has an exponential cutoff and above that a *condensate* occurs where a finite fraction of all edges is attached to an infinitely small fraction of nodes.

In an ensemble of connected tree graphs with a fixed, power law degree sequence, Burda et. al [44] have reported a phase transition as a function of two parameters: γ, the exponent of the degree distribution, and α, related to the probability of subgraphs (see Fig. 10). The analytic form of the dividing line between the identified *generic* and *crumpled* phases was computed, and numerical simulations were carried out using a Monte-Carlo sampling technique.

6 Summary

Graph models with energies provide a natural way to define microcanonical, canonical and grand canonical ensembles. These ensembles are often generated by equilibrium restructuring processes obeying detailed balance and ergodicity. Also, to describe a wider range of network models, it is useful to extend the definition and consider ensembles without energy as well. We have reviewed the main features of currently studied equilibrium graph ensembles, with a focus on degree-degree correlations, component sizes and spectral properties. We have also discussed continuous and discontinuous topological phase transitions in equilibrium graph ensembles. A solid basis of the equilibrium statistical mechanics of networks, as presented in this article, can facilitate the application of statistical physics tools in the field of networks, and can help to expand the analyses towards problems of high current interest, such as optimization and reverse engineering.

Acknowledgements

This work has been in part supported by the Hungarian Scientific Research Fund under grant No: OTKA 034995. I.F. acknowledges a scholarship from the Communication Networks Laboratory at ELTE.

References

1. L. Euler: Comm. Acad. Sci. Imp. Petrop. **8**, 128–140 (1736)
2. P. Erdős and A. Rényi: Publ. of the Math. Inst. of the Hung. Acad. of Sci. **5**, 17–61 (1960)
3. E. N. Gilbert: Ann. of Math. Stat. **30**, 1141 (1959)
4. B. Bollobás: *Random graphs*, 2nd ed. (Cambridge University Press, Cambridge, 2001)
5. D. J. Watts and S. H. Strogatz: Nature **393**, 440 (1998)
6. A.-L. Barabási and R. Albert: Science **286**, 509 (1999);
7. R. Albert and A.-L. Barabási: Rev. Mod. Phys. **74**, 47 (2002)
8. J. F. F. Mendes and S. N. Dorogovtsev: *Evolution of Networks: From Biological Nets to the Internet and WWW* (Oxford University Press, Oxford, 2003)
9. *Handbook of Graphs and Networks, From the Genome to the Internet* Ed. by S. Bornholdt and H. G. Schuster (Wiley-VCH, Berlin, 2002)
10. J. Berg and M. Lässig: Phys. Rev. Lett. **89**, 228701 (2002)
11. M. Baiesi and S. S. Manna: Phys. Rev. E **68**, 047103 (2003)
12. S. N. Dorogovtsev, J. F. F. Mendes, and A. N. Samukhin: Nucl. Phys. B **666**, 396 (2003)
13. I. Derényi, I. Farkas, G. Palla, and T. Vicsek: cond-mat/0306170 (to appear in Physica A)
14. G. Palla, I. Farkas, I Derényi, and T. Vicsek: cond-mat/0309556 (to appear in Phys. Rev. E)
15. G. Szabó, M. Alava, and J. Kertész: Phys. Rev. E **67**, 056102 (2003)
16. G. Bianconi and A.-L. Barabási: Phys. Rev. Lett. **86**, 5632 (2001)
17. P. L. Krapivsky, S. Redner, and F. Leyvraz: Phys. Rev. Lett. **85** 4629-4632 (2000)
18. B. Bollobás and O. Riordan: to appear in: *Combinatorica*.
19. P. Bialas and Z. Burda: Phys. Lett. B **384**, 75 (1996)
20. D. Stark and B. Vedres: Santa Fe Inst. working paper no. 01-12-081 (2001)
21. J.-P. Onnela, A. Chakraborti, K. Kaski, J. Kertész, and A. Kanto: Phys. Rev. E **68**, 056110 (2003)

22. M. E. J. Newman: *Phys. Rev. E* **64**, 016131 (2001); M. E. J. Newman: *Phys. Rev. E* **64**, 016132 (2001).
23. J. P. K. Doye: *Phys. Rev. Lett.* **88**, 238701 (2002).
24. G. Tusnády: *Fák evolúciója*. (The evolution of trees, in Hungarian.) Inaugural talk at the Hung. Acad. of Sci. (2001)
25. Z. Burda, J. Jurkiewicz, and A. Krzywicki: cond-mat/0312494
26. A. V. Goltsev, S. N. Dorogovtsev, and J. F. F. Mendes: *Phys. Rev. E* **67**, 026123 (2003)
27. S. N. Dorogovtsev, A. V. Goltsev, and J. F. F. Mendes: Phys. Rev. E **66**, 016104 (2002)
28. G. Bianconi: *Phys. Lett. A* **303**, 166 (2002)
29. A. Barrat and M. Weigt: Eur. Phys. J. B **13**, 547 (2000)
30. B. J. Kim, H. Hong, P. Holme, G. S. Jeon, P. Minnhagen, and M. Y. Choi: Phys. Rev. E **64**, 056135 (2001)
31. S. N. Dorogovtsev, A. V. Goltsev, and J. F. F. Mendes: cond-mat/0310693
32. M. E. J. Newman: Phys. Rev. E **66**, 016128 (2002)
33. D. H. Zanette and M. Kuperman: Physica A **309**, 445 (2002)
34. N. Zekri and J. P. Clerc: Phys. Rev. E **64**, 056115 (2001)
35. M. Anghel, Z. Toroczkai, K. E. Bassler, and G. Korniss: cond-mat/0307740
36. S.-H. Yook, H. Jeong, and A.-L. Barabási: Proc. Natl. Acad. Sci. USA **99**, 13382 (2002)
37. R. Pastor-Satorras, A. Vazquez, and A. Vespignani: Phys. Rev. Lett. **87**, 258701 (2001)
38. R. Guimerá, A. Díaz-Guilera, F. Vega-Redondo, A. Cabrales, and A. Arenas: Phys. Rev. Lett. **89**, 248701 (2002)
39. L. A. Braunstein, S. V. Buldyrev, and R. Cohen: S. Havlin, and H. E. Stanley: Phys. Rev. Lett. **91**, 168701 (2003)
40. K. W. Kohn: Mol. Biol. Cell **10**, 2703 (1999)
41. G. Simonyi: Graph entropy: a survey. In: *Combinatorial Optimization*, DIMACS Series in Discrete Mathematics and Computer Science, vol 20, ed by W. Cook, L. Lovasz, and P. Seymour (Center for Discrete Mathematics & Theoretical Computer Science, New Jersey, 1995) pp 399–441
42. I. Csiszár, J. Körner, L. Lovász, K. Marton, and G. Simonyi: Combinatorica **10**, 27 (1990)
43. M. Bauer and D. Bernard: cond-mat/0206150
44. Z. Burda, J. D. Correia, and A. Krzywicki: Phys. Rev. E **64**, 046118 (2001)
45. S. N. Dorogovtsev, J. F. F. Mendes, and A. N. Samukhin: cond-mat/0206131
46. J. Dall and M. Christensen: Phys. Rev. E **66**, 016121 (2002)
47. E. Ben-Naim, P. L. Krapivsky, and S. Redner, Extremal Properties of Random Structures, Lect. Notes Phys. **650**, 211–233 (2004)
48. A.-L. Barabási, R. Albert, and H. Jeong: Physica A**272**, 173 (1999)
49. S. Maslov and K. Sneppen: Science **296**, 910 (2002)
50. R. Xulvi-Brunet, W. Pietsch, and I. M. Sokolov: Phys. Rev. E **68**, 036119 (2003)
51. D. Bessis, C. Itzykson, and J. Zuber: Adv. Appl. Math. **1**, 109 (1980)
52. M. E. J. Newman: *Phys. Rev. E* **67**, 026126 (2003)
53. M. Boguña and R. Pastor-Satorras: Phys. Rev. E **68**, 036112 (2003)
54. S. N. Dorogovtsev: cond-mat/0308444
55. M. E. J. Newman, S. H. Strogatz, and D. J. Watts: Phys. Rev. E **64**, 026118 (2001)
56. M. Molloy and B. Reed: Random Struct. and Algorithms **6**(2-3), 161 (1995)
57. F. Chung and L. Lu: Ann. Comb. **6**, 125 (2002)

58. E. P. Wigner: The Ann. of Math. **67**, 325 (1958)
59. M. L. Mehta: *Random Matrices*, 2nd ed. (Academic, New York, 1991)
60. A. Crisanti, G. Paladin, and A. Vulpiani: *Products of Random Matrices in Statistical Physics*, vol 104, Springer Series in Solid-State Sciences (Springer, Berlin, 1993)
61. Z. Füredi and J. Komlós: Combinatorica **1**, 233 (1981)
62. F. Hiai and D. Petz: *The Semicircle Law, Free Random Variables and Entropy* (Am. Math. Soc., 2000)
63. F. Juhász: On the spectrum of a random graph. In: *Colloq. Math. Soc. J. Bolyai* (1978) pp 313–326
64. D. Cvetkovic and P. Rowlinson: Lin. and Multilin. Alg. **28**, 3 (1990)
65. I. J. Farkas, I. Derényi, A.-L. Barabási, and T. Vicsek: Phys. Rev. E **64**, 026704 (2001)
66. M. Bauer and O. Golinelli: J. Stat. Phys. **103**, 301 (2001)
67. B. V. Bronx: J. Math. Phys. **5**, 215 (1964)
68. M. Mihail and C. H. Papadimitriou: On the Eigenvalue Power Law. In: *Lecture Notes In Computer Science*. Proceedings of the 6th International Workshop on Randomization and Approximation Techniques (Springer-Verlag London, UK, 2002) pp 254–262
69. F. Chung, L. Lu, and V. Vu: Proc. Natl. Acad. Sci. USA **100**, 6313 (2003)
70. S. N. Dorogovtsev, A. V. Goltsev, J. F. F. Mendes, and A. N. Samukhin: Phys. Rev. E **68**, 046109 (2003)
71. M. Faloutsos, P. Faloutsos, and C. Faloutsos: Comput. Commun. Rev. **29**, 251 (1999)
72. K.-I. Goh, B. Kahng, and D. Kim: Phys. Rev. E **64**, 051903 (2001)
73. F. Chung, L. Lu, Linyuan, and V. Vu: Ann. Comb. **7**, 21 (2003)
74. L. Lovász: Random Walks on Graphs: A Survey. In: *Combinatorics, Paul Erdős is Eighty* vol 2 (Keszthely, Hungary, 1993) pp 1–46

Information Theory of Complex Networks: On Evolution and Architectural Constraints

Ricard V. Solé and Sergi Valverde

[1] Complex Systems Lab-ICREA, Universitat Pompeu Fabra (GRIB),
   Dr Aiguader 80, 08003 Barcelona, Spain
[2] Santa Fe Institute, 1399 Hyde Park Road, Santa Fe NM 87501, USA

Abstract. Complex networks are characterized by highly heterogeneous distributions of links, often pervading the presence of key properties such as robustness under node removal. Several correlation measures have been defined in order to characterize the structure of these nets. Here we show that mutual information, noise and joint entropies can be properly defined on a static graph. These measures are computed for a number of real networks and analytically estimated for some simple standard models. It is shown that real networks are clustered in a well-defined domain of the entropy-noise space. By using simulated annealing optimization, it is shown that optimally heterogeneous nets actually cluster around the same narrow domain, suggesting that strong constraints actually operate on the possible universe of complex networks. The evolutionary implications are discussed.

1 Introduction

Many complex systems are to some extent describable by the network of interactions among its components. Beyond the specific features displayed by each net, it has been shown that a number of widespread properties are common to most of them. One is the presence of the small-world phenomenon and the second the observation that in many cases they are highly heterogeneous in their connectivity patterns [1-4].

Heterogeneity can be easily identified by looking at the so called degree distribution P_k, which gives the probability of having a node with k links. Most complex networks (both natural and artificial) can be described by a degree distribution $P_k \sim k^{-\gamma}\phi(k/\xi)$ where $\phi(k/\xi)$ introduces a cut-off at some characteristic scale ξ [5]. An example of such scale-free networks is provided by the architecture of digital electronic circuits (Fig. 1). It has been shown [6] that these systems exhibit long tail distributions of links, where the nodes are electronic components and the links are physical wires between units. Most elements are connected to a few others (for circuits this usually means nearest neighbors) but some are connected to many others. In Fig. 2 several examples of the observed distributions for both analog (a-b) and digital (c-d) systems are shown. Although analog systems are closer to an exponential distribution (i. e. ξ small) digital, large-scale systems[3] exhibit scaling behavior, with $\gamma \sim 3$.

---

[3] Similar results have been obtained by looking at large-scale systems, such as VLSI networks. See for example: http://citeseer.nj.nec.com/450707.html

R.V. Solé and S. Valverde, Information Theory of Complex Networks: On Evolution and Architectural Constraints, Lect. Notes Phys. **650**, 189–207 (2004)
http://www.springerlink.com/

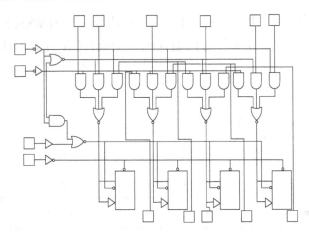

Fig. 1. Heterogeneity is a widespread feature of most (but not all) complex networks. An example from technology graphs are electronic circuits (upper plot) which have been shown to display scale-free distributions of links.

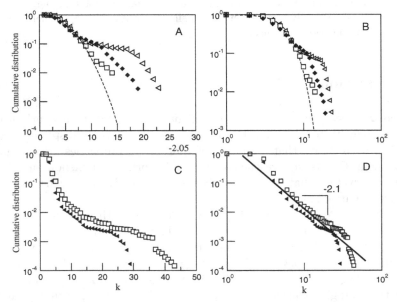

Fig. 2. Cumulative degree distributions for several examples of analogic (a-b) and digital (c-d). Although the analogic systems are less heterogeneous, analogic circuits (particularly large systems) display scaling in their degree distributions (Left plots are linear-log and right plots are in log-log scale).

Scale-free nets have been shown to be obtainable through a number of mechanisms, including preferential attachment [2,3,7,8], optimization [9,10], duplication and divergence [11,12] or fitness-dependent, rich-gets-richer mechanisms [13] or the "copying" model [14]. Beyond the common qualitative architecture shared by these systems, the dynamical patterns and their time scales that take place on top of these webs differs from system to system, although in a way or another deals with information propagation and/or processing. Moreover, the response to node removal differs from system to system. Although genetic and metabolic networks seem to be fairly robust against perturbations of different types, a totally different situation arises in electronic circuits. In biological nets failure of highly connected components will typically end in system's failure (for example, at the cellular level). But failure (by mutation or transient change) of a gene is often buffered by the rest of the system. This is not the case for electronic circuits and, to a similar extent, by software networks. Failure of any component typically leads to system's failure, no matter how much linked is the given unit.

Several quantitative measures can be used in order to characterize a given network. The first step is to define an appropriate representation in terms of a graph Ω, defined by a pair $\Omega = (W, E)$, where $W = \{s_i\}, (i = 1, ..., N)$ is the set of N nodes (species, proteins, neurons, etc) and $E = \{\{s_i, s_j\}\}$ is the set of edges/connections between nodes. The *adjacency matrix* ξ_{ij} indicates that an interaction exists between two nodes $s_i, s_j \in \Omega_p$ ($\xi_{ij} = 1$) or that the interaction is absent ($\xi_{ij} = 0$). Several statistical properties, such as average degree, clustering or diameter can be defined from the adjacency matrix.

But the universe of possible networks (Fig. 3), although not arbitrarily diverse, displays a number of structural variations that cannot be compressed by the previous average quantities. Real networks are not only typically heterogeneous, but they also involve other types of features, such as hierarchical organization [15].

In Fig. 3 we qualitatively summarize the basic types of network organization by using a generic, qualitative parameter space. Here heterogeneity, modularity and randomness define three axes. Assuming that such three parameters can be properly defined, different real and model graphs can be located at different locations. The current knowledge of network architecture in many different systems strongly indicates that the domain of random networks with long tailed degree distributions and some amount of modular structure are rather densely occupied. In spite that the evolutionary process leading to these different systems are rather diverse, it is interesting to see that there is a strong convergence towards this type of architectures. Here we will explore this problem by using information-based statistical measures.

2 Measuring Correlations

Beyond the degree distribution and average statistical measures, correlation measures offer considerable insight into the structural properties displayed by com-

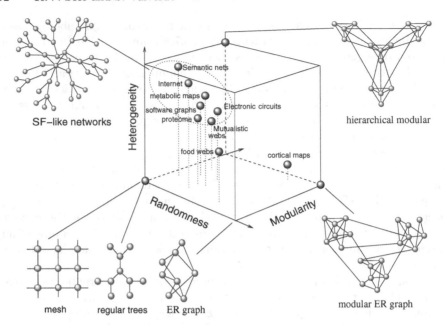

Fig. 3. A zoo of complex networks. In this qualitative space, three relevant character-istics are included: randomness, heterogeneity and modularity. The first introduces the amount of randomness involved in the process of network's building. The second mea-sures how diverse is the link distribution and the third would measure how modular is the architecture. The position of different examples are only a visual guide. The domain of highly heterogeneous, random hierarchical networks appears much more occupied than others. Scale-free like networks belong to this domain.

plex networks. One particularly interesting is network asortativeness [16]. Some networks show *assortative mixing* (AM): high degree vertices tend to attach to other high-degree vertices. At the other extreme there are graphs displaying *dis-sortative mixing* (DM), thus involving anticorrelation. The later are common in most biological nets, whereas the former are common in social and collaboration networks. It has been suggested that the presence and sign of assortativeness in this nets can have deep implications to their resilience under node removal or disease propagation.

Following a previous analysis [16] we will be interested here not in the degree distribution P_k but instead in the *remaining degree*: the number of edges leaving the vertex other than the one we arrived along (Fig. 4). This new distribution $q(k)$ is obtained from:

$$q(k) = \frac{(k+1)P_{k+1}}{\langle k \rangle} \tag{1}$$

where $\langle k \rangle = \sum_k k P_k$. In a network with no assortative (or disassortative) mixing $q_c(j,k)$ takes the value $q(j)q(k)$. If there is assortative mixing, $q_c(j,k)$ will differ from this value and the amount of assortative mixing can be quantified by the

Fig. 4. Computing correlations in a network. Here two given, connected nodes s_i, s_j are shown, displaying different degrees k_i, k_j. Since we are interested in the remaining degrees, a different value needs to be considered (here indicated as q_i, q_j).

connected degree-degree correlation function

$$\langle jk \rangle - \langle j \rangle \langle k \rangle = \sum_{jk} jk q_c(j,k) - \left(\sum_j j q(j) \right)^2 \qquad (2)$$

where $\langle \ldots \rangle$ indicates an average over edges.

The correlation function is zero for no assortative mixing and positive or negative for assortative or disassortative mixing respectively. In order to compare different networks, normalization is obtained by dividing it with by its maximal value, which it achieves on a perfectly assortative network, i.e., one with $q_c(j,k) = q(k)\delta_{jk}$. This value is equal to the variance $\sigma_q^2 = \sum_k k^2 q(k) - \left[\sum_k k q(k) \right]^2$ of the distribution $q(k)$, and hence the normalized correlation function is

$$r = \frac{1}{\sigma_q^2} \left[\sum_{jk} jk q_c(j,k) - \left(\sum_j j q(j) \right)^2 \right] \qquad (3)$$

As defined from the previous equation, we have $-1 < r < 0$ for DM and $0 < r < 1$ for AM. Both biological ad technological nets tend to display DM, whereas social webs are clearly assortative.

Correlation functions have been widely used both in statistical physics [17] and nonlinear dynamics [18]. A closely related, and more general approach involves the use of information-based measures [19-21]. One specially important quantity is the so called *mutual information*, which is a general measure of dependence between two variables [19,22]. Correlation functions measure linear relations, whereas mutual information measures the general dependence and is thus a less biased statistic. The relevance of this difference is illustrated by the analysis of chaotic dynamical systems: the second allows to determine the independent variables for (re-) constructing phase trajectories [23]. This cannot be done from linear correlation functions. Additionally, the definition of mutual information within the context of communication channels implies additional statistical quantities (such as channel entropy and noise) that provide a detailed characterization of system's complexity. Here we show how these quantities can be properly defined for complex networks, how they correlate with other statistical measures and what is their meaning and implications.

3 Entropy and Information

By using the previous distribution $\mathbf{q} = (q(1), ..., q(i), ..., q(N))$, an entropy measure $H(\mathbf{q})$ can be defined:

$$H(\mathbf{q}) = -\sum_{k=1}^{N} q(k) \log(q(k)) \tag{4}$$

The entropy of a network will be a measure of uncertainty [19]. Within the context of complex nets, it provides an average measure of network's heterogeneity, since it measures the diversity of the link distribution. The maximum is $H_{\max}(\mathbf{q}) = \log N$ is obtained for $q(i) = 1/N (\forall i = 1, ..., N)$ and $H_{\min}(\mathbf{q}) = 0$ which occurs when $\mathbf{q} = (1, 0., , , 0)$. In an information channel, there is a distinction between source and destination. Given the symmetric character of our system, no such distinction is made here. In Fig. 5 we can see the impact of heterogeneity on entropy. Specifically, we computed the entropy $H(\mathbf{q}; \gamma, \xi)$ for $\gamma \in (2, 3)$ and $\xi \in (0, 50)$ for a distribution $P_k \sim k^{-\gamma} \phi(k/\xi)$ using different scaling exponents γ and cut-offs ξ. The impact of diversity (long tails) is obvious, increasing the uncertainty. As the scaling exponent increases or the cut-off decreases, the network becomes less heterogeneous and as a result a lower entropy is observed.

Similarly, the joint entropy can be computed by using the previous joint probabilities:

$$H(\mathbf{q}, \mathbf{q}') = -\sum_{k=1}^{N} \sum_{k'=1}^{N} q_c(k, k') \log q_c(k, k') \tag{5}$$

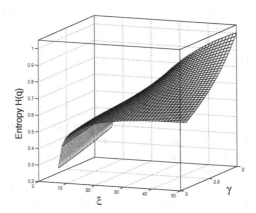

Fig. 5. Entropy of the remaining degree distribution obtained from a network with degree distribution $P_k \sim k^{-\gamma} \phi(k/\xi)$. Here $H(\mathbf{q}) = -\int P_k \log P_k dk$ is shown against the scaling exponent γ and the cut off ξ. Here we have used an exponential cut-off, i. e. $\phi(k/\xi) = \exp(-k/\xi)$ As expected, the entropy becomes larger for smaller γ and decreases as ξ is reduced.

Here $q_c(k, k')$ is the joint probability, and it is normalized, i. e.:

$$\sum_{k=1}^{N} \sum_{k'=1}^{N} q_c(k, k') = 1 \tag{6}$$

Since it considers all possible pairs of edges, this entropy provides a measure of the average uncertainty of the network. As before, it can be understood in terms of a measure of the diversity of linked pairs with given remaining degrees.

The mutual information $I(\{q_k\})$ of a given system is defined by means of the difference:

$$I(\mathbf{q}) = H(\mathbf{q}) - H_c(\mathbf{q}|\mathbf{q}') \tag{7}$$

where the last term $H_c(\mathbf{q}|\mathbf{q}')$ is the conditional entropy that involves a different set of *conditional* probabilities $\pi(k|k')$ [19]. They give the probability of observing a vertex with k edges leaving it provided that the vertex at the other end of the chosen edge has k' leaving edges. This entropy (the "noise" in our graph) is defined as:

$$H_c(\mathbf{q}|\mathbf{q}') = -\sum_{k=1}^{N} \sum_{k'=1}^{N} q(k)\pi(k|k') \log \pi(k|k') \tag{8}$$

Since the conditional and joint probabilities are related through:

$$\pi(k|k') = \frac{q_c(k, k')}{q(k')} \tag{9}$$

the conditional entropy can actually be computed in terms of the two previous distributions:

$$H_c(\mathbf{q}|\mathbf{q}') = -\sum_{k=1}^{N} \sum_{k'=1}^{N} q_c(k, k') \log \frac{q_c(k, k')}{q(k')} \tag{10}$$

we thus have, from the previous expressions,

$$I(\mathbf{q}) = H(\mathbf{q}) - H_c(\mathbf{q}|\mathbf{q}') \tag{11}$$

$$= -\sum_{k=1}^{N} \sum_{k'=1}^{N} q_c(k, k') \log q(k) + \sum_{k=1}^{N} \sum_{k'=1}^{N} q_c(k, k') \log \pi(k|k')$$

$$= -\sum_{k=1}^{N} \sum_{k'=1}^{N} q_c(k, k') \log \frac{q(k)}{\pi(k|k')}$$

which gives a final form for the information transfer function[4]:

---

[4] The previous measures can be extended (with some care) into continuous distributions. In this case, we must assume that the continuous counterparts of the pre-

$$I(\mathbf{q}) = \sum_{k=1}^{N} \sum_{k'=1}^{N} q_c(k, k') \log \frac{q_c(k, k')}{q(k)q(k')} \tag{12}$$

Some limit cases are of interest here. The first corresponds to the maximum information transfer, which is obtained, for a given $\{q_k\}$, when $H_c(\mathbf{q}|\mathbf{q}') = 0$, i. e. when the conditional probabilities are such that $\pi(k|k') = 1$ or 0 for all $k, k' = 1, ..., N$. Another is given by $\pi(k|k') = \delta_{k,k'}$. This case corresponds to a deterministic channel in standard information theory [19]. This implies that $q_c(k, k') = q(k')\delta_{k,k'}$ which is precisely the case of perfectly assortative network [16].

In analogy with information channels, we can find a maximum value of the information, which we call the *network's capacity* $C = \max_{\{q_k\}} I(\mathbf{q})$. There is no general method to compute C for an arbitrary channel. It can only be computed in some specific cases.

By using the previous functions, we will measure three key quantities: (a) the amount of correlation between nodes in the graph, as measured by the information; (b) the noise level, as defined by the conditional entropy, which will provide a measure of assortativeness and (c) the entropy of the $q(k)$ distribution. Since the total information involves the two last terms in a linear fashion, a noise-entropy space will be constructed and the distribution of real nets on this space will be analysed.

4 Model Networks

In the following sub-sections some simple, limit cases will be considered. Different types of architectures are represented by some standard networks exhibiting different degrees of heterogeneity and randomness. The list is far from exhaustive but provides an idea of what are the effects of each ingredient on information transfer and entropies.

4.1 Lattices and Trees

Lattice-like networks are common in some man-made architectures, particularly parallel computers [24-26]. These nets represent the highest degree of homogeneity and have no randomness. For a lattice, we have $P_k = \delta_{k,z}$, where z is a fixed number of links per node and δ_{ij} the Kronecker's delta function. For this ordered graph Ω_L, we have

---

vious degree distributions can be defined. The new distributions are such that the normalization conditions: $\int q(k)dk = 1$ and $\int\int q_c(k, k')dkdk' = 1$ are at work. Provided that the distributions are well behaved, the information transfer is now given by $I(\mathbf{q}) = \int\int q_c(k, k') \log\left(\frac{q_c(k,k')}{q(k)q(k')}\right) dkdk'$. Accordingly, entropy and noise would be obtained from: $H(\mathbf{q}) = -\int q(k) \log(q(k))dk$ and $H_c(\mathbf{q}|\mathbf{q}') = -\int\int q_c(k, k') \log \pi(k|k')dkdk'$

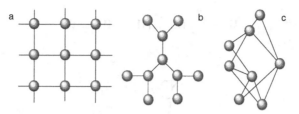

Fig. 6. Homogeneous networks: here two examples of a lattice (a) and a regular tree (b) are shown as examples of deterministic nets. In both cases each node has the same degree and thus both the entropy and the noise are zero. In (c) a random, Erdös-Renyi graph is shown. Here some amount of heterogeneity is at work, but the variance equals the mean and both noise and entropy are very close, giving as a result a small information i. e. no correlations (in the $N \to \infty$ limit).

$$q(k) = \delta_{k,z-1} \tag{13}$$

$$q_c(k, k') = \delta_{k,z-1}\delta_{k',z-1} \tag{14}$$

and thus

$$I(\mathbf{q}) = H(\mathbf{q}) = H_c(\mathbf{q}|\mathbf{q}') = 0 \tag{15}$$

This is a trivial case, since the homogeneous character of the degree distribution implies zero uncertainty. The same situation arises for a Cayley tree (Bethe lattice), where each node has exactly the same degree. Tree-like architectures are also common in designed systems, such as small-sized software graphs [27] and communication networks.

4.2 Erdös-Renyi Graphs

Erdös-Renyi graphs $\Omega_{N,p}$ are random graphs such that two nodes are joined with some probability p. These types of graphs have been widely used as the backbone of null models of genetic [28] ecological [29] and neural [30] networks. It seems also appropriate in describing the topology of species-poor ecosystems [31]. The distributions are single-scaled and thus low uncertainty and high randomness are at work. The average degree will be $\langle k \rangle \approx pN$, and it can be easily shown that the probability P_k that a vertex has a degree k follows a Poisson distribution $P_k = e^{\langle k \rangle}\langle k \rangle^{-k}/k!$, and thus

$$q(k) = \frac{(k+1)e^{\langle k \rangle}}{\langle k \rangle(k+1)!}\langle k \rangle^{(k+1)} = P_k \tag{16}$$

For this random graph, the independence associated to the link assignment implies (for N large) $q_c(k, k') = q(k)q(k')$ and thus information transfer is zero.

An interesting extension of the standard ER graph allows to introduce modularity into the graph structure [15]. In general, the graph Ω is partitioned into m subgraphs $\{\Omega_i\}, (i = 1, ..., m)$ of relative size $\eta_i = |\Omega_i|$ such that

$$W = \bigcup_{i=1}^{m} W_i \quad (W_j \cap W_k = \emptyset) \tag{17}$$

and such that

$$\sum_i \eta_i = 1 \tag{18}$$

All nodes $s_j \in \Omega_i$ are connected with probability q and additionally we have a probability p of connecting two nodes belonging to different modules. The average degree of this system is $<k> = pN/m + (m-1)Nq/m$. Given the random wiring, it is not difficult to show that for large N information will be typically very small.

4.3 Star Graph

Star graphs define another extreme within the universe of complex nets. Although no real network is likely to be described in terms of a pure star graph, it is certainly a common motif in many graphs. They are largely responsible for the short distances achieved in SF networks. Besides, a star graph can be shown to be optimal for low-cost communication [32].

This graph Ω_* is characterized by a degree distribution:

$$p(k) = \frac{n-1}{n}\delta_{k,1} + \frac{1}{n}\delta_{k,n-1} \tag{19}$$

The corresponding distribution $q(k)$ is:

$$q(k) = \frac{1}{2}[\delta_{k,0} + \delta_{k,N-1}] \tag{20}$$

and the joint probabilities are reduced to:

$$q_c(k,k') = \delta_{k,N-2}\delta_{k',0} \tag{21}$$

The entropy is maximal, given by:

$$H(\mathbf{q}) = -q(0)\log q(0) - q(N-1)\log q(N-1) \tag{22}$$

which gives $H(\mathbf{q}) = \log 2$. The noise term is $H_c(\mathbf{q}|\mathbf{q}') = 0$, since $\pi(k|k') = \delta_{kk'}$. The information is thus maximal, with $I(\mathbf{q}) = H(\mathbf{q}) = \log 2$. The star graph displays maximum information, as expected given the deterministic character of the conditional probabilities.

5 Real Networks

In this section we present some analysis of the information measures as applied to real networks. A large set of both technological and biological graphs has been studied. Specifically, three groups of data sets were used in our analysis, all of them known to be highly heterogeneous displaying scale-free architecture:

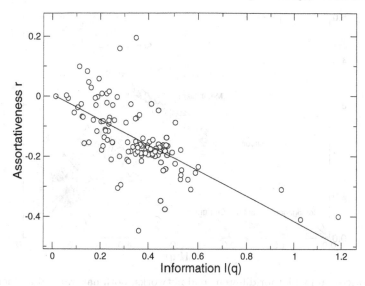

Fig. 7. Information transfer and assortativeness appear to be roughly correlated in a negative way. Here several systems have been used (all those analysed in this paper) and a linear interpolation has been displayed. Although the trend is clear, considerable variance can be appreciated, probably due to the underlying nonlinear mapping between both measures.

1. Metabolic networks: a graph theoretic representation of the biochemical reactions taking place in a given metabolic network can be easily constructed. In this representation, a metabolic network is built up of nodes, the substrates, that are connected to one another through links, which are the actual metabolic reactions [33].

2. Software class diagrams: Nodes are software components and links are relationships between software components. *Class diagrams* constitute a well-known example of such graphs [34,35].

3. Electronic circuits: they can be viewed as networks in which vertices (or nodes) are electronic components (e.g. logic gates in digital circuits and resistors, capacitors, diodes and so on in analogic circuits) and connections (or edges) are wires in a broad sense [6].

In Table 1 we also show a list of selected networks obtained from very different systems and ordered from the higher to the lower information. The system's size N, average connectivity $< k >$, information measures and the assortative mixing coefficient r are provided. We can see that most nets are disassortative, as predicted in [16]. Actually, information and r appear to be negatively correlated. This is shown in Fig. 7, where r is shown against $I(q)$ for different systems.

It is important to see that, in spite of the roughly negative correlation (a linear interpolation has been used) a large variance is observable, and a range of r values is associated to each information transfer. Such a variable plot is likely

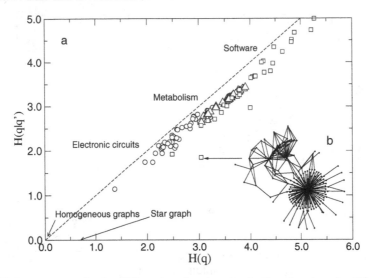

Fig. 8. Noise-entropy plot for different real networks, both natural and artificial. Here electronic circuits (open circles), metabolic (triangles) and software maps (squares) are shown to be close to the zero-information line, i. e. when entropy equals to noise. An example of a software graph that significantly deviates from the $H = H_c$. The network is small and has a rather particular shape, involving a large hub plus another cluster of connected classes.

to be the result of the nonlinear character of the information transfer, not shared by the (linear) correlation defined by assortative mixing measures.

By displaying noise against entropy, the general picture that emerges is that the set of complex networks analysed here displays typically uncorrelated structure. This is clear from the strongly linear dependence shown between noise and entropy (Fig. 8). If two given, randomly chosen nodes with remaining degrees k, k' are typically connected with some probability, roughly irrespective of their mutual degree (i. e. low assortativeness is present) we should expect:

$$q_c(k, k') \approx q(k)q(k') \tag{23}$$

and thus we would have

$$\pi(k|k') \approx q(k) \tag{24}$$

in this case, the noise will be given by:

$$H_c(\mathbf{q}|\mathbf{q}') = -\sum_{k=1}^{N}\sum_{k'=1}^{N} q(k, k') \log \pi(k|k') \tag{25}$$

$$= -\sum_{k=1}^{N}\sum_{k'=1}^{N} q(k, k') \log q(k) = H(\mathbf{q}) \tag{26}$$

Statistical independence among node degree thus gives a predicted straight line $H_c(\mathbf{q}|\mathbf{q}') = H(\mathbf{q})$ which seems to be suggested by our data. One first conclu-

Table 1. Information-based measures computed for different real and theoretical systems. For each subset the list is ordered from higher to lower information transfer.

Network type	N	$< k >$	$I(\mathbf{q})$	$H(\mathbf{q})$	$H_c(\mathbf{q}|\mathbf{q}')$	r
Technological networks						
Software 1	168	2.81	1.19	3.04	1.85	-0.39
Software 2	159	4.19	1.03	3.99	2.97	-0.41
Internet AS	3200	3.56	0.50	4.77	4.27	-0.22
Software 3	1993	5.00	0.30	4.82	4.51	-0.08
Circuit TV	320	3.17	0.23	1.37	1.14	0.010
Circuit EC05	899	4.14	0.15	2.98	2.82	-0.15
Software linux	5285	4.29	0.12	4.47	4.35	-0.06
Powergrid	4941	2.67	0.06	3.01	2.95	0.003
Biological networks						
Silwood park	154	4.75	0.94	4.09	3.14	-0.31
Ythan estuary	134	8.67	0.53	4.74	4.21	-0.24
p53 subnetwork	139	5.09	0.46	4.00	3.54	-0.24
Metabolic map	1173	4.84	0.39	3.58	3.19	-0.17
Neural net (C.elegans)	297	14.5	0.37	5.12	4.74	-0.16
Metabolic map	821	4.76	0.37	3.46	3.09	-0.18
Romanian syntax	5916	5.65	0.31	5.45	5.14	-0.18
Proteome map	1458	2.67	0.24	3.85	3.61	-0.21
Theoretical systems						
Star graph	17	1.88	1.00	1.00	0.00	-1.00
Barabási-Albert	3000	3.98	0.25	4.12	3.85	-0.078
Erdös-Renyi	300	6.82	0.06	3.31	3.25	-0.005
Modular E-R	500	10.3	0.04	3.67	3.62	-0.001

sion from this analysis is that network correlations in real graphs are small, being the diversity of pairs of linked nodes a direct consequence of the heterogeneous character of the degree distribution and nothing else. In spite that the cloud of points deviates from the straight line, these deviations might result from finite-size effects. Actually, if we plot information measures $I(\mathbf{q}; N)$ against system's size N, it can be shown that they follow a scaling $I(\mathbf{q}; N) \sim N^{-1}$.

Two points clearly deviate from the general pattern displayed by the majority of networks analysed here. Both are small systems and correspond to software graphs, and one of them is shown in Fig. 8b. As we can see this is a rather peculiar system, involving a large hub connected to a small module. It is thus a small structure dominated by the star graph component together with a homogeneous component. Such a nonuniform structure is likely to result from an process dealing with engineered, small-sized systems but unlikely to result from a natural process or from artificial evolution when some complexity thresholds are reached.

One possible explanation for the previous result is that correlations simply do not play any particular role in shaping network architecture[5]. However, it could be also argued that such a lack of correlation has been either chosen or selected for some underlying reason. But there's also another (more likely) scenario: that the observed structures are actually the only possible choices, at least when some complexity threshold is reached.

6 Simulated Annealing Search

The spread of real networks close to the zero-information boundary suggests that the possible structures allowed to occur (with a given heterogeneity and a given correlation) is rather constrained. This might be a consequence of the irrelevance of correlations for these systems but it would also be the case that some selective pressure is made towards heterogeneous networks with small correlations (i. e. no assortativeness).

In order to test the previous idea we can perform a Monte Carlo search in network space. Specifically, we explore the space of possible pairs entropy-noise available to candidate graphs Ω, i. e. $\Gamma = \{H(\mathbf{q}), H_c(\mathbf{q}|\mathbf{q}')\}$, which is constrained by two well-defined boundaries[6]:

$$\partial_1 \Gamma = \{(H(\Omega), H_c(\Omega)) \mid H_c(\Omega) = 0\} \tag{27}$$
$$\partial_2 \Gamma = \{(H(\Omega), H_c(\Omega)) \mid H(\Omega) = H_c(\Omega)\} \tag{28}$$

where $H(\Omega)$ and $H_c(\Omega)$ indicate the entropy and noise associated to a given graph Ω. It is not difficult to show that only two points occupy the lower boundary, i. e. $\partial_1 \Gamma = \{(0,0), (\log 2, 0)\}$. These correspond to purely homogeneous graphs and the star graph. The second boundary has already been studied.

For every random sample point (H, H_c), an optimizing searching process looks for candidate networks that minimize the error term or potential function $U(\Omega)$:

$$U(\Omega) = \sqrt{(H - H(\Omega))^2 + (H_c - H_c(\Omega))^2} \tag{29}$$

Here, we use the Boltzmann strategy presented in [36,37]. The algorithm explores the search space defined by all possible networks of N nodes. We assume that every possible state visited by the search process can be properly characterized by the scalar U_i. In the stationary limit (for a large number of searchers)

---

[5] This conclusion is reached under our specific, quasi-local definition of remaining degree. Other approaches, considering instead shortest paths among nodes might reveal important differences

[6] Strictly speaking, we are considering the entropies associated to the remaining degree distribution of a graph sampled from some graph ensemble by a stochastic process

define the occupation probability $p_i(t)$ of certain state i at time t. We require the optimization process to increase the occupation probability for the state of minimal potential. In general, many local minima exist and the search could be trapped in one of these states, which is undesirable.

A dynamics that finds the minimum is given by:

$$\frac{dp_i(t)}{dt} = \sum_{i \neq j} A_{ij} p_i(t) - A_{ji} p_j(t) \tag{30}$$

where

$$A_{ij} = A_{ij}^0 * \begin{cases} 1 & ; U_i < U_j \\ \exp\left(-(U_i - U_j)/T(t)\right) & ; U_i \geq U_j \end{cases} \tag{31}$$

is the transition probability for the searcher to move from state i to state j. The term A_{ij}^0 is 1 if and only if the state j can be reached by a little change or mutation and 0 otherwise. Here, the valid changes involve edge addition, edge removal and edge rewire, which are all equally selected with the same probability. The number of nodes of the network is always fixed. Transitions to lower energy states are always accepted but local minima is avoided because thermal fluctuations like in simulated annealing . As the search progresses, the temperature $T(t)$ is decreased following a power law rule:

$$T(t) = \frac{T_0}{1 + at} \tag{32}$$

where T_0 is the initial temperature (or starting degree of disorder) and a is the cooling rate. This allows the optimization process to perform a smooth transition from coarse to detailed search. The process starts from a random graph of N nodes with a given connectivity $< k >$ and lasts a given number of simulation steps.

By measuring the final error $\varepsilon(\Omega) = U(\Omega)$ for a large number of Monte Carlo samples it is possible to approximate the likelihood of a particular candidate network. Here we have used $\epsilon = 0.003$ and the optimization parameters are: $N = 500, < k >= 3, T_0 = 0.01, a = 0.002$ and 350000 steps. Our results indicate that the potential is effectively minimized only for a domain of pairs (H, H_c) along the second boundary $\partial_2 \Gamma$. In Fig. 9a the (smoothed) probability density $\mathcal{P}(H, H_c)$ of optimized networks is shown (for the upper part of the parameter space, $\mathcal{P}(H, H_c) = 0$). The distribution is peaked around a domain of Γ that fits very well the range of values satisfied by most real networks (compare with Fig. 8). There is also a clearly empty zone outside this domain, indicating that networks are difficult or simply impossible to find. An example of the optimized graphs is shown in Fig. 9b. This particular graph is scale-free, with an exponent $\gamma \approx 2.26$ and a cut-off at $\xi \sim 50$. The other networks in this domain are also SF, with an average scaling exponent $< \gamma > \sim 2.5$. By searching candidate networks that simultaneously fit the two requirements of given entropy and noise, the only possible solutions to be found are scale-free graphs with small levels of

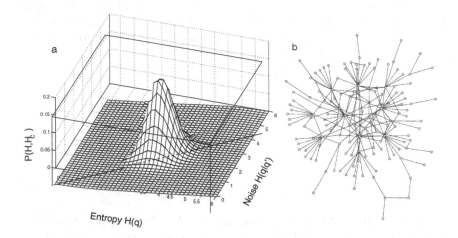

Fig. 9. (a) Noise-entropy probability plot obtained by exploring the Γ space using a Monte Carlo sampling. Different pairs of noise and entropy are generated and a simulated annealing search is performed looking for candidate networks. Here the smoothed probability distribution obtained from this algorithm is shown in (a). The highest density of observed networks appears to be close to the same domain observed for real networks. In (b) an example of a small sized network ($N = 142, < k >= 2.64$) is shown, together with its degree distribution (c). The cumulative degree distribution follows power law with exponent -1.26 (i. e. $\gamma = 2.26$). The graph has been obtained close to the boundary $H = H_c$ (with $H(\mathbf{q}) = 3.69, H(\mathbf{q}|\mathbf{q}') = 3.09$).

correlations. Interestingly, software networks deviate from this rule and are to be found along the upper region of the boundary ($H > 4$), where potential is not minimum. This might be a signature of frustrated optimization in software design processes [10].

7 Discussion

Complex networks display heterogeneous structures that result from different mechanisms of evolution [38]. Some are created through multiplicative processes (such as preferential attachment) while others seem to be well described in terms of optimization mechanisms [9]. Our study indicates that the possible universe of complex networks is actually rather constrained. Networks display scale-free architecture but also small assortativeness. The search algorithm, instead of assuming the presence of a given predefined mechanism of network growth, simply searches for candidate solutions to an optimization algorithm trying to approach simultaneously some amount of network heterogeneity and correlations. The result is that indeed networks are scale-free and involve low degree of correlations, but such situation is constrained to a well-defined domain. This domain is remarkably similar to the one inhabited by real graphs. Outside this domain, it is not feasible to find graphs simultaneously satisfying the two requirements.

The impact of SF architecture on biological and artificial networks is clearly different. Although the first can take advantage of the high homeostasis provided by scaling laws the second are completely dependent on the correct functioning of all units. Failure of a single diode in a circuit or of a single component in a software system leads to system collapse. Thus, homeostasis can not be a general explanation for scaling. We have conjectured that the leading force here is an optimization process where reliable communication at low cost shapes network architecture in first place [38]. The need of a sparse graph can be a consequence of different requirements. In an electronic circuit, saving wire is a strong constraint. In metabolic or genetic networks, it might be important in order to reduce the impact of unstable positive feedbacks. This can be satisfied by means of sparse graphs displaying scale-free architecture. What is the role of correlations? For the systems analysed here correlations don't seem to be of relevance to network performance. But what is more important: the lack of networks outside the densely populated domain is not due to some relevant, perhaps adaptive trait. It is actually a consequence of higher-level limitations imposed to network architecture.

Such a constrained set of possibilities fits very well the view of evolution as strongly dominated by intrinsic constraints [39-41] (see also [42] for a critical discussion). Under this view, the outcome of evolutionary searches would be not *any* possible architecture from the set of possible patterns but a choice from a narrow subset of attainable structures.

Acknowledgments

The authors thank the members of the Complex Systems Lab for useful discussions and to an anonymous referee for valuable comments. This work was supported by a grant BFM2001-2154, FET Open Project IST DELIS and by the Santa Fe Institute.

References

1. R. Albert and A.-L. Barabási. Statistical Mechanics of Complex Networks. Rev. Mod. Phys. **74**, 47-97 (2002a).
2. S. N. Dorogovtsev andJ. F. F. Mendes. Evolution of networks. *Adv. Phys.* **51**, 1079-1187 (2002).
3. S. N. Dorogovtsev andJ. F. F. Mendes. *Evolution of Networks: from biological nets to the Internet and WWW* Oxford U. Press, Oxford (2003).
4. S. Bornholdt and H. G. Schuster, eds. *Handbook of Graphs and Networks: From the Genome to the Internet.* Springer, Berlin (2002).
5. L. A. N. Amaral, A. Scala, M. Barthélemy and H. E. Stanley. Classes of behavior of small-world networks. Proc. Nat. Acad. Sci. USA **97**, 11149-11152 (2000).
6. R. Ferrer, C. Janssen and R. V. Solé. Topology of Technology Graphs: Small World Patterns in Electronic Circuits Physical Review E 64, 32767 (2001).
7. A.-L. Barabási and R. Albert. Emergence of scaling in random networks. Science **286**, 509-512 (1999).

8. S. N. Dorogovtsev andJ. F. F. Mendes. Accelerated growth of networks, in: *Handbook of Graphs and Networks: From the Genome to the Internet*, eds. S. Bornholdt and H.G. Schuster. pp. 320-343 Wiley-VCH, Berlin (2002).
9. R. Ferrer and R. V. Solé. Optimization in Complex Networks, Lect. Notes Phys. **625**, 114–125 (2003).
10. S. Valverde, R. Ferrer and R. V. Solé. Scale free networks from optimal design Europhys. Lett. **60**, 512-517 (2002).
11. R. V. Solé, R. Pastor-Satorras, R., Smith, E.D. and Kepler, T. A model of large-scale proteome evolution. *Adv. Complex Systems* **5**, 43-54 (2002).
12. A. Vazquez, A. Flammini, A. Maritan and A. Vespignani. Modeling of protein interaction networks. *Complexus*, **1**, 38-44 (2002).
13. G. Caldarelli, A. Capocci, P. De Los Rios and M. A. Muñoz. Scale-free networks from varying vertex intrinsic fitness. Phys. Rev. Lett. **89**, 258702 (2002).
14. F. Menczer. Growing and navigating the small world web by local content. Proc. Nat. Acad. Sci. USA **99**, 14014-14019 (2002).
15. E. Ravasz, A. L. Somera, D. A. Mongru, Z. N. Oltvai and A.-L. Barabási. Hierarchical Organization of Modularity in Metabolic Networks. Science 297, 1551-1555 (2002).
16. M. E. J. Newman. Assortative mixing in networks. Phys. Rev. Lett. **89**, 208701 (2002).
17. E. H. Stanley, S. V. Buldyrev, A. L. Goldberger, Z. D. Goldberger, S. Havlin, R. N. Mantegna, S. M. Ossadnik, C. K. Peng and M. Simon, Statistical mechanics in biology: how ubiquitous are long-range correlations? Physica A**205**, 214-253 (1996).
18. H. D. Abarbanel, R. Brown, J. L. Sidorowich and L. S. Tsimring. The analysis of observed chaotic data in physical systems. Rev. Mod. Phys. **65**, 1331-1392 (1993).
19. R. B. Ash. *Information Theory*, Dover, London (1965).
20. C. Adami. *Introduction to Artificial Life*. Springer, New York (1998).
21. W. Li. Mutual information versus correlation functions. J. Stat. Phys. **60**, 823-837 (1990).
22. W. Li. On the relationship between complexity and entropy for Markov chains and regular languages. Complex Syst. **5**, 381-399 (1991).
23. A. Fraser and H. Swinney. Independent coordinates for strange attractors from mutual information. Phys. Rev. A**33**, 1134-1140 (1986).
24. C. Germain-Renaud and J. P. Sansonnet. *Ordinateurs massivement paralleles*, Armand Colin, Paris (1991).
25. V. M. Milutinovic. *Computer Architecture*, North Holland, Elsevier (1988).
26. W. D. Hillis. *The Connection Machine*, MIT Press (Cambridge, MA, 1985).
27. S. Valverde, R. Ferrer and R. V. Solé, Scale-free networks from optimal design. Europhys. Lett. **60**, 512-517 (2002).
28. S. A. Kauffman. *Origins of Order*. Oxford U. Press, New York (1993).
29. R. M. May. *Stability and complexity in model ecosystems*. Princeton U. Press, New York (1973).
30. S. Amari. Characteristics of random nets of analog neuron-like elements. IEEE Trans. Man and Cybernetics **2**, 643-657 (1972).
31. J. M. Montoya and R. V. Solé. Topological properties of food webs: from real data to community assembly models Oikos **102**, 614-622 (2003).
32. R. Ferrer and R. V. Solé, Optimization in Complex Networks, Lect. Notes Phys. **625**, 114–125 (2003).
33. H. Jeong, S. Mason, A.-L. Barabási and Z. N. Oltvai. Lethality and centrality in protein networks Nature **411**, 41 (2001).

34. S. Valverde and R. V. Solé. Hierarchical small worlds in sotfware architecture. Santa Fe Institute Working Paper 03-07-044.
35. C. R. Myers. Software systems as complex networks: structure, function, and evolvability of software collaboration graphs, Phys. Rev. E **68**, 046116 (2003).
36. F. Schweitzer, W. Ebeling, H. Rose and O. Weiss. Network Optimization Using Evolutionary Strategies, in: *Parallel Problem Solving from Nature - PPSN IV*, (Eds. H.-M. Voigt, W. Ebeling, I. Rechenberg, H.-P. Schwefel), Lecture Notes in Computer Science, vol. 1141, Springer, Berlin (1996) pp. 940-949.
37. F. Schweitzer. *Brownian Agents and Active Particles*. Springer, Berlin (2002).
38. R. V. Solé, R. Ferrer-Cancho, J. M. Montoya and S. Valverde. Selection, tinkering and emergence in complex networks. Complexity **8**(1), 20-33 (2002).
39. F. Jacob. Evolution as tinkering. Science **196**, 1161-1166 (1976).
40. P. Alberch. The logic of monsters: evidence for internal constraint in development and evolution. Geobios **19**, 21-57 (1989).
41. B. C. Goodwin. *How the Leopard Changed Its Spots: the Evolution of Complexity*. Charles Scribner's Sons, New York (1994).
42. S. J. Gould. *The structure of evolutionary theory*. Belknap, Harvard (2003).

Part II

Network Dynamics

Extremal Properties of Random Structures

Eli Ben-Naim[1], Paul L. Krapivsky[2], and Sidney Redner[2]

[1] Theoretical Division and Center for Nonlinear Studies, Los Alamos National Laboratory, Los Alamos, NM 87545
[2] Center for Polymer Studies and Department of Physics, Boston University, Boston, MA 02215

Abstract. The extremal characteristics of random structures, including trees, graphs, and networks, are discussed. A statistical physics approach is employed in which extremal properties are obtained through suitably defined rate equations. A variety of unusual time dependences and system-size dependences for basic extremal properties are obtained.

1 Introduction

The goal of this article is to show that methods of non-equilibrium statistical physics are very useful for analyzing extreme properties of random structures. Extremes are compelling human curiosities — we are naturally drawn to compilations of various pinnacles of endeavor, such as lists of the most beautiful people, the richest people, the most-cited scientists, athletic records, *etc* [1]. More importantly, extremes often manifest themselves in catastrophes, such as the failure of space shuttles, the breaching of dams in flood conditions, or stock market crashes. The theory of extreme statistics [2–4] is a powerful tool for describing the extremes of a set of independent random variables; however, much less is known about extremes of correlated variables [5–7]. Such an understanding is crucial, since complex systems are composed of many subsystems that are highly correlated.

While estimates for the failure probability of a nuclear plant or a space shuttle still involve guesswork, understanding the extremes of certain correlated random variables is a hard science. Below we demonstrate this thesis for various extremal characteristics of geometrical features in basic evolving structures, such as randomly growing trees, graphs, and networks. In each case, the growth process of the structure induces correlations in the variables whose extremes are the focus of this review. We shall illustrate how the statistical physics of classical irreversible processes can be naturally adapted to elucidate both typical and extremal statistics.

We obtain new scaling laws for extreme properties and consequently give new insights for a variety of applications. For example, random trees arise naturally in data storage algorithms [8–10], an important branch of computer science, and the maximal branch height yields the worst-case performance of data retrieval algorithms. Random trees also describe various non-equilibrium processes, such as irreversible aggregation [11,12] and collisions in gases [13]. Random graphs

E. Ben-Naim, P.L. Krapivsky, and S. Redner, Extremal Properties of Random Structures, Lect. Notes Phys. **650**, 211–233 (2004)
http://www.springerlink.com/

[14,15] have numerous applications to computer science and to physical processes such as polymerization [16]. Random growing networks are used to model the distributions of biological genera, word frequencies, and income [17,18], the structure of the Internet [19], the World-Wide Web [20], and social networks [21, 22].

As a subtext to this review, it is worth mentioning that problems at the interface of statistical physics and computer science have been fruitful and symbiotic. Algorithms and methods developed in one area have found application in the other field; important examples include the Monte Carlo method, simulated annealing, and the Dijkstra algorithm. Statistical physics concepts such as criticality, scaling, universality, and techniques such as replicas have proved useful in diverse interdisciplinary applications such as algorithmic complexity, combinatorial optimization, error correction, compression algorithms, and image restoration; a review of these topics can be found in [23–28].

We will focus on three ubiquitous random structures — trees, graphs, and networks. Random trees (Sect. 2) can be viewed as the space-time diagram of irreversible aggregation with a size-independent merging rate. This connection allows to apply well-known results in aggregation to elucidate the growth of the largest component (the leader) and the number of changes in its identity. The number of lead changes grows quadratically with logarithm of the system size. The time-dependent number of lead changes becomes asymptotically self-similar, following a scaling form in which the scaling variable involves a logarithmic, rather than an algebraic ratio, between the typical size and the system size. Qualitatively similar properties also characterize the smallest component in the system.

Another characteristic of random trees is their height. The corresponding branch height distribution is Poissonian, reflecting the random nature of the merger process that underlies tree growth. The growth of the tree height (the maximal branch height) has an interesting relation to traveling wave propagation. The velocity of this wave yields typical and extremal height statistics as a corollary.

Random graphs (Sect. 3) are also equivalent to an aggregation process in which the merging rate of two components is proportional to the product of their sizes. This system undergoes a gelation transition in which a giant component, that contains a finite fraction of the entire mass in the system, arises. Near this transition, the size distribution of graph components follows a self-similar behavior. Despite the differences with the size distribution of random trees, leadership statistics in these two systems are remarkably robust.

Random networks (Sect. 4) can be grown by adding nodes and attaching the new node to a pre-existing node with a rate that depends on the degree of the target node. A hallmark of such systems is the statement that "the rich get richer"; that is, the more popular nodes tend to remain so. This adage is in keeping with human experience — a person who is rich is likely to stay rich, as evidenced by the continued appearance of the same individuals on lists of wealthiest individuals [29]. We examine whether the rich really get richer in growing networks by studying properties of the nodes with the largest degree. In keeping with analysis of random trees and graphs, we focus on the identity of the most popular node as a function of time, the expected degree of this most popular node, and the number of lead changes in the most popular node as a function of time.

2 Random Trees

Random trees underlie physical processes such as coagulation, collisions in gases [13], and fragmentation [30,31]. They are also important in computer science algorithms such as data storage and retrieval [8,9,32–35]. Different extremal characteristics may be important in different contexts. In aggregation processes, the maximal aggregate size is of interest. In other cases, the maximal or the minimal branch height are of interest. In Lorentz gases, the maximum branch height is related to the largest Lyapunov exponent, while in data storage, extremal heights yield best-case and worst-case algorithm performances.

Consider a forest of random trees that is generated randomly as follows (Fig. 1). Starting with N single-branch trees, two trees are picked randomly and merged. This process is repeated until a single tree containing all N branches is generated. We treat the merger process dynamically. Let s be the number of trees. The transition $s \to s-1$ occurs with rate r_s proportional to the total number of pairs. Choosing $2/N$ as the merger rate for each pair (i.e., $r_s = s(s-1)/N$) is convenient as in the thermodynamic limit $N \to \infty$, the normalized density $c = \langle s \rangle / N$ evolves according to $\frac{d}{dt}c = -c^2$. Given the initial condition $c(0) = 1$, the density is

$$c(t) = \frac{1}{1+t}. \tag{1}$$

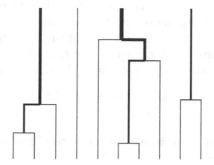

Fig. 1. Random trees. A forest of random growing trees is equivalent to the space-time evolution of irreversible aggregation with a size-independent merging rate. Each branch corresponds to the world line of a cluster. The thickness of each branch is proportional to the size of the cluster. The sizes of the 4 trees are (left to right) 3, 1, 4, 2 and their heights are 2, 0, 3, 1.

The number of trees is therefore $\mathcal{N} = \langle s \rangle = N(1+t)^{-1}$. Moreover, conservation of the total number of branches yields the average tree size $m = 1 + t$. The results are stated in terms of the physical time t, but can be easily re-expressed in terms of the intrinsic quantities $\mathcal{N}$ or m.

2.1 Size Statistics

Let $n_k(t)$ be the number of trees with k branches at time t. The normalized density $c_k(t) = n_k(t)/N$ evolves according to the Smoluchowski rate equation [11,12,36]

$$\frac{dc_k}{dt} = \sum_{i+j=k} c_i c_j - 2cc_k \tag{2}$$

with the monodisperse initial conditions $c_k(0) = \delta_{k,1}$. This evolution equation reflects the fact that trees merge randomly, independent of their size. The well-known solution to this equation is

$$c_k(t) = \frac{t^{k-1}}{(1+t)^{k+1}}. \tag{3}$$

Taking the long time limit $t \to \infty$ while keeping the variable k/t fixed, the size distribution approaches the asymptotic form $c_k(t) \to t^{-2} e^{-k/t}$. More generally, this can be recast as the scaling form

$$c_k(t) \simeq k_*^{-2} \Phi(k/k_*), \tag{4}$$

with the scaling function $\Phi(z) = e^{-z}$ and the typical tree size $k_* \simeq t$.

2.2 The Leader

Extremal characteristics, such as the size of the largest tree — the leader — and the number of lead changes, follow directly from the size distribution. We focus on the asymptotic time regime[3], where most of the lead changes occur, and use the scaled size distribution (4). Let $l(t, N)$ be the average size of the leader at time t. The basic criterion used to determine the size of the leader is

$$U_l(t) \equiv \sum_{j \geq l} n_j \simeq Nt^{-1} e^{-l/t} = 1. \tag{5}$$

This simply states that there is one cluster whose size exceeds $l(t, N)$. Solving for the leader size gives

$$l(t, N) \simeq t \ln \frac{N}{t}. \tag{6}$$

---

[3] The behavior in the early time regime, $t \ll 1$, can be obtained by using the exact time dependence (3).

This expression holds in the asymptotic time regime $t \gg 1$. For short times the leader size grows logarithmically with system size $l(t \approx 1) \sim \ln N$. Finally, at times of the order N, the leader becomes of the order of the system size. The final leader, that is, the ultimate winner, emerges on a time scale of the order N. This is consistent with the fact that the average "final" time for a single tree to remain in the system t_f, is given by $t_f = N - 1$ as follows from $\mathcal{N} = 1$.

We now consider the quantity $L(t, N)$, defined as the average number of lead changes during the time interval $(0, t)$. Lead changes occur when two trees (neither of which is the leader) merge and overtake the leader. The flux of probability to surpass the leader is simply the rate of change of the cumulative distribution. Thus $\frac{d}{dt} L(t, N) = \frac{\partial}{\partial t} U_k \big|_{k=l}$. Using $U_l = 1$ yields $\frac{d}{dt} L(t, N) \simeq lt^{-2} \simeq t^{-1} \ln \frac{N}{t}$. Therefore, the time-dependent number of lead changes is [37]

$$L(t, N) \simeq \ln t \ln N - \frac{1}{2}(\ln t)^2, \tag{7}$$

which can be recast in the self-similar form

$$L(t, N) \simeq (\ln N)^2 F(x), \qquad x = \frac{\ln t}{\ln N}, \tag{8}$$

with the quadratic scaling function $F(x) = x - \frac{1}{2}x^2$. Notice the unusual scaling variable — a ratio of logarithms — in contrast to the ordinary scaling variable $z = k/k_*$ underlying the size distribution (4). The scaling variable still involves the typical size, $x = \ln k_*/\ln N$. Note also that the leader size (6) can be expressed in terms of same scaling variable $l(t, N) \simeq t \ln N f(x)$ with $f(x) = \frac{d}{dx} F(x) = 1 - x$. Numerical simulations confirm this scaling behavior [37]. However, the convergence to these asymptotics is slow due to the logarithmic functional dependences on the system size and time.

The total number of lead changes $L(N)$ as a function of system size N follows from the time dependent behavior (7). The eventual winner emerges at time of order N. Using $L(N) \cong L(t \propto N, N)$ we obtain

$$L(N) \simeq A(\ln N)^2 \tag{9}$$

with $A = F(1) = 1/2$ (see Fig. 2). The correction to this leading asymptotic behavior is of the order $\ln N$. The logarithmic dependence implies that lead changes are relatively infrequent.

Both the size of the leader and the number of lead changes grow logarithmically in the early time regime, $l(t \approx 1, N) \propto L(t \approx 1, N) \propto \ln N$. The first relation implies that initially the leader size predominantly grows in increments of one and every leader is a new leader. When $t \gg 1$, the size of the leader greatly exceeds the number of lead changes as the increments of the leader size grow roughly linearly with time.

The distribution of the number of lead changes $P_n(t, N)$, i.e., the probability that n lead changes occur by time t, can be determined by noting that lead changes occur by a random process in which the average flux of probability to surpass the leader is $\frac{d}{dt} L$. Hence, the probability distribution obeys

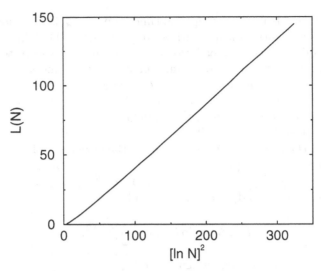

Fig. 2. The total number of lead changes $L(N)$ versus the system size N. The simulation data represents an average over 10^4 independent realizations of the random tree generation process with N up to 10^8.

$\frac{d}{dt}P_n = (\frac{d}{dt}L)\,[P_{n-1} - P_n]$ with the initial condition $P_n(0, N) = \delta_{n,0}$. Therefore, the distribution of the number of lead changes is Poissonian and it is characterized solely by the average number of lead changes

$$P_n(t, N) = \frac{[L(t, N)]^n}{n!}\, e^{-L(t,N)}. \tag{10}$$

As a result, the ultimate number of lead changes is also Poissonian distributed, $P_n(N) = \frac{L^n}{n!}\,e^{-L}$, with $L \equiv L(N)$ given by (9). Asymptotically, the Poissonian distribution approaches a Gaussian in the proximity of the peak:

$$P_n(N) \simeq \frac{1}{\sqrt{2\pi L}} \exp\left[-\frac{(n-L)^2}{2L}\right]. \tag{11}$$

The number of lead changes is a self-averaging quantity; however, the system size should be huge to ensure that relative fluctuations $\frac{\delta n}{n} \sim \frac{\sqrt{L}}{L} \sim (\ln N)^{-1}$ are small. Hence in a given realization for a system of size $N = 10^8$ (the maximum size in our simulations), lead changes are still relatively erratic.

Another interesting quantity is $S(N)$, the probability that no lead change ever occurs. This is obviously the "survival" probability that the first leader, whose size is initially $k = 2$, never relinquishes the lead. This survival probability is given by $S(N) \equiv P_0(N) = \exp(-L)$, so it decays faster than a power-law but slower than a stretched exponential (Fig. 3)

$$S(N) \simeq \exp\left[-A(\ln N)^2\right]. \tag{12}$$

The above formalism extends to the statistics of the rth-largest tree. Using $U_l = r$, the average size of the rth-largest tree grows according to $l_r \simeq t \ln \frac{N}{rt}$.

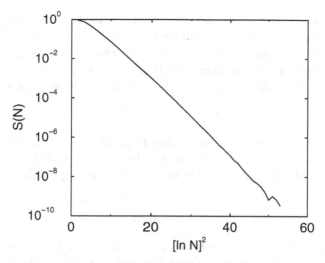

Fig. 3. The survival probability of the first leader $S(N)$ versus the system size N obtained from an average over 10^{10} independent realizations. The slope $A = 1/2$ is in accordance with (12).

Moreover, the total number of changes in the group of r-largest trees grows linearly with r according to $\frac{r}{2}(\ln N)^2$.

Among several open problems we mention just two: What is the size of the winner (the last emerging leader)? At what time does the winner emerge? The averages of both these random quantities grow linearly with N, but we do not know the proportionality factors. The computation of these factors, and the determination of the distribution of these random quantities, are interesting open problems.

2.3 The Laggard

At the opposite end of the size spectrum sits the laggard, the smallest component in the system. Unlike the leader, the laggard does not change its size for a relatively long period. The total number of monomers is $n_1 = N(1+t)^{-2}$. Thus, monomers are depleted from the system only when the time becomes of the order of $N^{1/2}$. Until this time, the laggard size remains unity. To investigate laggard statistics in the interesting regime $N^{1/2} \ll t \ll N$ we employ the same approach as for the leader. First, we estimate the cumulative distribution $u_k = \sum_{j=1}^{k} n_j$ and find $u_k \simeq t^{-1}(1 - e^{-k/t})$. Then we use the criterion $u_\ell = 1$ and get the average laggard size

$$\ell(t, N) \simeq -t \ln \left(1 - \frac{t}{N}\right). \tag{13}$$

In the time regime $N^{1/2} \ll t \ll N$, the above expression simplifies to $\ell(t, N) \simeq t^2/N$. As in the leader case, the laggard size is proportional to the typical size, but modified by a logarithmic correction.

The number of changes in the identity of the laggard, $\mathcal{L}(t, N)$, is given by $\frac{d}{dt}\mathcal{L}(t, N) = -\frac{\partial}{\partial t}u_k\big|_{k=l}$. Using asymptotics for u and ℓ, we simplify the right-hand side and obtain $\frac{d}{dt}\mathcal{L} = t^{-1} - Nt^{-2}(1 - t/N)\ln(1 - t/N) \simeq 2t^{-1}$ for $t \ll N$. Integrating over time and recalling that the first laggard change occurs at time of the order $N^{1/2}$ we obtain $\mathcal{L}(t, N) \simeq \int_{N^{1/2}}^{t} ds\, 2s^{-1}$. Consequently,

$$\mathcal{L}(t, N) \simeq 2\ln t - \ln N. \tag{14}$$

This behavior can be recast in the scaling form $\mathcal{L}(t, N) \simeq (\ln N)F(x)$ with the same scaling variable as in the leader problem, $x = \ln t/\ln N$, and the linear scaling function $F(x) = 2x - 1$. The total number of laggard changes saturates at

$$\mathcal{L}(N) \simeq \ln N. \tag{15}$$

Numerical simulations confirm this behavior. Thus, the total number of laggard changes is much smaller compared with the leader. This behavior is intuitive: it is more difficult to catch up with the rest of the pack than it is to remain ahead of the pack.

The distribution of the number of laggard changes is also Poissonian, as in (10). Moreover, the survival probability still decays exponentially with the total number of changes $\mathcal{S}(N) = \exp[-\mathcal{L}(N)]$. However, the growth of the average is only logarithmic in this case, so the survival probability decays as a power law

$$\mathcal{S}(N) \sim N^{-1}, \tag{16}$$

i.e., much slower than in the leader case. This can be understood by considering the probability that the laggard remains a monomer until the very last merger event between the final two subtrees. Interestingly, the size distribution of these final two trees is uniform as can be seen immediately by considering the time-reversed merger process. The probability that the laggard in the last merging event is a monomer is simply $2/(N - 1)$. This lower bound for the survival probability is indeed consistent with (16). An interesting open question is the size distribution of the loser (the final laggard).

2.4 Height Statistics

The height (or depth) of a tree branch provides another fundamental size characterization. It is defined as the number of different-width line segments between a branch and the tree root (see Fig. 1). Thus, different heights correspond to different branches in the tree. It is therefore natural to ask: What is the typical branch height? What is the typical tree height (the maximal branch height)? What is the maximal tree height?

First, consider the distribution of branch heights. Each time two branches merge, the distance to the root increases by one (the branch height can also be viewed as the generation number). Let $h(t)$ be the average number of merger events experienced by a given branch up to time t. The rate of growth of the average height is proportional to the number density of trees and since the merger rate equals 2, we have $\frac{d}{dt}h(t) = 2c(t)$, with $h(0) = 0$. Therefore, the average branch height is $h(t) = 2\ln(1+t)$, or, in terms of the average tree size m,

$$h = 2\ln m. \tag{17}$$

Thus, the branch height grows logarithmically with its size. Because the merger process is random, the probability $P_n(t)$ that the branch height equals n is Poissonian

$$P_n(t) = \frac{[h(t)]^n}{n!}e^{-h(t)}, \tag{18}$$

with $h(t)$ the average height.

The height of a tree is defined as the maximal branch height. For example, the (left-to-right) trees in Fig. 1 have heights of 2, 0, 3, and 1, respectively. Based on the branch height behavior, we anticipate that the tree height grows logarithmically, $H_{\max} \simeq v_{\max}\ln m$. Similar to the calculation of the maximal size from the cumulative distribution, the tree height can be obtained heuristically from the properly normalized branch height distribution $c^{-1}P_n$ via $\sum_{n \geq H_{\max}} c^{-1}P_n = 1$. Estimating the tails of the Poisson distribution (18) by using the Stirling formula leads to the transcendental equation [38]

$$v\ln\frac{2e}{v} = 1. \tag{19}$$

The larger root of this equation yields the growth of the tree height

$$H_{\max} \simeq v_{\max}\ln m, \qquad v_{\max} \cong 4.31107. \tag{20}$$

This value was obtained in different contexts, including fragmentation processes [30,31], and collision processes in gases, where this value is related to the largest Lyapunov exponent [13].

Each tree carries a height k. The result of a merger between trees with heights i and j is a tree with height $\max(i,j) + 1$. The number density of trees with height k, $H_k(t)$, evolves according to the master equation (the initial conditions are $H_k(0) = \delta_{k,0}$)

$$\frac{dH_k}{dt} = H_{k-1}^2 - 2c\,H_k + 2H_{k-1}\sum_{j=0}^{k-2} H_j. \tag{21}$$

The rate equations (21) are more complicated than the *recursive* Smoluchowski equations (2) for the tree size distribution. Fortunately, one can extract analytically almost all relevant information without explicitly solving (21). Figure 4 shows that the normalized distribution $c^{-1}H_k$ approaches a traveling wave in the large time limit. This suggests seeking an asymptotic solution of the traveling wave; this construction therefore greatly simplifies our analysis. The traveling

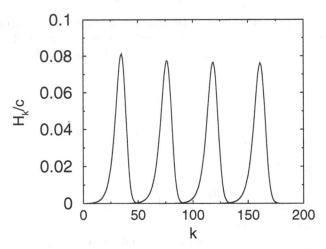

Fig. 4. The traveling wave. Shown is the normalized distribution $c^{-1}H_k$ vs. k at different times $\tau = 10, 20, 30, 40$ obtained from numerical integration of the rate equation (22).

wave form has significant qualitative implications for the tree height statistics, *e.g.*, fluctuations with respect to the mean saturate to some fixed value.

The equations simplify using the cumulative fractions $A_k = c^{-1}\sum_{j=0}^{k} H_j$ and the time variable $\tau = \int_0^t dt'\, c(t') = \ln(1 + t)$. With these transformations, (21) become

$$\frac{dA_k}{d\tau} = A_{k-1}^2 - A_k \tag{22}$$

with the initial conditions $A_k(0) = \delta_{k,0}$. Substituting the traveling wave solution, $A_k(\tau) \to A(k - v\tau)$, into (22) we find that $A(x)$ satisfies the nonlinear difference-differential equation

$$v A'(x) = A(x) - A^2(x - 1) \tag{23}$$

with the boundary conditions $A(-\infty) = 0$ and $A(\infty) = 1$. This nonlinear and nonlocal equation appears insoluble; however, important physical features can now be established analytically. For example, both extreme tails of $A(x)$ are exponential:

$$A(x) \sim \begin{cases} e^{x/v} & x \to -\infty; \\ 1 - e^{-\lambda x} & x \to \infty. \end{cases} \tag{24}$$

Consequently, the distribution of both very large and very small (compared with the typical) heights are exponential. The propagation velocity of the wave, which characterizes the typical behavior, follows from the large-k tail. Substituting $1 - A(x) \sim e^{-\lambda x}$ into (23) gives a dispersion relation, *i.e.*, a relation between the velocity v and the decay constant λ:

$$v = \frac{2e^\lambda - 1}{\lambda}. \tag{25}$$

Out of the spectrum of possible v only one value, the maximal possible velocity, is selected[4]. From (25) we find $v_{\max} \cong 4.31107$, corresponding to $\lambda \cong 0.768039$. This velocity satisfies (19) and is identical to the one obtained heuristically (20). Numerical integration shows that a traveling wave is indeed approached (Fig. 4) and the predicted propagation velocity is confirmed to within 0.1%. The choice of the extremal velocity is the fundamental selection principle that applies to classical reaction-diffusion equations [39–43] and to numerous difference-differential equations [44].

The traveling wave form of the height distribution implies that the height — the elemental random variable — is highly concentrated near the average; more precisely, each moment $\langle (H_k - \langle H_k \rangle)^n \rangle$ is finite. Thus, accurate determination of the average is especially important. We already know that $\langle H_k \rangle \cong v\tau$; a more sophisticated traveling wave technique yields the leading (logarithmic) correction: $\langle H_k \rangle \cong v\tau - \frac{3}{2\lambda} \ln \tau$ [38].

Similar analysis can also be performed for the minimal branch height [38]. The resulting velocity $v_{\min} \cong 0.373365$ is the smaller root of the transcendental equation (19).

2.5 The Tallest and the Shortest

The tallest tree is defined as the one with largest height and similarly for the shortest tree. The tallest and the shortest are merely the height leader and laggard, respectively. The number of changes in the identity of these extremal trees throughout the evolution process follows from the tails of the height distribution.

Consider the height distribution $H_k(t)$ and the corresponding cumulative distribution $U_k = N \sum_{j \geq k} H_j(t)$. Both of these distributions have exponential tails[5], $U_k(t) \sim N H_k(t) \sim N t^{\lambda v - 1} \exp(-\lambda k)$, as follows from the large-x tail of the traveling wave (24). The criterion $U_l = 1$ yields the average height of the tallest tree

$$l(t, N) \simeq \lambda^{-1} \ln[N t^{\lambda v - 1}]. \tag{26}$$

Indeed, the height of the tallest tree saturates at a time scale of the order N consistent with the saturation value $l_{\text{final}}(N) \simeq v \ln N$. This is also an upper bound for the total number of lead changes since the height of the tallest tree grows by increments of unity. Similar to the leader, $L(t \approx 1) \sim l(t \approx 1) \sim \ln N$. However, at later times the rate of change is slower, $\frac{d}{dt} L(t) \sim t^{-1}$, as follows from the flux criterion $\frac{d}{dt} L = \frac{\partial}{\partial t} U_k \big|_{k=l}$. The overall number of changes now grows slower than in the leader case $L(N) \simeq \varphi \ln N$ with $\varphi \leq v_{\max}$ and consequently, the survival probability of the first tallest tree decays algebraically

---

[4] This actually happens for a wide class of initial conditions including all that vanish for sufficiently small k.

[5] The proportionality factor is tacitly ignored as it is irrelevant asymptotically. The determination of its value requires a nonlinear analysis of the traveling wave.

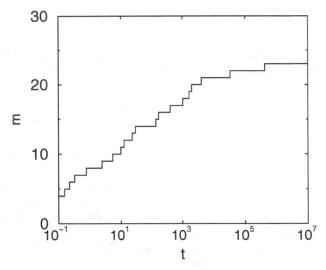

Fig. 5. Number of times m the tallest tree changes versus time t in a single realization with $N = 10^7$.

$$S(N) \sim N^{-\varphi} \qquad (27)$$

with an apparently non-trivial exponent φ. Determination this constant is challenging since the number of lead changes in the early and the late time regimes are comparable. Nevertheless, this heuristic approach successfully yields extremal statistics of an extremal tree characteristic, namely, the maximal branch height. The irregular nature of the lead changing process is manifest when a single realization is considered (Fig. 5).

Extremal statistics of the shortest tree follow from the cumulative distribution $u_k = N \sum_{j \leq k} H_j \sim Nt^{-2} \exp(k/v)$ and the criteria $u_\ell = 1$ and $\frac{d}{dt}\mathcal{L} = -\frac{\partial}{\partial t} u_k \big|_{k=\ell}$. The size of the smallest tree thus grows according to

$$\ell(t, N) \simeq v \ln \frac{t^2}{N} \qquad (28)$$

for times $t \gg N^{1/2}$ (at earlier times the shortest tree is a monomer). Even though the shortest tree has a different growth law than the laggard (13), the time dependent number of changes grows according to (14). Thus the total number of changes $\mathcal{L}(N) \simeq \ln N$ and the survival probability $\mathcal{S}(N) \sim N^{-1}$ are as in the laggard case.

We conclude that leadership statistics generally exhibit logarithmic dependences on the system size. However, they are not universal. Different behaviors may characterize leaders and laggards and the behavior may depend on the type of geometric feature, $i.e.$, size or height. We have observed both linear and quadratic growth with $\ln N$. A third possibility, saturation at a finite value, is found for random networks, as will be shown below.

3 Random Graphs

Random graphs are fundamental in theoretical computer science [9,14,10,15]. They have been used to model social networks [21,22], and physical processes such as percolation [45] and polymerization [16]. We discuss size statistics only. The size distribution is derived and then used to obtain leader statistics.

3.1 The Size Distribution

A random graph is grown from an initially disconnected graph with N nodes. Two nodes are then selected at random and are connected (Fig. 6). This process occurs at a constant rate, that we set equal to unity without loss of generality. This linking is repeated indefinitely until all N nodes form a single connected component.

Let n_k be the number of components of size k. The normalized density $c_k = n_k/N$ evolves according to the Smoluchowski equation

$$\frac{dc_k}{dt} = \frac{1}{2} \sum_{i+j=k} ij c_i c_j - k\, c_k. \tag{29}$$

The initial conditions are $c_k(0) = \delta_{k,1}$. In writing (29), the conservation law $\sum_k k c_k = 1$ is employed. Equations (29) reflect that components are linked with a rate proportional to the product of their sizes.

The generating function $F(z,t) = \sum_k k c_k(t) e^{kz}$, evolves according to $\frac{\partial F}{\partial t} = (F-1)\frac{\partial F}{\partial z}$ with the initial condition $F(z,0) = e^z$. Writing the derivatives through Jacobians, $\frac{\partial F}{\partial t} = \frac{\partial(F,z)}{\partial(t,z)}$ and $\frac{\partial F}{\partial z} = \frac{\partial(F,t)}{\partial(z,t)}$, and using the relation $\frac{\partial z}{\partial t} = \frac{\partial(z,F)}{\partial(t,F)}$, the nonlinear equation for $F(z,t)$ is recast into the linear equation $\frac{\partial z}{\partial t} = 1 - F$, from which we get[6] $z(t) = (1 - F)t + \ln F$. Exponentiating this equality gives an implicit relation for the generating functions

$$F(z,t)\, e^{-tF(z,t)} = e^{z-t}. \tag{30}$$

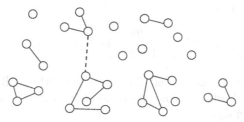

Fig. 6. A random graph. The dashed line indicates a newly-added link that joins two randomly-selected nodes. The probability of joining together two components that contain i and j nodes is proportional to ij.

---

[6] The integration constant $\ln F$ follows from the initial condition $F(z,0) = e^z$.

The Lagrange inversion formula[7] conveniently yields the size distribution [47,48]

$$c_k(t) = \frac{(kt)^{k-1}}{k \cdot k!} e^{-kt}. \tag{31}$$

The system undergoes a gelation transition at time $t_g = 1$. At this point a giant component arises that eventually engulfs the entire mass in the system. Close to the gelation time, the size distribution attains the scaling behavior

$$c_k(t) \simeq k_*^{-5/2} \Phi(k/k_*), \tag{32}$$

with the scaling function $\Phi(z) = \frac{1}{\sqrt{2\pi}} z^{-5/2} e^{-z/2}$. The typical size diverges, $k_* \simeq (1-t)^{-2}$, as $t \to t_g$. Beyond the gelation point, there exists an infinite sequence of transitions at times $t_k \simeq k^{-1} \ln N$ beyond which components of size k disappear. At the last such transition time t_1, the system consists of the giant component and a few surviving monomers. The smallest component is always a monomer and the laggard problem is trivial.

3.2 The Leader

The size of the giant component (the last emerging leader) follows from the size distribution. Exactly at the gelation time, the large-size tail of the size distribution is algebraic, $c_k(t=1) \sim k^{-5/2}$, so that the cumulative distribution is $u_k \sim N k^{-3/2}$. The criterion $u_{l_w} \sim 1$ gives the average size of the giant component $l_w \sim N^{2/3}$ [14] and the time at which it emerges is $1 - t_w \sim N^{-1/3}$.

Consider the size of the leader, $l(t, N)$, and the number of lead changes $L(t, N)$. At early times ($t \ll 1$), the behavior is the same as for random trees: the size of the leader $l(t, N)$, the number of lead changes $L(t, N)$, as well as the number of distinct leaders are all of the order $\ln N$. The asymptotic time regime in this case is $t \to 1$, as suggested by the size distribution. The tail of the size distribution together with $u_l = 1$ yield an implicit relation for the size of the leader, $l \simeq 2(1-t)^{-2} \ln N - 3(1-t)^{-2} \ln l$. Substituting the zeroth order approximation $l^{(0)} = 2(1-t)^{-2} \ln N$ into $\ln l$ and ignoring subdominant $\ln \ln N$ terms gives the leader size

$$l(t, N) \simeq \frac{2}{(1-t)^2} \ln[N(1-t)^3]. \tag{33}$$

At early stages ($t \ll 1$) the leader size grows logarithmically with the system size. Moreover, the leader size is proportional to the typical size but with a logarithmic enhancement.

The rate of leadership change is estimated as in the random tree case and we find $\frac{d}{dt} L(t, N) = (1-t) l(t, N)$, so that the time dependence of the number of lead changes is

---

[7] The series $v = \sum_{n \geq 1} \frac{n^{n-1}}{n!} u^n$ is a solution of the equation $v e^{-v} = u$ [46].

$$L(t, N) \simeq 2 \ln N \ln \frac{1}{1-t} - 3 \left[\ln \frac{1}{1-t} \right]^2. \qquad (34)$$

It follows that the scaling form is

$$L(t, N) \simeq (\ln N)^2 \, F(x) \qquad x = \frac{\ln \frac{1}{1-t}}{\ln N}, \qquad (35)$$

with the scaling function $F(x) = 2x - 3x^2$. This scaling function is related to the leader size: $l(t, N) \simeq k_* \ln N f(x)$, with $f(x) = \frac{d}{dx} F(x) = 2 - 6x$. The scaling behavior is obeyed until the giant component emerges, $i.e.$, up to a time t_w, with $1 - t_w \sim N^{-1/3}$. We neglected extremely slowly growing terms that are of the order $\ln \ln N / \ln N$ to obtain the scaling behavior. Thus, the approach to the scaling behavior may be very slow.

The total number of lead changes, $L(N) \simeq \frac{1}{3} (\ln N)^2$, is similar to the random tree case[8]. Furthermore, the distribution of lead changes is Poissonian, as in (10), and the survival probability decays according to (12).

Random trees and random graphs show very different size characteristics. Gelation occurs in one case but not in the other. Nevertheless, leadership statistics in these two systems are remarkably similar. In both cases, the total number of lead changes grows as $L(N) \sim (\ln N)^2$. Moreover, the seemingly different scaling variables underlying (8) and (35) can be both related to the typical size $x = \ln k_* / \ln N$.

4 Random Networks

In the case of sequentially growing networks, the basic quantity is the degree distribution N_k, defined as the number of nodes of degree k when the network contains N total nodes. In this section, we investigate extremal properties of the degree distribution. We are again interested in the leader, namely, the node with the highest degree and its associated statistical properties.

4.1 Identity of the Leader

We characterize the Jth node that enters the network as having an index J (Fig. 7). To start with an unambiguous leader node, we initialize the system to have $N = 3$ nodes, with the initial leader having degree 2 (and index 1) and the other two nodes having degree 1. A new leader arises when its degree exceeds that of the current leader.

For a constant attachment rate ($A_k = 1$), the average index of the leader grows algebraically, $J_{\text{lead}}(N) \sim N^\psi$, with $\psi \approx 0.41$. The leader is typically an early node (since $\psi < 1$), but not necessarily one of the very earliest. For example, a node with index greater than 100 has a probability of approximately

---

[8] The relation $1 - t_w \sim N^{-1/3}$ shows that $x \leq 1/3$, and the prefactor is obtained from the scaling function: $A = F(1/3) = 1/3$.

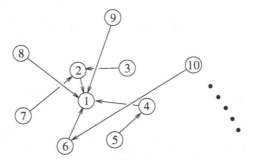

Fig. 7. A random network. The network is grown by adding links sequentially. A new node has a single outgoing link that joins to an earlier node of degree k with an attachment rate A_k. Each node is labeled by its index J.

10^{-2} of being the leader in a graph of $N = 10^5$ nodes. Thus the order of node creation plays a significant but not deterministic role in the identity of the leader node for constant attachment rate — there is partial egalitarianism.

We can understand this behavior analytically from the joint index-degree distribution. Let $C_k(J, N)$ be the average number of nodes of index J and degree k. For constant attachment rate, this joint distribution obeys the rate equation

$$\frac{\partial C_k}{\partial N} = \frac{\partial C_k}{\partial J} + \frac{C_{k-1} - C_k}{N} + \delta_{k1}\delta(N - J). \tag{36}$$

This is a slight generalization of the rate equation for the degree distribution itself [49]. The new feature is the first term on the right that accounts for node "aging".

The homogeneous form of this equation implies a self-similar solution. Thus, we seek a solution as a function of the single variable J/N rather than two separate variables [49]

$$C_k(J, N) = f_k(x) \qquad \text{with} \quad x = \frac{J}{N}. \tag{37}$$

This turns (36) into the ordinary differential equation

$$-x \frac{df_k}{dx} = f_{k-1} - f_k. \tag{38}$$

We have omitted the delta function term, since it merely provides the boundary condition $c_k(J = N, N) = \delta_{k,1}$, or $f_k(1) = \delta_{k,1}$. The solution is simply the Poisson distribution in the variable $\ln x$, *i.e.*,

$$C_k(J, N) = \frac{J}{N} \frac{|\ln(J/N)|^{k-1}}{(k-1)!}, \tag{39}$$

from which the average index of a node of degree k is

$$J_k(N) = \frac{\sum_{1 \le J \le N} J\, C_k(J, N)}{\sum_{1 \le J \le N} C_k(J, N)} = N \left(\frac{2}{3}\right)^k. \tag{40}$$

Thus the index of the leader is $J_{\text{lead}}(N) = N(2/3)^{k_{\max}}$. The maximum degree is estimated from the extreme value criterion $\sum_{k \geq k_{\max}} N_k(N) \approx 1$ and using $N_k(N) = N/2^k$ [49] gives $k_{\max} \sim \ln N / \ln 2$. Therefore [50]

$$J_{\text{lead}}(N) \sim N^{\psi}, \qquad \text{with} \qquad \psi = 2 - \frac{\ln 3}{\ln 2} \cong 0.415\,037,$$

in excellent agreement with numerical results.

For the linear attachment rate, $A_k = k$, numerical simulations indicate that a rich gets richer phenomenon arises, as the average index of the leader $J_{\text{lead}}(N)$ saturates to a finite value of approximately 3.4 as $N \to \infty$. With probability ≈ 0.9, the leader is among the 10 earliest nodes, while the probability ≈ 0.99 the leader is among the 30 earliest nodes [50]. In general, we find similar behavior for the more general case of the shifted linear attachment rate $A_k = k + \lambda$.

We can understand these results analytically through the joint index-degree distribution. For the linear attachment rate one has [49]

$$C_k(J, N) = \sqrt{\frac{J}{N}} \left(1 - \sqrt{\frac{J}{N}} \right)^{k-1}, \tag{41}$$

from which $J_k(N) = 12N/[(k+3)(k+4)]$. Since $N_k(N) \simeq 4N/k^3$ for linear attachment [51,49], the extreme statistics criterion now gives $k_{\max} \sim N^{1/2}$. Therefore $J_{\text{lead}}(N) \simeq 12N/k_{\max}^2 = \mathcal{O}(1)$ indeed saturates to a finite value. Thus the leader is one of the first few nodes in the network.

4.2 Number of Lead Changes

In contrast with random trees and random graphs, the average number of lead changes $L(N)$ grows only logarithmically in N for both the attachment rates $A_k = 1$ and $A_k = k$. While the average number of lead changes appears to be universal, there is a significant difference in the distribution of the number of lead changes, $P_n(N)$, at fixed N. For $A_k = 1$, this distribution is sharply localized, while for $A_k = k$, $P_n(N)$ has a significant large-n tail. This tail stems from repeated lead changes among the two leading nodes. Related to lead changes is the number of *distinct* nodes that enjoy the lead over the history of the network. Simulations indicate that this quantity also grows logarithmically in N.

This logarithmic behavior can be easily understood for the attachment rate $A_k = 1$. Here the number of lead changes cannot exceed an upper bound given by the maximal degree $k_{\max} \sim \ln N / \ln 2$. To establish the logarithmic growth for the general attachment rate $A_k = k + \lambda$, we first note that when a new node is added, the lead changes if the leadership is currently shared between two (or more) nodes and the new node attaches to a co-leader. The number of co-leader nodes (with degree $k = k_{\max}$) is $N/k_{\max}^{3+\lambda}$, while the probability of attaching to a co-leader is $k_{\max}/N$. Thus the average number of lead changes satisfies

$$\frac{dL(N)}{dN} \sim \frac{k_{\max}}{N} \frac{N}{k_{\max}^{3+\lambda}}. \tag{42}$$

Since $k_{\text{max}} \sim N^{1/(2+\lambda)}$, (42) reduces to $dL/dN \sim N^{-1}$ and thus gives the logarithmic growth $L(N) \sim \ln N$.

4.3 Fate of The First Leader

We now turn to the probability that the first leader retains the lead throughout the network growth. For the linear attachment rate $A_k = k + \lambda$ (rich get richer systems), the initial leader has a finite chance to remain in the lead forever. However, for the egalitarian attachment rate $A_k = 1$, the initial leader is eventually replaced by another leader. Here, the probability that the initial leader retains the lead decays very slowly in time with an unusual decay law.

To understand the fate of the initial leader, we need to understand the degree distribution of the first node. We can straightforwardly determine this degree distribution analytically for the constant and linear attachment rates [50,52]. Let $P(k, N)$ be the probability that the first node has degree k in a network of N links[9]. For $A_k = k$, this probability obeys [50]

$$P(k, N + 1) = \frac{k - 1}{2N} P(k - 1, N) + \frac{2N - k}{2N} P(k, N). \tag{43}$$

The first term on the right accounts for the case that the earliest node has degree $k - 1$. Then a new node attaches to it with probability $(k - 1)/2N$, thereby increasing the probability for the node to have degree k. Conversely, with probability $(2N - k)/2N$ a new node does not attach to the earliest node, thereby giving the second contribution to $P(k, N + 1)$.

The solution to (43) for the "dimer" initial condition o—o is

$$P(k, N) = \frac{1}{2^{2N-k-1}} \frac{(2N - k - 1)!}{(N - k)!\,(N - 1)!}. \tag{44}$$

For $N \to \infty$, this simplifies to the Gaussian distribution

$$P(k, N) \simeq \frac{1}{\sqrt{\pi N}} e^{-k^2/4N} \tag{45}$$

for finite values of the scaling variable $k/N^{1/2}$. Thus the typical degree of the first node is of the order of $N^{1/2}$; this is the same scaling behavior as the degree of the leader node. For the trimer initial condition (which we typically used in simulations) we obtain the degree distribution of the first node as a series of ratios of gamma functions in which $P(k, N)$ has an $e^{-k^2/4N}$ Gaussian tail, independent of the initial condition. The degree of the first node also approximates that of the leader node more and more closely as the degree of the first node in the initial state is increased [53].

---

[9] The normalized attachment probability is A_k/A, with $A = \sum A_j N_j$. For the linear attachment rate, A is twice the total number of links. Hence formulae are neater if we denote by N the total number of links.

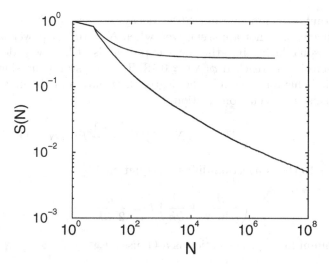

Fig. 8. The probability that the first node leads throughout the evolution obtained from 10^5 realizations of up to size $N = 10^7$ for $A_k = k$ (upper), and up to $N = 10^8$ for $A_k = 1$ (lower).

Although $P(k, N)$ contains all information about the degree of the first node, the behavior of its moments $\langle k^a \rangle_N = \sum k^a P(k, N)$ is simpler to appreciate. To determine these moments, it is more convenient to construct their governing recursion relations directly, rather than to calculate them from $P(k, N)$. Using (43), the average degree of the first node satisfies the recursion relation $\langle k \rangle_{N+1} = \langle k \rangle_N \left(1 + \frac{1}{2N}\right)$ whose solution is

$$\langle k \rangle_N = \Lambda \frac{\Gamma\left(N + \frac{1}{2}\right)}{\Gamma\left(\frac{1}{2}\right)\Gamma(N)} \simeq \frac{\Lambda}{\sqrt{\pi}} N^{1/2}. \tag{46}$$

The prefactor Λ depends on the initial conditions, with $\Lambda = 2, 8/3, 16/5, \ldots$ for the dimer, trimer, tetramer, *etc.*, initial conditions.

This multiplicative dependence on the initial conditions means that the first few growth steps substantially affect the average degree of the first node. For example, for the dimer initial condition, the average degree of the first node is, asymptotically, $\langle k \rangle_N \simeq 2\sqrt{N/\pi}$. However, if the second link attaches to the first node, an effective trimer initial condition arises and $\langle k \rangle_N \simeq (8/3)\sqrt{N/\pi}$. Thus, small initial perturbations lead to huge differences in the degree of the first node.

An intriguing manifestation of the rich get richer phenomenon is the behavior of the survival probability $S(N)$ that the first node leads throughout the growth up to size N (Fig. 8). For the linear attachment rate, $S(N)$ saturates to a finite non-zero value of approximately 0.277 as $N \to \infty$; saturation also occurs for the general attachment rate $A_k = k + \lambda$. We conclude that for popularity-driven systems, the rich get richer holds in a strong form—the lead never changes with a positive probability.

For constant attachment rate, $S(N)$ decays to zero as $N \to \infty$, but the asymptotic behavior is not apparent even when $N = 10^8$. A power law $S(N) \sim N^{-\phi}$ is a reasonable fit, but the local exponent is still slowly decreasing at $N \approx 10^8$ where it has reached $\phi(N) \approx 0.18$. To understand the slow approach to asymptotic behavior, we study the degree distribution of the first node. This quantity satisfies the recursion relation

$$P(k, N) = \frac{1}{N} P(k - 1, N - 1) + \frac{N - 1}{N} P(k, N - 1) \tag{47}$$

which reduces to the convection-diffusion equation

$$\left(\frac{\partial}{\partial \ln N} + \frac{\partial}{\partial k} \right) P = \frac{1}{2} \frac{\partial^2 P}{\partial k^2} \tag{48}$$

in the continuum limit. The solution is a Gaussian

$$P(k, N) \simeq \frac{1}{\sqrt{2\pi \ln N}} \exp \left[-\frac{(k - \ln N)^2}{2 \ln N} \right]. \tag{49}$$

Therefore the degree of the first node grows as $\ln N$, with fluctuations of the order of $\sqrt{\ln N}$. On the other hand, the maximal degree grows faster, as $\ln N / \ln 2$, with negligible fluctuations.

We now estimate the large-N behavior of $S(N)$ as $\sum_{k \geq k_{\max}} P(k, N)$. This approximation gives

$$S(N) \sim \int_{v \ln N}^{\infty} \frac{dk}{\sqrt{\ln N}} \exp \left[-\frac{(k - \ln N)^2}{2 \ln N} \right] \sim N^{-\phi} (\ln N)^{-1/2}, \tag{50}$$

with $\phi = [(\ln 2)^{-1} - 1]^2/2 \cong 0.097989$. The logarithmic factor leads to a very slow approach to asymptotic behavior.

The above estimate is based on a Gaussian approximate for $P(k, N)$ which is not accurate for $|k - \ln N| \gg \sqrt{\ln N}$. However, we can determine $P(k, N)$ exactly because its defining recursion formula, (47), is closely related to the Stirling numbers $\begin{bmatrix} N \\ k \end{bmatrix}$ of the first kind [54]. For the dimer initial condition, the solution reads $P(k, N) = \begin{bmatrix} N \\ k \end{bmatrix}/N!$. The corresponding generating function is [54]

$$S_N(x) = \sum_{k=1}^{N} P(k, N) x^k = \frac{x(x + 1) \dots (x + N - 1)}{N!}. \tag{51}$$

Using the Cauchy theorem, we express $P(k, N)$ in terms of the contour integral $S_N(x)/x^{k+1}$. When $N \to \infty$, this contour integral is easily computed using the saddle point technique. Finally, we arrive at (50) with the same logarithmic prefactor but with the slightly smaller *exact* transcendental exponent $\phi = 1 - \frac{1 + \ln \ln 2}{\ln 2} \cong 0.08607$. The remarkably small exponent value and the logarithmic correction are the reasons why simulations with $N = 10^8$ observed an exponent that was more that twice larger.

5 Summary and Discussion

Extremal properties provide an important statistical characterization of random structures and these properties yield many insights and surprises. Generally, extremes involve logarithmic dependences on system size. The practical consequences are numerous: slow convergence to asymptotic behavior, significant statistical fluctuations, erratic changes in extremal characteristics, and sensitive dependence on the initial conditions. Such behavior is consistent with our experience. For example, changes in athletic records are rare and unpredictable. As another example, the number of changes in the composition of the bellwether Dow Jones stock index (the 30 largest companies) ranged from a high of 11 in the 1990's to a low of 0 in the 1950's [55].

Leadership statistics of random graphs and random trees are quite similar: lead changes are infrequent; their total number increases logarithmically with the system size. The time-dependent number of lead changes approaches a self-similar form. The convergence to the asymptotic behavior is much slower for extremal statistics compared with size statistics because of the presence of various logarithmic dependences. Hence, the asymptotic behavior is difficult to detect in practice, especially for random graphs.

The most elementary leadership characteristic is the overall number of lead changes as a function of system size. This quantity can be measured simply by counting the number of changes until the process ends, making no reference to time. We have seen that introducing the time variable and treating the merger process dynamically not only produces this quantity, but also reveals an important self-similar behavior throughout the growth process.

Lead changes are also rare in popularity-driven network growth processes, where leadership is restricted to the earliest nodes. With finite probability, the first node remains the leader throughout the evolution. For growth with no popularity bias, leadership is shared among a somewhat larger cadre of nodes. As a consequence, the average index of the leader node grows algebraically with the network size. The possibility of sharing the lead among a larger subset of nodes gives a rich dynamics in which the probability that the first node retains the lead decays algebraically with the system size.

Extremal height properties of random trees can be obtained by analyzing the underlying nonlinear evolution equations. The cumulative distributions of tree heights approach a traveling wave form and the mean values grow logarithmically with the tree size. The corresponding growth coefficients can be obtained using either an elementary probabilistic argument or using an extremum selection criteria on the traveling wave. The same formalism used to analyze the leader and the laggard extends naturally to extremal statistics of extremal characteristics such as the heights of the tallest and the shortest trees.

To obtain leader or laggard characteristics, we employed the scaling behavior of the size distribution outside the scaling regime, namely, at sizes much larger than the typical size where, at least formally, statistical fluctuations can no longer be ignored. Nevertheless, the size dependences for these various leadership statistics appear to be asymptotically exact. Further analysis is needed to illuminate the role of statistical fluctuations, for example, by characterizing corrections to the leading behavior [56–58].

Acknowledgements

We are thankful to our collaborator Satya Majumdar. This research was supported by DOE(W-7405-ENG-36) and NSF(DMR0227670).

References

1. See, *e. g.*, *The Guiness Book of World Records* (Downtown Book Center, Inc., 2001).
2. E. J. Gumbel, *Statistics of Extremes* (Columbia University Press, New York, 1958).
3. J. Galambos, *The Asymptotic Theory of Extreme Order Statistics* (R.E. Krieger Publishing Co., Malabar, 1987).
4. R. E. Ellis, *Entropy, Large Deviations, and Statistical Mechanics* (Springer-Verlag, New York, 1985).
5. D. Carpentier and P. Le Doussal, Phys. Rev. E **63**, 026110 (2001)
6. D.S. Dean and S.N. Majumdar, Phys. Rev. E **64**, 046121 (2001)
7. S. Raychaudhuri, M. Cranston, C. Pryzybla, and Y. Shapir, Phys. Rev. Lett. **87**, 136101 (2001)
8. H. M. Mahmoud, *Evolution of Random Search Trees* (John Wiley & Sons, New York, 1992).
9. D. E. Knuth, *The Art of Computer Programming, vol. 3, Sorting and Searching* (Addison-Wesley, Reading, 1998).
10. W. Szpankowski, *Average Case Analysis of Algorithms on Sequences* (John Wiley & Sons, New York, 2001).
11. M. V. Smoluchowski, Z. Phys. Chem. **92**, 215 (1917).
12. S. Chandrasekhar, Rev. Mod. Phys. **15**, 1 (1943).
13. R. van Zon, H. van Beijeren, and Ch. Dellago, Phys. Rev. Lett. **80**, 2035 (1998).
14. B. Bollobás, *Random Graphs* (Academic Press, London, 1985).
15. S. Janson, T. Łuczak, and A. Rucinski, *Random Graphs* (John Wiley & Sons, New York, 2000).
16. P. J. Flory, *Principles of Polymer Chemistry* (Cornell University Press, Ithaca, 1953).
17. G. U. Yule, Phil. Trans. Roy. Soc. B **213**, 21 (1924); *The Statistical Study of Literary Vocabulary* (Cambridge University Press, Cambridge, 1944).
18. H. A. Simon, Biometrica **42**, 425 (1955); Infor. Control **3**, 80 (1960).
19. W. Willinger, R. Govindan, S. Jamin, V. Paxson, and S. Shenker, Proc. Natl. Acad. Sci. **99**, 2573 (2002).
20. See *e.g.*, S. H. Strogatz, Nature **410**, 268 (2001).
21. B. Skyrms and R. Pemantle, Proc. Natl. Acad. Sci. **97**, 9340 (2000).
22. M. Girvan and M. E. J. Newman, Proc. Natl. Acad. Sci. **99**, 7821 (2002).
23. O. C. Martin, R. Monasson, and R. Zecchina, Theor. Comput. Sci. **265**, 3 (2001).
24. H. Nishimori, *Statistical Physics of Spin Glasses and Information Processing* (Oxford University Press, New York, 2001).
25. M. Mezard, G. Parisi, and G. Zecchina, Science **297**, 812 (2002).
26. J. Baik, *math-PR/0310347*

27. P. L. Ferrari, M. Prähofer, H. Spohn, *math-ph*/0310053.
28. C. Knessl and W. Szpankowsky, SIAM J. Comput. **30**, 923 (2000).
29. See *e.g.*, the annual list of the 400 wealthiest individuals in *Fortune Magazine*.
30. T. Hattori and H. Ochiai, preprint (1998).
31. P. L. Krapivsky and S. N. Majumdar, Phys. Rev. Lett. **85**, 5492 (2000).
32. B. Pittel, J. Math. Anal. Appl. **103**, 461 (1984).
33. L. Devroye, J. ACM **33**, 489 (1986).
34. P. Noguiera, Disc. Appl. Math. **109**, 253 (2001).
35. S. N. Majumdar and P. L. Krapivsky, Phys. Rev. E **65**, 036127 (2001).
36. For a recent review, see D. Aldous, Bernoulli **5**, 3 (1999).
37. E. Ben-Naim and P. L. Krapivsky, Europhys. Lett. **65**, 151 (2004).
38. E. Ben-Naim, P. L. Krapivsky, and S. N. Majumdar, Phys. Rev. E **64**, 035101(R) (2001).
39. J. D. Murray, *Mathematical Biology* (Springer-Verlag, New York, 1989).
40. M. Bramson, *Convergence of Solutions of the Kolmogorov Equation to Travelling Waves* (American Mathematical Society, Providence, R. I., 1983).
41. W. van Saarloos, Phys. Rev. A **39**, 6367 (1989).
42. E. Brunet and B. Derrida, Phys. Rev. E **56**, 2597 (1997).
43. U. Ebert and W. van Saarloos, Phys. Rev. Lett. **80**, 1650 (1998); Physica D **146**, 1 (2000).
44. S. N. Majumdar and P. L. Krapivsky, Physica A **318**, 161 (2003).
45. D. Stauffer, *Introduction to Percolation Theory*, (Taylor & Francis, London, 1985).
46. H. S. Wilf, *Generatingfunctionology* (Academic Press, Boston, 1990).
47. J. B. McLeod, Quart. J. Math. Oxford **13**, 119 (1962); *ibid* **13**, 193 (1962); *ibid* **13**, 283 (1962).
48. E. M. Hendriks, M. H. Ernst, and R. M. Ziff, J. Stat. Phys. **31**, 519 (1983).
49. P. L. Krapivsky and S. Redner, Phys. Rev. E **63**, 066123 (2001).
50. P. L. Krapivsky and S. Redner, Phys. Rev. Lett. **89**, 258703 (2002).
51. A.-L. Barabási and R. Albert, Science **286**, 509 (1999).
52. S. N. Dorogovtsev, J. F. F. Mendes, and A. N. Samukhin, Phys. Rev. E **63**, 062101 (2001).
53. A. A. Moreira, J. S. de Andrade Jr., and L. A. N. Amaral, Phys. Rev. Lett. **89**, 268703 (2002).
54. J. Stirling, *Methodus Differentialis* (London, 1730); see also R. L. Graham, D. E. Knuth, and O. Patashnik, *Concrete Mathematics: A Foundation for Computer Science* (Reading, Mass.: Addison-Wesley, 1989).
55. http://indexes.dowjones.com.
56. A. A. Lushnikov, J. Colloid Inter. Sci. **65**, 276 (1977).
57. J. L. Spouge, J. Colloid Inter. Sci. **107**, 38 (1985).
58. P. G. J van Dongen and M. H. Ernst, J. Stat. Phys. **49**, 879 (1987).

On the Analysis of Backtrack Procedures
for the Colouring of Random Graphs

Rémi Monasson

[1] CNRS-Laboratoire de Physique Théorique de l'ENS, 24 rue Lhomond, 75005 Paris, France
[2] CNRS-Laboratoire de Physique Théorique, 3 rue de l'Université, 67000 Strasbourg, France

Abstract. Backtrack search algorithms are procedures capable of deciding whether a decision problem has a solution or not through a sequence of trials and errors. Analysis of the performances of these procedures is a long-standing open problem in theoretical computer science. I present some statistical physics ideas and techniques to attack this problem. The approach is illustrated on the colouring of random graphs, and some current limitations and perspectives are presented.

1 Introduction

1.1 Why Studying Backtrack Algorithms?

Many computational tasks e.g. constrained satisfaction, scheduling, ... amount to a search for the optimum of a cost function which depends upon a set of variables taking values in a huge space of possible configurations. Finding the true optimum with certainty is not easy from an algorithmic point of view and can be done through 'clever' exhaustive search only. Indeed, as stated by D. Knuth in 1975,

"*the majority of all combinatorial computing applications can apparently be handled only by what amounts to an exhaustive search through all possibilities. Such searches can readily be performed by using a well-known "depth-first" procedure (...) called backtracking*" [1].

Three decades later, Knuth's statement still holds. It is a fundamental conjecture of theoretical computer science that exhaustive search is essentially the only way to solve many combinatorial problems, called NP-complete [2]. For such problems, backtracking is among the most efficient solving procedures, and often the only one. Unfortunately, the running time of backtrack-based algorithms is hardly predictable, as explained by the same author:

"*Sometimes a backtrack program will run to completion in less than a second, while other applications of backtracking seem to go on forever. The author once waited all night for the output from such a program, only to discover that the answers would not be forthcoming for about 10^6 centuries. A "slight increase" in one of the parameters of a backtrack routine might slow down the total running time by a factor of a thousand; conversely, a "minor improvement" to the algorithm might cause a hundredfold improvement in speed; and a sophisticated "major improvement" might actually make the program ten times slower*" [1].

R. Monasson, On the Analysis of Backtrack Procedures for the Colouring of Random Graphs, Lect. Notes Phys. **650**, 235–254 (2004)
http://www.springerlink.com/

The study of backtrack algorithms has a long and rich story in theoretical computer science [3]. Recently, the use of out-of-equilibrium ideas stemming from statistical physics has led to some improvement in our understanding of when and why a backtrack procedure is very fast, or slow to solve computational problems [4]. The purpose of this article is to illustrate this approach on the colouring problem, discuss its limitations, and propose some ways to circumvent those.

1.2 The Colouring Problems and Some Definitions

An example of combinatorial problem that can be solved by a backtrack procedure is the colouring of graphs (COL). An input of the K-COL decision problem consists in a graph G. The problem consists in finding a mapping from the set of vertices to the set of K colours such that no two neighbouring vertices (connected by an edge) have the same colour, or proving there exists no such mapping. K-COL is a NP-complete problem for any $K \geq 3$ [2], and we choose $K = 3$ in most of what follows. The operation of the backtrack procedure, called Davis-Putnam-Logemann-Loveland (DPLL) [5] on an input of the 3-COL problem is illustrated in Fig. 1.

In order to study in a quantitative way the performances of DPLL, we need first to define in a precise way the notion of running time, and then the features of the input graphs we want to colour.

Running Time and Search Tree

It is convenient to represent the history of the search process followed by DPLL, that is, the sequence of trials and errors by a search tree. Examples of search trees are given in Fig. 2. Nodes in the tree are attached to assignment of variables, while edges represent logical consequences (elimination of satisfied constraints, simplification of other constraints) resulting from these assignments. A good computer-independent measure of the complexity of resolution is the size of the search tree generated by DPLL. This search tree varies with the input of the problem under consideration *i.e.* the graph to be coloured, and the sequence of assignments carried out by the search procedure.

Random Graphs

In spite of being NP-complete, 3-COL is not always hard. Deciding whether a given graph is 3-colourable or not may sometimes be very easy. For instance, it is immediate to recognise that a square lattice is 3-colourable, irrespectively of its size (number of nodes), while a complete graph with 4 vertices (or more) is not. To obtain a reliable estimate of the performances of DPLL on 3-COL, we want to discard such instances.

A possibility is to estimate resolution complexity for some underlying probability distribution of instances. This 'average-case' behaviour depends, of course,

UNCOLORABLE   COLORABLE

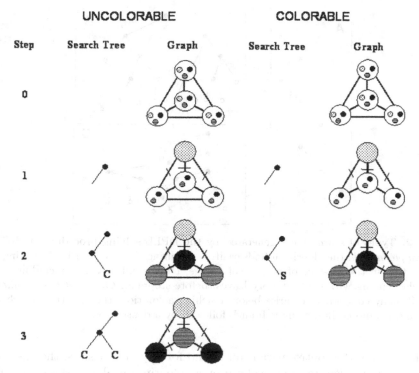

Fig. 1. Two examples which demonstrate how the DPLL algorithm acts onto a uncolourable (left side) and an colourable (right side) graph. The figure illustrates how the search tree grows with the operation of the algorithm. Available colours at each step are denoted by the patterns of the filled circles attached to vertices. When a vertex is coloured, it is removed from the graph, together with all its attached edges. In addition, the chosen colour is removed from the neighbours' sets of available colours. On the right side of the figure, a colourable graph is coloured by the algorithm. No contradiction is encountered, and the algorithm finds a solution without backtracking. On the left side, the algorithm tries to colour an uncolourable graph. When it first hits a contradiction (step 2) *i.e.* when two 1-colour vertices connected by an edge are left with the same available colour, the algorithm backtracks to the last-coloured vertex, and tries to colour it with the second available colour. When a contradiction is hit again, the algorithm terminates. Note, that in principle, it could backtrack to the first-coloured node, and try other colour options. However, due to colour gauge symmetry, this will not yield a solution.

on the input distribution and on the resolution algorithm considered [6,7]. Such distributions are usually unrealistic compared to structured instances from the real world, but are simple enough to allow for some analytical treatment. A popular input distribution for 3-COL are random graphs G *à la* Erdös-Renyì *i.e.* drawn with uniform probability among all the graphs having N vertices and E edges. The limit of interest is $N, E \to \infty$ at fixed ratio $c = 2E/N$ of edges per vertex [8–10]. Random 3-COL exhibits a phase transition phenomenon. For

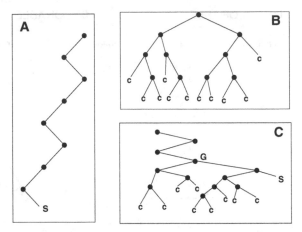

Fig. 2. Types of search trees generated by the DPLL solving procedure on 3-COL. **A.** *simple branch:* the algorithm finds easily a colouring without ever backtracking. **B.** *dense tree:* in the absence of proper colourings, DPLL builds a tree, including many branches ending with contradictory leaves, before stopping. **C.** *mixed case, branch + tree:* if many contradictions arise before reaching a solution, the resulting search tree can be decomposed into a single branch followed by a dense tree.

small values of the control parameter c, and for large input sizes, the answer to the decision problem (existence of a proper colouring) is almost surely yes. This holds as long as c remains smaller than a critical value $c_S \simeq 4.7$ called threshold, see [11] for a recent estimate of c_S with statistical mechanics techniques. Above the threshold, the answer is no with high probability.

1.3 Average Performances of DPLL. An Overview

A rigorous framework has been developed over the past years in theoretical computer science to understand the performances of DPLL in the yes region, at small enough ratios. There, decision is easily reached through a search tree as in Fig. 2A. The complexity of search is linear in the graph size, and essentially no backtracking takes place. This property allows for a rigorous analysis. It enormously simplifies the search by turning it into a Markovian process where a node in the graph is never seen twice, and the quenched character of the random graph to be coloured is not essential. The underlying structure (graph) is dynamically annealed. An informal account of the analysis of the operation of DPLL at small graph degree is presented in Sect. 2.

The situation is totally different above the threshold c_S. Proving the absence of solution requires the building up of a search tree like the one of Fig. 2B[3]. Though it is proven that the size of the search tree is exponentially large (in N) [12], not much more is known. In particular, no quantitative characterisation

---

[3] Massive backtracking is also present slightly below threshold where proper colourings are found at the price of intense computational effort, see search tree in Fig. 2C.

of this tree has been obtained by theoretical computer scientists so far. Due to massive backtracking, DPLL attempts to colour a node many different times at different stages of the search process. Furthermore, the Markovian character is lost: partial colourings of the graph, associated to nodes in the search tree, have to be stored to allow for the search to resume after a contradiction terminates a branch.

The understanding of DPLL operation in presence of massive backtracking is a formidable task from a probabilistic point of view. Recently, studies inspired from out-of-equilibrium statistical mechanics have permitted to tackle this problem to some extent. We present in Sect. 3 some of the ideas and results obtained along these lines. Finally, criticisms and remarks are given in Conclusion.

2 Colouring in the Absence of Backtracking

We first briefly review the 'rigorous' analysis of DPLL acting on an random input of 3-COL with small average vertex degree.

2.1 Main Features of the Search Heuristic

The action of the colouring procedure, illustrated in Fig. 1, is described as follows:

- *List of available colors:* while running, the algorithm maintains for each uncoloured vertices, a list of available colours, which consists of all the colours that can be assigned to this vertex. A node with $j(= 1, 2, 3)$ available colors is called j–colour node.
- *List-Updating:* to ensure that no adjacent vertices have the same colour, whenever a vertex is assigned a colour, this colour is removed from the lists (if present) attached to each of the uncoloured neighbours.
- *Colouring Order:* most constrained vertices *i.e.* with the least number of available colours are coloured first. At each step, a vertex is chosen among the most constrained vertices, and a colour is selected from the list of available colours. Both choices are done according to some heuristic rule, which can be unbiased (no preference is made between colours), or biased (following a hierarchy between colours), see next section.
- *Contradictions and Backtracking:* a contradiction occurs as soon as one of the lists becomes empty. Then, the algorithm backtracks to the most recently chosen vertex, which have more than one available colour (the closest node in the search tree - see definition below).
- *Termination Condition:* the algorithm stops when all vertices are coloured, or when all colouring possibilities have been tried.

Let us call Greedy heuristic the incomplete version of the above algorithm, obtained when the algorithm stops if a colouring is found (and outputs "Colourable"), or just after the first contradiction instead of backtracking (and outputs "Don't know if colourable or not"). In contrast to the algorithm with backtracking, the Greedy heuristic is not able to prove the absence of solution, but is amenable to rigorous analysis [13,9]. In the simplest case, vertices and colours are chosen purely randomly without any bias between colours (Colouring Order step described above). This Greedy heuristic enjoys two key properties. The first one is a statistical invariance: throughout the execution of the algorithm, the uncoloured part of the graph is uniformly randomly distributed, with an average vertex degree equal to $c(1-t)$ where t is the fraction of coloured vertices. The second property is colour symmetry: the search heuristic is symmetric with respect to the different colours, and the initial conditions are symmetric as well. Hence, the evolution of the algorithm can be monitored by tracking of the three numbers $N_j(T)$ of j-colour nodes ($j = 1, 2, 3$) without distinction between the colours available to each of these nodes.

2.2 Dynamics, Concentration, and Fluctuations of Node Populations

The evolution of these numbers in the course of the colouring was analysed by Achlioptas and Molloy [13]. evolution equations for the three populations of vertices read,

$$
\begin{aligned}
N_3(T+1) &= N_3(T) - w_2(T), \\
N_2(T+1) &= N_2(T) + w_2(T) - w_1(T) - \delta N_1(T), \\
N_1(T+1) &= N_1(T) + w_1(T) - (1 - \delta N_1(T)),
\end{aligned}
\tag{1}
$$

where $\delta N_1(T) = 1$ if $N_1(T) = 0$ (a 2-colour vertex is coloured) and $\delta N_1(T) = 0$ if $N_1(T) \neq 0$ (a 1-colour vertex is coloured). Quantities $w_2(T)$ and $w_1(T)$ are the 'flows' of vertices from $N_3(T)$ to $N_2(T)$, and from $N_2(T)$ to $N_1(T)$ respectively (Fig. 3). These are stochastic numbers depending on the graph under consideration and on the random choices made by the Greedy heuristic.

As a result of the additivity of (1), some concentration phenomenon takes place in the large size limit. The numbers of j-colour nodes do not fluctuate too much,

$$
N_j(T) = n_j(T/N) \, N + o(N).
\tag{2}
$$

where the n_j's are the population densities averaged over the graph (quenched disorder) and the choices of colours ("thermal" disorder). In other words, the densities of j-colour nodes are self-averaging quantities and we shall attempt at calculating their mean values only. Note that, in order to prevent the occurrence of contradictions, the number of 1-colour nodes must remain small and the density n_1 has to vanish.

Formula (1) also illustrates another essential feature of the dynamics of populations. Two time scales are at play. The short time scale, of the order of the unity, corresponds to the fast variations of the numbers of clauses $N_j(T)$ ($j = 1, 2, 3$). When time increases from T to $T + O(1)$ (with respect to the size N), all N_j's vary by $O(1)$ amounts. Consequently, the densities n_j of nodes, that

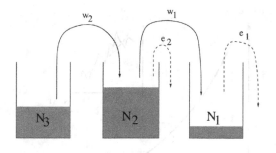

3–color nodes   2–color nodes   1–color nodes

Fig. 3. Schematic view of the dynamics of the Greedy heuristic. Nodes are sorted into three recipients according to the number of available colours. Each time a node is coloured by DPLL, populations N_1, N_2, N_3 are modified, resulting in a dynamics of the recipients populations (lines with arrows). Bold lines represent the reduction of 3-nodes into 2-nodes, or 2-nodes into 1-nodes with flows denoted by w_2, w_1 respectively. A solution is found when all recipients are empty. The level of the rightmost recipient coincides with the number of 1-colour nodes. If this level is low (*i.e.* $O(1)$), the probability that two 1-colour nodes (with the same available colour) are adjacent on the graph is vanishingly small. When the level is high (*i.e.* $O(\sqrt{N})$), contradictions will occur with high probability. Flow e_1 (respectively e_2) is equal to unity if the node coloured by the Greedy heuristic is chosen from the 1-colour (resp. 2-colour) recipient.

is, their numbers divided by N, are changed by $O(1/N)$ only. The densities n_js evolve on a long time scale of the order of N and depend on the reduced time $t = T/N$ only.

Due to the concentration phenomenon underlined above, the densities $n_j(t)$ will evolve in a deterministic way with the reduced time t. On the short time scale, the relative populations $\Delta N_j(T) = N_j(T) - N\, n_j(T/N)$ fluctuate (with amplitude $\ll N$) and are stochastic variables. As said above the evolution process for these relative numbers of clauses is Markovian and the probability rates (master equation) are functions of slow variables only, *i.e.* of the reduced time t and of the densities n_2 and n_3. On intermediary time scales, much larger than unity and much smaller than N, the ΔN_js reach some stationary distribution that depend upon the slow variables.

To sum up, the dynamical evolution of the clause populations may be seen as a slow and deterministic evolution of the j–colour nodes densities to which are superimposed fast, small fluctuations. The equilibrium distribution of the latter adiabatically follows the slow trajectory. This scenario is sketched in Fig. 4.

2.3 Resolution Trajectories and Percolation

Due to the statistical invariance property, the average flows of vertices $w_2(T)$ and $w_1(T)$ can be easily calculated. Each time a node is coloured, its 3-colour neighbors are turned into 2-colour vertices. The average number of neighbours is $c(1 - t)$, and the probability that a neighbour is a 3-colour nodes (prior to

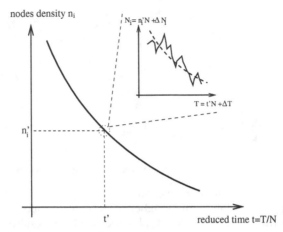

Fig. 4. Deterministic versus stochastic dynamics of the nodes population N_i as a function of the number of steps T of the algorithm. On the slow time scale (reduced time $t = T/N$), the density $n_i = N_i/N$ of (2- or 3-colour) nodes varies smoothly according to a deterministic law. Blowing up of the dynamics around some point t', n'_i shows the existence of small and fast fluctuations around this trajectory. Fluctuations are stochastic: their distribution depends upon the slow variables t', n'_i.

coloring) equals $N_3(T)/(N(1-t))$, leading to $w_2(T) = c\,N_3(T)/N$. A similar argument gives $w_1(T) = 2\,c\,N_2(T)/(3\,N)$. Thus, the evolution equations for the densities are

$$\frac{dn_3(t)}{dt} = -c\,n_3(t), \qquad \frac{dn_2(t)}{dt} = c\,n_3(t) - 1. \tag{3}$$

The solution of these differential equations, with initial conditions $n_3(0) = 1$, $n_2(0) = 0$, is $n_3(t) = e^{-ct}$, $n_2(t) = 1 - t - e^{-ct}$. Eqs. (3) were obtained under the assumption that $n_2(t) > 0$ and hold until time $t = \tau$ defined through,

$$1 - \tau = e^{-c\tau}, \tag{4}$$

at which the density n_2 of 2-colour nodes vanishes. For $t > \tau$, 2-colour vertices do not accumulate anymore. They are coloured as soon as they are created. 1-colour vertices are almost never created, and the vertices coloured by the algorithm are either 2-, or 3-colour vertices. Thus, when $\tau < t < 1$, $n_2(t) = 0$, and $n_3(t) = 1 - t$ decreases to zero. A proper colouring is found at $t = 1$ i.e. when all nodes have been coloured, see trajectory in Fig. 5. Notice that, if $c < 1$, $\tau = 0$ and n_2 vanishes at all times.

The interpretation of the change taking place at time τ is simple. Let us first assume that $c < 1$, then $\tau = 0$. The random graph to be coloured is essentially made of small trees[4]. Colouring of such a tree is straightforwardly done by the Greedy procedure. Since the graph is essentially a tree[5], colouring starts from one

---

[4] There may also exist small odd cycles e.g. triangles easily colourable with 3 colours.

[5] The presence of unicycle does not affect the property of being 3-colorable.

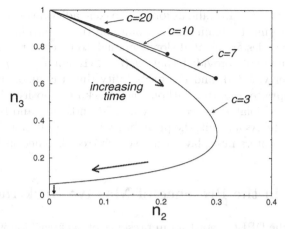

Fig. 5. Trajectories of dominant search branches generated by DPLL in the un-colourable phase ($c > c_3 \simeq 4.7$) compared to a search trajectory in the easy colourable phase ($c < c_L \simeq 3.85$). Horizontal and vertical axis represent the densities n_2 and n_3 of 2- and 3-colour nodes respectively. Trajectories are depicted by solid curves, and the arrows indicate the direction of motion (increasing depth of the search tree); they originate from the left top corner, with coordinates ($n_2 = 0, n_3 = 1$), since all nodes in the initial graph are 3-colour nodes. Dots at the end of the uncolourable trajectories ($c = 7, 10, 20$) symbolise the halt point at which condition $n_2 < 3\ln 2/c$ ceases to be fulfilled, and the search tree stops growing. Note that as the initial connectivity increases, the trajectories halt at earlier stage, implying the early appearance of contradictions as the problem becomes over-constrained (large connectivity values). The colourable trajectory (shown here for $c = 3$) represents the under-constrained region of the problem, where the very first search branch is able to find a proper colouring (bottom left corner with coordinates ($n_2 = 0, n_3 = 0$)).

node and then proceeds, branch after branch. The lists of available colours of not-yet-coloured nodes contain at least two colours. During this process, n_3 decreases linearly with time T, while $N_2(T)$ is bounded from above by the number of nodes in the tree. Hence the density of 2-colour nodes vanishes. Assume now that $c > 1$. A randomly drawn graph is now made of a giant percolating component including a fraction of the vertices equal to τ defined in eqn (4), while the fraction of remaining nodes, $1 - \tau$, almost surely belong to small trees. During the initial stage of the colouring process e.g. $1 \ll T \ll N$, the probability that a node belonging to the giant component is chosen by the Greedy heuristic tends to one. On the average, the number of its neighbours equals c, and this coincides with the number N_2 of 2-colour nodes created. Next, only one of these nodes is coloured, resulting in an average net creation rate of $c - 1$ 2-colour nodes. Hence N_2 initially grows, and the density n_2 becomes finite. This goes on until all the nodes in the percolating component have been assigned some colour *i.e.* up to time τ. Later on, the Greedy algorithm is left with the colouring of the remaining small trees (resolution trajectory at zero 2-colour density in Fig. 5).

Motion equations (3) are valid as long as no contradiction occurs. The number of 1-colour vertices must remain small throughout the execution of the algorithm. Clearly, 1-colour nodes are created slowly enough to be coloured ans eliminated, and do not accumulate provided that $w_1(t) < 1$. For $c < c_L \approx 3.847$, this condition is never violated, and the probability that the algorithm succeeds in finding an appropriate colouring without backtracking is positive [6]. For $c_L < c < c_S$, the condition is violated at $t = t_d(c)$ which depends on c, and 1-colour vertices start to accumulate. As a result, the probability for contradictions becomes large, and backtracking enters into play with a search tree sketched in Fig. 2C.

3 Colouring in the Presence of Massive Backtracking

The analysis of the DPLL algorithm in presence of backtracking was initiated on the random SAT problem. The case of 3-COL is qualitatively similar. Hereafter, we study the average complexity of showing that a random graph G with average degree c is not 3-colourable (as happens with high probability if $c > c_S$).

3.1 From Depth-First to Breadth-First Search:
The Markovian Evolution Matrix

The probabilistic analysis of DPLL in the uncolourable regime appears to be a formidable task since the search tree of Fig. 2B is the output of a complex, sequential process: nodes and edges are added by DPLL through successive descents and backtrackings (depth-first search). We have imagined a different, breadth-first building up of the refutation tree, which results in the same complete tree but can be mathematically analysed. In our imaginary process, the tree grows in parallel, layer after layer (Fig. 6). At time $T = 0$, the tree reduces to a root node, to which is attached the initial colouring E consisting in assigning to one randomly chosen vertex, say vertex number 1, a colour, say, Red, and an attached outgoing edge. Nothing else is known at the beginning of the search process. We suppose that the graph G is connected, and not 3-colourable. At time T, that is, after having coloured T vertices of the graphs attached to each branch, the tree is made of $B(T)$ ($\leq 2^T$) branches, each one carrying a partial colouring. At next time step $T \to T + 1$, a new layer is added by colouring, according to DPLL heuristic, one more node along every branch. As a result, a branch may keep growing through 1-colour node colouring, get hit by a contradiction and die out, or split if the colouring proceeds through 2-colour node colouring.

This parallel growth process is Markovian, and can be encoded in an instance–dependent evolution operator $\mathbf{H}$. A detailed definition and construction of $\mathbf{H}$ for the SAT problem is presented in [15]. We hereafter expose the main steps:

---

[6] See ref. [14] for a recent study of the occurrence of exponentially hard resolutions with massive backtracking in this range of connectivities.

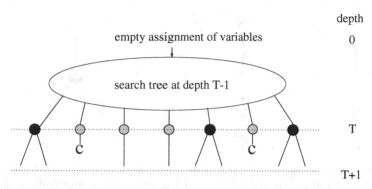

Fig. 6. Imaginary, parallel growth process of a search tree proving uncolourability used in the theoretical analysis. Colouring proceeds along the DPLL rule, but branches evolve in parallel. T denotes the depth in the tree, that is the number of nodes coloured by DPLL along each branch. At depth T, one node is chosen on each branch among 1-colour nodes (grey circles not represented in Fig. 2), or 2-, 3-colour nodes (splitting, black circles as in Fig. 2). If a contradiction occurs, the branch gets marked with C and dies out. The growth of the tree proceeds until all branches carry C leaves. The resulting tree is identical to the one built through the usual, sequential operation of DPLL.

– A 4^N dimensional–vector space $\mathbf{V}$ is introduced. Each vector $|S\rangle$ in the spanning basis is in one–to–one correspondence with a partial colouring $S = (s_1, s_2, \ldots, s_N)$ of the N vertices, where $s_i = R, G, B$ if vertex i is coloured with colour R, G, B respectively, or $s_i = U$ (Unknown) if vertex i has not been assigned any colour yet.

– Let S be a partial colouring for the graph under consideration with $s_j = U$ for some vertex j. Then $S^{(j,x)}$ is the partial colouring obtained from S by assigning colour $x(= R, G, B)$ to vertex j. Call $h_n(j|S)$ and $h_v(x|S, j)$ the probabilities that the Greedy heuristic respectively chooses vertex j when presented the graph with partial colouring S, and then colour it with colour x.

– For a partial colouring S, we denote by $N_i(S)$ the number of uncoloured nodes having i available colours *i.e.* having coloured neighbours of $3 - i$ different colours.

– The evolution operator $\mathbf{H}$ encodes the action of DPLL. Its matrix elements in the spanning basis are, see Fig. 7,

1. if S is an improper colouring of G, that is if two adjacent vertices have the same colour in S, $\langle S'|\mathbf{H}|S\rangle = 1$ if $S' = S$, 0 otherwise.
2. if S is a partial but not improper colouring of G, $\langle S'|\mathbf{H}|S\rangle = h_n(j|S) \times h_v(x|S, j)$ if $N_1(S) \geq 1$ and $S' = S^{(j,x_1)}$, $h_n(j|S)$ if $N_1(S) = 0$ and $(S' = S^{(j,x_2)}$ or $S' = S^{(j,x_2')})$, 0 otherwise. Here S, S' are the partial colourings attached to $|S\rangle, |S'\rangle$, x_1 (respectively x_2, x_2') denote the colour(s) available to the 1-colour (resp. 2-colour) node[7].

---

[7] Notice that, under our assumption that the graph is connected, there will always be at least one node with two or one available colours.

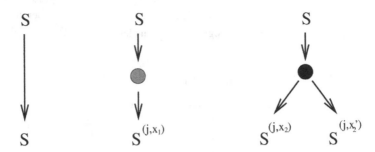

Fig. 7. Transitions allowed by the heuristic-induced evolution operator. Grey and black nodes correspond to the case of colouring of 1-colour and 2-colour nodes respectively, as in Fig. 6. **A.** If partial colouring S is improper for G, it is left unchanged. **B.** If the partial colouring has not led to any contradiction yet and there is at least 1-colour node, a 1-colour node is chosen, say j, and coloured as required, say with colour $s_j = x$ (grey node). The output partial colouring is $S^{j,x}$. **C.** If the partial colouring is not improper and there is no 1-colour node, a 2-colour node is coloured through splitting (black node). Two partial colourings are generated, S^{j,x_2} and $S^{j,x_2'}$ where $s_j = (x_2, x_2')$.

3.2 Gluing Partial Proofs of Uncolourability and Time Reversal

In this section, we show that the expectation value over the random colourings of variables of the size (number of leaves) of the search tree produced by DPLL to prove the uncolourability of G is equal to

$$B = \sum_S \langle S | \mathbf{H}^N | E \rangle, \tag{5}$$

where $\mathbf{H}^N$ denotes the Nth matrix power of $\mathbf{H}$, the sum runs over all 4^N partial colourings S, and the rightmost vector $|E\rangle = |R, U, U, \ldots, U\rangle$ corresponds to the initial colouring of G [15].

Let S be a partial colouring. We call refutation tree built from S a complete search tree that proves the uncolourability of G conditioned to the fact that DPLL is allowed to colour only nodes which are uncoloured in S. The height of the search tree is the maximal number of colourings leading from the root node (attached to partial colouring S) to a contradictory leaf. Let $b_T(S)$ be the average size (number of leaves) of refutation trees of height $\leq T$ that can be built from partial colouring S. Let us call W the set of improper colourings of G *i.e.* of partial colourings S assigning the same colour to two adjacent vertices. Clearly, $b_T(S) = 1$ if $S \in W$ and $b_T(S) \geq 2$ otherwise.

Assume now T is an integer larger or equal to 1, S a partial colouring with $N_1(S)$ 1-colour nodes. Our parallel representation of DPLL allows us to write simple recursion relations:

1. if $S \in W$, $b_T(S) = 1 = b_{T-1}(S)$.
2. if $S \notin W$ and $N_1(S) \geq 1$,

$$b_T(S) = \sum_{j=1}^{N} \sum_{x_1} h_n(j|S)\, h_v(x_1|S,j)\, b_{T-1}\big(S^{(j,x_1)}\big). \tag{6}$$

3. if $S \notin W$ and $N_1(S) = 0$,

$$b_T(S) = \sum_{j=1}^{N} h_n(j|S)\, \Big[b_{T-1}\big(S^{(j,x_2)}\big) + b_{T-1}\big(S^{(j,x_2')}\big) \Big]. \tag{7}$$

In the above equations, x_1, x_2, x_2' denote colours as defined in Sect. 3.1. These three different cases are symbolised on Fig. 7A, B and C respectively. From the definition of $\mathbf{H}$, these recursion relations are equivalent to

$$b_T(S) = \sum_{S'} \langle S' | \mathbf{H} | S \rangle\, b_{T-1}(S'), \tag{8}$$

for any partial colouring S. Let $|b_T\rangle$ be the vector of $\mathbf{V}$ whose coefficients on the spanning basis $\{|S\rangle\}$ are the $b_T(S)$'s. In particular, $|b_0\rangle$ is the sum of all improper colourings in W. Then identity (8) can be written as $|b_T\rangle = \mathbf{H}^\dagger\, |b_{T-1}\rangle$ where $\mathbf{H}^\dagger$ is the transposed of the evolution operator. The apparition of the time reversal operator $\mathbf{H}^\dagger$ is very natural since we glue partial refutation trees to build bigger and bigger ones to finally refute G from the initial (almost) empty colouring E.

The average size of refutation trees of height T obtained without any *a priori* knowledge on G is simply $b_T(E) = \langle E|b_T\rangle$. Since refutation trees cannot have height larger than the number of nodes N, it is easy to show that the average size of the proof of uncolourability of G generated by DPLL is

$$B \equiv b_N(E) = \langle E|(\mathbf{H}^\dagger)^N|b_0\rangle = \sum_S \langle E|(\mathbf{H}^\dagger)^N|S\rangle = \sum_S \langle S|\mathbf{H}^N|E\rangle \tag{9}$$

as claimed in equation (5).

3.3 Dynamical Annealing and the Search Growth Process

Calculation of the expectation value of the Nth power of $\mathbf{H}$, and of its average over the instance distribution is a hard task. We therefore turned to a simplifying approximation, called dynamical annealing. Call population vector $\mathbf{N}(S)$ of a partial colouring S the three dimensional vector $\mathbf{N} = (N_1, N_2, N_3)$ where N_j is the number of nodes with j available colours. The quantity we focus on is $\bar{B}(\mathbf{N}; T+1)$, the expectation number of branches at depth T in the search tree (Fig. 6) carrying partial colourings with population vector $\mathbf{N} = (N_1, N_2, N_3)$. Within the dynamical annealing approximation, the evolution of the $\bar{B}$'s is Markovian,

$$\bar{B}(\boldsymbol{N};T+1) = \sum_{\boldsymbol{N}'} \bar{\mathbf{H}}\left[\boldsymbol{N},\boldsymbol{N}';T\right]\bar{B}(\boldsymbol{N}';T). \tag{10}$$

The entries of the evolution matrix $\bar{\mathbf{H}}[\boldsymbol{N},\boldsymbol{N}';T]$ can be calculated from the definition of the evolution matrix $\boldsymbol{N}$ [15]. They can be interpreted as the average number of branches with population vector $\boldsymbol{N}$ that DPLL will generate through the colouring of one node from a partial colouring with population vector $\boldsymbol{N}'$. We find

$$\bar{\mathbf{H}}(\boldsymbol{N},\boldsymbol{N}';T) = \sum_{w_2=0}^{N_3'} \binom{N_3'}{w_2}\left(\frac{c}{N}\right)^{w_2}\left(1-\frac{c}{N}\right)^{N_3}\delta_{N_3'-N_3-w_2}\times \tag{11}$$

$$\left\{\{(1-\delta_{N_1'})\sum_{w_1=0}^{N_2'}\binom{N_2'}{w_1}\left(\frac{2c}{3N}\right)^{w_1}\left(1-\frac{2c}{3N}\right)^{N_2'-w_1}\delta_{N_2-N_2'-(w_2-w_1)}\delta_{N_1-N_1'-w_1+1}+ \right.$$

$$\left. 2\,\delta_{N_1'}\sum_{w_1=0}^{N_2'-1}\binom{N_2'-1}{w_1}\left(\frac{2c}{3N}\right)^{w_1}\left(1-\frac{2c}{3N}\right)^{N_2'-w_1-1}\delta_{N_2-N_2'-(w_2-w_1-1)}\delta_{N_1-N_1'-w_1}\right\}$$

where δ_N is the Kronecker delta function. Note that (11) is written under the condition that no 3-colour nodes are chosen by the algorithm throughout the growth process. This condition is consistent with the assumption that the graph G is connected, and with the fact that initially one node is coloured.

Let us examine how a step of the algorithm affects the size of the three populations N_1, N_2, N_3. Since the average connectivity is $O(1)$ *i.e.* each vertex is connected on average only to $O(1)$ vertices, when a vertex is coloured, the number of vertices whose status (the number of available colours) is subsequently changed is bounded from above by the number of neighbours of the coloured vertex. Hence a reasonable assumption is that the densities $n_i = N_i/N$ change by $O(1)$ after $T = t \times N$ vertices are coloured. In addition, we expect that, as soon as $N_1(T)$ becomes very large, contradictions are very likely to occur, and the growth process stops. Throughout the growth process, $N_1 = O(1)$ almost surely. Thus $n_1 = 0$ with high probability. The corresponding Ansatz for the number of branches is,

$$\tilde{B}(\boldsymbol{N};T) = e^{N\,\omega(n_2,n_3;t)+o(N)} \tag{12}$$

where non-exponential terms in N depend on the populations of i-colour nodes $(i = 1, 2, 3)$. At the initial stage of the tree building up, there is a single outgoing branch from the root node, carrying a fully uncoloured graph. Thus, $\bar{B}(\boldsymbol{N};T = 0) = 1$, and

$$\omega(n_2,n_3;t=0) = \begin{cases} 0 & \text{if } (n_2,n_3) = (0,1)\ , \\ -\infty & \text{if } (n_2,n_3) \neq (0,1)\ . \end{cases}$$

Insertion of Ansatz (12) into evolution equation (10) leads to the following partial differential equation for the logarithm $\omega(n_2,n_3;t)$ of the average number of branches with densities n_2, n_3 of 2- ,3-colours nodes [16],

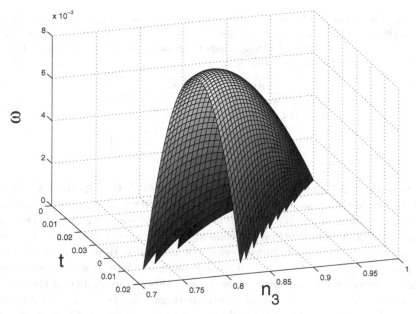

Fig. 8. Function ω (log. of number of branches with densities $n_2 = 1 - t - n_3$, n_3 of 2- and 3-colour nodes at depth t in the search tree) as a function of n_3 and t for $c = 10$. The top of the curve at given time t, $\omega^*(t)$, is reached for the dominant branch 3-colour density $n_3^*(t)$. The evolution of ω is shown till $t = t_h$ at which dominant branches in the search tree stop growing (die from the onset of contradictions). The maximal ω at t_h, $\omega^*(t_h)$, is the theoretical prediction for the complexity.

$$\frac{\partial \omega}{\partial t} = \frac{\partial \omega}{\partial n_2} + \ln 2 - \frac{c}{3} n_2 + c\, n_3 \left[\exp\left(\frac{\partial \omega}{\partial n_3} - \frac{\partial \omega}{\partial n_2} \right) - 1 \right]. \qquad (13)$$

The surface ω, growing with "time" t above the plane n_2, n_3 describes the whole distribution of branches. Here, this distribution simplifies due to nodes conservation. The sum $n_2 + n_3$ of 2- and 3-colour nodes densities necessarily equals the fraction $1 - t$ of not-yet coloured nodes. Therefore, ω is a function of n_3 and t only, whose expression is obtained through the exact resolution of (13) with the above initial condition (see Sect. 4 for comments on the resolution),

$$\omega(n_3; t) = \frac{c}{6} t \left(1 - 2t - 4 n_3\right) - n_3 \ln n_3 - (1 - n_3) \ln(1 - n_3) -$$
$$(1 - t - n_3) \ln 2 + (1 - n_3) \ln \left[3 \left(1 - e^{-2tc/3} \right) \right]. \qquad (14)$$

Figure 8 exhibits $\omega(n_3, t)$ for $c = 10$.

The maximum $\omega^*(t)$ over n_2, n_3 of $\omega(n_2, n_3; t)$ at depth t in the tree gives the logarithm of the average number of branches at depth t divided by N. The search tree keeps growing as long as no contradictions are encountered *i.e.* as long as 1-colour vertices do not accumulate. This amounts to say that dominant branches are not suppressed by contradictions and become more and more nu-

Table 1. Analytical results and simulation results of the complexity ω for different connectivities c in the uncolourable phase. The analytical values of ω_{THE} are derived from theory; ω_{EXP} is obtained through experimental measures of the search tree size and averages over input graphs [16].

c	ω_{THE}	ω_{NOD}
20	$2.886 * 10^{-3}$	$3 * 10^{-3} \pm 3 * 10^{-4}$
15	$5.255 * 10^{-3}$	$5.8 * 10^{-3} \pm 5 * 10^{-4}$
10	$1.311 * 10^{-2}$	$1.5 * 10^{-2} \pm 1 * 10^{-3}$
7	$2.135 * 10^{-2}$	$3. * 10^{-2} \pm 3.6 * 10^{-3}$

merous through 2-colour nodes colouring, $d\omega^*/dt > 0$. Call t_h the halt time at which this condition ceases to be fulfilled. The logarithm $\omega^*(t_h)$ of the number of dominant branches at $t = t_h$, when divided by $\ln 2$, yields our analytical estimate for the complexity of resolution. Agreement between theory and numerics is very good at large c (see Table 1) but deteriorates at small c. However, the high computational complexity of the algorithm for small c values, does not allow us to obtain numerical results for large sizes N, and affects the quality of the large N extrapolation of ω. As c increases, contradictions emerge in an earlier stage of the algorithm, the probability that the same vertex appears in different branches reduces, and the analytical prediction becomes exact. As a consequence of the early appearance of contradictions, the complexity ω decreases with c. At very large c, we find

$$\omega(c) \asymp \frac{3 \ln 2}{2} \frac{1}{c^2} \simeq \frac{1.040}{c^2} \quad , \tag{15}$$

and therefore that the (logarithm of the) complexity exhibits a power law decay with exponent 2 as a function of connectivity c.

3.4 More Than Three Colours

The scaling exponent appearing in eqn (15) strongly depends on the number of colours, here three. The whole procedure described above can be extend to the study of K-COL, where K is the number of colours. Let us concentrate on the case of large degrees c, where the output of the dynamical annealing procedure is conjectured to be exact. We find that the logarithm of the average complexity scales as

$$\omega(c) \asymp \Omega(K) \, c^{-(K-1)/(K-2)} \tag{16}$$

where the constant Ω depends on the details of the Greedy heuristic. For the heuristic analysed in Sect. 1.2 with preferential colouring of nodes with the smallest number of available colours,

$$\Omega(K) = \frac{K(K-2)}{K-1} \left[\frac{2 \ln 2}{K-1} \right]^{1/(K-2)} \quad . \tag{17}$$

A less sophisticated Greedy heuristic corresponds to colouring nodes with one available color if any, any node otherwise. The multiplicative factor entering eqn (16) is then,

$$\Omega(K) = \frac{K(K-2)}{K-1} \ln K \left[\frac{K \ln K}{(K-1)^2} \right]^{1/(K-2)}. \tag{18}$$

Recent works suggest that the correctness of the decay exponent in eqn (16) could be rigorously established [17].

4 Conclusions: What Is Missing?

In this article, we have presented a quantitative study of the search tree growth process accompanying the backtrack resolution of the random graph colouring problem, especially in presence of massive backtracking. Here are some remarks regarding the approach:

1. From a mathematical point of view, it is worth noticing that monitoring the growth of the search tree requires a partial differential equation, while ordinary differential equations are sufficient to account for the evolution of a single branch [18]. Yet, the partial differential equation is of the first-order[8], and can be solved using the characteristics method based on the use of an appropriate set of ordinary differential equations only. Dominant branches at time $t' > 0$ are indeed completely described by the set of coupled ordinary differential equations, see eqn (13),

$$\frac{dn_3(t)}{dt} = -c\, n_3(t)\, \exp\left(\psi(t)\right), \tag{19}$$

$$\frac{d\psi(t)}{dt} = c\, \exp\left(\psi(t)\right) - \frac{2}{3}\, c, \tag{20}$$

for all intermediate times $0 < t < t'$ with the boundary conditions $n_3(0) = 1$ and $\psi(t') = 0$. The new field ψ is related to the derivative of the surface ω at density n_3. Notice the equivalence of equation (19) when $\psi = 0$ and the dynamical equation (3) describing a single branch. In other words, the equations for the evolution of a search tree look like to the ones of a single branch in the absence of backtracking up to the presence of a statistical bias imposed by the field ψ.

Remarkably, the idea of an effective branch was at the base of the phenomenological approach proposed by Knuth three decades ago [1]. The statistical physics approach is, to some extent, an analytical realization of this idea.

---

[8] This statement is correct in the large size limit only. Finite size corrections would introduce second derivative terms with $1/N$ multiplicative coefficients. See [19] for a similar situation.

2. It is likely from the above discussion and numerical experiments that our theory is exact at large connectivity c but requires some corrections for small connectivities. These corrections come from the fact that we have neglected statistical correlations between branches in the search tree resulting from the average over the random graph to be colored. Though these correlations are expected to vanish at large c, they should give some contribution for finite c to the equations describing the dynamical evolution of the search tree. What is the structure of this contribution? A precise answer to this question will come from a detailed study of operator $\mathbf{H}$ along the lines recently initiated for another out-of-equilibrium system called the Contact Process [20]. It may be guessed that the output of such a calculation will be the emergence a systematic $1/c$ expansion for n_3 with non Markovian retarded terms on the right hand side of equation (19) involving the value of the density of 3-color nodes at all times $< t$. This structure is expected from the similarity of the present problem and the one treated in [20], and the fact that neglecting correlations between branches in the search tree precisely amounts to neglecting all non-Markovian terms in the evolution of the search tree[9].

3. From a qualitative point of view, the average complexity of DPLL for the 3-COL problem is linear for $c < c_L$ with finite probability, and exponential for $c > c_L$. Right at the location of the cross-over, that is, for connectivity $c = c_L$, we expect an average complexity growing as a stretched exponential of the size of the graph. A precise determination of the exponent is under way.

4. The present study is part of a general effort to understand the extremal statistics of correlated variables, and is related to studies on random binary trees to which statistical mechanics ideas have recently been applied [21]. From this point of view, it would be interesting to see to what extent the results presented in this article are changed when the input graph distribution is modified e.g. for random graphs with preferential attachment and power-law degree distributions.

5. Last of all, the study of the operator $\mathbf{H}$ is interesting regardless of its computer science interpretation. $\mathbf{H}$ is a non Hermitean evolution operator, and the halt of dominant branches is deeply related to a localization vs. delocalization transition of its dominant eigenvector [4].

In view of the above remarks, one may be quite confident that a major understanding of the average-case performances of backtracking algorithms will be obtained in the next future. To what extent these results will find rigorous support is however not clear.

---

[9] To be more precise, as stated in Sect. 3.1, the parallel coloring process is Markovian for a fixed graph but is not any longer once the average over the underlying graph has been carried out.

Acknowledgements

This work is the fruit of a long standing collaboration with S. Cocco on the Satisfiability problem. Results on graph Colouring have been obtained in collaboration with L. Ein-Dor. The discussion on the $1/c$ expansion is based on a recent work with C. Deroulers. I am grateful to O. Dubois and C. Moore for stimulating and very useful discussions. Partial support from the ACI Jeunes Chercheurs "Algorithmes d'optimisation et systèmes désordonnés quantiques" is acknowledged.

References

1. Knuth, D.E. Estimating the efficiency of backtrack programs, *Math. Comp.* **29**, 12-136 (1975).
2. Garey, M. R. and Johnson, D. S, Computers and Intractability: A Guide to the Theory of NP-Completeness, W.H. Freeman and Company, San Fransico (1979).
3. Knuth, D.E. Selected Papers on Analysis of Algorithms, Center for the Study of Language and Information Lecture Notes 102, Stanford CA (2000).
4. Cocco, S. and Monasson, R. Trajectories in phase diagrams, growth processes and computational complexity: how search algorithms solve the 3-Satisfiability problem, *Phys. Rev. Lett.* **86**, 1654 (2001); Analysis of the computational complexity of solving random satisfiability problems using branch and bound search algorithms, *Eur. Phys. J. B* **22**, 505 (2001).
5. Davis, M., Logemann, G., Loveland, D. A machine program for theorem proving. *Communications of the ACM* **5**, 394-397 (1962).
6. Mitchell, D., Selman, B. and Levesque, H. Hard and Easy Distributions of SAT Problems, *Proc. of the Tenth Natl. Conf. on Artificial Intelligence (AAAI-92)*, 440-446, The AAAI Press / MIT Press, Cambridge, MA (1992).
7. Turner, J. S. Almost All k-Colorable Graphs Are Easy to Color, *Journal of Algorithms* **9**, 63–82 (1988).
8. Achlioptas, D. and Friedgut, E. A sharp threshold for k-colorability, *Random Structures and Algorithms* **14(1)**, 63–70 (1999).
9. Achlioptas, D. and Moore, C. Almost all graphs with average degree 4 are 3-colorable *Proc. on 34th Annual ACM Symposium on Theory of Computing, May 19-21*, Montreal, Quebec, Canada, ACM, Montreal, 199–208 (2002)
10. Culbersome, J. C. and Gent, I. P. Frozen development in graph coloring, *Theor. Comp. Sci.* **265(1-2)**, 227–264 (2001).
11. Mulet, R., Pagnani, A., Weigt, M. and Zecchina R. Coloring random graphs. *Phys. Rev. Lett.* **89**, 268701 (2002).
12. Chvàtal, V. and Szmeredi, E. Many hard examples for resolution, *Journal of the ACM* **35**, 759–768 (1988).
13. Achlioptas, D. and Molloy, M. Analysis of a List-colouring Algorithm on a Random Graph, *Proc. of FOCS 97* 204 (1997).
14. Jia, H., and Moore, C. How much backtracking does it take to color sparse random graphs? Rigorous results on heavy tails. *preprint* (2003).
15. Cocco, S. and Monasson R. Heuristic average-case analysis of backtrack resolution of random 3-Satisfiability instances, to appear in *Theoretical Computer Science* (2004).

16. Ein-Dor, L. and Monasson, R. The dynamics of proving uncolorability of large random graphs. I. symmetric colouring heuristic, *J. Phys. A* **36**, 11055 (2003).
17. Beame, P., Culberson, J., Mitchell, D. and Moore, C. The resolution complexity of random graph k-colorability, *preprint* (2004).
18. Achlioptas, D. Lower bounds for random 3-SAT via differential equations, *Theor. Comp. Sci.* **265**, 159–185 (2001).
19. Griffiths, R.B., Weng, C-H. and Langer, J.S. Relaxation times for metastable states in the mean-field model of a ferromagnet, *Phys. Rev.* **149**, 301 (1966).
20. Deroulers, C. and Monasson, R. Field theoretic approach to metastability in the contact process. *Phys. Rev. E* **69** 016126 (2004).
21. Majumdar, S.N. and Krapivsky, P.I. Extreme value statistics and traveling fronts: an application to computer science. *Phys. Rev. E* **65**, 036127 (2002).

Small-World Synchronized Computing Networks for Scalable Parallel Discrete-Event Simulations

Hasan Guclu[1], György Korniss[1], Zoltán Toroczkai[2], and Mark A. Novotny[3]

[1] Department of Physics, Applied Physics, and Astronomy, Rensselaer Polytechnic Institute, 110 8th Street, Troy, NY 12180, USA
[2] Theoretical Division and Center for Nonlinear Studies, Los Alamos National Laboratory, MS B258 Los Alamos, NM 87545, USA
[3] Department of Physics and Astronomy and ERC Center for Computational Sciences, Mississippi State University, P.O. Box 5167, Mississippi State, MS 39762, USA

Abstract. We study the scalability of parallel discrete-event simulations for arbitrary short-range interacting systems with asynchronous dynamics. When the synchronization topology mimics that of the short-range interacting underlying system, the virtual time horizon (corresponding to the progress of the processing elements) exhibits Kardar-Parisi-Zhang-like kinetic roughening. Although the virtual times, on average, progress at a nonzero rate, their statistical spread diverges with the number of processing elements, hindering efficient data collection. We show that when the synchronization topology is extended to include quenched random communication links between the processing elements, they make a close-to-uniform progress with a nonzero rate, without global synchronization. We discuss in detail a coarse-grained description for the small-world synchronized virtual time horizon and compare the findings to those obtained by "simulating the simulations" based on the exact algorithmic rules.

1 Introduction

Synchronization is a fundamental problem in natural or artificial coupled multi-component systems [1]. To achieve it in an autonomous fashion can be a particularly challenging task from a system design viewpoint. In this chapter we discuss such a problem in the context of scalable Parallel Discrete-Event Simulations (PDES) [2–4]. Examples of PDES applications include dynamic channel allocation in cell phone communication network [4,5], models of the spread of diseases [6], battle-field simulations [7], and dynamic phenomena in highly anisotropic magnetic systems [8–10]. In these examples the discrete events are call arrivals, infections, troop movements, and changes of the orientation of the local magnetic moments, respectively. We focus on the basic algorithm suitable for simulating large spatially extended systems with short-range interactions and asynchronous dynamics [11,12].

In discrete-event simulations, the instantaneous local updates (discrete-events) occur in continuous time. The algorithm must faithfully and reproducibly keep track of the asynchrony of the local updates in the system's configuration. For example standard random-sequential Monte Carlo simulations naturally produce Poisson asynchrony. In fact, such continuous-time simulations (e.g., sin-

H. Guclu, G. Korniss, Z. Toroczkai, and M.A. Novotny, Small-World Synchronized Computing Networks for Scalable Parallel Discrete-Event Simulations, Lect. Notes Phys. **650**, 255–275 (2004)
http://www.springerlink.com/   © Springer-Verlag Berlin Heidelberg 2004

gle spin-flip Glauber dynamics) were long believed to be inherently serial until Lubachevsky's illuminating work [11,12] on the parallelization of these simulations *without* altering the underlying dynamics. The essence of the problem is to algorithmically parallelize "physically" non-parallel dynamics of the underlying system. This requires some kind of synchronization to ensure causality. The two basic ingredients of PDES are the set of local simulated times (or virtual times [13]) and a synchronization scheme. First, a scalable PDES scheme must ensure that the average progress rate of the simulation approaches a nonzero constant in the long-time limit as the number of Processing Elements (PEs) N goes to infinity. Second, the "width" of the simulated time horizon (the spread of the progress of the individual PEs) should be bounded as N goes to infinity [14]. The second requirement is crucial for the measurement phase of the simulation to be scalable: a large width of the virtual time horizon hinders scalable data management. Temporarily storing a large amount of data on each PE (being accumulated for "on-the-fly" measurements) is limited by available memory while frequent global synchronizations can get costly for large N. Thus, one aims to devise a scheme where the PEs make a nonzero and close-to-uniform progress without global synchronization. In such a scheme, the PEs autonomously learn the global state of the system (without receiving explicit global messages) and adjust their progress rate accordingly.

As the number of PEs available on parallel architectures increases to hundreds of thousands [15], or grid-computing networks proliferate the internet [16, 17] fundamental questions of the scalability of the underlying algorithms must be addressed. The center of our interest here is to understand the effects of the "microscopic dynamics" (corresponding to the algorithmic synchronization rules) and the effects of the underlying communication network among the PEs on the evolution and the morphological properties of the virtual time horizon. We achieve this by looking at the parallel simulation itself as a complex interacting system. A similar approach was also successful to establish connection [18] between rollback-based (or optimistic) schemes [13] and self-organized criticality [19]. Our main finding is that extending the basic conservative synchronization rules [11,12] to a small-world-like [20] communication topology among the PEs results in both a finite width of the time horizon and a nonzero progress rate of the simulation [21]. Performing additional synchronizational steps through the random links at a very small rate can only reduce the average progress rate infinitesimally while the width is reduced from infinity (in the limit of an infinite number of PEs) to some manageable finite value.

2 The Basic Conservative Scheme

The basic notion of discrete-event simulations is that time is continuous and the discrete events occur instantaneously. Between events, the state (configuration) of the system remains unchanged. If the events occur at random instants of time, the dynamics can be referred to as asynchronous. In *conservative* PDEs schemes [22], only those PEs that are guaranteed not to violate causality are allowed to process their events and increment their local time. The rest of the PEs must "idle". For simplicity we consider an arbitrary but one-dimensional underlying system (the "physical" system to be simulated) with nearest-neighbor interactions in which discrete events (update attempts in the local configuration) exhibit Poisson asynchrony. Further, we focus on the one site-per-PE scenario where each PE has its own local simulated time $h_i(t)$, constituting the virtual time horizon $\{h_i(t)\}_{i=1}^N$. Here t is the number of parallel steps executed by all PEs (proportional to the wall-clock time) and N is the number of PEs. By construction, $h_i(t)$ is the progress of PE i after parallel step t. In the following, we will use the terms "height", "simulated time", or "virtual time" interchangeably, since we refer to the same observable.

According to the basic conservative synchronization scheme, first introduced by Lubachevsky, [11,12], at each parallel step t, only those PEs for which the local simulated time is not greater then the local simulated times of their virtual neighbors, can increment their local time by an exponentially distributed random amount. (Without loss of generality we assume that the mean of the local time increment is one in simulated time units [stu].) Thus, for the one-site-per-PE, one-dimensional regular virtual topology, if $h_i(t) \leq \min\{h_{i-1}(t), h_{i+1}(t)\}$, PE i can update the configuration of the underlying site it carries and determine the time of the next event. Otherwise, it idles. Despite its simplicity, this rule preserves unaltered the asynchronous causal dynamics of the underlying system [11,12]. (More general PDES schemes, where events to be processed by a PE are initiated (or generated) by the same PE (such as the basic conservative scheme above), are also referred to as self-initiating discrete-event schemes [23,24].) In the original algorithm, the virtual communication topology between PEs mimics the interaction topology of the underlying system [11,12,25]. When "simulating the simulations" based on the above simple "microscopic" rules for the evolution of the time horizon, we implemented periodic boundary conditions, i.e., the PEs are placed on a ring. In analyzing the performance of the above scheme, it is enormously helpful that the progress of the simulation itself is decoupled from the possibly complex behavior of the underlying system. This is contrary to optimistic approaches, where the evolution of the underlying system and the progress of the PDES simulation are strongly entangled [18], making scalability analysis a much more difficult task.

To understand the scalability and performance of the basic conservative scheme we study two basic observables: the average utilization $\langle u \rangle$ (the fraction of non-idling PEs), which directly corresponds to the average rate of progress of the simulation, and the average width of the virtual time horizon, which probes the complexity of data management during the simulation. On a *regular* one-dimensional lattice the utilization is the density of local minima

$$\langle u \rangle = \langle \Theta(h_{i-1} - h_i)\Theta(h_{i+1} - h_i)\rangle = \langle \Theta(-\phi_{i-1})\Theta(\phi_i)\rangle , \qquad (1)$$

where $\phi_i \equiv h_{i+1} - h_i$ is the local slope, $\Theta(\dots)$ is the Heaviside step function, and $\langle \dots \rangle$ denotes an ensemble average over the stochastic, exponentially distributed local simulated time increments. For a system of identical PEs (implying translational invariance), the above quantity is independent of i. The width, characterizing the spread of the time horizon, is defined as

$$\langle w^2 \rangle = \left\langle \frac{1}{N} \sum_{i=1}^{N} (h_i - \bar{h})^2 \right\rangle \tag{2}$$

where $\bar{h} = (1/N) \sum_{i=1}^{N} h_i$ is the mean-height.

Here we use a coarse-grained description for the virtual time horizon to perform the scalability analysis [25,26]. It was shown [25] that the virtual time horizon exhibits Kardar-Parisi-Zhang (KPZ)-like [27] kinetic roughening [28] and the steady-state behavior in one dimension is governed by the Edwards-Wilkinson (EW) Hamiltonian [29]. The evolution of the simulated time horizon is effectively governed by the Langevin equation

$$\partial_t h_i(t) = \nabla^2 h_i - \lambda (\nabla h_i)^2 + \ldots + \eta_i(t) , \tag{3}$$

where $\eta_i(t)$ is a delta correlated Gaussian noise $\langle \eta_i(t)\eta_j(t') \rangle = 2D\delta_{ij}\delta(t-t')$, and ∇ and ∇^2 are the discrete gradient and discrete Laplacian operators on a regular lattice, respectively. The $\ldots$ in (3) stands for infinitely many irrelevant terms in the long-time, large-N limit. Being primarily interested in the steady-state properties of the algorithm, we consider the equal-time height-height correlations, or alternatively, its Fourier transform, the corresponding structure factor $S^{(h)}(k,t)$, defined through

$$S^{(h)}(k,t) N \delta_{k,-k'} \equiv \langle \tilde{h}_k(t)\tilde{h}_{k'}(t) \rangle . \tag{4}$$

Here $\tilde{h}_k = \sum_{j=1}^{N} e^{-ikj} h_j$ is the Fourier transform of the virtual times with the wave number $k = (2\pi n)/N$, $n = 0, 1, 2, \ldots, N-1$. In the long time limit in one dimension (EW stationary state), one has [26]

$$S^{(h)}(k) \equiv \lim_{t\to\infty} S^h(k,t) = \frac{D}{2[1 - \cos(k)]} . \tag{5}$$

The structure factor essentially contains all the "physics" needed to describe the scaling behavior of the time horizon. Figure 1(a) shows the measured structure factor, obtained by simulating the PDES simulation itself, based on the exact rules for the evolution of the local times. It confirms the $\sim 1/k^2$ coarse-grained prediction for small k values. Using the steady-state structure factor, one can express the width as

$$\langle w^2 \rangle = \frac{1}{N} \sum_{k \neq 0} S^{(h)}(k) . \tag{6}$$

The above summation can be carried out for the structure factor given by (5), yielding

$$\langle w^2 \rangle_N \simeq \frac{D}{12} N \sim N^{2\alpha} , \tag{7}$$

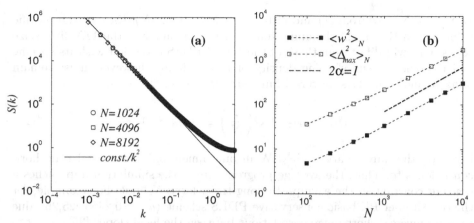

Fig. 1. (a) Steady-state height-height structure factor for various system sizes for the regular one-dimensional lattice, one-site-per-PE basic conservative PDES time horizon. The solid straight line indicates the theoretical $\sim 1/k^2$ behavior, (5), for small k values. (b) Steady-state width and extreme-height fluctuations for the same scheme as a function of the number of PEs. The dashed straight line corresponds to the exact KPZ (EW in one dimension) roughening (7).

(corresponding to a roughness exponent $\alpha=1/2$) in the limit of large N. Figure 1(b) shows the measured width, asymptotically approaching the above scaling form. For later calculations, we will also need the slope-slope steady-state structure factor

$$S^{(\phi)}(k) = 2[1 - \cos(k)]S^{(h)}(k) = D \qquad (8)$$

and the corresponding correlation function

$$C^{(\phi)}(l) = \langle \phi_i \phi_{i+l} \rangle = \frac{1}{N} \sum_{k \neq 0} e^{ikl} S^{(\phi)}(k) \qquad (9)$$

to study the density of local minima. From (8) and (9) it trivially follows that $C^{(\phi)}(l)=D\delta_{l,0}$ (i.e., the local slopes become independent) in the infinite system-size limit. Then the probability that two neighboring local slopes form a local minima is 1/4. Hence, the density of local minima and the utilization $\langle u \rangle$ [see (1)] approaches 1/4. (The steady state is governed by the EW Hamiltonian where the local slopes are independent.)

For more general two-point functions (but still within the coarse-grained Gaussian picture (5), we utilize a simple relationship between the density of local minima and the slope-slope correlation function [26]

$$\langle \Theta(-\phi_{i-1})\Theta(\phi_i) \rangle = \frac{1}{2\pi} \arccos\left(\frac{C^{(\phi)}(1)}{C^{\phi}(0)} \right). \qquad (10)$$

The above formula can be used, e.g., to extract finite-size corrections to the utilization [26]. From (8) and (9), for a *finite* system, one finds that $C^{(\phi)}(l) =$

$D(\delta_{l,0} - 1/N)$ and from (1) and (10), $\langle u \rangle \simeq 1/4 + 1/(2\pi N)$. Clearly, the specific value $1/4$ in the thermodynamic limit and the prefactor of the $1/N$ finite-size corrections will differ from those of the actual PDES evolution with its specific "microscopic dynamics". The density of local minima, however, must remain *nonzero* and it displays universal finite-size effects [25,26,30–34],

$$\langle u \rangle_N \simeq \langle u \rangle_\infty + \mathcal{O}\left(\frac{1}{N}\right) \, , \quad \langle u \rangle_\infty \neq 0 \, , \tag{11}$$

based on the universality class (EW in one dimension) the virtual time horizon belongs to. Thus, the average progress rate of the simulation approaches a nonzero constant in the asymptotic long-time, large-N limit. For example, for the one-site-per PE basic conservative PDES scheme $\langle u \rangle_\infty \simeq 0.2464$ [25,26], due to non-universal short-range correlations between the local slopes [35].

The average width of the virtual time horizon, however, diverges as $N \to \infty$ [see (7)], making the measurement phase of the PDES scheme (data collection) *not* scalable [30]. Since the effect of very large fluctuations in the progress of the individual PEs is also important (after all, delays will be caused by state-saving difficulties on the individual nodes, where extreme events occur), we investigated the properties of the extremal-height fluctuations. We considered the average of the largest height fluctuations above the mean $\Delta_{\max} \equiv h_{\max} - \bar{h}$. The average or typical extreme-height fluctuations in the basic conservative PDES scheme exhibit the same scaling behavior as the width itself, $\langle \Delta_{\max}^2 \rangle \sim N$ [Fig. 1(b)]. This is not particularly surprising in that the extreme fluctuations emerge through the dominating collective long-wavelength modes of the "critical" surface. This finding was also observed [36] for other surface growth models belonging to KPZ universality class.

Finally, we note that, in an attempt to construct an analytically tractable model for PDES, Greenberg et al. [14] introduced the K-random model. Here at each update attempt, PEs compare their local simulated times to the local simulated times of K *randomly* chosen PEs (rechosen at every update attempt). They showed that in the $t \to \infty$, $N \to \infty$ limit the average rate of progress of the simulation converges to a *non-zero* constant, $1/(K+1)$. Further, they also showed that the evolution of the time horizon converges to a traveling wave solution described by a *finite width* of the distribution of the local times. Finally, they suggested that the qualitative properties of the K-random model are universal and hold for regular lattice models as well. As we have shown above, their latter conjecture for the width does not hold, thus, the basic conservative PDES scheme for *regular lattices* cannot be equivalently described by K-random model (at least not below the critical dimension of the KPZ universality class [28,37]). Nevertheless, their "annealed" random connection model is highly inspiring in that the underlying connection topology can have crucial effects on the universal behavior of the evolution of the virtual time horizon, and in turn, on the synchronizability of PDES schemes.

3 The Small-World Synchronized Conservative PDES Scheme

3.1 Motivation and Properties for the Synchronization Network

The divergent width and extreme-height fluctuations (with increasing N), discussed in the previous section, are the result of the divergent lateral correlation length $\xi^{(h)}$ of the virtual time surface, which reaches the system size N in the steady state [28,30]. To de-correlate the simulated time horizon, first, we modify the virtual communication topology of the PEs. The resulting communication network *must include* the original short-range (nearest-neighbor) connections to faithfully simulate the dynamics of the underlying system. In the modified network, the connectivity of the nodes (the number of neighbors) should remain *non-extensive* (i.e., only a finite number of virtual neighbors per node is allowed). This is in accordance with our desire to design a PDES scheme where no global "intervention" or synchronization is employed (PEs can only have $\mathcal{O}(1)$ communication exchanges per step). It is clear that the added synchronization links (or at least some of those) have to be long range. (Only short range links would not change the universality class and the scaling properties of the width of the time horizon). Also, fluctuations in the individual connectivity should be avoided for load balancing purposes, i.e., requiring the same number of added links (e.g., one) for each node is a reasonable constraint.

One may wonder how the collective behavior of the PDES scheme would change if each node was connected to the one located at the "maximum" possible distance away from it ($N/2$ on a ring) [Fig. 2(a)] [38]. Consider a linear coarse-grained Langevin equation with Gaussian noise where the effective strength of the added long-range links is Σ,

$$\partial_t h_i(t) = (h_{i+1} + h_{i-1} - 2h_i) - \Sigma(h_i - h_{i+N/2}) + \eta_i(t) , \qquad (12)$$

with periodic boundary conditions. After elementary calculations one obtains for the width

$$\langle w^2 \rangle = \frac{1}{N} \sum_{k \neq 0} S^{(h)}(k) = \frac{1}{N} \sum_{k \neq 0} \frac{D}{2[1 - \cos(k)] + 2\Sigma[1 - \cos(kN/2)]} , \qquad (13)$$

where $k = (2\pi n)/N$, $n = 0, 1, 2, \ldots, N\text{--}1$ as before (and N is even for simplicity). Separating the terms with even and odd n values above, we find

$$\langle w^2 \rangle = \frac{1}{N} \sum_{n=\text{odd}} \frac{D}{2[1 - \cos(2\pi n/N)] + 4\Sigma}$$
$$+ \frac{1}{N} \sum_{n=\text{even}} \frac{D}{2[1 - \cos(2\pi n/N)]} . \qquad (14)$$

The first sum yields a finite N independent value in the $N \to \infty$ limit. The second sum, on the other hand, is identical to the width of the EW model on

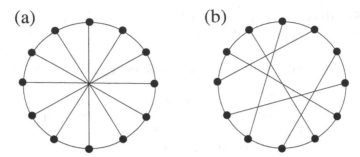

Fig. 2. Schematic diagrams for the PDES synchronization networks. (a) Maximal-distance connected network as described in the text. (b) Small-world network where each PE has exactly one quenched random neighbor.

a regular network of size $N/2$. Thus, in the large N limit the width for the "maximal-distance" connected network [Fig. 2(a)] diverges as $\langle w^2 \rangle_N \simeq DN/24$. Indeed, one can realize, that such regularly patterned long-range links make the network equivalent to a $2 \times (N/2)$ quasi one-dimensional system with only nearest-neighbor interactions and helical boundary conditions. The above extreme case suggests, that the purely maximum-range synchronization cannot work either.

We then choose the extra synchronization links in such a way that they cover all lengthscales with equal weight [21]. With the one extra link per PE constraint, we employ *quenched* random bidirectional links, i.e., each PE is connected to exactly one other PE, as illustrated on Fig. 2(b). That is, pairs of sites selected at random, and once they are linked they cannot be selected again. The resulting network resembles a (constrained) small-world-like network [20]. It differs from both the original ("rewiring") [20,39] and the "soft" version [40,41] of the Small-World (SW) network (where an Erdős-Rényi random graph is thrown on top of a regular lattice). Our construction too, however, exhibits a well balanced coexistence among short- and long-range links (random links are placed on the top of a regular substrate), and we will refer to it as a SW network in what follows. When explicit distinction is needed among the above versions of the SW networks, we will refer to our construction as the "hard" version of the SW network. This terminology is motivated by the eigenvalue spectrum of the Laplacian on the different variations of the SW networks [42,43], discussed in more detail in [43,44] and in the chapter by Hastings and Kozma in this book.

As one can expect, the average path length $\langle l \rangle_N$ (the average minimum number of links connecting two randomly chosen nodes) for our synchronization network scales *logarithmically* with the system size N [Fig 3], i.e., like most other random networks [45], it too exhibits the "small-world" character (or low-degree of separation).

We now describe the modified algorithmic steps for the SW connected PEs [21]. In the modified conservative PDES scheme, at every parallel step each PE with probability p compares its local simulated time with its *full* virtual

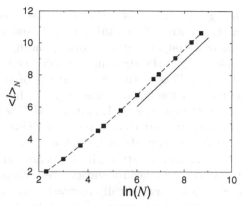

Fig. 3. Average shortest path as a function of the logarithm of the number of nodes (PEs) for our small-world synchronization network [Fig. 2(b)]. The straight line represents the slope of the asymptotic large N behavior of the average shortest path $\langle l \rangle_N \simeq 1.42 \ln(N)$.

neighborhood and can only advance if it is a neighborhood minimum, i.e., if $h_i(t) \leq \min\{h_{i-1}(t), h_{i+1}(t), h_{r(i)}(t)\}$, where $r(i)$ is the random connection of PE i. With probability $(1 - p)$ each PE follows the original scheme, i.e., the PE then can advance if $h_i(t) \leq \min\{h_{i-1}(t), h_{i+1}(t)\}$. Note that the occasional extra checking of the simulated time of the random neighbor is *not* needed for the faithfulness of the simulation. It is merely introduced to control the width of the time horizon.

3.2 Coarse-Grained Equation of Motion
for the Small-World-Coupled Conservative PDES Scheme

We now obtain a coarse-grained description for the evolution of the virtual time horizon. The occasional checking of the virtual time (at every $1/p$ parallel steps on average) through the random links introduces an effective strength $\tilde{p}$ for these links. Note that this is a dynamic "averaging" process, controlled by the parameter p, the probability of checking the random neighbor as well. The only properties we assume about $\tilde{p}(p)$ is that it is a monotonically increasing function of p and is only zero when $p=0$. The effective Langevin equation then becomes

$$\partial_t h_i(t) = (h_{i+1} + h_{i-1} - 2h_i) - \sum_{j=1}^{N} J_{ij}(h_i - h_j) + \ldots + \eta_i(t) , \qquad (15)$$

where $\eta_i(t)$ is delta-correlated Gaussian noise as in (3) and J_{ij} is proportional to the symmetric adjacency matrix of the random part of the network with exactly one non-zero element (being equal to $\tilde{p}$) in each row and column. The former property implies that $\sum_l J_{il} = \tilde{p}$ for all i, which is related to our construction that there are no fluctuations in the individual connectivity. The ... in (15) stand for *all* non-linear terms (involving non-linear interactions through the random links as well). "Phenomenological" results of simulating the simulation (Sect. 3.3) suggest that the dynamic control of the link strength and non-linearities only give rise to a renormalized coupling and a corresponding renormalized mass (in a field theory sense). Thus, the dynamics is effectively governed by EW relaxation in a small world. This motivates the study of the EW model on a SW network, i.e., keeping only the linear terms in (15). That problem is studied in detail in [43,44] and in the chapter by Hastings and Kozma in this book. A disorder-averaged systematic perturbation expansion yields an effective "mass" $\Sigma(\tilde{p}) \sim \tilde{p} + \mathcal{O}(\tilde{p}^{3/2})$ in the asymptotic small-$\tilde{p}$ limit. In our case, when non-linearities are indeed present and the strength of the random links is controlled by the relative frequency p of the synchronization steps through those links, we will only assume that $\Sigma(p)$ is a monotonically increasing function of p and is only zero when $p=0$. In the following, for brevity, $\langle \ldots \rangle$ will denote the double average: ensemble average based on the stochastic dynamics [e.g., over the noise in (15)], and disorder average over the random network realizations. The resulting steady-state structure factor (or propagator) for (15) then reads as [44]

$$S^{(h)}(k) = \frac{1}{N} \langle \tilde{h}_k \tilde{h}_{-k} \rangle = \frac{D}{2[1 - \cos(k)] + \Sigma} . \tag{16}$$

The above structure factor contains the essential properties of the SW synchronized PDES scheme at the coarse grained level. In particular, the SW links induce a finite correlation length $\xi^{(h)}$ for the surface fluctuations. In the following we will only discuss the infinite-system small-Σ behavior, when the finite-size effects vanish and the discrete-lattice effects become negligible. In this limit, $\xi^{(h)} \simeq 1/\sqrt{\Sigma}$. Also from (16), for the width of the time horizon one obtains

$$\langle w^2 \rangle = \frac{1}{N} \sum_{k \neq 0} S^{(h)}(k) \simeq \frac{1}{2\sqrt{\Sigma}} . \tag{17}$$

i.e., the width remains finite in the $N \to \infty$ limit. (Note that $\Sigma(p)$ is only zero when $p=0$.) The implication of this result for the SW synchronized PDES scheme is that the spread of the virtual time horizon will approach a finite value in the limit of infinite number of PEs for any nonzero value of p.

We now discuss some general considerations for the the utilization $\langle u \rangle$ (the average progress rate) for the SW synchronized PDES scheme. From the algorithmic rules it follows that

$$\langle u \rangle = (1 - p)\langle \Theta(-\phi_{i-1})\Theta(\phi_i) \rangle + p\langle \Theta(-\phi_{i-1})\Theta(\phi_i)\Theta(h_{r(i)} - h_i) \rangle , \tag{18}$$

where p is the probability to include the random neighbor as well in the synchronization step. Note that the disorder averaging makes the right hand side independent of i. For general p (with the random links present) it is hard to carry out quantitative approximations for the utilization. Since the height fluctuations *become* short-range correlated (16) and the local slopes *remain* short-range correlated [see discussion below, (21)], it is guaranteed that both terms in (18), and subsequently $\langle u \rangle$, *remain non-zero* for any $0 \leq p \leq 1$ [35,46]. Rearranging the terms in (18) one obtains

$$\langle u \rangle = \langle \Theta(-\phi_{i-1})\Theta(\phi_i) \rangle$$
$$- p \left[\langle \Theta(-\phi_{i-1})\Theta(\phi_i) \rangle - \langle \Theta(-\phi_{i-1})\Theta(\phi_i)\Theta(h_{r(i)} - h_i) \rangle \right] . \qquad (19)$$

The first term, $\langle \Theta(-\phi_{i-1})\Theta(\phi_i) \rangle = \langle \Theta(h_{i-1} - h_i)\Theta(h_{i+1} - h_i) \rangle$, is an increasing function of p, as the heights become less correlated [46]. For example, it would be $1/4$ for completely independent slopes, and it would be $1/3$ for completely independent heights. The actual values differ for the PDES time horizon (the slopes and heights exhibit some short-range correlations), but the above trend remains and $\langle \Theta(-\phi_{i-1})\Theta(\phi_i) \rangle$ saturates rapidly as a function of p [46]. The quantity in $[\dots]$ in (19) is always positive, bounded from zero, so it will eventually lead to the *decrease* in $\langle u \rangle$ as $\mathcal{O}(p)$, *once* $\langle \Theta(-\phi_{i-1})\Theta(\phi_i) \rangle$ saturates. For very small values of p, however, the leading order correction to $\langle \Theta(-\phi_{i-1})\Theta(\phi_i) \rangle$ may become more dominant than $\mathcal{O}(p)$. In this case, as it is clear from (19), the small-p behavior of $\langle \Theta(-\phi_{i-1})\Theta(\phi_i) \rangle$ alone yields the asymptotic small-p behavior of $\langle u \rangle$.

We now continue to discuss the density of local minima and the utilization for the coarse-grained linear model with Gaussian noise [(15) and (16)], which may capture some of the small-p features of the actual PDES time horizon. We also make the mean-field assumption that the structure factor and correlation functions are self-averaging in the large system-size limit. First, from (16) we find for the slope-slope structure factor

$$S^{(\phi)}(k) = 2[1 - \cos(k)]S^{(h)}(k) = D \left\{ 1 - \frac{\Sigma}{2[1 - \cos(k)] + \Sigma} \right\} , \qquad (20)$$

which yields

$$C^{(\phi)}(l) \simeq D \left\{ \delta_{l,0} - \Sigma \frac{e^{-l\sqrt{\Sigma}}}{2\sqrt{\Sigma}} \right\} = D \left\{ \delta_{l,0} - \frac{\sqrt{\Sigma}}{2} e^{-l\sqrt{\Sigma}} \right\} \qquad (21)$$

for the slope correlation function in the infinite system-size, small-Σ limit. The above equation shows explicitly, that the local slopes remain short-range correlated for the SW-synchronized time horizon. Using (10) and the above form of the slope correlations, in the small-Σ limit we obtain

$$\langle \Theta(-\phi_{i-1})\Theta(\phi_i) \rangle \simeq \frac{1}{4} + \frac{\sqrt{\Sigma}}{4\pi} - \frac{\Sigma}{8\pi} + \dots \qquad (22)$$

This implies that increasing the effective mass increases the density of local minima. This is not surprising, in that increasing Σ reduces the correlation length $\xi^{(h)}$ for the height fluctuations, as discussed above. Using (19) and (22), we obtain for the utilization

$$\langle u \rangle \simeq \frac{1}{4} + \frac{\sqrt{\Sigma(p)}}{4\pi} - \frac{\Sigma(p)}{8\pi} + \dots + \mathcal{O}(p) , \qquad (23)$$

where we now explicitly indicated the p-dependence of Σ. If $\Sigma(p)$ is known, more precisely, if $\Sigma(p) \sim p^s$ with $s < 2$ for small p values, the above equation becomes useful to extract the asymptotic small-p behavior of the utilization. It is instructive to consider the mean-field case when the effective strength of the random links [in (15)] scales as p, at least for small p values. Then $\Sigma(p) \sim p$ [44], and to leading order in p, one finds

$$\langle u \rangle \simeq \frac{1}{4} + \frac{\sqrt{p}}{4\pi} + \mathcal{O}(p) \ . \tag{24}$$

The above counterintuitive behavior of increasing $\langle u \rangle$ by actually synchronizing "more" is the result of the gain in $\langle \Theta(-\phi_{i-1})\Theta(\phi_i) \rangle$ $[\mathcal{O}(\sqrt{p})]$ winning over the loss due to the occasional extra random synchronizations $[\mathcal{O}(p)]$, for *asymptotically small p* values. As p is increased, (24) will not be valid anymore; $\langle \Theta(-\phi_{i-1})\Theta(\phi_i) \rangle$ starts to saturate, so the change in $\langle u \rangle$ will be dominated by the $-\mathcal{O}(p)$ factor in (19).

Some analogy between the evolution of the virtual time horizon with quenched random links added and the sliding state of charge-density waves with "no-passing" rule [47] suggests [48] that the above mean-field coarse-graining argument may break down and the average rate of progress of the SW-synchronized conservative scheme for arbirary small p is bounded by that of the $p=0$ case.

3.3 Comparison with the Simulated Small-World Synchronized PDES Results

We now turn to discussing the results obtained by simulating the actual PDES scheme, based on the specific update rules for the local simulated times (see the end of Sect. 3.1). The fundamental difference between the original and the SW-synchronized conservative scheme is illustrated on Fig. 4. The snapshots of the virtual times indicate that, indeed, the large-amplitude long-wavelength fluctuations, present in the original time horizon [Fig. 4(a)] are suppressed when the extra synchronizations through the quenched random links are implemented [Fig. 4(b)].

Since all steady-state characteristics are "encoded" in the structure factor of the virtual times, we measured this quantity, and compared it to

$$S^{(h)}(k) \propto \frac{1}{k^2 + \Sigma(p)} \ , \tag{25}$$

the small k limit of (16). Figure 5(a) shows the disorder-averaged structure factors, as well as individual realizations for various system sizes. As one can observe, finite-size effects become small and $S^{(h)}(k)$ approaches a *finite* value as $k \rightarrow 0$. Thus, there are no large-amplitude, long-wavelength modes in the virtual time horizon. Further, the inset of Fig. 5(a) confirms the "massive" behavior (25) for small k values. It is important to note that for the actual PDES time horizon, the effective mass $\Sigma(p)$ may depend non-trivially on p as a result of the dynamic control of the "link strength" and renormalization by nonlinear effects in the specific "microscopic dynamics". The form of $S(k)$, however, seems to follow the linear theory, discussed in Sect. 3.2. By plotting $1/S^{(h)}(k)$ vs. k^2

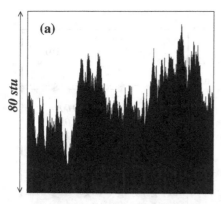
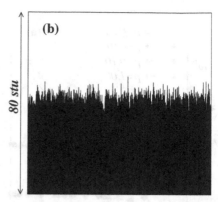

Fig. 4. Snapshots of the virtual time horizon in the steady state (a) for the original regular lattice (p=0) and (b) for the small-world synchronized (p=0.10) PDEs scheme for N=10000. The vertical scale is the same in (a) and (b).

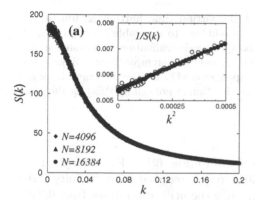
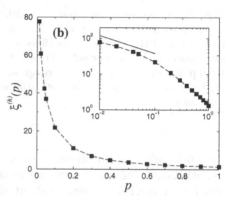

Fig. 5. (a) Steady-state structure factor of the virtual time horizon for p=0.10. In addition to ensemble averages over 100 realizations of the random links (filled symbols), single realizations (the same open symbols) are also shown. The inset shows a magnified view of $1/S(k)$ versus k^2 for small k for the largest system. The solid straight line is the best linear fit used to determine the correlation length. (b) Correlation length of the virtual time horizon as a function of the random synchronization frequency p. For reference, the solid straight line corresponds to the power-law divergence with the power $-1/2$.

[inset of Fig. 5(a)] for the measured structure factors, we can "phenomenologically" determine the correlation length of the fluctuations of the virtual times, $\xi^{(h)}(p) \simeq 1/\sqrt{\Sigma(p)}$. The results are shown in Fig. 5(b)]. The inset implies, that the asymptotic small-p scaling regime has not been reached yet at our smallest value p=0.01.

The results for the width and for the utilization also agree with the basic prediction of the coarse-grained linear model. Even for small p values, the width saturates, and approaches a system-size independent value for large N

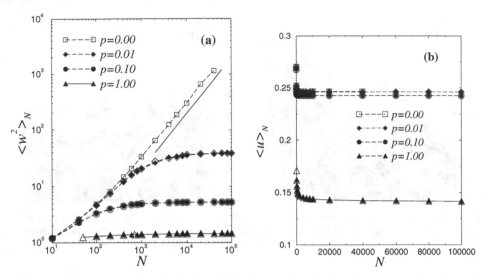

Fig. 6. (a) Average steady-state width of the virtual time horizon as function of the number of PEs for various values of p. In addition to ensemble averages over 10 realizations of the random links (filled symbols), a single realization is also shown (the same open symbols). The solid straight line represents the asymptotic one-dimensional KPZ power-law divergence with roughness exponent $\alpha=1/2$ for the $p=0$ case. Note the log-log scales. (b) The steady-state utilization (fraction of non-idling PEs) for the same cases as in (a).

[Fig. 6(a)] while the utilization remains non-zero [Fig. 6(b)]. For example, for a hypothetically infinite system, for $p=0.01$, $\langle w^2 \rangle$ is reduced from "infinity" (the width for the KPZ surface) to about 40, while the utilization drops from 0.2464 only to about 0.2460. For $p=0.10$, the width is further reduced to about 5, while the utilization is down only to 0.242. One can also observe the clear self-averaging property for both global observables (the width and the utilization), i.e., their values become independent of the realization of the SW network for large enough N.

To extract the asymptotic small-p behavior of the width and the utilization, one would need larger system sizes, longer steady-state PDES time series, and more network realizations to obtain reliable statistics, and to compare all aspects of the linearized coarse-grained model with the actual PDES simulations. In particular, it would be interesting to see, whether the utilization increases initially for sufficiently small p values (the subtle prediction of the coarse-grained linear theory). We have not observed this, but the systems we simulated have not yet reached their asymptotic scaling regime [inset of Fig. 5(b)]. Further, finite-size corrections and error bars may become comparable to this possible asymptotically small effect in $\langle u \rangle$.

3.4 Extremal Fluctuations of the Virtual Time Horizon

In addition to the average value of the fluctuations of the local field variables (such as the height in the context of surface growth models), the typical value of the largest fluctuations can also be of great importance [49–51] in a number of applications. For example, in load balancing networks [52] or state-saving schemes for PDES schemes [30,53], extreme (load or accumulated data) fluctuations on an *individual* node will cause the delays. Thus, in interacting multi-component systems such as the above examples, failures or delays are triggered by extreme-events occurring on the individual components [51].

Relationship between extremum statistics and universal fluctuations in *correlated* systems have been discussed intensively in recent years. [36,54–60]. For the original PDES scheme (p=0, regular lattice synchronization) exhibiting a KPZ-like rough (or critical) surface, we illustrated (Sect. 2) that the extremal fluctuations of the time horizon diverge in the same fashion as the width itself [Fig. 1(b)]. We now discuss to what extent SW synchronizations lead to the suppression of the extreme-height fluctuations in the virtual time horizon [53], closely related to the measurement scalability of the conservative PDES scheme.

First, consider N *independent* identically distributed stochastic variables with a complementer cumulative distribution $P_>(x)$ (the probability that the individual stochastic variable is greater than x). Then the cumulative distribution $P_<^{\max}(x)$ for the largest of the N events (the probability that the extremal value is less than x) can be approximated as [60,61]

$$P_<^{\max}(x) = [P_<(x)]^N = [1 - P_>(x)]^N = e^{N \ln[1 - P_>(x)]} \simeq e^{-NP_>(x)} , \qquad (26)$$

where one typically assumes that the dominant contribution to the statistics of the extremes comes from the tail of the individual distribution $P_>(x)$. For example, for exponentially-tailed individual variables, $P_>(x) \simeq e^{-cx}$, the above equation yields

$$P_<^{\max}(x) \simeq e^{-e^{-cx+\ln(N)}} . \qquad (27)$$

Thus, the sequence of scaled variables $\tilde{x} = c(x - \ln(N)/c)$ asymptotically approaches the standard Fisher-Tippett-Gumbel (FTG) distribution [49,50]

$$\tilde{P}_<^{\max}(\tilde{x}) \simeq e^{-e^{-\tilde{x}}} \qquad (28)$$

with mean $\langle \tilde{x} \rangle = \gamma$ ($\gamma = 0.577 \ldots$ being the Euler constant) and variance $\langle \tilde{x}^2 \rangle - \langle \tilde{x} \rangle^2 = \pi^2/6$. It immediately follows that the average value of the largest of the N original random variables then scales as

$$\langle x^{\max} \rangle = \gamma/c + \ln(N)/c \simeq \ln(N)/c \qquad (29)$$

for large N values. When comparing with simulation or experimental data, it is often convenient to use the scaled variables $\hat{x} = (x - \langle x \rangle)/\sigma_x$ which for the above case yields the FTG limit distribution with zero mean and unit variance

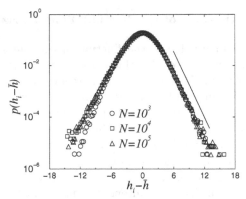

Fig. 7. Disorder-averaged probability density (histogram) for the individual simulated time fluctuations for various system sizes at p=0.10 (log-normal scales). The solid straight line indicates a pure exponential tail.

$$\hat{P}_<^{\max}(\hat{x}) \simeq e^{-e^{-(a\hat{x}+\gamma)}} , \tag{30}$$

where $a=\pi/\sqrt{6}$. Note that with *appropriately* chosen scaled variables, the convergence to the FTG distribution holds not only for exponential variables, but also for more general ones with "exponential-like" tails $P_>(x) \simeq e^{-cx^\delta}$ (i.e., decaying faster than any power law) [49,50,60,61]. For any $\delta \neq 1$, however, the convergence to (28) or (30) is *extremely* (logarithmically) slow [61].

For the SW synchronized PDES scheme with N PEs we showed that a finite correlation length $\xi^{(h)}(p) \simeq 1/\sqrt{\Sigma(p)}$ effectively decouples the local simulated times. Then, the extreme-value limit theorems can be applied [60,61] using the number of independent blocks $N/\xi^{(h)}$ in the system. Further, we found [62] that the tail of disorder-averaged distribution of the individual relative height fluctuations (independent of the site i) are simple exponentials $P_>(h_i - \bar{h}) \simeq \exp[-c(h_i - \bar{h})/w]$ with $w \equiv \sqrt{\langle w^2 \rangle}$. The histogram for the corresponding probability *density* function, $p(h_i - \bar{h})$, is shown in Fig. 7.

From the general extreme-value limit theorems, discussed above, it follows that the scaled extreme-height fluctuations are governed by the FTG distribution (30) (if scaled to zero mean and unit variance). Further, from (29), the average maximum relative height, $\Delta_{\max} = h_{\max} - \bar{h}$, will scale as

$$\langle \Delta_{\max} \rangle \simeq \frac{w}{c} \ln(N/\xi^{(h)}) \simeq \frac{w}{c} \ln(N) , \tag{31}$$

where we dropped all N-independent terms. Note that both w and $\xi^{(h)}$ approach their finite N-independent values for any non-zero p, and the only N dependent factor is $\ln(N)$ for large N values.

Agreement between the simulated PDES extremal fluctuations and the above considerations are rather convincing. In Fig. 8(a) we show the scaled histograms (to zero mean and unit variance) for the extreme-height fluctuations together with the probability *density*, corresponding to (30)

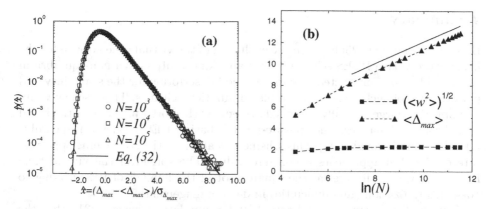

Fig. 8. Extreme fluctuations in the the small-world synchronized PDES time horizon for $p=0.10$. (a) Probability density of the scaled extremal fluctuations and comparison with the FTG density equation (32). (b) Average of the extremal fluctuations (also shown is the width for comparison). The solid straight line indicates the logarithmic divergence.

$$\hat{p}(\hat{x}) \simeq ae^{-(a\hat{x}+\gamma)-e^{-(a\hat{x}+\gamma)}} \; , \tag{32}$$

We note again, that the underlying reason for the fast convergence to the FTG density of the simulated time horizon is that the local relative height distributions exhibit *pure* exponential tails. Also, for the more general distribution $P_>(h_i - \bar{h}) \simeq \exp[-c((h_i - \bar{h})/w)^\delta]$, the approach to the FTG limit distribution would be very slow and the corresponding maximum fluctuations would scale as $\sim[\ln(N)]^{1/\delta}$ [53,61,62], as opposed to (31).

In Fig. 8(b) we show the average of the largest fluctuations above the mean, $\Delta_{\max}$, for the simulated PDES time horizon. The figure confirms that for large enough N (when $\langle w^2 \rangle$ essentially becomes system-size independent) $\Delta_{\max}$ increases *logarithmically* with the system size, according to (31) [Fig. 8(b)]. Simulation results for the actual PDES scheme also indicate [62] that the largest deviations below the mean, $\Delta_{\min} = \bar{h} - h_{\min}$, and the maximum separation, $\Delta = h_{\max} - h_{\min}$, scale the same way as $\Delta_{\max}$, i.e., diverge logarithmically with the system size. Note, that similar to the width and the utilization, the extremal height fluctuations are also self-averaging [53,62].

The implication of these findings is that while the width becomes finite for SW-synchronized virtual times, any node can exhibit fluctuations of size $\mathcal{O}(\ln(N))$ in its local simulated time (related to the local memory need). We refer to this property as "marginally" scalable for the measurement phase (due to the weak logarithmic divergence). This property still ensures synchronization in a practical sense for the SW-synchronized PDES scheme with millions of PEs. Note, that this logarithmic system-size dependence of the extreme fluctuations is generic to coupled multi-component system, where some local relaxational dynamics is extended to a SW network [53,62].

4 Summary

Based on a mapping [25] between the evolution of the virtual time horizon for the basic conservative PDES scheme [11,12] and kinetically grown non-equilibrium surfaces [28], we constructed a coarse-grained description for the scalability and performance of such large-scale parallel simulation schemes. These schemes can be applied to large spatially extended systems with short-range interactions and asynchronous dynamics. The one-site-per PE basic PDES was shown to exhibit KPZ-like kinetic roughening. This scheme is scalable in that the average progress rate of the PEs approaches a non-zero value. The spread of the virtual time horizon, however, diverges as the square root of the number of PEs, leading to "de-synchronization" and difficulties in data management.

Universality arguments, and actual PDES simulations suggest [31], that the above characteristics generically hold for any underlying system with short-range interactions for any finite number of sites per PE implementations. Possible idling due to the conservative synchronization rules and actual communication times can be greatly suppressed by each PE carrying a large block of sites [11, 12], yielding encouraging values for the utilization for actual implementations [8]. When the PEs carry many sites, however, the saturation value of the width can become extremely large. More precisely, there is an additional fast-roughening phase for early times when the evolution of the time horizon corresponds to random deposition [31]. Subsequently, it will cross over to the KPZ growth regime and finally saturate. This further motivates the need for some sort of extra synchronizations to suppress the roughness of the time horizon.

Our goal here was to achieve synchronization without any global intervention. We constructed a specific version of the SW network, where each PE was connected to exactly one other randomly chosen PE. The extra synchronizational steps through the random links are merely used to control the width. The virtual time horizon for the SW-synchronized PDES scheme becomes "macroscopically" smooth and essentially exhibits mean-field like characteristics. The random links, on top of a regular lattice, generate an effective "mass" for the propagator of the virtual time horizon, corresponding to a nonzero correlation length. The width becomes finite, for an arbitrary small rate of synchronization through the random links, while the utilization remains nonzero, yielding a fully scalable PDES scheme. The former statement is only marginally weakened by observing that the extreme fluctuations in the time horizon can exhibit logarithmically large values as a function of the total number of PEs. The above predictions of the coarse-grained PDES model were confirmed by actually "simulating the simulations".

The generalization when random links are added to a higher-dimensional underlying regular lattice is clear: since our construction of the SW network ("hard" version) on a one-dimensional regular substrate is already mean-field, in higher dimensions it will be even more so [44] (i.e., the critical dimension of our SW network is less than one). Note that synchronizability on scale-free networks [45,63] was also studied recently. The results indicate that a PDES scheme is marginally scalable if the communication topology between the PEs is a scale-free network [35]. The implication of this finding is that the internet, which is already exploited for distributed computing for mostly "embarrassingly parallel" problems through existing GRID-based schemes [16,17], may have the potential to accommodate efficient complex system simulations (such as asynchronous PDES) where the nodes frequently have to synchronize with each other.

The above construction of a fully scalable algorithm for simulating large systems with asynchronous dynamics and short-range interactions is an example for the enormous "computational power and synchronizability" [20] that can be achieved by SW couplings. The suppression of critical fluctuations of the virtual time horizon is also closely related to the emergence of mean-field-like phase transitions and phase ordering in *non-frustrated* interacting systems [1,64–70].

Recent theoretical work also supports [43,44,71] that systems without inherent frustration exhibit strict or anomalous mean-field characteristics when the original short-range interaction topology is modified to a SW network.

Acknowledgements

Discussions with G. Györgyi, M.B. Hastings, G. Istrate, B. Kozma, B.D. Lubachevsky, Z. Rácz, and P.A. Rikvold, are gratefully acknowledged. G.K. and M.A.N thank CNLS, Los Alamos National Laboratory for their hospitality during their stays in summer 2002 and 2003 where part of this work was completed. We acknowledge the support of US NSF through Grant No. DMR-0113049 and the support of the Research Corporation Grant No. RI0761. Z.T. is supported by the US DOE under contract W-7405-ENG-36. H.G. was also supported in part by the Los Alamos summer student program in 2002 and 2003, DOE W-7405-ENG-36.

References

1. S.H. Strogatz, Nature **410**, 268 (2001).
2. R. Fujimoto, Commun. ACM **33**, 30 (1990).
3. D.M. Nicol, R.M. Fujimoto, Ann. Oper. Res. **53**, 249 (1994).
4. B.D. Lubachevsky, Bell Labs Tech. J. **5** April-June 2000, 134 (2000).
5. A.G. Greenberg et al. in *Proc. 8th Workshop on Parallel and Distributed Simulation (PADS'94), Edinburgh, UK, 1994* (SCS, San Diego, CA, 1994), p. 187.
6. E. Deelman, B.K. Szymanski, T. Caraco, in *Proc. 28th Winter Simulation Conference*, (ACM, New York, 1996), p. 1191.
7. D.M. Nicol, *Proc. 1988 SCS Multiconference*, Simulation Series, SCS, Vol. 19, p. 141 (1988).
8. G. Korniss, M.A. Novotny, P.A. Rikvold, J. Comput. Phys. **153**, 488 (1999).
9. G. Korniss, C.J. White, P.A. Rikvold, M.A. Novotny, Phys. Rev. E **63**, 016120 (2001).
10. G. Korniss, P.A. Rikvold, M.A. Novotny, Phys. Rev. E **66**, 056127 (2002).
11. B.D. Lubachevsky, Complex Syst. **1**, 1099 (1987).
12. B.D. Lubachevsky, J. Comput. Phys. **75**, 103 (1988).
13. D.R. Jefferson, ACM Trans. Prog. Lang. and Syst. **7**, 404 (1985).

14. A.G. Greenberg, S. Shenker, and A.L. Stolyar, Performance Eval. Rev. **24**, 91 (1996).
15. *Blue Gene/L project*, partnership between IBM and DoE, announced Nov.9, 2001; 64K processors, expected scale 200 teraflops, a step towards a petaflop scale; expected completion 2005; IBM Research Report, RC22570 (W0209-033) September 10, 2002.
16. See, e.g., www.gridforum.org and setiathome.ssl.berkeley.edu.
17. S. Kirkpatrick, Science **299**, 668 (2003).
18. P.M.A. Sloot, B.J. Overeinder, A. Schoneveld, Comput. Phys. Commun. **142**, 76 (2001).
19. P. Bak, C. Tang, K. Wiesenfeld, *Phys. Rev. Lett.* **59**, 381 (1987).
20. D.J. Watts and S.H. Strogatz, Nature **393**, 440 (1998).
21. G. Korniss, M.A. Novotny, H. Guclu, Z. Toroczkai, and P.A. Rikvold, Science **299**, 677 (2003).
22. K.M. Chandy, J. Misra, *Commun. ACM* **24**, 198 (1981).
23. R.E. Felderman and L. Kleinrock, ACM Trans. Model. Comput. Simul. **1**, 386 (1991).
24. D.M. Nicol, ACM Trans. Model. Comput. Simul. **1**, 24 (1991).
25. G. Korniss, Z. Toroczkai, M.A. Novotny, and P.A. Rikvold, Phys. Rev. Lett. **84**, 1351 (2000).
26. Z. Toroczkai, G. Korniss, S. Das Sarma, and R.K.P. Zia, Phys. Rev. E **62**, 276 (2000).
27. M. Kardar, G. Parisi, Y.-C. Zhang, Phys. Rev. Lett. **56**, 889 (1986).
28. A.-L. Barabási and H.E. Stanley, *Fractal Concepts in Surface Growth* (Cambridge University Press, Cambridge, 1995).
29. S.F. Edwards and D.R. Wilkinson, Proc. R. Soc. London, Ser A **381**, 17 (1982).
30. G. Korniss, M.A. Novotny, A.K. Kolakowska, and H. Guclu, SAC 2002, *Proceedings of the 2002 ACM Symposium on Applied Computing*, pp. 132-138, (2002).
31. A. Kolakowska, M. A. Novotny, and G. Korniss, Phys. Rev. E **67**, 046703 (2003).
32. A. Kolakowska, M. A. Novotny, and P.A. Rikvold, Phys. Rev. E **68**, 046705 (2003).
33. A. Kolakowska and M. A. Novotny, Phys. Rev. B **69**, 075407 (2004).
34. J. Krug and P. Meakin, J. Phys. A **23**, L987 (1990).
35. Z. Toroczkai, G. Korniss, M. A. Novotny, and H. Guclu, in *Computational Complexity and Statistical Physics*, edited by A. Percus, G. Istrate, and C. Moore, Santa Fe Institute Studies in the Sciences of Complexity Series (Oxford University Press, 2004, in press); arXiv:cond-mat/0304617 (2003).
36. S. Raychaudhuri, M. Cranston, C. Przybyla, and Y. Shapir, Phys. Rev. Lett. **87**, 136101 (2001).
37. E. Marinari, A. Pagnani, G. Parisi, and Z. Rácz, Phys. Rev. E **65**, 026136 (2002).
38. Z. Rácz, private communications.
39. D.J. Watts, *Small Worlds* (Princeton Univ. Press, Princeton, 1999).
40. M.E.J. Newman, J. Stat. Phys. **101**, 819 (2000).
41. M.E.J. Newman and D.J. Watts, Phys. Lett. A **263**, 341 (1999).
42. R. Monasson, Eur. Phys. J. B **12**, 555 (1999).
43. B. Kozma and G. Korniss, in *Computer Simulation Studies in Condensed Matter Physics XVI*, edited by D.P. Landau, S.P. Lewis, and H.-B. Schüttler, Springer Proceedings in Physics (Springer, Berlin, 2004, in press); arXiv:cond-mat/0305025 (2003).
44. B. Kozma, M.B. Hastings, and G. Korniss, Phys. Rev. Lett. **92**, 108701 (2004).

45. Réka Albert and Albert-László Barabási, Rev. Mod. Phys. **74**, 47 (2002).
46. H. Guclu, G. Korniss, M.A. Novotny, and Z. Toroczkai, in preparation.
47. A.A. Middleton, Phys. Rev. Lett. **68**, 670 (1992).
48. M.B. Hastings, private communications (2003).
49. R.A. Fisher and L.H.C. Tippett, Proc. Camb. Philos. Soc. **24**, 180 (1928)
50. E.J. Gumbel, *Statistics of Extremes* (Columbia University Press, New York, 1958).
51. *Extreme Value Theory and Applications*, edited by J. Galambos, J. Lechner, and E. Simin (Kluwer, Dordrecht, 1994).
52. Y. Rabani, A. Sinclair, and R. Wanka, *Proceedings of the 39th Annual Symposium on Foundations of Computer Science* (IEEE Comput. Soc, Los Alamitos, CA, 1998) pp. 694-703.
53. H. Guclu and G. Korniss, arXiv:cond-mat/0311575 (2003).
54. S.T. Bramwell, P.C.W. Holdsworth, and J.-F. Plinton, Nature **396** 552 (1998).
55. S.T. Bramwell *et al.*, Phys. Rev. Lett. **84** 3744 (2000).
56. S.T. Bramwell *et al.*, Phys. Rev. E **63** 041106 (2001).
57. T. Antal, M. Droz, G. Györgyi, and Z. Rácz, Phys. Rev. Lett. **87**, 240601 (2001).
58. S. C. Chapman, G. Rowlands, and N. W. Watkins, Nonlinear Processes in Geophysics **9**, 409 (2002); arXiv:cond-mat/0106015.
59. V. Aji and N. Goldenfeld, Phys. Rev. Lett. **86**, 1007 (2001).
60. J.-P. Bouchaud and M. Mézard, J. Phys. A **30**, 7997 (1997).
61. A. Baldassarri, *Statistics of Persistent Extreme Events*, Ph.D. Thesis, De l'Université Paris XI Orsay (2000).
62. H. Guclu and G. Korniss, in preparation.
63. A.-L. Barabási and R. Albert, Science **286** 509 (1999).
64. A. Barrat, M. Weigt, Eur. Phys. J. B **13**, 547 (2000).
65. M. Gitterman, J. Phys. A **33**, 8373 (2000).
66. B.J. Kim et al, Phys. Rev. E **64**, 056135 (2001).
67. H. Hong, B.J. Kim, M.Y. Choi, *Phys. Rev. E* **66**, 018101 (2002).
68. H. Hong, M.Y. Choi, and B.J. Kim, Phys. Rev. E **65**, 047104 (2002).
69. C.P. Herrero, Phys. Rev. E **65**, 066110 (2002).
70. M.A. Novotny and Shannon M. Wheeler, Braz. J. Phys., *Proceedings for the III Brazilian Meeting on Simulational Physics* (2003); arXiv:cond-mat/0308602 (2003).
71. M.B. Hastings, Phys. Rev. Lett. **91**, 098701 (2003).

Critical Phenomena in a Small World

Matthew B. Hastings[1] and Balázs Kozma[2]

[1] Theoretical Division and Center for Nonlinear Studies, Los Alamos National Laboratory, Los Alamos, NM, 87545 USA
[2] Department of Physics, Applied Physics, and Astronomy, Rensselaer Polytechnic Institute, 110 8th Street, Troy, NY 12180 USA

Abstract. We consider the behavior of various systems on a small-world network near a critical point. Our starting point is a different, nonrandom system with combined short- and long-range interactions. We analyze this model and find that the critical behavior is mean-field in general, with mean-field amplitudes that depend in an anomalous way on the strength of the long-range interaction. We then compare this model to the original small-world model, and derive a general criterion which determines when the two models have the same scaling behavior. The criterion can be applied to a variety of equilibrium statistical mechanics models as well as to various non-equilibrium processes. Finally, we apply these results to the specific case of the Edwards-Wilkinson equation. There, we find that the mean-field behavior is valid for $d > 2$ dimensions, but that for $d \leq 2$ dimensions there is anomalous scaling.

1 Introduction

The study of systems on networks [1] is a very general field. The systems studied have included both equilibrium systems, such as Ising or XY models or percolation problems, and non-equilibrium processes, such as epidemic spreading. The networks themselves can be any general collection of nodes and vertices. However, one particularly important class of networks is the small-world network model [2]. The purpose of this paper is simply to show that a wide variety of systems on the small-world can be described by a combination of mean-field theory with known results in scaling theory.

The small-world network model is a particular network model that combines both long-range and short-range aspects, and interpolates between regular lattices and random graphs. In this model, a regular lattice is modified by either randomly "re-wiring" links or else by simply randomly adding long-range links. We will choose in general the latter case: adding links to a regular lattice. The model has some notion of locality, as most of the links remain the same as that of the original underlying lattice. This leads to clustering in the system: with an underlying square lattice, for example, if there is a path from node i to node j to node k to node l, then there is often a direct path from node i to node l, in the case that the 4 given nodes lie around a plaquette on the lattice. However, this model also includes the "small world effect", that the average path length between sites on the network scales only as the logarithm of the network size. While many other networks also possess this small world effect, throughout this

M.B. Hastings and B. Kozma, Critical Phenomena in a Small World, Lect. Notes Phys. **650**, 277–297 (2004)

paper we consider only the specific case with long-range links added to a regular lattice, referring for brevity to this network, and to various systems on this network, simply as the small-world model.

To define the small-world model, one considers all pairs of nodes. Then, with probability p/V, where V is the total number of nodes in the system, an additional long-range link is added connecting those two nodes. Thus, in a region of linear size l on the original lattice, such that $pl^d \approx 1$, one expects to find a long-range link. This already gives some indication that at short distances the model is described by the local behavior of the underlying lattice, while on long length scales the model becomes highly connected and some kind of mean-field theory will take over. However, we will see below a crucial difference: for many models of equilibrium statistical mechanics the length scale at which this crossover occurs is much larger than l, as the number of long-range links in the given length scale will be much larger than unity.

Equilibrium statistical systems have been well studied on small-world systems [3–6], with traditional mean-field critical exponents being typically observed. This contrasts with the behavior on scale-free networks [7,8]; for scale-free networks the absence of critical fluctuations still leads to mean-field behavior, but the mean-field exponents in this case typically differ from the traditional mean-field exponents found in systems with finite connectivity. We will consider only the first case, the small-world case.

However, the presence of quenched randomness makes the small-world model difficult to treat analytically. Thus, in the next section, we thus first consider a different model which is easier to handle, lacking quenched randomness [9]. Instead of a fixed set of long-range links connecting pairs of sites with probability p/V, we will connect every site to every other site by weak links of strength p/V. We will refer to this model as the long-range model. In many cases, depending on inequality (11) below, the long-range model gives the same universal critical behavior as the small-world model, thus explaining the critical phenomena in the small-world system. This inequality will play the role of a modified Harris criterion [10] for the small-world system. We show the existence of mean-field critical behavior in the long-range model, albeit with anomalous exponents describing the width of the critical region and various mean-field amplitudes. These anomalous exponents can complicate the interpretation of numerical data. Finally, we will consider a non-equilibrium case. We will find that the stationary statistics of the non-equilibrium case are described by a very similar theory to that in equilibrium, while the relaxation to a stationary state can be described by a branching process.

Further, in some cases, the long-range model may be more appropriate than the usual small-world model. In the spread of a disease, for example, people tend to spread the disease to those geographically nearby (the regular lattice). There is a chance of a long-range spread of the disease, if, for example, a given person travels a long distance, typically by air. However, these links are not fixed. A given person may spread a disease from one geographical region to another geographical region by travelling by air. However, the next time a contact is formed between people in these two different geographic regions, it is very unlikely that it will be the same person responsible. Thus, a slight probability of long-range contact between any two people may be a better description than a set of fixed long-range links between people, so that at the level of people, the long-range model is more appropriate. At the level of larger social groupings, such as cities, fixed links may become a more appropriate description than the long-range model: certain cities are major hubs for air travel and thus are exposed more to long-range contact than others.

The results in the first section will apply to a wide variety of specific systems. In the next section, we will consider a specific system: the Edwards-Wilkinson equation. The study of this equation will exemplify our criterion (11) for the equivalence of the small-world and long-range models, as we will see that it provides a rare case in which the criterion fails [11]. There are a number of reasons to study this equation: on the theoretical side, the equation is linear, simplifying the analysis, but the randomness still leads to interesting properties. On the practical side, the scaling properties of the EW model on regular or random networks are directly relevant to the scalability and synchronizability of parallel discrete-event simulations (PDES) [12,13]. In PDES schemes each processing element performs a series of update attempt. A successful update leads to an advance of the processing elements simulated time [14,15]. There is some randomness in the rate at which the simulated time advances compared to real time. Further, neighboring processing elements synchronize with each other: a given processing element cannot get too far ahead of the elements with which it interacts. This need for synchronization between neighboring processing elements leads to a Kardar-Parisi-Zhang (KPZ)-like kinetic roughening [16,17] of the simulated time horizon [12,18]. For a one-dimensional chain or ring of processing elements, the steady-state performance of the PDES scheme is governed by the EW Hamiltonian [12]. In particular, for V processing elements, the width of the simulated time horizon *diverges* as $V^{1/2}$, seriously hindering efficient data collection and state saving [13,19]. Since scalable data management crucially depends on the finiteness of the width of the time horizon (the spread of the progress of the individual processing elements), one must suppress the diverging fluctuations of the simulated time horizon. As an alternative to costly and frequent global synchronizations among the processing elements, an autonomous small-world synchronization scheme was demonstrated to work [13]. In this scheme synchronization also takes place between processing elements which are connected by long-range links on the small-world topology. This finding provided another concrete example for synchronizability in generalized multi-agent systems facilitated by a small-world network [2,20].

2 Long-Range Versus Small-World

2.1 Long-Range Model

The basis for many results in this paper is the long-range model. To define this model, we again start with a regular lattice of V sites in d dimensions. Rather than adding long-range links with probability p, we give each site of the lattice a weak coupling, of order p/V, to every other site in the lattice [21]. We will find that many results can be obtained on this system by combining mean-field with standard renormalization group techniques. It is interesting to compare this model to a related mean-field solution of path lengths on a small-world network [22].

In many cases this model will enable us to extract the scaling properties of the small-world model. Our general strategy will be to study the case of small p. Locally, there is a clear distinction between the small-world and long-range models: a given site in the small-world network either has a long-range link or it does not. However, looking at large regions containing many links we may expect that the random fluctuations in the number of links in the small-world network become unimportant. As p becomes small, we will find that there is a universal behavior controlled by these large regions, such that the details of the local interaction become unimportant, and such that the small-world and long-range models often coincide.

2.2 Equilibrium Statistics

In this subsection, we consider models from equilibrium statistical mechanics. Generally, we will consider models with uniform, ferromagnetic couplings, such as Ising models, XY models, etc..., as our approach will be suited to cases in which the long-range interactions can be handled by introducing a single mean-field, corresponding to an average magnetization. However, it may be possible to handle certain frustrated models. For example, a model with ferromagnetic short-range interactions and antiferromagnetic long-range interactions can again be handled by introducing a mean-field. Models with spin-glass interactions may also be amenable to these techniques using the replica trick. These possibilities will not be considered below.

The models we consider can be represented by introducing a field $\phi(x)$, where x labels lattice sites and where ϕ has $n = 1, 2, \ldots$ components, with a partition function

$$Z = \sum_{\{\phi\}} \exp\left[-S[\phi]\right], \tag{1}$$

where $S = E/kT$ is a statistical weight for a configuration of energy E at temperature T.

For a model on a regular d-dimensional lattice, $S[\phi] = S_{\text{local}}[\phi]$, where $S_{\text{local}}[\phi]$ includes only short-range interactions. We refer to this as the local system. We choose instead for the long-range model a statistical weight with additional long-range couplings of strength p/V:

$$S[\phi] = S_{\text{local}}[\phi] - \frac{p}{2V} \sum_{x_1,x_2} \phi(x_1) \cdot \phi(x_2). \tag{2}$$

We neglect the temperature dependence of the second term on the right-hand side of (2) in what follows, as it leads to corrections to physical quantities which are higher order in p. The temperature dependence of $S[\phi]$ will be contained completely within $S_{\text{local}}[\phi]$.

Now, decouple the long-range interaction to find

$$Z = \int_{R^n} d^n h \, \exp\left[-\frac{Vh^2}{2p}\right] Z(h), \tag{3}$$

where h has n components and where

$$Z(h) = \sum_{\{\phi\}} \exp\left[-S_{\text{local}}[\phi] + h \cdot \sum_x \phi(x)\right]. \tag{4}$$

Here, $Z(h)$ is equal to the partition function of the local system in the presence of a magnetic field, h. Defining $h = |h|$, then $Z(h) = Z(h)$, as the value of the partition function does not depend on the direction of h.

For p large, the long-range interaction outweighs the short-range interaction, and the system can be approximately solved by mean-field theory. As mentioned above, we consider precisely the opposite case: when p is small. This is the only case in which we expect the dependence of physical quantities, such as the specific heat, on p to be universal. Then, the critical point of the long-range system, $\tilde{T}_c$, is close to the critical point of the local system, T_c. Thus, we can use scaling laws for the local system to find the behavior of various physical quantities. Define the magnetization $m = |m|$, where $m = (1/V) \sum_x \phi(x)$. Then, the magnetization, m, obeys $m(T = T_c, h) = A_m h^{1/\delta}$, defining the critical exponent δ. For $T > T_c$, the susceptibility χ obeys $\chi(T, h = 0) = A_\chi^+ |T - T_c|^{-\gamma}$. In general, we can write a scaling function: $m = h^{1/\delta} f((T - T_c) h^{-1/(\delta\beta)})$. For $h >> (T - T_c)^{\delta\beta}$, we use the first result $m(T = T_c, h) = A_m h^{1/\delta}$, while for $h << (T - T_c)^{\delta\beta}$, we use the second $m(T = T_c, h) = A_\chi^+ h |T - T_c|^{-\gamma} - B_\chi h^3 |T - T_c|^{-\gamma - 2\delta\beta}$, where we have added the h^3 term in the expansion of m.

The magnetization is defined by $\partial \ln Z(h)/\partial h = mV$. Thus,

$$Z(h) = Z(0) \exp\left[\int_0^h dh' \, m(h')V\right]. \tag{5}$$

We now consider $Z(h)$ in two different limits, depending on whether h is much greater or much smaller than $(T - T_c)^{\delta\beta}$.

We first consider the case of T near T_c so that $h >> (T - T_c)^{\delta\beta}$. Then, the dominant contribution to (5) arises from $h' >> (T - T_c)^{\delta\beta}$ and thus (3) gives

$$Z = Z(0) \int\limits_{R^n} d^n h \exp\left[-\frac{Vh^2}{2p} + \frac{A_m V h^{1+1/\delta}}{1+1/\delta}\right]. \tag{6}$$

Since V is taken large, we can use a saddle point to arrive at $h = (pA_m)^{\delta/(\delta-1)}$, or

$$m = A_m (pA_m)^{1/(\delta-1)}. \tag{7}$$

The correlation length ξ of the local system in the presence of this field is proportional to $m^{-\nu/\beta}$, and hence diverges as $p \to 0$. This correlation length sets the scale at which the local lattice is important. Beyond the correlation length, the correlations in the long-range system are controlled by the averaged field h, so that $\langle \phi(0) \cdot \phi(x) \rangle$ approaches m^2 for $x \gg \xi$. Up to this length, the short-range system controls the correlations, and the correlation function $\langle \phi(0) \cdot \phi(x) \rangle$ has a power law decay.

We have seen that at $T = T_c$, the system has a net magnetization, and thus $T_c < \tilde{T}_c$. To study the transition itself, we now consider the second case, $h \ll (T - T_c)^{\delta\beta}$. Now, (3) and (5) give

$$Z = Z(0) \int\limits_{R^n} d^n h \exp\left[-\frac{Vh^2}{2p} + \frac{A_\chi^+ V h^2 |T - T_c|^{-\gamma}}{2} - \frac{B_\chi V h^4 |T - T_c|^{-\gamma-2\delta\beta}}{4}\right]. \tag{8}$$

This equation can again be studied using a saddle point. The term quadratic in h changes sign at the critical point $\tilde{T}_c$, determined by

$$\tilde{T}_c - T_c = (pA_\chi^+)^{1/\gamma}. \tag{9}$$

One finds that $m = h = 0$ for $T > \tilde{T}_c$. This formal determination of the modified critical temperature can be expressed in a much more physical manner: the uniform long-range links are equivalent to studying the system in a net field h, where h is determined self-consistently by $h = pm$. For a small field $|h|$, the magnetization is determined by the susceptibility: $m = \chi h + \cdots$, where the $\cdots$ represent terms of order h^3 and higher. Combining these two results we find $h = p\chi h + \cdots$. For $p\chi < 1$, this equation cannot be satisfied self-consistently with non-zero m: the field h does not produce sufficient m to lead to the given h. For $p\chi > 1$, however, the response to the applied field h exceeds the field h itself, leading to a non-vanishing m. Thus, we identify the critical point by the criterion $p\chi = 1$. This self-consistent calculation is valid in the large V limit when m has no fluctuations; thus, it is equivalent to the saddle-point approximation above.

Slightly below the critical point we find $h = \sqrt{\tilde{T}_c - T}\sqrt{\frac{\gamma A_\chi^+}{B_\chi}}(\tilde{T}_c - T_c)^{\delta\beta-1/2}$, and a magnetization given by (using the scaling law $\delta\beta - \gamma = \beta$ [23])

$$m = \sqrt{\tilde{T}_c - T} A_\chi^+ \sqrt{\frac{\gamma A_\chi^+}{B_\chi}}(\tilde{T}_c - T_c)^{\beta-1/2}. \tag{10}$$

Thus, the magnetization behaves as $m = \tilde{A}\sqrt{\tilde{T}_c - T}$, with $\tilde{A} \propto p^{(\beta-1/2)/\gamma}$. If the local system is described by mean-field theory, then $\beta = 1/2$ and A does not diverge as $p \to 0$. In other cases, $\beta < 1/2$, and the mean-field amplitude $\tilde{A}$ diverges for small p. The specific heat of the system can be obtained by differentiating the partition function (8) twice with respect to temperature. The partition function $Z(0)$ is analytic in T at $T = \tilde{T}_c$, while the integral over h is not, leading to a specific heat jump at $T = \tilde{T}_c$ equal to $(\gamma A_\chi^+ |\tilde{T}_c - T_c|^{-\gamma-1})^2/(2B_\chi|\tilde{T}_c - T_c|^{-\gamma-2\delta\beta}) \propto |\tilde{T}_c - T_c|^{2\delta\beta-\gamma-2} = |\tilde{T}_c - T_c|^{-\alpha}$. Thus, the jump in specific heat is of order $|\tilde{T}_c - T_c|^{-\alpha}$; the specific heat of the local system at temperature $\tilde{T}_c$ is at the same order. Therefore, if we examine the behavior of the specific heat approaching the critical point, we find that initially the specific heat increases in a manner controlled by the local system. Then, the specific heat jumps by an amount which *is comparable to the specific heat immediately above the transition.* This behavior of the specific heat jump is seen in numerical studies [6].

We now consider the width of the mean-field critical region. Let us first check that the mean-field critical region extends at least to T_c. To check this, we extrapolate the result (10) to $T = T_c$, and we observe, using various exponent equalities, that the power of p in the result is consistent with the power of p obtained in (7). Next, let us consider how far the mean-field critical region extends below T_c. For $T < T_c$, the average magnetization of the local system in the absence of a field behaves as $|T_c - T|^\beta$, and the susceptibility is given by $\chi = A_\chi^- |T_c - T|^{-\gamma}$. In the long-range system, this magnetization produces a field $h \propto p|T_c - T|^\beta$, which in turns feeds back and increases the magnetization an amount of order $\chi h = p|T_c - T|^{\beta-\gamma}$. For $|T_c - T|^\gamma >> p$, this effect is negligible compared to the averaged field itself, $|T_c - T|^\beta$. Thus, at such temperatures the long-range interactions have negligible effect on the magnetization and so the mean-field critical behavior only extends to $|T_c - T| \propto p^{1/\gamma} \propto |\tilde{T}_c - T_c|$. Therefore, for small p, the width of the mean-field critical region is small. The scaling arguments above all rely on this width becoming narrower than the width of the anomalous critical region in the local system, in which case both mean-field and anomalous scaling will be seen in the same system.

2.3 Effect of Randomness

In the small-world model one adds a quenched set of strong links, while the long-range model lacks randomness and has links of *strength* p/V. We now identify a criterion, (11), for when it is justified to ignore the quenched nature of the links in the small-world model, at least for determining the universal scaling of quantities with p, as in (9) and (10). The strategy is to consider the long-range model without randomness, and then to add the effects of randomness in the small-world model as a perturbation, determining when it is self-consistent to ignore randomness for small p.

At $\tilde{T}_c$, the correlation length of the local system is $\xi \propto (\tilde{T}_c - T)^{-\nu} \propto p^{-\nu/\gamma}$. Thus, within a correlation volume, there are $p^{-\nu d/\gamma}$ sites. In the small-world model, each site coupled with a long-range link feels an average field proportional to m, and also feels statistical fluctuations about this field. These effects lead to perturbations in the two relevant variables, the magnetic field and temperature, which must be treated carefully.

Consider first the average field acting on a correlation volume due to the long-range links. If sites in the correlation volume are chosen instead with probability p to have long-range links, then an average of $p^{1-\nu d/\gamma}$ sites are chosen. The exponent $1 - \nu d/\gamma$ can be rewritten using Fisher's equality: $\gamma = \nu(2 - \eta)$ [24]. Thus, the average number of sites scales as $p^{1-d/(2-\eta)}$. However, in general, the correlation function of the field ϕ between two sites separated by a distance r decays as $1/r^{d-2+\eta}$. In order for this indeed to describe a decaying correlation function, we need $d - 2 + \eta > 0$, which implies that $1 - d/(2 - \eta) = 1 - \nu d/\gamma < 0$. Thus, the number of sites chosen in a correlation volume, $p^{1-\nu d/\gamma}$ diverges as $p \to 0$. Then, there are a large number of sites with long-range links within each correlation volume and so the sample-to-sample fluctuation in the number of such sites within each correlation volume is negligible in the small-world model. Thus, the sample-to-sample fluctuation in the average field are negligible for most systems. There are some exceptions, however, in systems in which $1 - \nu d/\gamma \geq 0$. One example is the one- and two-dimensional Edwards-Wilkinson model, which will be discussed in more detail in the next section. Another example is the one-dimensional Ising model near its zero-temperature critical point, which will also be discussed below.

Next consider the statistical fluctuations in the field, which *reduce* the correlation of the given site with its neighbors. This effectively raises the temperature of a site with a long-range link. Consider the number of sites with long-range links. The root mean square sample-to-sample fluctuation in the number of such sites scales as $(p\xi^d)^{1/2}$, and thus the sample-to-sample fluctuation in the temperature averaged over a correlation volume scales as $(p/\xi^d)^{1/2} \propto p^{1/2+\nu d/(2\gamma)} \propto |\tilde{T}_c - T_c|^{\gamma/2+\nu d/2}$. Compare this to the difference in temperatures, $\tilde{T}_c - T_c$. As long as

$$\gamma/2 + \nu d/2 > 1, \tag{11}$$

the sample-to-sample fluctuation in temperature is negligible as $p \to 0$. Equation (11) resembles the Harris criterion [10] for the relevance of disorder, with an additional term $\gamma/2$ on the left-hand side. For any model where (11) holds (this includes most unfrustrated models), fluctuations in field and temperature are both negligible and the scaling of both $\tilde{T}_c - T$ and $\tilde{A}$ with p will be the same in the small-world and long-range models.

Intuitively, we expect that the transition temperature in the long-range model will be higher than that in the small-world model: this is definitely true if we ignore the local couplings, and consider only the long-range links. Therefore, in cases when (11) does not hold, $\tilde{T}_c - T_c$ should scale as at least as large a power of p in the small-world model as it does in the long-range model. This, in fact, is what is found below in the Edwards-Wilkinson model: the effective mass, defined below, scales as a larger power of p in the small-world model than in the long-range model. For a further study of the effect of randomness, it was shown elsewhere [9] that the long-range model provides an upper bound to the free energy of the small-world model.

2.4 Comparison to Numerics

An important work was a numerical calculation of some of these quantities, looking for the shift in the transition temperature [6]. In that paper, a different scaling argument was made for the shift, $\tilde{T}_c - T_c \sim p^{1/(\nu d)}$. This is the temperature at which a correlation volume includes roughly one long-range link. However, we have argued that the shift in transition temperature actually scales as $p^{1/\gamma}$, which is less than $p^{1/(\nu d)}$ as $p \to 0$. The difference arises since one long-range link is not sufficient to magnetize an entire correlation volume; several such links are required.

The numerical results in two dimensions are consistent with a shift in transition temperature scaling as $p^{1/\gamma} = p^{0.57\cdots}$. The numerical results in three dimensions indicate a shift scaling as $p^{0.96}$, while taking $\gamma = 1.2396$ from ϵ-expansion [25] gives $1/\gamma \approx 0.81$. This indicates some discrepancy with the numerical results. However, in the numerical study [6], it was argued that their results do not yet involve sufficiently large lattices to obtain accurate scaling; certainly, $p^{0.81}$ is closer to the observed scaling than $p^{1/(\nu d)} \approx p^{0.53}$ is.

These two- and three-dimensional Ising systems satisfy the inequality (11), as do all other $O(N)$, $N \geq 1$, models in two or more dimensions. In the next section, we consider the Edwards-Wilkinson equation which does not satisfy this inequality in one and two dimensions.

Finally, it is worth mentioning various systems with zero temperature phase transitions, such as the one-dimensional Ising model. In the absence of long-range links, this model has no phase transition at any finite temperature. However, the long-range links do lead to a finite temperature phase transition. Unfortunately, the long-range model does not provide a good approximation in this case. Consider a single long-range link between sites x_1, x_2 with spins $\phi(x_1), \phi(x_2) = \pm 1$. which adds a term $e^{\beta \phi(x_1) \phi(x_2)}$ to the partition function. In this case the term diverges at the critical point for small p ($p \to 0$ implies the critical $\tilde{\beta}_c \to \infty$), so that we cannot average the long-range interactions over large numbers of sites. However, there is still some form of mean-field theory available, as the phase transition temperature is sufficiently low that there are a large number of long-range links in a correlation length [3].

In another case, the case of the two-dimensional XY model, with $N = 2$, the system has no spontaneous magnetization at any non-vanishing temperature. However, there is a phase transition, the Kosterlitz-Thouless [26] transition, to a phase with a divergent susceptibility. This transition happens at a finite temperature, and in this case the scaling theory we have developed can be used. Approaching the phase transition from above, the susceptibility diverges as $\chi \propto \exp[c(T - T_c)^{-1/2}]$, where c is a constant [27]. Applying the criterion $p\chi = 1$ for the critical point, we find

$$T - T_c = c/\ln(p)^2. \tag{12}$$

2.5 Non-equilibrium Dynamics

We now consider the generalization to a non-equilibrium process, the contact process [28], in which each site is marked either infected or susceptible. An infected site becomes susceptible at unit rate, while an infected site can turn a neighboring susceptible site infected at a rate λ/q, with q the lattice coordination number. The state with all sites susceptible is absorbing. However, above a critical λ_c, if a single infected site is placed in an infinite lattice of susceptible sites, there is a non-zero probability of the epidemic persisting for all time. We modify the model as follows: each susceptible site can be infected by any other infected site, not necessarily a neighbor, at a rate equal to p/V.

We start by recalling some exponents in the local case. For $\lambda > \lambda_c$, there is an average density, $\rho \propto |\lambda - \lambda_c|^\beta$. In the presence of a source, where susceptible sites become infected at a rate h, the density $\rho(\lambda = \lambda_c, h) = A_\rho h^{1/\delta_h}$. Also, consider the infection spreading from a single source in the local model. At $\lambda = \lambda_c$, the survival probability of the infection after time t, $P(t)$ obeys $P(t) = A_p t^{-\delta}$, The number of infected sites is a random variable, $n(t)$; the average number of such sites obeys $\overline{n}(t) = A_n t^\eta$. The radius of the infection scales as $t^{z/2}$. For $\lambda < \lambda_c$, the infection dies out exponentially, with an asymptotic survival probability $P(t) \propto e^{-t/\tau}$, with $\tau \propto |\lambda_c - \lambda|^{-\nu_\parallel}$. This gives rise to a divergent susceptibility: in the presence of a source h at $\lambda \neq \lambda_c$, the susceptibility, $\chi \equiv \partial_h \rho$ at $\rho = 0$, obeys $\chi(\lambda, h = 0) = A_\chi |\lambda_c - \lambda|^{-\gamma}$.

Then, the general development in the stationary state of the long-range model will be very similar to the equilibrium case. The dynamics in the long-range model can be described by that of the local model with an additional time-dependent source $h(t) = p\rho(t)$. In the stationary state, in the large V limit, the time-dependence of $h(t)$ can be neglected, giving rise to a set of self-consistent equations for h.

First, we consider $\lambda = \lambda_c$. Then, the density obeys $h = pA_\rho h^{1/\delta_h}$, or $h = (pA_\rho)^{\delta_h/(\delta_h-1)}$ and thus

$$\rho = A_\rho (pA_\rho)^{1/(\delta_h-1)}. \tag{13}$$

Equation (13) should be compared to (7). It implies that the transition to a spreading epidemic happens at $\lambda = \tilde{\lambda}_c < \lambda_c$. We can find $\tilde{\lambda}_c$ by using the susceptibility and a self-consistent calculation of the density of particles: in the presence of a source h in the local model,

$$\rho = A_\chi |\lambda_c - \lambda|^{-\gamma} h - \cdots, \tag{14}$$

where the ... represent terms of order h^2 and higher. The source is due to the long-range interaction, and hence $h = p\rho$. Thus, for $pA_\chi |\lambda_c - \lambda|^{-\gamma} = 1$, (14) is obeyed to linear order. For larger λ, the response to the field exceeds the field at linear order, and thus the higher order terms in (14) are required, indicated that there is a non-vanishing net density. This identifies $\tilde{\lambda}_c$ by $\lambda_c - \tilde{\lambda}_c = (pA_\chi)^{1/\gamma}$. For λ slightly above $\tilde{\lambda}_c$, following the same steps as in the equilibrium case leads to the same result as (10), except that the role of magnetization m is replaced by density

ρ and $\sqrt{\tilde{T}_c - T}(\tilde{T}_c - T_c)^{\beta - 1/2}$ is replaced by $(\tilde{\lambda}_c - \lambda)(\tilde{\lambda}_c - \lambda_c)^{\beta - 1}$. The transition is again mean-field. Thus, the stationary results in this non-equilibrium model are described by the same scaling theory as in the equilibrium models. with one difference being that the mean-field density ρ has a linear, rather than square-root, dependence on the distance from the critical point. This is due to the appearance of terms of order h^2 in the average density as a function of field, while in the equilibrium case the magnetization has terms of order h, h^3 and higher.

However, the spread of infection starting from a single source is described by an interesting branching dynamics. This dynamics describes the approach to the stationary state. Consider the dynamics in the long-range model, with $\lambda = \lambda_c$, with a single source for an infection. This source grows as described, with the given $P(t), \overline{n}(t)$. However, the local outbreak starting from that source can produce other local outbreaks elsewhere, via the long-range links, at a rate equal to p times the number of infected sites. For p small, the number of infected sites $n(t)$ will be large before such an event, and thus the fluctuations in the $n(t)$ are described by a random process with a universal distribution. In the large V limit, at fixed t, each new local outbreak produced via a long-range link is well separated in space from the other local outbreaks. Thus, we can describe the dynamics of the spread from a single source simply: there is initially one local outbreak, created at time 0, which survives at time t with probability $P(t)$, and which produces additional local outbreaks at a rate equal to $pn(t)$. Each local outbreak, created at time t', evolves independently, surviving with probability $P(t-t')$, and producing additional local outbreaks with rate $pn(t-t')$. This fully describes the dynamics via a branching process. For $\lambda \neq \lambda_c$, this description of the dynamics remains valid with a changed $P(t)$ and distribution of $n(t)$.

At short times, the average number of infected sites in this dynamics is equal to $\overline{n}(t)$. At long times, the average number of infected sites grows exponentially [29]. To describe this exponential growth, realize that at long times the number of local outbreaks becomes large. If $s(t')$ describes the number of local outbreaks started at time t', then the average number of particles at time t is equal to $\int_0^t dt'\, s(t')n(t-t')$, and thus on average $s(t) = p\int_0^t dt'\, s(t')n(t-t')$. The ansatz $s(t) = e^{\alpha t}$ gives

$$\alpha = [A_\eta p \Gamma(1+\eta)]^{1/(1+\eta)}. \tag{15}$$

Each local outbreak takes a volume of order $t^{dz/2}$. Eventually, at sufficiently large time, such that $e^{\alpha t} \sim (V/t^{dz/2})$, the individual local outbreaks start to merge, and the dynamics of different local outbreaks become coupled. This time t is of order $\ln(V)$.

Up to this time t, the system is approaching its stationary state, and the long-range and small-world models describe definitely dynamics. Consider what happens after the first local outbreak produces a second local outbreak. At some point, the second local outbreak will again infect some other site via a long-range link. In the long-range model, it is unlikely (vanishing probability in the limit $V \to \infty$) that this site will be near the first local outbreak. However, in the small-world model there is a fixed link connecting the two outbreaks, and they can re-infect each other multiple times with high probability. Beyond this time t, one approaches a stationary state with density ρ given above.

3 Edwards-Wilkinson Equation: An Example

3.1 Hard and Soft Cases

The equilibrium Edwards-Wilkinson (EW) model [30] is an interesting case in that it does *not* satisfy the mean-field criterion (11) when each pair of sites is connected with probability p/V with a link of unit strength on top of a one-dimensional regular substrate [11]. The exponents ν, γ for this model can be found by considering it as a Gaussian approximation to a model at a continuous phase transition. We will see that adding a mass to the equation is equivalent to moving above the critical temperature. Thus, the EW model on this small-world network is expected to scaling properties (as a function of p) which differ drastically from the mean-field version of the model in one dimension. In two dimensions, the criterion (11) is marginal, and we will find logarithmic corrections to the mean-field results.

We will consider the EW model on two distinct variants of the small-world network model. The first is the "soft" variant. This is the version considered in the previous section, and the most commonly studied version. In this case, random links of unit strength are added to the one-dimensional substrate with probability p/V to each pair of sites [31–33]. In the "hard" version, each site has *exactly one* random long-range link (in addition to the nearest-neighbors). This link is connected to another randomly chosen site on the lattice, and the strength of the interaction through the random links is p. To construct this network, one picks a random pair of sites, and connects them. This process of selecting pairs of sites and connecting continues; once a site is selected it is never selected again. One motivation for studying the hard network is in its application to scalable PDES schemes, where all processing elements are chosen to have the same connectivity; the weak connection along the link (of strength p) corresponds to synchronization checks along the long-range links which are rare compared to local synchronizations [13]. A further motivation for studying the two different networks is due to their different scaling properties; the soft network, as we will see, has anomalous scaling, while the hard network does not. The terminology "soft" and "hard" is due to the absence or presence of a gap in the excitation spectrum, respectively [33], discussed below.

We consider the equation (for a single realization of the small-world)

$$\partial_t h_i = h_{i+1} + h_{i-1} - 2h_i - \sum_{j=1}^{V} \hat{J}_{ij}(h_i - h_j) + \eta_i(t) \,, \tag{16}$$

where h_i is the surface height, $\eta_i(t)$ is a delta-correlated Gaussian noise with variance 2 (without loss of generality), and we have dropped the t-dependence from the argument of h_i for brevity. The *symmetric* matrix $\hat{J}_{ij}$ represents the quenched random links on top of a one-dimensional lattice of length V with periodic boundary conditions. For the hard version of the small-world $\hat{J}_{ij}$ has exactly one non-zero element (being equal to p) in each row and column. This construction results in $\sum_l \hat{J}_{il} = p$ for *all i*. For the soft version of the small-world, each element of $\hat{J}_{ij}$ (e.g., above the diagonal) is 1 with probability p/V and zero otherwise. In this case $[\sum_l \hat{J}_{il}] = p$, where $[...]$ denotes the average over the network disorder: the *average* coordination number is p and each bond has unit strength. The number of long-range bonds emanating from a given node on the network has a Poisson distribution.

We write (16) as $\partial_t h_i = -\sum_j \hat{\Gamma}_{ij} h_j + \eta_i$, where $\hat{\Gamma} = \hat{\Gamma}^o + \hat{U}$. Here, $\hat{\Gamma}^o$ is minus the Laplacian of the original one-dimensional ring: $\hat{\Gamma}^o = -\delta_{i,j-1} - \delta_{i,j+1} + 2\delta_{i,j}$. The matrix $\hat{U}_{ij} = -\hat{J}_{ij} + \delta_{ij} \sum_l \hat{J}_{il}$ is minus the Laplacian on the random part of the network.

As a basic physical observable, we focus on the behavior of the width, which probes the generic collective properties of the underlying networks by providing a sensitive measure of synchronization [13]. The operator $\hat{\Gamma}$ has a zero mode, corresponding to a uniform shift in the surface: $h_i \to h_i + \text{const}$. It is useful to introduce the projection operator: for all i,j, define $P_{ij} = \delta_{ij} - 1/V$. This operator projects onto the vector space orthogonal to this zero mode. For a given realization of the small-world network the average surface width characterizing the roughness is equal to

$$\langle w^2 \rangle_V \equiv \left\langle \frac{1}{V} \sum_{i=1}^V (h_i - \bar{h})^2 \right\rangle = \frac{1}{V} \sum_{k=1}^{V-1} \frac{1}{\lambda_k} = \left\langle \frac{1}{V} h_i P_{ij} h_j \right\rangle. \tag{17}$$

Here $\bar{h} = (1/V) \sum_{i=1}^V h_i$ is the mean height, $\langle ... \rangle$ denotes an ensemble average over the noise in (16), and λ_k are the non-zero eigenvalues of the real symmetric coupling matrix $\hat{\Gamma}$. We will quote numerical results [11] which relied on the exact diagonalization [34] of the coupling matrix $\hat{\Gamma}$ for a given realization of the network. Then, from the eigenvalues of $\hat{\Gamma}$, the right side of (17) was used to obtain the width. Finally, the *disorder-averaged* (denoted by $[...]$) width $[\langle w^2 \rangle_V]$ was obtained by averaging over a large number (ranging from 100 to 1000) of realizations.

To connect to the previous section, we consider the criterion (11) for various dimensions d for the EW model on the soft version of the network. As an approximation to the soft network, we consider instead the equivalent long-range model, where in addition to the nearest-neighbor connections, each site is coupled to *all* others with strength p/V. This yields the equation of motion

$$\partial_t h_i = h_{i+1} + h_{i-1} - 2h_i - ph_i + p\bar{h} + \eta_i(t) \equiv -\hat{\Gamma}_{ij}^{lr} h_j + \eta_i(t). \tag{18}$$

The operator $\hat{\Gamma}^{kr}$ has a zero mode. However, in the space orthogonal to the zero mode, the equation is equivalent to a massive equation of motion: $\partial_t h_i = -(2h_i - h_{i+1} - h_{i-1}) - ph_i + \eta_i(t)$, where the projection to the subspace orthogonal to the zero mode permits us to ignore the term $p\bar{h}$. Introducing the mass p leads to a correlation length $\xi \sim p^{-1/2}$, and a correlation volume $\xi^d \sim p^{-d/2}$. This introduction of a mass corresponds to shifting away from the critical temperature, and thus determines the exponent $\nu = 1/2$. The exponent $\eta = 0$ since this is a Gaussian model, and so $\gamma = 1$. Now, let us check if this model is a valid description of the soft network. For the soft network, within this volume one would have on average $p\xi^d \sim p^{1-d/2}$ links. For $d > 2$ this number diverges as $p \to 0$; then for small p there are a large number of links leaving the volume and the sample-to-sample fluctuation in the mass in a correlation volume is negligible compared to the mass itself so that the trivial mean-field behavior is expected to be valid. For $d \leq 2$, the trivial behavior breaks down. The requirement $1 - \nu d/\gamma < 0$ is not satisfied in this case. The failure of this requirement is related to the fact that $1/r^{d-2+\eta}$ does *not* decay with r. We will see the resulting divergence in a perturbative calculation. In contrast, for the hard version of the network the density of random links is unity, resulting in trivial mean-field scaling for all d in the $p \to 0$ limit, as there will always be a divergent $(p^{-d/2})$ number of links leaving any correlation volume.

The model also fails the criterion (11). We have seen that the critical exponents are $\nu = 1/2, \gamma = 1$. The criterion $\nu d/2 + \gamma/2 > 1$ for the validity of mean-field behavior is thus violated for $d = 1$, while $d = 2$ is marginal.

3.2 Perturbation Theory

We define the propagator, or Green's function, G to be equal to $\hat{\Gamma}^{-1}$ in the space of non-zero eigenvalues of $\hat{\Gamma}$, while G vanishes when acting on the zero mode of $\hat{\Gamma}$. Thus, $G = P(\hat{\Gamma} + i\epsilon)^{-1}$ (the $i\epsilon$ is inserted so that $\hat{\Gamma} + i\epsilon$ will be invertible). The projection operator is inserted so that we do not need to worry about the effect of the zero mode in what follows. The zero mode always has the same eigenvector, independent of the disorder realization.

We now use the techniques of impurity averaged perturbation theory [35] to study the disorder-averaged propagator $[G]$. The perturbative expansion of $[G]$ can be obtained by

$$[G] = G^o - [G^o\hat{U}G] = G^o - [G^o\hat{U}G^o] + [G^o\hat{U}G^o\hat{U}G^o] - \cdots, \qquad (19)$$

where $G^o = P(\hat{\Gamma}^o + i\epsilon)^{-1}$ is the propagator of the Laplacian on the original one-dimensional lattice. To obtain $[G]$ it is necessary to average this expansion over the network disorder in $\hat{U}$.

To deal with only one-particle irreducible disorder-averaged diagrams, we calculate the self-energy $\Sigma = ([G])^{-1} - (G^o)^{-1}$ up to next-to-lowest order for both the soft and hard versions of the small-world. In Fig. 1, a single line denotes the propagator G^o while a double line denotes $[G]$. The relation between $[G], G^o$, and Σ is shown at the top of the figure. We will then represent Σ as a sum of different diagrams, some of which are shown in Fig. 1.

In these calculations, a cross with no dashed lines attached is used to denote the average over different realizations of the network of a diagonal term $\hat{U}_{ii}$ in

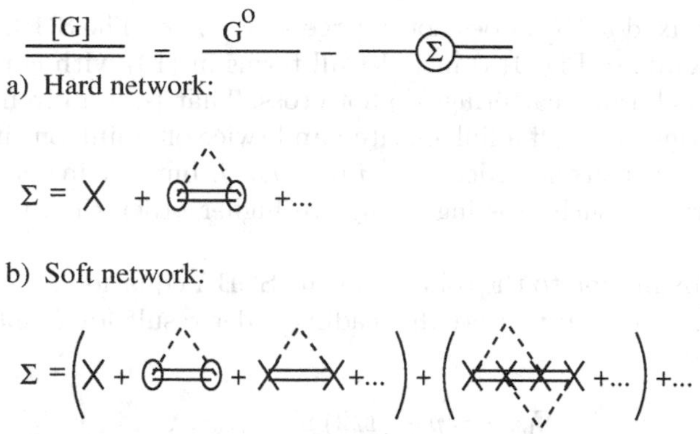

$$\overline{\underline{[G]}} = \frac{G^0}{} - \overline{\boxed{\Sigma}}$$

a) Hard network:

$$\Sigma = \times + \text{⊖⎯⎯⊖} + ...$$

b) Soft network:

$$\Sigma = \left(\times + \text{⊖⎯⎯⊖} + \times\!\!\longleftarrow\!\!\times + ... \right) + \left(\times\!\!\times\!\!\times\!\!\times + ... \right) + ...$$

Fig. 1. Diagrams for the calculation of the Green's function and the self-energy.

$\hat{U}$. A pair of crosses connected by a dashed line is used to denote an average $[\hat{U}_{ii}\hat{U}_{jj}] - [\hat{U}_{ii}][\hat{U}_{jj}]$, while three or more crosses connected by dashed lines are used to denote higher cumulants. Similarly, circles connected by dashed lines are used to denote averages of off-diagonal terms $-\hat{J}_{ij}$ in V. Dashed lines can connect both circles and crosses.

Terms in the expansion with a circle, not connected by dashed lines to other circles, vanish as $1/V$ for large system size V and so may be neglected. For the hard network, terms in the expansion with two or more crosses connected by dashed lines vanish since there are no random fluctuations in $\delta_{ij} \sum_l \hat{J}_{il}$ in this case, while for the soft network these terms do appear. In the hard network, each power of $\hat{U}$ comes with a power of p, while in soft network, each set of circles or crosses connected by dashed lines comes with a power of p.

In the hard version of the network, we can proceed by expanding $[G]$ in powers of $\hat{U}$. The first diagram in Fig. 1(a) yields the lowest order result $\Sigma = p$. Then, the representation of $[G]$ in terms of G^0, Σ gives $[G] = G^0 - G^0 p G^0 + G^0 p G^0 p G^0 - \cdots = (\Sigma + (G^0)^{-1})^{-1}$. Comparing to (19), we see that this lowest order expansion amounts to neglecting all the terms involving higher order cumulants: we have taken $[\hat{U}] = p, [\hat{U}_{ii}\hat{U}_{jj}] = p^2, \ldots$

Adding the second diagram yields $\Sigma = p - p^2[G]_{ii} = p - p^2/(2\sqrt{\Sigma})$, where we use the fact that, for large V, $[G]_{ii} = 1/(2\sqrt{\Sigma})$ plus terms of order $\sqrt{\Sigma}$ (see, e.g., [36]). Here we are using the so-called self-consistent Born approximation (SCBA). If the line in the second diagram were instead taken to be a single-line, we would have obtained $\Sigma = p - p^2 G^0_{ii}$. That approximation amounts to including the second order cumulant in the average of (19) $[\hat{U}_{ii}\hat{U}_{jj}] = p^2 + p^2\delta_{ij}$, *for scatterings off $\hat{U}$ which occur successively in the expansion (19), with no intervening scattering events.* However, using the single line leads to a serious problem: as $V \to \infty$, the propagator G^0_{ii} diverges. This problem can be traced to the failure to include intervening scattering events between the two scatterings off $\hat{U}$. This is why we use the double line (SCBA). The SCBA includes intervening scattering events, and resums a large class of diagrams. It avoids the divergence of G^o_{ii} since for fixed p, $[G]_{ii}$ does not diverge as $V \to \infty$. The SCBA as shown in the second diagram of Fig. 1(a) includes all terms in (19) with cumulants *up to second order*, such that scatterings do not cross. That is, if a given term in (19) involves scattering twice off a link on site i and twice off a link on site j, then the scattering can occur in the order i, j, j, i or j, i, i, j, but not in the order i, j, i, j or j, i, j, i. However, such crossing terms are higher order than those shown in Fig. 1(a).

As an approximation to the solution of the SCBA equation $\Sigma = p - p^2[G]_{ii} = p - p^2/(2\sqrt{\Sigma})$, we can just insert the leading order result for Σ into the second diagram to find

$$\Sigma_{\text{hard}} = p - (1/2)\,p^{3/2} + \dots \,, \tag{20}$$

so that the higher powers of $\hat{U}$ lead to corrections to Σ which are higher order in p as required.

In the soft version of the network, the above procedure does not work. The first diagram in Fig. 1(b) yields $\Sigma = p$. Inserting this result for Σ into the next two diagrams leads to $\Sigma = p - p^{1/2}$, so that the expansion in $\hat{U}$ does not lead to an expansion in p. To correct this, we start by summing up all the diagrams involving a single link. The first terms in this sum are shown in the first pair of parenthesis in Fig. 1(b). The infinite sum yields $\Sigma = p(1 - 2[G]_{ii} + 4[G]^2_{ii} - \dots) = p/(1 + 2[G]_{ii})$. Solving this equation self-consistently to lowest order in p yields $\Sigma = p^2$. Physically, since the density of links is small, of order p, this first step consists of exactly solving the interaction with a single link.

Having done this infinite summation, we can now consider another infinite series of diagrams, starting with the last one shown in Fig. 1(b), and adding additional diagrams where each single interaction with the given link is replaced by two, three, or more interactions with the link, as well as diagrams in which crosses are replaced by circles. Let us determine the order of this summation in p, where we will use the result from the previous summation for $[G]$. The infinite summation replaces the interaction with a given link with the sum: $1 - 2[G]_{ii} + 4[G]^2_{ii} - \dots = 1/(1 + 2[G]_{ii}) \propto p$. Thus, we have resummed the scattering off of a single link, and each link scatters only with strength p, now. The diagram has four such scatterings, leading to a result of order p^4. There is an additional factor of p^2, due to the appearance of two sets of circles/crosses connected by dashed lines; this is the probability of finding two such links. There is a summation over the spacing between the two impurities; this spacing is of order p^{-1}, as this is the scale on which $[G]$ decays, as determined from the solution of scattering off a single link. There is a further factor of $[G]^3_{ij}$ from the three Green's functions. For two links separated by a distance of order p^{-1}, we have $[G]^3_{ij} \propto p^{-3}$. As a result, this sum over diagrams yields a result of order $p^4 \cdot p^2 \cdot p^{-1} \cdot p^{-3}$, which is again of order p^2. More complicated diagrams continue to yield results of order p^2. As a result, after this resummation, we are able to determine only that Σ_{soft} scales as p^2, but not the exact coefficient, and not the higher order corrections:

$$\Sigma_{\text{soft}} \propto p^2 \tag{21}$$

3.3 Two Dimensions

For $d>2$, Σ is asymptotically of order p for both soft and hard cases, i.e, the systems are effectively mean-field. In the soft case, this result can be found by the resummed expansion.

For $d=2$, one finds $\Sigma_{\text{hard}} = p - \cdots$ using Fig. 1(a). For the soft case in $d = 2$ the resummation of the scattering off of a single link (the first set of diagrams in Fig. 1(b)) enables one to find a logarithmic correction to the trivial mean-field behavior:

$$\Sigma_{\text{soft}} = 2\pi p/|\log(p)| - \cdots . \tag{22}$$

Now consider the infinite sum of diagrams starting with the last one shown in Fig. 1(b). In $d = 1$, we found that this diagram was asymptotically the same order in p as the first set of diagrams. For $d = 2$, we find that the resummation of the scattering off a single link reduces that scattering strength to an amount of order $1/|\log(p)|$; there are four such scatterings. There is a factor of p^2 for the two links. The separation between the two links can be of order $\Sigma^{-1/2}$, leading to a total area of order $\Sigma^{-1} \propto |\log(p)|/p$. Finally, the Green's function G_{ij}^3 is of order unity, with no logarithmic divergence in p, if i, j are separated a distance of order $\Sigma^{-1/2}$. This gives the total power of the diagram equal to $(1/|\log(p)|)^4 \cdot p^2 \cdot \Sigma^{-1} \propto p/|\log(p)|^3$. As $p \to 0$, the contribution of this diagram is asymptotically vanishing compared to that of (22).

Thus, Σ_{soft} depends anomalously on p in $d = 2$. However, higher order corrections are small, so that the system is asymptotically mean-field, albeit with a renormalized p.

The absence of higher order corrections in $d = 2$ raises the possibility of an ϵ-expansion for $d = 2 - \epsilon$. This will be considered in future work [37].

3.4 Numerics

Then, the disorder-averaged width can be found from Σ by $[\langle w^2 \rangle] = [G]_{ii} \simeq \frac{1}{2\sqrt{\Sigma}}$. Thus, the *asymptotic* small-p behavior of the width in the thermodynamic limit is $[\langle w^2 \rangle]_{\text{hard}} \simeq 1/(2\sqrt{p})$ and $[\langle w^2 \rangle]_{\text{soft}} \propto 1/p$, for the respective versions of the small-world networks. These asymptotic small-p, infinite system-size behaviors are indicated with the two slopes next to the numerical data [11] in Fig. 2. In an attempt to match the behavior of the width for finite systems in the hard network, the finite-system version of the propagator (see, e.g., [36]) was used with the effective mass from (20). We ignored various finite-size corrections to Σ itself, as being too complicated to be worth calculating, and also as entering only at next-to-leading order. That is, we assumed that the system was described by (18) with the mass p replaced by Σ_{hard}, and obtained the width for this equation in a finite-size system. The results lead to good agreement for a wide range of p. The inset to Fig. 3 shows the improvement of the next-to-leading order for the hard network.

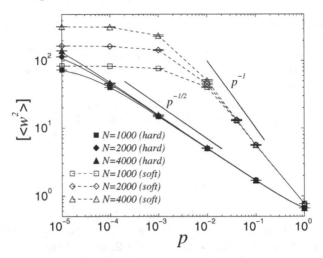

Fig. 2. Disorder-averaged width obtained by exact numerical diagonalization for the hard (filled symbols) and for the soft (open symbols) version of the small-world as a function p for system sizes indicated in the figure. The two slopes indicate the asymptotic small-p infinite system-size behavior. The solid line is obtained using the finite-system propagator with the effective mass equations (20).

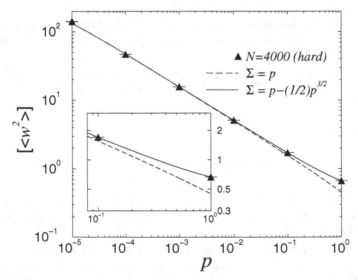

Fig. 3. Comparison of the numerically obtained width (symbols) for the hard version of the small-world network for $V=4000$ with the lowest (dashed lines) and next-to-lowest order (solid lines) perturbative results. The inset shows a magnified view of the improvement of the latter in the region where p is small enough so that the perturbative expansion becomes sensible, but is also sufficiently large so that the finite-size corrections of the mass are negligible.

These results indicate that for both small-world networks, the width approaches a finite value for any non-zero value of p as $V \to \infty$. In the hard network, the scaling of the width approaches that of the mean-field version of (16), while the soft network does not exhibit this scaling, as expected from the general criterion [9].

3.5 Non-perturbative Results

The effective mass obtained perturbatively, (20) and (21) describes the average behavior of the width of the system. However, it does not determine the eigenvalue spectrum of the system. Define the density of eigenvalues of the matrix Γ by $\rho(\lambda)$. Then, we can calculate $\rho(\lambda)$ by defining $[G(\lambda)] = P[(\Gamma - \lambda + i\epsilon)^{-1}]$ so that $\rho(\lambda) = -\text{Im}[\text{Tr}(G(\lambda))]/\pi V$. Then, one must compute a λ-dependent self-energy $\Sigma(\lambda)$. To do this, all the Green's functions appearing in the self-energy must be replaced with $G(\lambda)$. In the case of the hard network, the self-energy Σ is only weakly dependent: the leading order term, p, does not involve any Green's functions in it and hence does not depend on λ. Only at next order does one find a λ-dependence in Σ. Thus, one expects that there is a true gap in the spectrum which is equal to p, plus higher order corrections of order p^2.

To understand this better, we review some earlier results, studying the "soft" version of the small-world network [33]. Throughout, we assume that the system size, V is taken to infinity. Then, consider a segment of the chain of length l. Any given site has a probability e^{-p} of having no long-range links, and thus the segment has a probability e^{-pl} of having no such links. Although the probability of these quasilinear chain segments of length l is exponentially small, they have eigenvalues of order $1/l^2$ [38,39]. Summing up over large l values with the exponential weight above yields $\rho(\lambda) \sim (1/\sqrt{\lambda})e^{-cp/\sqrt{\lambda}}$ for small λ [33], where c is a constant. Therefore, the spectrum is gapless, but the density of states vanishes exponentially fast for small λ as a result of the essential singularity in the exponent. From the right side of (17) it follows that in the $V \to \infty$ limit, the disorder-averaged width can be expressed in terms of the density of states as $[\langle w^2 \rangle] = \int (1/\lambda)\rho(\lambda)d\lambda$. The small-$\lambda$ behavior of $\rho(\lambda)$ determines whether the width remains finite or diverges in the thermodynamic limit. Although there is no gap in the spectrum, the exponentially small $\rho(\lambda)$ above more than compensates for the term λ in the denominator, so that the integral converges as $\lambda \to 0$, yielding $[\langle w^2 \rangle]_{\text{soft}} \sim 1/p$. This construction can be extended to arbitrary dimension d, where we find that $\rho(\lambda) \sim \lambda^{d/2-1}e^{-cp\lambda^{-d/2}}$. Thus, even for $d > 2$ where mean-field scaling of the width prevails, the spectrum remains gapless for the soft system.

Instead, for the hard version of the network, we expect that there is indeed a true gap of order p. Each site has one link. Considering a fixed segment of the chain, as $V \to \infty$, there is a vanishing probability that any of these links connect two sites in the given segment. Thus, there is a vanishing probability of finding a segment with no long-range links leaving the segment.

4 Discussion

We have considered the behavior of a general class of systems on a small-world network. Our starting point was a model with combined short- and long-range interactions. This model was solved using a combination of scaling theory and mean-field theory. We find that the long-range interactions lead to mean-field behavior, but with a scaling region whose width vanishes as $p \to 0$, as well as with anomalous scaling of the mean-field amplitudes in p. We have then compared this model to the original small-world model and found that in a number of situations the same scaling behavior describes both. We have found that the description is applicable to both equilibrium and non-equilibrium processes in the stationary state, while the approach to the stationary state in the non-equilibrium case is describe by a branching process. Finally, we have illustrated some of the results with the specific example of the Edwards-Wilkinson equation.

Acknowledgements

We thank G. Korniss for collaborating on work in [11], and for permission to use his results in this work. We thank Z. Toroczkai for discussions and inspiration. MBH was supported by DOE W-7405-ENG-36 and BK was supported by NSF Grant No. DMR-0113049 and Research Corporation Grant No. RI0761.

References

1. R. Albert and A.-L. Barabasi, Rev. Mod. Phys. **74**, 47 (2002).
2. D. J. Watts and S. H. Strogatz, Nature **393**, 440 (1998).
3. A. Barrat and M. Weigt, Eur. Phys. J. B. **13**, 547 (2000).
4. M. Gitterman, J. Phys. A **33**, 8373 (2000).
5. B. J. Kim et. al., Phys. Rev. E **64**, 56135 (2001).
6. C. P. Herrero, Phys. Rev. E **65**, 66110 (2002).
7. A. V. Goltsev, S. N. Dorogovtsev, and J. F. F. Mendes, Phys. Rev. E **67**, 026123 (2003)
8. M. Leone, A. Vazquez, A. Vespignani, and R. Zecchina, Eur. Phys. J. B **28**, 191 (2002).
9. M.B. Hastings, Phys. Rev. Lett **91**, 098701 (2003), preprint cond-mat/0304530 (2003).
10. A. B. Harris, J. Phys. C **7**, 1671 (1974).
11. B. Kozma, M.B. Hastings, and G. Korniss, Phys. Rev. Lett. 92, 108701 (2004).
12. G. Korniss, Z. Toroczkai, M.A. Novotny, and P.A. Rikvold, Phys. Rev. Lett. **84**, 1351 (2000).
13. G. Korniss, M.A. Novotny, H. Guclu, and Z. Toroczkai, P.A. Rikvold, Science **299**, 677 (2003).
14. R. Fujimoto, Commun. ACM **33**, 30 (1990).
15. B.D. Lubachevsky, J. Comput. Phys. **75**, 103 (1988).
16. M. Kardar, G. Parisi, Y.-C. Zhang, Phys. Rev. Lett. **56**, 889 (1986).

17. A.-L. Barabási and H.E. Stanley, *Fractal Concepts in Surface Growth* (Cambridge University Press, Cambridge, 1995).
18. G. Korniss, M.A. Novotny, P.A. Rikvold, H. Guclu, and Z. Toroczkai, *Materials Research Society Symposium Proceedings Series* Vol. 700, pp. 297-308 (2002).
19. G. Korniss, M.A. Novotny, A.K. Kolakowska, and H. Guclu, SAC 2002, *Proceedings of the 2002 ACM Symposium on Applied Computing*, pp. 132-138, (2002).
20. S.H. Strogatz, Nature **410**, 268 (2001).
21. M. Kardar, Phys. Rev. B **28**, 244 (1983).
22. M. E. J. Newman, C. Moore, and D. J. Watts, Phys. Rev. Lett. **84**, 3201 (2000).
23. This and other scaling laws are covered in G. Parisi, *Statistical Field Theory* (Perseus Books, Advanced Book Classics, Reading, 1998).
24. M. E. Fisher, J. Math. Phys. **5**, 944 (1964).
25. J. Zinn-Justin, *Quantum Field Theory and Critical Phenomena*, (Oxford Science Publications, 2001).
26. J. M. Kosterlitz and D. J. Thouless, J. Phys. C **6**, 1181 (1973).
27. J. M. Kosterlitz, J. Phys. C **7**, 1046 (1974).
28. T. E. Harris, Ann. Probab. **2**, 969 (1974); J. Marro and R. Dickman, *Nonequilibrium Phase Transitions in Lattice Models*, (Cambridge University Press, 1999).
29. N. Zekri and J. P. Clerc, Phys. Rev. E **64**, 056115 (2001).
30. S.F. Edwards and D.R. Wilkinson, Proc. R. Soc. London, Ser A **381**, 17 (1982).
31. M.E.J. Newman, J. Stat. Phys. **101**, 819 (2000).
32. M.E.J. Newman and D.J. Watts, Phys. Lett. A **263**, 341 (1999).
33. R. Monasson, Eur. Phys. J. B **12**, 555 (1999).
34. W.H. Press, S.A. Teukolsky, W.T. Vetterling, B.P. Flannery *Numerical Recipes in C*, 2nd ed. (Cambridge Univ. Press, Cambridge, 1995), Sects. 11.2 and 11.3
35. J. Rammer, Rev. Mod. Phys. **63**, 781 (1991).
36. Z. Toroczkai, G. Korniss, S. Das Sarma, and R.K.P. Zia, Phys. Rev. E **62**, 276 (2000).
37. M. B. Hastings, to be published.
38. I.M. Lifshitz, Sov. Phys.-JETP **17**, 1159 (1963).
39. A.J. Bray and G.J. Rodgers, Phys. Rev. B **38**, 11461 (1988).

Attacks and Cascades in Complex Networks

Ying-Cheng Lai[1], Adilson E. Motter[2], and Takashi Nishikawa[3]

[1] Department of Mathematics and Statistics, Department of Electrical Engineering, Arizona State University, Tempe, AZ 85287, USA
[2] Max Planck Institute for the Physics of Complex Systems, Nöthnitzer Strasse 38, 01187 Dresden, Germany
[3] Department of Mathematics, Southern Methodist University, Dallas, TX 75275, USA

Abstract. This paper reviews two problems in the security of complex networks: *cascades of overload failures on nodes* and *range-based attacks on links*. Cascading failures have been reported for numerous networks and refer to the subsequent failure of other parts of the network induced by the failure of or attacks on only a few nodes. We investigate a mechanism leading to cascades of overload failures in complex networks by constructing a simple model incorporating the flow of physical quantities in the network. The second problem is motivated by the fact that most existing works on security of complex networks consider attacks on nodes rather than on links. We address attacks on links. Our investigation leads to the finding that many scale-free networks are more sensitive to attacks on short-range than on long-range links. Besides its importance concerning network security, our result has the unexpected implication that the small-world phenomenon in these scale-free networks is mainly due to short-range links.

1 Introduction

Complex networks [1] such as the Internet, the electrical power grid, and the transportation network, are an essential part of a modern society. The security of such a network under random or intentional attacks is of great concern. Recently, an interdisciplinary field among information science and engineering, statistical and nonlinear physics, applied mathematics, and social science has emerged, bringing novel concepts and approaches to the study of complex networks [2–5]. Issues such as the characterization of the network architecture, dynamics on complex networks, and the effect of attacks on network operation have begun to be addressed. A central point of this review is that the flow of information and other physical quantities in the network can be critically important for network security. This *dynamical* aspect of the security problem, despite its highly practical relevance, has been only partially understood in the context of complex networks. Here we shall review some of our initial results in this direction.

Most large natural and man-made networks are sparse and evolve in time. Two important properties displayed by many of these networks are the small-world [6] and scale-free [7] properties. Small-world networks are characterized by the clustering coefficient C and the average network distance L. The former is the probability that any two nodes are connected to each other, given that they are both connected to a common node. The latter measures the average minimal

Y.-C. Lai, A.E. Motter, and T. Nishikawa, Attacks and Cascades in Complex Networks, Lect. Notes Phys. **650**, 299–310 (2004)
http://www.springerlink.com/

number of links connecting any two nodes in the network. Many regular networks have high clustering coefficients and large network distances. Random networks, on the other hand, have small network distances and low clustering coefficients [8]. Small-world networks fall somewhere in between these two extremes as they have large clustering coefficients and small average network distances [6,9]. A small-world network is then locally similar to a regular network but globally similar to a random network. The scale-free property, on the other hand, is defined by an algebraic behavior in the probability distribution $P(k)$ of the number k of links at a node. Barabási and Albert [7] have presented a model which generates a class of scale-free networks. Their model incorporates two basic features in the evolution of the network: growth and preferential attachment. The former means that the number of nodes in the network increases with time and the latter stipulates that the probability for a new node to be connected to an existing node depends on the number of links that this node already has. A number of other models of scale-free networks have been proposed (see, for example, [10]).

Most existing works on the security of scale-free networks consider attacks on nodes rather than on links ([11,12] are among the few exceptions). We believe that attacks on links are as important for the network security as those on nodes, and therefore deserve a careful investigation. As we argue, studying the effect of attacks on links can provide an understanding to the fundamental question of why scale-free networks are typically highly efficient. Roughly, the efficiency of a scale-free network is determined by the average network distance between nodes. It has been assumed that long-range connections are responsible for the small average network distance observed in complex networks. In the Watts-Strogatz model of small-world networks, the small network distances are due to links connecting nodes that would otherwise be separated by a long distance, i.e. long-range links [6]. The range of a link l_{ij} connecting nodes i and j is defined to be the shortest distance between i and j when l_{ij} is removed [9]. The intuition is then that scale-free networks are much more sensitive to attacks on long-range than those on short-range links. We show that in fact, for many scale-free networks, the opposite is true. Thus, the small-world phenomenon in these scale-free networks is caused by short-range links.

This review is organized as follows. In Sect. 2, we will present an example of complex network that may be of broad interest: the conceptual network of English words. The topology of this network was recently studied by us [13] and we hope this example can serve to illustrate the interdisciplinary nature of research on complex networks, and how quantitative characterizations can be useful for a discipline that has traditionally been qualitative. In Sect. 3, we present a simple model to address the issue of attack-induced cascades in complex networks [14]. Ranged-based attacks on links and the origin of the small-world phenomenon in scale-free networks [15] are detailed in Sect. 4. A brief discussion is presented in Sect. 5.

2 Conceptual Network of Language

A language can be regarded as a network where words correspond to nodes of the network. We define two words in a language to be connected if they express similar concepts. The resulting network of connections among many thousands of words is potentially relevant not only for the study of the languages themselves, but also for cognitive science. This issue has recently been studied quantitatively [13] by mapping out the conceptual network of English language. In particular it has been shown that this network exhibits the small-world property.

To construct the network [13], we define the connections according to the entries of a Thesaurus dictionary. Such a dictionary gives for every entry a list of words that are conceptually similar to the entry word. For instance, for the word "nature" it lists "character", "world", "universe" etc. We define a network where each *entry* word is a node, and two nodes are connected if one of the corresponding words is listed as conceptually similar the other one, as depicted in Fig. 1. In our study we used online English Thesaurus that is available at [16], which has over 30,000 entries. The resulting network has an average of about 60 connections per node.

Despite being sparse, the conceptual network is expected to be highly clustered, because there are many sets of related words that are densely interconnected. Indeed, the numerical computation of the clustering coefficient C yields a number more than 250 times larger than the corresponding value for a random network with the same parameters (see Table I). On the other hand, because the network is sparsely connected and only words expressing similar concepts are linked, one might naively conclude that the average network distance L should be large. However, our numerical computation yields $L = 3.2$, which is very close to the value of about 2.5 of the corresponding random network (see Table I). This means that two words in the 30,000-words dictionary are connected by only *three degrees of separation*, on average. This surprisingly small L is due to words that correspond to two or more very different concepts and work as shortcuts, connecting regions of the network that would otherwise be separated by many links. In fact, less than 1 percent of the words require more than 4 steps to be reached from other words, on average. Words that require many links to be

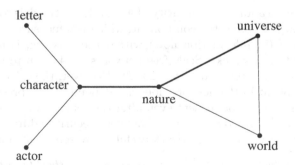

Fig. 1. Small part of the conceptual network of the English language.

Table 1. Comparison between the conceptual network defined by the Thesaurus dictionary and a random network with the same parameters. N is the total number of nodes (entry words) in the largest connected component, $\bar{k}$ is the average number of links per node, C is the clustering coefficient, and L is the average network distance.

	N	$\bar{k}$	C	L
Actual configuration	30,244	59.9	0.53	3.16
Random configuration	30,244	59.9	0.002	2.5

reached are usually very specialized, such as "appendectomy" which requires a path of length eight to be connected with "quadrillion"[4].

Therefore, the conceptual network English language is highly clustered and at the same time has a very small average network distance, *i.e.*, it is a *small-world network*. Although we have focused on a particular language (English) we expect similar results to hold of other languages as well because high clustering comes from the existence of concepts shared by more than two words and short average network distance comes from the existence of words that share meanings with otherwise unrelated words. Both features are seemingly present in many languages.

This result is potentially relevant for cognitive science. From the standpoint of retrieval of information in an associative memory, the small-world property of the network represents a maximization of efficiency. On the one hand, similar pieces of information are stored together; on the other hand, even very different pieces of information are never separated by more than a few links. The former makes searching by association possible, while the latter guarantees a fast search [17]. It is thus tempting to speculate that associative memory may have arisen partly because of a maximization of efficiency in the retrieval of information by natural selection.

For more details we refer to [13]. Different aspects of language networks have been addressed by other authors [18–22,3,23,24].

3 Attack-Induced Cascades in Complex Networks

A convenient way to address the security of a complex network is to examine how the size of the largest connected component, which is a measure of the efficiency of communication (or information flow) within the network, is reduced under random or intentional attacks. Scale-free networks are known to be sensitive to the removal of highly connected nodes [25–29]. However, the existence of a giant connected component in the network does not depend on the presence of highly connected nodes and can be present even after the removal of a significant number of nodes [29,30]. Previous studies on network security address mainly static properties. Our concern is that network architecture represents only one aspect

---

[4] *quadrillion* → *googol* → *infinity* → *holiness* → *purity* → *sterility* → *birth control* →
vasectomy → *appendectomy*

of the security problem. An important question for many real-world situations is how attacks affect the functions of a network when the flow of information or other physical quantity in the network are taken into consideration. In particular, the removal of nodes changes the balance of flows and may trigger a cascading failure [31–34], as the one that happened on August 10, 1996 in the western U.S. power grid [35,36]. A simple model has been recently introduced [14] for cascades of *overload* failures in complex networks. We show that for networks where loads can redistribute among the nodes, intentional attacks on highly loaded nodes can trigger a large-scale cascade of overload failures.

Our model is defined as follows [14]. Suppose that at each time step one unit of the relevant quantity is exchanged between every pair of nodes in the network and is transmitted along the shortest paths connecting them. The load at a node is then simply the betweenness centrality [37–39], i.e. total number of shortest paths passing through the node. The capacity of a node is the maximum load that the node can handle. Since capacity is costly, it is natural to assume that the capacity C_i of node i is proportional to the initial load L_i on that node,

$$C_i = (1 + \alpha)L_i, \quad i = 1, 2, ...N, \tag{1}$$

where $\alpha \geq 0$ is the tolerance parameter, and N is the initial number of nodes. When all the nodes are connected, the entire network operates insofar as $\alpha \geq 0$. But the removal of nodes in general changes the distribution of loads. The load at a particular node can then change. If it increases and becomes larger than the capacity, the corresponding node fails. Any failure leads to a new redistribution of loads and, as a result, subsequent failures can occur. Because of the global redistribution of load, new failures may be driven by events happening far away. This cascading process can stop after a few steps but it can also propagate and shutdown a considerable fraction of the network[5]. But under what conditions can such a global cascade happen?

Our result is that global cascades occur if the network exhibits a *highly heterogeneous distribution of loads* and the removed nodes are among those with *higher load*. Otherwise, cascades are not expected. In order to understand this result, consider the removal of a single node. If the node has small load, its removal will not cause major changes in the balance of loads. However, when the load at the node is large, its removal is likely to affect significantly the loads at other nodes and possibly starts a sequence of overload failures. In networks with some degree of randomness, the distribution of loads is highly correlated with the distribution of links. In particular, networks with heterogeneous distribution of links, such as scale-free networks, are expected to be heterogeneous with respect to load as well, so that nodes with larger number of links will have higher load [38,40], on average. This results reveals another aspect of the robust-yet-fragile property of heterogeneous networks, which was first observed for the attack on *several* nodes [25]. In the case of cascades, a large damage can be caused by the attack on one or very few nodes.

---

[5] A different model and mechanism for overload breakdown due to networks growth has been considered in [39].

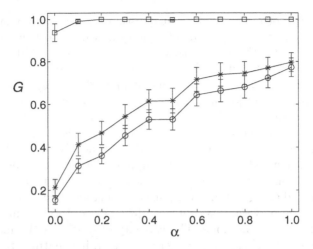

Fig. 2. Cascading failure in scale-free networks with scaling exponent $\gamma = 3$, as triggered by the removal of one node chosen at random (squares), or among those with largest connectivities (stars) or highest loads (circles). Each curve corresponds to the average over 5 triggers and 10 realizations of the network. The error bars represent the standard deviation. The number of nodes in the largest component is $5000 \leq N \leq 5100$.

We simulate cascades triggered by random failures and by intentional attacks. In the case of failures, we choose a trigger at random among all the nodes of the network. In the case of attacks, the targeted node is selected from those with highest loads or largest connectivities. We consider heterogeneous networks with scale-free distribution of links and compare them with an equivalent homogeneous configuration. To generate the networks, we start with a list of integers representing the connectivities of the nodes, i.e. the number of end-links of each node [41,15]. Next, we pick up pairs of end-links at random and connect them to form a link and repeat this process until the last pair is connected, prohibiting self- and repeated links. Let N denote the number of nodes in the largest connected component of the resulting network. The damage caused by a cascade is quantified in terms of the relative size G of the largest connected component $G = N'/N$, where N' are the number of nodes in the largest component after the cascade.

Figure 2 shows results for scale-free networks with scaling exponent $\gamma = 3$. On average, G remains close to unity in the case of random breakdowns but is significantly reduced under intentional attacks, even for α unrealistically large. This result is in agreement with intuition, because in the case of random breakdown the trigger is probably one of the many nodes with small load, while in the case of intentional attack it is a node with very large load. The damage is larger for smaller values of α, and the attack of nodes with highest loads is more destructive than the attack of nodes with largest connectivities. Figure 3 shows the corresponding results for a homogeneous network with the same number of nodes and exactly 3 links per node. In the inset we display results for scale-free

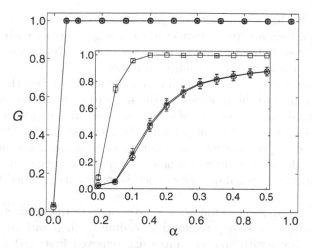

Fig. 3. Cascading failures in homogeneous networks with degree $k = 3$ and $N = 5000$. Inset: the same for scale-free networks with $\gamma = 3$, $N = 5000$, and $k \geq 2$ (different from the networks in Fig. 1, here each node has 2 or more links). The resulting average connectivity is $\langle k \rangle \approx 3.1$. The legends and other parameters are the same as in Fig. 1.

networks with about the same average number of links per node. The homogeneous network does not experience cascading failures due either to random breakdown or to intentional attacks for α as small as 0.05. For the scale-free (heterogeneous) network, cascades triggered by the attack on a key node can drastically reduce the size of the the the largest connected component, as shown in the inset. Therefore, networks with homogeneous distribution of load appear to be more robust against attacks than the heterogeneous ones. This conclusion does not rely on the particular properties of these models, as the same was also observed for other classes of networks.

These findings are expected to be important for real-world networks. Indeed, many infrastructure networks have heterogeneous distribution of load and as such are expected to undergo large-scale cascades if some vital nodes are attacked, but rarely in the case of random breakdown. For details see [14].

4 Range-Based Attacks on Links in Complex Networks

The Watts and Strogatz [6] model of small-world networks identifies the small shortest paths observed in locally structured, sparse networks as being due to long-range connections, while short-range links are responsible for high clustering. This observation matches with the known results for the Erdös-Rényi model of random networks [42], where almost all links are long-range connections and the average network distance increases only logarithmically with the number N of nodes [8]. In most regular networks, on the other hand, all the links have small range and the average shortest path increases with a power of N. All these models display a relatively homogeneous distribution of connectivities. Many real networks having very small average network distance have been identified as scale-free [7,3]. Scale-free networks are heterogeneous as their connectivity can vary significantly from node to node and a considerable number of links can be associated with a few highly connected nodes.

A recent paper [15], which we shall review here, has studied the contribution of short-range links to the shortness of the node-to-node distances in scale-free networks, by analyzing the impact of attacks on short-range links versus those on long-range links. Our results contrast with the tacit assumption that long-range connections are responsible for the small average network distance exhibited by these networks. Our findings are based on the observation that the average network distance is a global quantity which is mainly determined by links with large load.

Our attack strategy is as follows [15]. We measure the *efficiency* of the network as links are successively removed according to their ranges: (*i*) for short-range attacks, links with shorter ranges are removed first; (*ii*) for long-range attacks, links with longer ranges are removed first. The efficiency is measured by the shortest paths between pairs of nodes. A convenient quantity to characterize the efficiency is [43]

$$E = \frac{2}{N(N-1)} \sum \frac{1}{d_{ij}}, \tag{2}$$

where d_{ij} is the length of the shortest path between nodes i and j and the sum is over all $N(N-1)/2$ pairs of nodes. The network is more efficient when it has small shortest paths, which according to our definition corresponds to large E.

To be specific we consider the network model described in the previous section, where the nodes are connected randomly for a given scale-free distribution with scaling exponent γ, and self- and repeated links are prohibited. In order to have nontrivial networks in the limits of small and large γ, we bound the connectivity so that $k_{min} \leq k_i \leq k_{max}$ for $i = 1, 2, \ldots, N$, where k_{min} and k_{max} are constant integers. For $\gamma \to \infty$, the network is homogeneous as all the nodes have the same connectivity k_{min}. The distribution of connectivities becomes increasingly more heterogeneous as γ is decreased.

In randomly generated networks, nodes with larger connectivity are expected to be on average closer to each other than those with smaller connectivity [15]. More specifically, the distance d_{ij} between nodes i and j is expected to be highly correlated with the product of the connectivities k_i and k_j. This suggests that the range is also correlated with the product of the connectivities so that short-range links tend to link together highly connected nodes, while long-range links tend to connect nodes with very few links. Moreover, links between nodes with large connectivities are expected to be passed through by a large number of shortest paths (see [12]). That is, on average these links should possess a higher load [12] than those connected to nodes with few links, where the load of a link is defined as the number of shortest paths passing through the link [37,38]. These have been confirmed numerically, as shown in Fig. 4 for $\gamma = 3$. As a result, high load should be associated mainly with short-range links. Since links with higher load are expected to contribute more to the shortness of the paths between nodes,

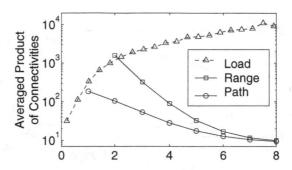

Fig. 4. Averaged product of connectivities as a function of the shortest path, range, and load for $\gamma = 3$, where the load is binned and normalized by 10^4. Each curve corresponds to the average over 10 realizations for $N = 5000$, $k_{min} = 3$, and $k_{max} = 500$.

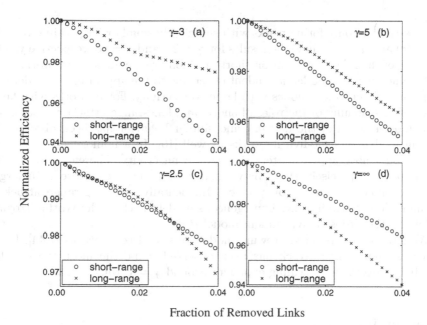

Fraction of Removed Links

Fig. 5. Efficiency for short- and long-range attacks (normalized by the initial value) as a function of the fraction of removed links. All the parameters other than γ are the same as in Fig. 5.

such a correlation between load and range implies that attacks on short-range links are more destructive than those on long-range links.

In Fig. 5 we show the efficiency for both short- and long-range attacks, for different values of γ. Short-range attacks are clearly more destructive than long-range ones for intermediate values of γ, as shown in Figs. 5(a) and 5(b) for $\gamma = 3$ and $\gamma = 5$, respectively. The corresponding relation between the average load and range, plotted in Fig. 6 for $\gamma = 3$, confirms that higher load on links with shorter range is the mechanism underlying this phenomenon. Long-range attacks become

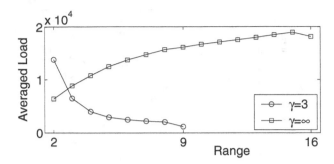

Fig. 6. Averaged load as a function of the range for $\gamma = 3$ and $\gamma = \infty$. All the parameters other than γ are the same as in Fig. 4.

more destructive only for networks with sufficiently small or large values of γ. In Figs. 5(c) and 5(d) we show the results for $\gamma = 2.5$ and $\gamma = \infty$, respectively. The exchange of the roles of attacks on short- and long-range links for networks with small values of γ is a model dependent effect due to the appearance of a densely connected subnetwork of nodes with large connectivity. For networks with large values of γ, switching of the roles of short- and long-range attacks is caused by the homogenization of the network and similar behavior has been observed in growing models of scale-free networks as well [15]. In a homogeneous network all the nodes have approximately the same connectivity. Therefore, links with higher load are precisely those between distant nodes, i.e., those with larger range, as shown in Fig. 6 for $\gamma = \infty$. Incidentally, the long-range attack is also more destructive in other homogeneous models, such as the Watts-Strogatz model and the Erdös-Rényi random model [15].

We have also considered growing models of scale-free networks [7,44]. In all the cases, short-range attack has been observed to be the most effective for scale-free networks with scaling exponent around $\gamma = 3$ [15].

5 Discussion

In this paper, we have reviewed two problems concerning attacks on and security of complex networks. The study of attacks on complex networks is important in order to identify the vulnerabilities of real-world networks, which can be used either for protection (e.g., of infrastructures) or for destruction (e.g., in the control of epidemic diseases). Additionally, it can provide guidance in designing more robust artificial networks (e.g., communication networks).

Our result on cascades in complex networks indicates that while the scale-free property makes many natural and man-made networks quite robust against to random failure of nodes, the presence of a few nodes with very large load may make the network vulnerable to a cascade of overload failures capable of disrupting the network into small fragments. Such a global cascade represents a serious threat because it may be triggered by relatively small events and prevents an efficient communication between most nodes in the network.

We have also shown that for a wide interval of the scaling exponent around $\gamma = 3$, fairly random scale-free networks are more vulnerable to short-range attacks than long-range ones. This property results from a higher concentration of load on short-range links. Our findings have the important implication that short-range links are more important than long-range links for an efficient communication between nodes, which is the opposite to what one might expect from other classes of small-world networks. This result is potentially relevant for the spread of sexual diseases, which has been argued to take place in a scale-free network [45]. Although we have focused on scale-free networks, similar results are expected to hold for other classes of heterogeneous networks.

More details about the content of this review can be found in [13–15].

Acknowledgements

This work was supported by NSF under Grant No. ITR-0312131 and by AFOSR under Grant No. F49620-01-1-0317.

References

1. S. H. Strogatz, Nature (London) **410**, 268 (2001).
2. L. A. N. Amaral, A. Scala, M. Barthélémy, and H. E. Stanley, Proc. Natl. Acad. Sci. U.S.A. **97**, 11149 (2000).
3. R. Albert and A.-L. Barabási, Rev. Mod. Phys. **74**, 47 (2002).
4. S. N. Dorogovtsev and J. F. F. Mendes, Adv. Phys. **51**, 1079 (2002).
5. M. E. J. Newman, SIAM Rev. **45**, 167 (2003).
6. D. J. Watts and S. H. Strogatz, Nature (London) **393**, 440 (1998).
7. A.-L. Barabási and R. Albert, Science **286**, 509 (1999).
8. B. Bollobás, *Random Graphs* (Academic Press, London, 1985).
9. D. J. Watts, *Small Worlds: The Dynamics of Networks between Order and Randomness* (Princeton University Press, Princeton, 1999).
10. K. Klemn and V. M. Eguíluz, Phys. Rev. E **65**, 057102 (2002).
11. M. Girvan and M. E. J. Newman, Proc. Natl. Acad. Sci. U.S.A. **99**, 8271 (2002).
12. P. Holme, B. J. Kim, C. N. Yoon, and S. K. Han, Phys. Rev. E **65**, 056109 (2002).
13. A. E. Motter, A. P. S. de Moura, Y.-C. Lai, and P. Dasgupta, Phys. Rev. E **65**, 065102 (2002).
14. A. E. Motter and Y.-C. Lai, Phys. Rev. E **66**, 065102 (2002).
15. A. E. Motter, T. Nishikawa, and Y.-C. Lai, Phys. Rev. E **66**, 065103 (2002).
16. ftp://ibiblio.org/pub/docs/books/gutenberg/etext02/mthes10.zip
17. A. P. S. de Moura, A. E. Motter, and C. Grebogi, Phys. Rev. E **68**, 036106 (2003).
18. M. Steyvers and J. B. Tenenbaum, cond-mat/0110012 (2001).
19. R. F. I. Cancho and R. V. Solé, Proc. Royal Soc. London B **268**, 2261 (2001).
20. S. N. Dorogovtsev and J. F. F. Mendes, Proc. Royal Soc. London B **268**, 2603 (2001).
21. O. Kinouchi, A. S. Martinez, G. F. Lima, G. M. Lourenço, and S. Risau-Gusman, Physica A **315**, 665 (2002).

22. M. Sigman and G. A. Cecchi, Proc. Natl. Acad. Sci. U.S.A. **99**, 1742 (2002).
23. L. F. Costa, cond-mat/0309266 (2003).
24. P. Allegrini, P. Grigolini, and L. Palatella, cond-mat/0310648 (2003).
25. R. Albert, H. Jeong, and A.-L. Barabási, Nature (London) **406**, 378 (2000).
26. R. Cohen, K. Erez, D. ben-Avraham, and S. Havlin, Phys. Rev. Lett. **85**, 4626 (2000).
27. D. S. Callaway, M. E. J. Newman, S. H. Strogatz, and D. J. Watts, Phys. Rev. Lett. **85**, 5468 (2000).
28. R. Cohen, K. Erez, D. ben-Avraham, and S. Havlin, Phys. Rev. Lett. **86**, 3682 (2001).
29. A. Broder, R. Kumar, F. Maghoul, P. Raghavan, S. Rajagopalan, R. Stata, A. Tomkins, and J. Wiener, Comput. Netw. **33**, 309 (2000).
30. A. P. S. de Moura, Y.-C. Lai, and A. E. Motter, Phys. Rev. E **68**, 017102 (2003).
31. D. J. Watts, Proc. Natl. Acad. Sci. USA **99**, 5766 (2002).
32. Y. Moreno, J. B. Gómez, and A. F. Pacheco, Europhys. Lett. **58**, 630 (2002).
33. K.-I. Goh, D.-S. Lee, B. Kahng, and D. Kim, Phys. Rev. Lett. **91**, 148701 (2003).
34. Y. Moreno, R. Pastor-Satorras, A. Vázquez, and A. Vespignani, Europhys. Lett. **62**, 292 (2003).
35. B. A. Carreras, D. E. Newman, I. Dolrou, and A. B. Poole, in: *Proceedings of Hawaii International Conference on System Sciences*, January 4-7, 2000, Maui, Hawaii.
36. M. L. Sachtjen, B. A. Carreras, and V. E. Lynch, Phys. Rev. E **61**, 4877 (2000).
37. M. E. J. Newman, Phys. Rev. E **64**, 016132 (2001).
38. K.-I. Goh, B. Kahng, and D. Kim, Phys. Rev. Lett. **87**, 278701 (2001).
39. P. Holme and B. J. Kim, Phys. Rev. E **65**, 066109 (2002).
40. M. Barthélemy, Phys. Rev. Lett. **91**, 189803 (2003).
41. M. E. J. Newman, S. H. Strogatz, and D. J. Watts, Phys. Rev. E **64**, 026118 (2001).
42. P. Erdös and A. Rényi, Publ. Math. Inst. Hung. Acad. Sci. **5** , 17 (1960).
43. V. Latora and M. Marchiori, Phys. Rev. Lett. **87**, 198701 (2001).
44. S. N. Dorogovtsev and J. F. F. Mendes, Phys. Rev. E **62**, 1842 (2000).
45. F. Liljeros, C. R. Edling, L. A. N. Amaral, H. E. Stanley, and Y. Aberg, Nature (London) **411**, 907 (2001).

Part III

Information Networks & Social Networks

Scholarly Information Network

Paul Ginsparg

Departments of Physics and Computing & Information Science, Cornell University, Ithaca, NY 14853, USA

Abstract. I review the background and some recent trends of a particular scholarly information network, arXiv.org, and discuss some of its implications for new scholarly publication models. If we were to start from scratch today to design a quality-controlled archive and distribution system for scientific and technical information, it could take a very different form from what has evolved in the past decade from pre-existing print infrastructure. Near-term advances in automated classification systems, authoring tools, and document formats will facilitate efficient datamining and long-term archival stability, and I discuss how these could provide not only more efficient means of accessing and navigating the information, but also more cost-effective means of authentication and quality control. Finally, I illustrate the use of machine learning techniques to analyze, structure, maintain, and evolve a large online corpus of academic literature. An emerging field of research can be identified as part of an existing corpus, permitting the implementation of a more coherent community structure for its network of practitioners.

1 arXiv Background and Lessons

I will discuss here some aspects of arXiv.org, a research/social network in which many researchers have been participating for more than a decade. While primarily providing research communications infrastructure, it is also a significant resource for model building and algorithmic experiments in networks. I'll also provide some "cultural enlightenment", in the form of a lighting review of some basic text classification methods (the "Vector Space Model" and "Support Vector Machine"), and their potential use in cleaning, building and maintaining the arXiv subject class structure. The specific experiment I discuss is an example of extracting an emerging research area, "Quantitative Biology", from a larger scale resource. The issue of disentangling sub-networks and their associated subcommunities from the global network also has applications in the broader worldwideweb arena.

The arXiv[1] is an automated repository of over 250,000 full-text research articles[2] in physics and related disciplines (mathematics, non-linear sciences, computer science) going back over a decade and growing at a rate of 40,000 new submissions per year. New submissions are received at a rate of over 175 per

---

[1] See `arXiv.org`. For general background, see [1,2].
[2] as of mid-Oct 2003

weekday from scientists all over the world and the submission rate is increasing at about 10% per year.

The arXiv began in 1991 as an e-mail interface to create, maintain, and access a set of documents for specialists in a particular subject area, a subset of theoretical High Energy Physics. It quickly grew to encompass other subject areas, added a web interface in 1993, and its expansion in use and breadth of coverage continued throughout the 1990's, and continues today. It now serves over 10 million requests per month [3], including tens of thousands of search queries per day, and over 20 million full-text downloads during calendar year '02. It is a significant example of a Web-based service that has changed the practice of research in a major scientific discipline. It now provides nearly comprehensive coverage of large areas of physics, and serves as an on-line seminar system for those areas.

Usage data has been collected since 1991, including Web usage logs beginning in 1993. With this unique dataset, arXiv.org provides a fertile ground for postulating hypotheses and evaluating them. It also serves as a testbed for identifying optimal formats and protocols for rendering, indexing, linking, querying, accessing, mining, and transmitting information, and to identifying sociological, legal and financial obstacles to realization of ideal research communication systems.

When the arXiv was initiated in 1991, no physics journals were yet on-line. Figure 1 shows the continued growth in the monthly submission rate during the past 12 years, now corresponding to 50% more new submissions per month than just four years ago. The original intent was not to supplant journals, but to provide equal and uniform global access to prepublication materials (originally it was only to have had a three month retention time). Due to the multi-year period from 1991 until established journals did come on-line en masse, the arXiv de facto took on a much larger role, by providing the unique on-line platform for near-term (5–10 year) "archival" access. Electronic offerings have of course become commonplace since the early 1990's, and conventional publishers are also set up to provide superior services wherever manual oversight, at additional cost, can improve on the author's product: e.g., correcting bibliographic errors and standardizing the front- and back-matter for automated harvesting. Some of these costs will ultimately decline or disappear, however, with a more standardized "next-generation" document format, and improved authoring tools to produce it — developments from which automated distribution systems will benefit equally.

What are the overlapping roles of the arXiv and the current journal system? Primarily, the arXiv provides instant pre-review dissemination, aggregated on a field-wide basis, a breadth far beyond the capacity of any one journal. The journals augment this with some measure of authentication of authors, and a certain amount of quality control of the research content, providing at least the minimum certification of "not obviously incorrect, not obviously uninteresting". In many cases they provide more than that, e.g., those journals known to have higher selectivity convey an additional measure of short-term prestige. Both the arXiv and the journals provide access to past materials. It is occasionally argued

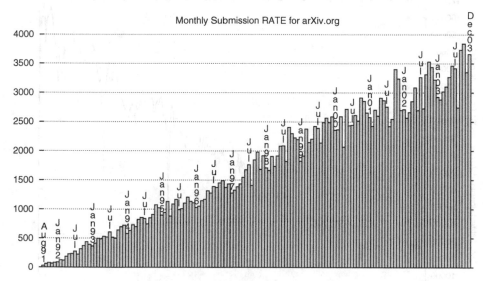

First 12.4 years (31 Dec '03 total = 259,888)

Fig. 1. (See `arXiv.org/show_monthly_submissions` for update)
The number of archive submissions deposited per month, from August 1991 through the end of 2003. Submissions are currently growing at a rate of roughly 10% per year, and there were just under 40,000 new submissions during calendar year 2003. The total number of submissions since 1991 was just under 260,000 at the end of 2003, over half of which were received since the beginning of the year 2000. The majority of growth is in areas of physics other than High Energy Physics (see `arXiv.org/Stats/hcamonthly.html`), which already saturated near 100% participation during the 1990's.

that organized journals may be able to provide a greater degree of long-term archival stability, both in aggregate and for individual items, though looking a century or more into the future this is difficult to project one way or the other.

With conventional overlapping journals having made so much on-line progress, does there remain a continued role for the arXiv, or is it on the verge of obsolescence? Informal polls of researchers suggest that it remains unthinkable to discontinue the resource, that it would simply have to be reinvented because it plays some essential role not fulfilled by any other resource. Hard statistics substantiate this: as mentioned earlier over 20 million full-text downloads during calendar year 2002, on average the full-text of each submission downloaded over 300 times in the 7 years from 1996–2002, and some downloaded in the tens of thousands of times. Figure 2 shows the steady growth in numbers of full-text downloads from the main site. The usage is significantly higher than comparable on-line journals in the field, and, most importantly, the access numbers have accelerated upwards as the conventional journals have come on-line over the past seven years. (Note also that the usage per user has also increased over this period, signaling a measurable change in user behavior over this period.) This is

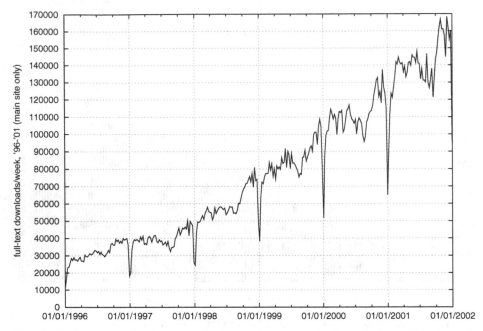

Fig. 2. Full-text downloads per week at the arXiv.org main site only from 1996 through the end of 2001. We continue to see a slightly greater than linear increase in the number of downloads and estimate a total of 27 million full-text downloads from the full arXiv network during 2003. (The yearly chasms result from a mysterious disappearance of readers during New Year holiday periods.)

not to suggest, however, that physicist users are in favor of rapid discontinuation of the conventional journal system either.

What then is so essential about the arXiv to its users? The immediate answer is "It gives instant communication, without having to wait a few months for the peer review process." Does that mean that items should then be removed after some fixed time period? The answer is still "No, it remains incredibly useful as a comprehensive archival aggregator," i.e., a place where for certain fields instead of reading any particular journal, or set of journals, one can browse or search and be certain that the relevant article is there, and if it's not there it's because it doesn't exist.

It has been remarked [4] that physicists use the arXiv site and do not appear concerned that the papers on it are not refereed. The vast majority of submissions are nonetheless submitted in parallel to conventional journals (at no "cost" to the author), and those that aren't are most frequently items such as theses or contributions to conference proceedings that nonetheless have undergone some effective form of review. Moreover, the site has never been a random UseNet newsgroup-like free-for-all. From the outset, a variety of heuristic screening mechanisms have been in place to ensure insofar as possible that submissions are at least *of refereeable quality*. That means they satisfy the minimal criterion that they would not be peremptorily rejected by any competent journal editor as nutty, offensive, or otherwise manifestly inappropriate, and would instead at least in principle be suitable for review (i.e., without the risk of alienating or wasting the time of a referee, that essential unaccounted resource). These mechanisms are an important — if not essential — component of why readers find the site so useful: though the most recently submitted articles have not yet necessarily undergone formal review, the vast majority of the articles can, would, or do eventually satisfy editorial requirements somewhere. Virtually none are entirely useless to active physicists. That is probably why expert arXiv readers are eager and willing to navigate the raw deposited material, and greatly value the accelerated availability over the filtering and refinement provided by the journal editorial processes (even as little as a few months later).

The arXiv operates as a pure dissemination system, i.e., without the editorial operations associated with peer review, at a factor of 100 to 1000 times lower in cost than a conventionally peer-reviewed system [2]. This is the real lesson of the move to electronic formats and distribution: not that everything should somehow be free, but that with many of the production tasks automatable or off-loadable to the authors, the editorial costs will then dominate the costs of an unreviewed distribution system by many orders of magnitude. This is the subtle difference from the paper system, in which the expenses directly associated with print production and distribution were roughly the same order of magnitude as the editorial costs. When the two were comparable in cost, it wasn't as essential to ask whether the production and dissemination system should be decoupled from the intellectual authentication system. Now that the former may be feasible at a cost of less than 1% of the latter, the unavoidable question is whether the utility provided by the latter, in its naive extrapolation to electronic form, continues to justify the associated time and expense. Since many communities rely in an essential way on the structuring of the literature provided by the editorial process, a first related question is whether some hybrid methodology might provide all of the benefits of the current system, but for a cost somewhere in between the greater than $1000/article cost of current editorial methodology and the less than $10/article cost of a pure distribution system. A second question is whether a hybrid methodology might also be better optimized for the differing needs, on differing timescales, of expert readers on the one hand and neophytes on the other.

Before considering these questions in more detail, some logistical and policy issues facing the system are worth mentioning. Ever-increasing automation is necessary so that a constant small staff can maintain current standards of service and appropriateness despite the constantly increasing rate of new submissions. The labor per submission is a skewed distribution: there are subsets, such as the original hep-th (High Energy Physics – Theory), which operate according to the original "fully automated" design, with users requiring no assistance at all. Indeed the vast majority of submissions require zero labor time and only a very small number of new users and problematic submissions are responsible for all labor time spent. This has to be the case, since there are upwards of 200 new submissions and replacements per weekday — if each took even just 15 minutes of human labor at the arXiv end, that would mean over 50 hours of work per day, i.e., at least 7 full-time employees. The current tiny percentage of problematic submissions, and smattering of other user questions, in reality requires less than a single full-time equivalent.

The system also faces policy issues regarding what constitutes the legitimate participating author community. This was not so pressing back in the early '90s, before the non-academic world had discovered the internet. It is clear there needs to be a line drawn somewhere, since readers are not served by submissions that are either manifestly non-science or contain offensive content (pornographic or otherwise). It is likely the system will turn to some form of "sponsorship" system, in which new registrants need to be endorsed by existing submitters before permitted to make new submissions. This is one way in which the on-line community could reflect the physical research community: comprised of people at known institutions, and the people they know and trust, iterated. This web of relationships defined by this networked community is related to the co-authorship web [5] reported on elsewhere at this meeting, and could be analyzed in similar ways.

2 New Scholarly Publication Models

The question for our scholarly research communications infrastructure is: if we were not burdened with the legacy print system and associated methodology, what sort of network would we design for our scholarly communications infrastructure? Do the technological advances of the past decade suggest a new methodology that provides greater utility to the research enterprise at the same or lower cost? (Parts of this section are adapted from [3].)

2.1 Open Access

There has been much recent discussion of free access to the on-line scholarly literature. It is argued that this material becomes that much more valuable when freely accessible [6], and moreover that it is in public policy interests to make the results of publicly funded research freely available as a public good [7]. It is also suggested that this could ultimately lead to a more cost-efficient scholarly publication system. The response of the publishing community has been that their editorial processes provide an essential service to the research community, that these are labor-intensive and hence costly, and that even if delayed, free access could impair their ability to support these operations. (Or, in the case of commercial publishers, reduce revenues to below the profit level necessary to satisfy their shareholders or investors.) Informal surveys (e.g., [2]) of medium- to large-scale publishing operations suggest a wide range in revenues per article published, from the order of $1000/article to more than $10,000/article. The smaller numbers typically come from non-profit operations that provide a roughly equivalent level of service, and hence are more likely representative of actual cost associated with peer reviewed publication. Even some of these latter operations are more costly than might ultimately be necessary, due to the continued need to support legacy print distribution, but the savings from eliminating print and going to an all-electronic in-house work-flow are estimated for a large non-profit publisher to be *at most* on the order of 30%.[3] The majority of the expenses are for the non-automatable editorial oversight and production staff: labor expenses that are not only unaffected by the new technology but that also increase faster than the overall inflation rate in developed countries.

It is also useful to bear in mind that much of the current entrenched methodology is largely a post World War II construct, including both the largescale entry of commercial publishers and the widespread use of peer review for mass production quality control. It is estimated that there are well over $8 billion/year in revenues in STM (Scientific, Technical, and Medical) primary publishing, for somewhere on the order of 1.5-2 million articles published/year. If non-profit operations had the capacity to handle the entirety, and if they could continue to operate in the $500-$1500 revenue per published article range, then with no other change in methodology there might be an immediate 75% savings in the system, releasing well over $5 billion globally.

One proposal to continue funding the current peer-review editorial system is to move entirely from the subscription model to an "author-subsidy" model, in which authors or their institutions pay for the material, either when submitted or when accepted for publication, and the material is then made freely available to readers. While such a system may prove workable in the long-run, it is difficult to impress upon authors the near-term advantages of moving in that direction. It would have the very useful effect of making more manifest directly to authors not only what the minimum real costs are, but also what are the cost hierarchies within the system. This could help to bring market forces to bear on a system that currently operates on a monopolistic basis. A few examples of recently created journals experimenting with this mode are the New Journal of Physics (www.njp.org), the BioMedCentral journals (www.biomedcentral.com), and the Public Library of Science journals (www.plos.org).

From the institutional standpoint, it would also mean that institutions that produce a disproportionate amount of quality research would pay a greater *percentage* of the costs. Some could consider this unfair, though in the long-term a fully reformed and less expensive scholarly publication system should nonetheless offer real savings to those institutions, since they already carry the highest costs in the subscription model. Another short-term difficulty with implementing such a system is the global nature of the research enterprise, in which special

---

[3] This estimate is for the American Physical Society, which publishes over 14,000 articles per year, and derives from figures discussed with its publications oversight committee. The percentage estimated for other publishing operations will vary, especially when editorial time and overhead is differentially accounted. In the discussion that follows, however, it matters only that there will be no *windfall* savings to publishers from going all-electronic, while employing the same overall labor-intensive methodology.

dispensation might be needed to accommodate researchers in developing countries, operating on lower funding scales. Correcting this problem could entail some form of progressive charging scheme and a proportionate increase in the charges to authors in developed countries, increasing the psychological barrier to moving towards an author-subsidy system. A system in which editorial costs are truly compensated equitably would also involve a charge for manuscripts that are rejected (sometimes these require even more editorial time than those accepted), but implementing that is also logistically problematic.

2.2 Peer Review

Many participants in the current peer review system regard it as the only possible quality control mechanism for the literature, signalling important contributions to readers, and necessary for deciding job and grant allocations. But this viewpoint relies on two very strong implicit assumptions: a) that the necessary signal results directly from the peer review process itself, and b) that the signal in question could *only* result from this process. The question is not whether we still need to facilitate *some* form of quality control on the literature; it is instead whether given the emergence of new technology and dissemination methods in the past decade, is the current implementation of peer review still the most effective and efficient means to provide the desired signal?

Appearance in the peer-reviewed journal literature certainly does not provide sufficient signal: otherwise there would be no need to supplement the publication record with detailed letters of recommendation and other measures of importance and influence. On the other hand, the detailed letters and citation analyses *would* be sufficient for the above purposes, even if applied to a literature that had not undergone that systematic first editorial pass through a peer review system. This exposes one of the hidden assumptions in the above: namely that peer-reviewed publication is a prerequisite to entry into a system that supports archival availability and other functions such as citation analysis. In the electronic world, that is no longer necessarily the case.

My own experience as a reader, author, and referee in Physics suggests that current peer review methodology in this field strives to fulfill roles for two different timescales: to provide a guide to expert readers (those well-versed in the discipline) in the short-term, and to provide a certification imprimatur for the long-term. The attempt to perform both functions in one step necessarily falls short on both timescales: too slow for the former, and not stringent enough for the latter.

The observed behavior of expert readers indicates that they don't value the extra level of filtering provided by the current review process above their preference for instant availability of material "of refereeable quality." Non-expert readers typically don't need the availability on the timescale of a few months, but do eventually need a much higher level of selective filtering than is provided on the short timescale. Expert readers as well could benefit on a longer timescale (say a year or longer) from more stringent selection criteria, for the simple reason that the literature of the past decade is always much larger than the "instantaneous" literature. More stringent criteria on a longer timescale would also aid significantly in the job and grant evaluation functions, for which signal on the year or more timescale remains sufficiently timely. More stringent evaluation could potentially play a far greater role than peer-reviewed publication currently does, as compared with external letters and citation analyses.

The simplest modification proposal is thus a two-tier system[4], in which on a first pass only some cursory examination or other pro forma certification is given for acceptance into a standard tier. This could be minimally labor-intensive, perhaps relying primarily on an automated check of author institutional affiliation, prior publication record, research grant status, or other related background; and involve human labor primarily to adjudicate incomplete or ambiguous results of an automated pass. The standard tier availability could also be used to collect confidential commentary from interested readers so that eventual referees would have access to a wealth of currently inaccessible information held by the community, and help to avoid duplication of effort. Then at some later point (which could vary from article to article, perhaps with no time limit), a much smaller set of articles would be selected for the full peer review process. The initial selection criteria for this smaller set could be any of a variety of impact measures, to be determined, and based explicitly on their prior widespread and systematic availability and citability: e.g., reader nomination or rating, citation impact, usage statistics, editorial selection, ...

The precise criteria would depend on the architectural details of the repositories. In a federation of institutionally and disciplinarily held repositories, the institutional repositories (e.g. Dspace, www.dspace.org) could rely on some form of internal endorsement, while the disciplinary aggregates could rely either on affiliation or on prior established credentials (termed "career review" in [12], as opposed to "peer review"). Alternate entry paths for new participants, such as referrals from prior credentialed participants or direct appeal for cursory editorial evaluation (not full-fledged peer review), would also be possible. The essential idea is to facilitate communication within the recognized research community, without excessive noise from the exterior [9]. While multiple logically independent (though potentially overlapping [2]) upper tiers could naturally evolve, only a single globally held standard tier is strictly necessary, with of course any necessary redundancy for full archival stability. Suitable licensing procedures or copyright retention [7] to facilitate such a system are consistent with the spirit of copyright law, "To promote the Progress of Science and useful Arts" (for a recent discussion, see [13]).

---

[4] For more details, see, e.g., [3]. The idea of using prior electronic distribution to augment the referee process goes back at least to [8]. Proposals along the lines of decoupling peer review from arXiv distribution can be found in [9] and the notion of "overlay" journals is further discussed in [1]. A review of various "decoupling" and "author subsidy" models proposed in the mid to late 1990's, taking advantage of new technology to implement improvements in research communication, can be found in [10]. In particular, the "eprint moderator model" [11] was intended to reduce costs by reducing the amount of material distributed in a commercial manner.

Recent experience in Physics and related disciplines continues to reinforce the desirability of experimentation within this model space, with the expectation that similar implementations will prove feasible in other disciplines.

3 Novel Corpus Navigation Tools

The size of the arXiv corpus, over 250,000 documents, presents navigation challenges to researchers using the system and provides an excellent test-bed for experimentation with navigation tools. For example, very preliminary experiments in the use of a burst detection algorithm [14] applied to word occurrences in titles provide tantalizing results. Figure 3 shows the title word bursts from this analysis of hep-th submissions from 1991 until the end of 2001. Physicist practitioners in the area recognize that this coarse first-pass version of the algorithm, coupled with equally primitive visualization, already provides a credible map to significant trends in theoretical High Energy Physics over this period.

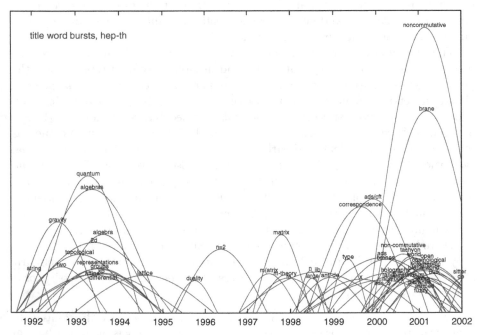

Fig. 3. Results from burst analysis of the words in the titles of submissions to the hep-th archive. The curves indicate both the intensity of onset (height) and the rough duration (width) of the bursts. (The vertical scale is somewhat arbitrary in this context, proportional to the "cost function" of [14], rather than more intuitive notions of the "strength" of the burst. The shape of the bursts, including the left–right symmetry, is also an artifact of this primitive visualization.)

Eventually this work will be extended to produce an on-line navigational tool available to users of arXiv.org that will allow such analyses to be performed over any portion of the archive. The intent is to implement intuitive and informative visualization methods that permit users to identify the most important temporal patterns, and then navigate the underlying intellectual trends via hypertext links annotating the visualization. This will require refinement of the basic burst detection algorithm to the textual case at hand, including proper handling of both common word multi-grams and synonymy. Use of citation tree data, together with typical co-occurring terms, will permit a characterization of the intellectual content of the burst, and a mapping of its prior and post intellectual development. Such maps can be imagined as the analog of an annotated genome for the field.

arXiv.org usage logs go back 12 years and embody a huge amount of information that has yet to be mined systematically for useful signal. Figure 4 shows the results of a straightforward analysis of the numbers of abstract and full-text downloads at the main Cornell site and the LANL mirror site during 2002. Care has been taken to avoid over-estimating the number of downloads (by screening for rapidly repeated or partial downloads and robotic activity).

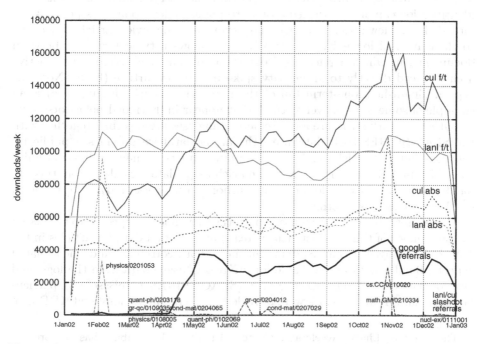

Fig. 4. Overview of download patterns at the main site (CUL) and the LANL mirror site during 2002. The ratio of abstract views ('abs') to the number of full-text ('f/t') downloads was roughly constant for many years until Google was permitted to full-text index in early 2002, and provided external referrers directly to the texts.

A number of interesting features stand out in this graph. First, the arXiv.org main site was moved from LANL to Cornell in Dec 2001, and the figure shows the transition in usage between the two sites. The downloads from the two sites still remained roughly comparable, in part due to the persistence of hard-coded links throughout the web to the old arXiv address, xxx.lanl.gov, which now points to the LANL mirror lanl.arXiv.org. (This has the coincidentally beneficial effect of providing load-balancing between the sites, though by the end of 2003 — a year after the end of the graph in Fig. 4 — the distribution had shifted to the main Cornell site handling more than 2.5 times the volume of the LANL mirror.) Second is the onset of referrals from Google searches starting in Apr 2002. Google was permitted to harvest full texts from the arXiv for indexing starting in Mar 2002. Third, the "spikes" at the foot of the graph show downloads for individual submissions resulting from Slashdot (slashdot.org) referrals, some a sufficiently large percentage of overall activity to be visible in the aggregate download graphs.

Automated characterization of documents based on their full-text usage patterns is another current research area. For example, review articles are known to have a particular usage "signature": a long tail in the usage pattern over time. Figure 5 ((a) and (b)) shows this signature in the usage patterns for a review paper and a textbook. The similarity in patterns between the review articles and the textbook confirms the intuition that they serve similar functions. The identification can also be turned around to point readers to articles perhaps not intended as review articles, but that nonetheless accrue review article status due to their archival longevity. Figure 5 (a) also shows a typical relation between documents, in which a reference from a non-review article (red) in an active area can lead directly to an activity spike in a review article (blue). Note the typical diminished long-term tail of the non-review article. Figure 5 (c) is an example of some of the novel features that appear in download behaviors.

Other network structure and prediction analyses are possible: together with the SLAC SPIRES-HEP database[5], the arXiv database provides a public resource of full-text articles and associated citation tree of many millions of links, with a focused disciplinary coverage, and rich usage data. The data from this site has been used, for example, in the co-authorship network analysis [5] reported on elsewhere at this meeting.

Recent research comparing usage patterns (downloads, expected to be related to the number of times a document is read) and citation data for articles in the NASA Astrophysical Data System [16] shows both correlated and uncorrelated aspects. Similar data are available for arXiv.org: usage data from the logs, and citation data extracted by other services (SLAC-SPIRES, CiteBase[6]). In particular, such data from arXiv.org were provided in the recently completed KDD Cup

---

[5] The Stanford Linear Accelerator Center SPIRES-HEP database has comprehensively catalogued the High Energy Particle Physics (HEP) literature online since 1974, and indexes more than 500,000 high-energy physics related articles including their full citation tree (see [15]).

[6] See citebase.eprints.org

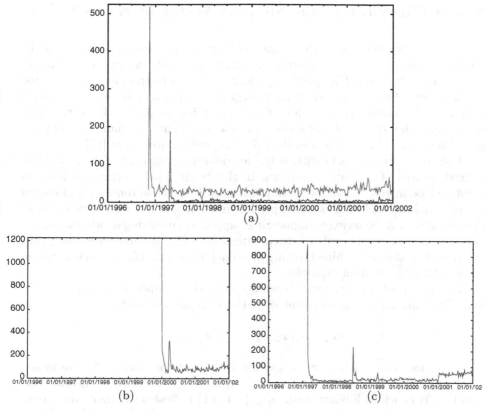

Fig. 5. Graphs showing the number of downloads per week as a function of time. (The horizontal scale on all of (a,b,c) covers the same time period from the beginning of 1996 to the end of 2001, and the retrievals shown are from the main site only.) (a) Retrievals of a review article (light gray) submitted in Nov 1996. A secondary spike in Apr 1997 coincides with the appearance of a non-review article (darker gray) that referred to it. Note also the characteristic long tail for the review article. (b) Retrievals of a textbook submitted in Dec 1999. (c) Usage pattern for an article submitted in Feb 1997 with some atypical features, including an event in early 2001 shifting it to a higher "excited state", and effective review article status.

2003 data mining competition[7] (see www.cs.cornell.edu/projects/kddcup), and used as the basis for a network prediction task: *predicting* the number of citations articles would receive in the future based on the network structure annotated by activity levels and other associated metadata. It is intriguing that such predictions can be reliably made more or less independent of the intellectual content of the actual articles.

---

[7] held in conjunction with the Ninth ACM SIGKDD International Conference on Knowledge Discovery and Data Mining

4 Text Classification and Support Vector Machines

In what follows, we will use arXiv data to illustrate how machine learning methods can be used to analyze, structure, maintain, and evolve a large online corpus of academic literature. The specific application will be to train a support vector machine text classifier to extract an emerging research area from a larger-scale resource. The automated detection of such subunits can play an important role in disentangling other sub-networks and associated sub-communities from the global network. This section and the following are adapted from [17].

The goal of text classification is the automatic assignment of documents to a fixed number of semantic categories. In the "multi-label" setting, each document can be in zero or one or more categories. Efficient automated techniques are essential to avoid tedious and expensive manual category assigment for large document sets. A "knowledge engineering" approach, involving hand-crafting accurate text classification rules, is surprisingly difficult and time-consuming [18]. We therefore take a machine learning approach to generating text classification rules automatically from examples.

The machine learning approach can be phrased as a supervised learning problem. The learning task is represented by the training sample S_n

$$(\boldsymbol{x}_1, y_1), (\boldsymbol{x}_2, y_2), \dots , (\boldsymbol{x}_n, y_n) \tag{1}$$

of size n documents, where $\boldsymbol{x}_i$ represents the document content. In the multi-label setting, each category label is treated as a separate binary classification problem. For each such binary task, $y_i \in \{-1, +1\}$ indicates whether a document belongs to a particular class. The task of the learning algorithm $\mathcal{L}$ is to find a decision rule $h_{\mathcal{L}} : \boldsymbol{x} \longrightarrow \{-1, +1\}$ based on S_n that classifies new documents $\boldsymbol{x}$ as accurately as possible.

Documents need to be transformed into a representation suitable for the learning algorithm and the classification task. Information Retrieval research suggests that words work well as representation units, and that for many tasks their ordering can be ignored without losing too much information. This type of representation is commonly called the "bag-of-words" model, an attribute–value representation of text. Each text document is represented by a vector in the lexicon space, i.e., by a "term frequency" feature vector $\mathrm{TF}(w_i, x)$, with component values equal to the number of times each distinct word w_i in the corpus occurs in the document x. Figure 6 shows an example feature vector for a particular document. This is a particular form of the "Vector Space Model" for representing text.

This basic representation is ordinarily refined in a few ways:

TF×IDF Weighting: Scaling the components of the feature vector with their *inverse document frequency* $\mathrm{IDF}(w_i)$ [19] often leads to improved performance. In general, $\mathrm{IDF}(w_i)$ is some decreasing function of the word frequency $\mathrm{DF}(w_i)$, equal to the number of documents in the corpus which contain the word w_i. For example,

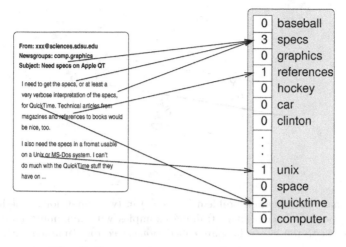

Fig. 6. Representing text as a feature vector.

$$\text{IDF}(w_i) = \log\left(\frac{n}{\text{DF}(w_i)}\right) \tag{2}$$

where n is the total number of documents. Intuitively, the inverse document frequency assumes that rarer terms have more significance for classification purposes, and hence gives them greater weight. To compensate for the effect of different document lengths, each document feature vector x_i is normalized to unit length: $\|x_i\| = 1$.

Stemming: Instead of treating each occurrence form of a word as a different feature, stemming is used to project the different forms of a word onto a single feature, the word stem, by removing inflection information [20]. For example "computes", "computing", and "computer" are all mapped to the same stem "comput". The terms "word" and "word stem" will be used synonymously in the following.

Stopword Removal: For many classification tasks, common words like "the", "and", or "he" do not help discriminate between document classes. Stopword removal describes the process of eliminating such words from the document by matching against a predefined list of stop-words. We use a standard stoplist of roughly 300 words.

SVMs [21] were developed by V. Vapnik et al. based on the structural risk minimization principle from statistical learning theory. They have proven to be a highly effective method for learning text classification rules, achieving state-of-the-art performance on a broad range of tasks [22,23]. Two main advantages of using SVMs for text classification lie in their ability to handle the high dimensional feature spaces arising from the bag-of-words representation. From a statistical perspective, they are robust to overfitting and are well suited for the statistical properties of text. perspective, they can be trained efficiently despite

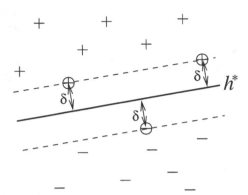

Fig. 7. A binary classification problem (+ vs. −) in two dimensions. The hyperplane h^* separates positive and negative training examples with maximum margin δ. The examples closest to the hyperplane are called *support vectors* (marked with circles).

the large number of features. A detailed overview of the SVM approach to text classification, with more details on the notation used below, is given in [24].

In their basic form, SVMs learn linear decision rules

$$h(\boldsymbol{x}) = \text{sgn}\{\boldsymbol{w} \cdot \boldsymbol{x} + b\} \ , \tag{3}$$

described by a weight vector $\boldsymbol{w}$ and a threshold b, from an input sample of n training examples $S_n = ((\boldsymbol{x}_1, y_1), \cdots, (\boldsymbol{x}_n, y_n))$, $\boldsymbol{x}_i \in \mathbf{IR}^N$, $y_i \in \{-1, +1\}$. For a linearly separable S_n, the SVM finds the hyperplane with maximum Euclidean distance to the closest training examples. This distance is called the margin δ, as depicted in Fig. 7. Geometrically, the hyperplane is defined by its normal vector, $\boldsymbol{w}$, and its distance from the origin, $-b$. For non-separable training sets, the amount of training error is measured using slack variables ξ_i.

Computing the position of the hyperplane is equivalent to solving the following convex quadratic optimization problem [21]:

Optimization Problem 1 (SVM (primal))

$$\text{minimize:} \quad V(\boldsymbol{w}, b, \boldsymbol{\xi}) = \frac{1}{2}\boldsymbol{w} \cdot \boldsymbol{w} + C \sum_{i=1}^{n} \xi_i \tag{4}$$

$$\text{subj. to:} \quad \forall_{i=1}^{n} : y_i[\boldsymbol{w} \cdot \boldsymbol{x}_i + b] \geq 1 - \xi_i \tag{5}$$

$$\forall_{i=1}^{n} : \xi_i > 0 \tag{6}$$

The margin of the resulting hyperplane is $\delta = 1/\|\boldsymbol{w}\|$.

The constraints (5) require that all training examples are classified correctly up to some slack ξ_i. If a training example lies on the "wrong" side of the hyperplane, we have the corresponding $\xi_i \geq 1$, and thus $\sum_{i=1}^{n} \xi_i$ is an upper bound on the number of training errors. The factor C in (4) is a parameter that allows trading off training error vs. model complexity. The optimal value of this parameter depends on the particular classification task and must be chosen via cross-validation or by some other model selection strategy. For text classification, however, the default value of $C = 1/\max_i ||\boldsymbol{x}_i||^2 = 1$ has proven to be effective over a large range of tasks [24].

OP1 has an equivalent dual formulation:

Optimization Problem 2 (SVM (dual))

$$\text{maximize:} \quad W(\boldsymbol{\alpha}) = \sum_{i=1}^{n} \alpha_i - \frac{1}{2}\sum_{i=1}^{n}\sum_{j=1}^{n} y_i y_j \alpha_i \alpha_j (\boldsymbol{x}_i \cdot \boldsymbol{x}_j) \tag{7}$$

$$\text{subj. to:} \quad \sum_{i=1}^{n} y_i \alpha_i = 0 \tag{8}$$

$$\forall i \in [1..n] : 0 \leq \alpha_i \leq C \tag{9}$$

From the solution of the dual, the classification rule solution can be constructed as

$$\boldsymbol{w} = \sum_{i=1}^{n} \alpha_i y_i \boldsymbol{x}_i \quad \text{and} \quad b = y_{\text{usv}} - \boldsymbol{w} \cdot \boldsymbol{x}_{\text{usv}} , \tag{10}$$

where $(\boldsymbol{x}_{\text{usv}}, y_{\text{usv}})$ is some training example with $0 < \alpha_{\text{usv}} < C$. For the experiments in this paper, SVM^{Light} [24] is used for solving the dual optimization problem[8]. More detailed introductions to SVMs can be found in [25,26].

5 arXiv q-bio Extraction

Before using the machine learning framework to identify new subject area content, we first assessed its performance on the existing (author-provided) category classifications. Roughly 180,000 titles and abstracts were fed to model building software which constructed a lexicon of roughly 100,000 distinct words and produced training files containing the TD×IDF document vectors for SVM^{Light}. (While the SVM machinery could easily be used to analyze the full document content, previous experiments [24] suggest that well-written titles and abstracts provide a highly focused characterization of content, at least as effective for our document classification purposes.) The set of support vectors and weight parameters output by SVM^{Light} was converted into a form specific to the linear SVM, (3): a weight vector $\boldsymbol{w}_c$ and a threshold b_c, where c is an index over the categories.

As seen in Fig. 8, the success of the SVM in classifying documents improves as the size of a category increases. The SVM is remarkably successful at identifying documents in large ($> 10{,}000$ documents) categories and less successful on smaller subject areas (< 500 documents). A cutoff was imposed to exclude subject areas with fewer than 100 documents.[9]

---

[8] SVM^{Light} is available at `svmlight.joachims.org`

[9] Some of the smaller subject areas are known to be less topically focused, so the difficulty in recall, based solely on title/abstract terminology, was expected.

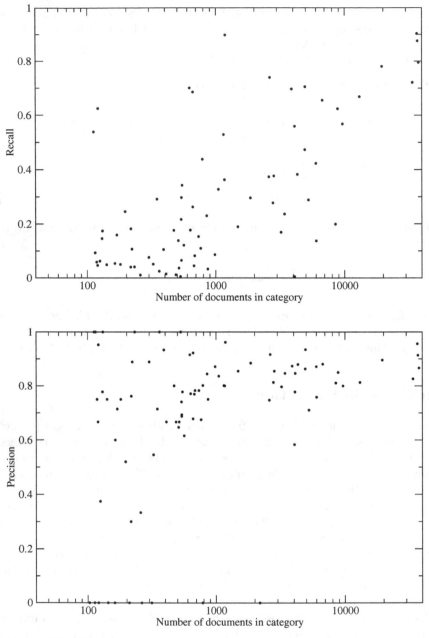

Fig. 8. Recall and Precision as functions of category size for 78 arXiv major categories and minor subject classes. 2/3 of the sample was used as a training set and 1/3 as a test set. The four largest categories, each with over 30,000 documents are cond-mat, astro-ph, hep-th, and hep-ph.

Experiments showed that the use of TF×IDF weighting as in (2) improved accuracy consistently over pure TF weighting, so TF×IDF weighting was used in the experiments to follow. We also used a document frequency threshold to exclude rare words from the lexicon, but found little difference in accuracy between using a document occurrence threshold of two and five.[10] Increasing the weight of title words with respect to abstract words, on the other hand, consistently worsened accuracy, indicating that words in a well-written abstract contain as much or more content classification import as words in the title. Changes in the word tokenization and stemming algorithms did not have a significant impact on overall accuracy. The default value of $C = 1$ in (9) was preferred.

There has been recent anecdotal evidence of an intellectual trend among physicists towards work in biology, ranging from biomolecules, molecular pathways and networks, gene expression, cellular and multicellular systems to population dynamics and evolution.[11] This work has appeared in separate parts of the archive, particularly under "Soft Condensed Matter", "Statistical Mechanics", "Disordered Systems and Neural Networks", "Biophysics, and "Adaptive and Self-Organizing Systems" (abbreviated cond-mat.soft, cond-mat.stat-mech, cond-mat.dis-nn, physics.bio-ph, and nlin.AO). A more coherent forum for the exchange of these ideas was requested, under the nomenclature "Quantitative Biology" (abbreviated "q-bio").

To identify first whether there was indeed a real trend to nurture and amplify, and to create a training set, volunteers were enlisted to identify the q-bio content from the above subject areas in which it was most highly focused. Of 5565 such articles received from Jan 2002 through Mar 2003, 466 (8.4%) were found to have one of the above biological topics as its primary focus. The total number of distinct words in these titles, abstracts, plus author names, was 23558, of which 7984 were above the $DF = 2$ document frequency threshold. (Author names were included in the analysis since they have potential "semantic" content in this context, i.e., are potentially useful for document classification. The SVM algorithm will automatically determine whether or not to use the information by choosing suitable weights.)

A data-cleaning procedure was employed, in which SVM^{Light} was first run with $C = 10$. We recall from (4) and (9) that larger C penalizes training errors and requires larger α's to fit the data. Inspecting the "outlier" documents [27] with the largest $|\alpha_i|$ then permitted manual cleaning of the training set. 10 were moved into q-bio, and 15 moved out, for a net movement to 461 q-bio (8.3%) of

---

[10] Words that appeared in fewer than two documents constituted roughly 50% of the lexicon, and those that appeared in fewer than five documents roughly 70%. Ignoring rare and consequently uninformative words hence reduces the computational needs.

[11] This is really a long tradition, going back to prominent examples such as M. Delbrück, F. Crick, W. Gilbert, ..., and including J. Hopfield at the "Complex Networks" conference at which the current presentation was given. So the question is really whether there has been a recent acceleration of the trend. Note also that with few recent examples of biologists working on string theory or elementary particle physics, it appears to be something of a one-way street, with no detailed balance in the diffusion process.

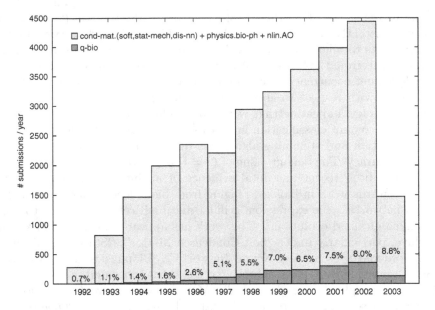

Fig. 9. The number of submissions per year from 1992 through April 2003 in particular subsets of arXiv subject areas of cond-mat, physics, and nlin most likely to have "quantitative biology" content. The percentage of q-bio content in these areas grew from roughly 1% to nearly 10% during this timeframe, suggesting a change in intellectual activity among arXiv-using members of these communities.

the 5565 total. Some of the others flagged involved word confusions, e.g., "genetic algorithms" typically involved programming rather than biological applications. Other "q-bio" words with frequent non-biological senses were "epidemic" (used for rumor propagation), "evolution" (used also for dynamics of sandpiles), "survival probabilities", and extinction. "Scale-free networks", were sometimes used for biological applications, and sometimes not (just as in other contributions to this volume). To help resolve some of these ambiguities, the vocabulary was enlarged to include a list of most frequently used two-word phrases with semantic content different from their constituent words .

With a training set fully representative of the categories in question, it was then possible to run the classifier on the entirety of the same subject area content received from 1992 through Apr 2003, a total of 28,830 documents. The results are shown in Fig. 9. A total of 1649 q-bio documents was identified, and the trend towards an increasing percentage of q-bio activity among these arXiv users is evident: individual authors can be tracked as they migrate into the domain. Visibility of the new domain can be further enhanced by referring submitters in real time, via the automated classifier running behind the submission interface.

Some components of the weight vector generated by the SVM for this training set are shown in Fig. 10. Since the distance of a document to the classifying hyperplane is determined by taking the dot product with the normal vector, its

protein	+8.57		:	point	-1.04
dna	+7.08	forward	+0.00	equation	-1.05
biological	+5.06	minimalist	+0.00	boundary	-1.06
neuron	+5.05	region	+0.00	social	-1.06
rna	+3.30	confinement	+0.00	n	-1.09
gene	+3.21	implies	+0.00	relaxation	-1.14
mutation	+3.15	96	+0.00	fluid	-1.15
population	+3.11	y_togashi	+0.00	indian	-1.15
epidemic	+3.05	n_wingreen	+0.00	spin	-1.17
biology	+3.02	mean_free	+0.00	spin_glass	-1.17
disease	+2.93	narrower	+0.00	traffic	-1.18
cell	+2.90	shot	-0.00	system	-1.30
neural	+2.89	repton	-0.00	polymer	-1.33
brain	+2.83	kyoto	-0.00	class	-1.35
ecosystem	+2.56	regular	-0.00	emerge	-1.36
tissue	+2.52	generalisation	-0.00	gradient	-1.39
sequence	+2.51	d_saakian	-0.00	quantum	-1.43
genetic	+2.51	conformity	-0.00	surface	-1.43
bacterial	+2.48	aware	-0.00	synchronization	-1.45
blood	+2.43	even_though	-0.00	market	-1.47
genome	+2.37	practitioner	-0.00	particle	-1.52
peptide	+2.37	permittivity	-0.00	polyelectrolyte	-1.53
infection	+2.34		:	world	-1.57

Fig. 10. Shown above are the most positive, a few intermediate, and most negative components of the q-bio classifying weight vector.

component values can be interpreted as the classifying weight for the associated words. The approach here illustrates one of the major lessons of the past decade: the surprising power of simple algorithms operating on large datasets. In some sense we have also come full circle, with the modern SVM technology of the late 90's providing a formal framework for a heuristic binary "peer review" classifier originally proposed over a decade ago (see Fig. 11).

The q-bio extraction described above was not just a thought experiment, but a prelude to an engineering experiment. The new category went on-line in mid-September 2003 (see arXiv.org/new/q-bio.html), at which time past submitters flagged by the system were asked to confirm the q-bio content of their submissions, and to send future submissions to the new category. The activity levels at the outset corresponded precisely to the predictions of the SVM text classifier, and later began to show indications of growth catalyzed by the public existence of the new category (see (arxiv.org/year/q-bio/03)). The subject classes range from the small scale of biomolecules, genomics, and molecular networks, to the intermediate scale of cellular and multicellular systems, to the large scales of population dynamics and evolution.

The experience gained from applying machine learning and text classification techniques to extract the q-bio archive will be used in the future to develop an enhanced SVM tool to facilitate better structuring of the arXiv.org corpus. Together with document clustering algorithms, the tool will permit dynami-

What's new on the <u>arXiv.org</u> (xxx) e-print archives

(messages from '91-'94: http://arXiv.org/new/91-94.html)

. . .

Nov 12 1991

```
In response to popular demand to limit the flux of papers,
This resource has adopted a referee system designed as a
faithful emulation of that employed, e.g., by the Physical Review
(except that papers will typically be evaluated within a few milliseconds).

The evaluation is based on a preprogrammed list of keywords, each with an
assigned positive or negative integer (reflecting respectively desirable
and undesirable words). The overall value of a paper is determined by the
signed sum of the number of occurrences of each keyword times its
associated value. Papers with value < or =0 are rejected. Keywords include
names of current and past fads, authornames (for both paper and references),
macropackages, etc. The assigned integers are programmed to vary randomly
within given preset ranges to emulate individual variation among referees,
and for additional realism are chosen entirely at random 20% of the time.
The preset ranges evolve dynamically in a neural net that determines positive
and negative contexts of word appearances via a rudimentary syntactic analysis.

Submitters of rejected papers will receive a detailed account of the
calculation, so they will know precisely why their paper was rejected.
Rejected papers may not be revised and resubmitted.
Objections, questions, and complaints will be automatically ignored.
```

Fig. 11. A binary classifier for "peer review", originally proposed in Nov 1991 (see arXiv.org/new/91-94.html). Compare methodology with the results of the SVM "q-bio" classifier in Fig. 10.

cal subdivisions of the existing document collection, and perform data cleaning and "outlier" detection. Once these text classification techniques are refined to sufficient robustness, a text classification system will be deployed to provide additional facilities such as real time classification suggestions to submitters at upload-time, automated alerts of mis-classified articles, and keyword-based alerting systems for arXiv.org subscribers (useful in finding articles otherwise dispersed within the classification scheme).

6 Conclusion

While implemented as a passive dissemination system, the arXiv database has also played a social engineering role in scholarly information network evolution, with active research users developing an affinity to the system and adjusting their behavior accordingly. They scan new submissions on a daily basis, assume others in their field do so and are consequently aware of anything relevant that has appeared there (while anything that doesn't may as well not exist), and use it to stake intellectual priority claims in advance of journal publication. arXiv has also played a major role in pointing the way towards new scholarly publication models, as described in Sect. 2. It serves moreover as a testbed in the development of new corpus navigational tools, as described in Sect. 3. Finally, the machine learning tools of Sect. 4 were seen in Sect. 5 to be able to characterize a subdomain and thereby help accelerate its growth, via the interaction of an information resource with the research network of its practitioners.

Acknowledgements

These notes are based on a talk given at the CNLS 23rd annual conference "Networks: Structure, Dynamics and Function" (Santa Fe, NM, 12–16 May 2003). I thank the organizers for creating a most stimulating meeting, and for inviting me to give the closing talk to help participants decompress.

References

1. P. Ginsparg. Winners and losers in the global research village. In Sir R. Elliot and D. Shaw, editors, *Electronic Publishing in Science I, Proceedings of joint ICSU Press/UNESCO conference, Paris*, 1996 (copy at http://arXiv.org/blurb/pg96unesco.html). URL: http://users.ox.ac.uk/~icsuinfo/ginsparg.htm.
2. P. Ginsparg. Creating a global knowledge network. In Sir R. Elliot and D. Shaw, editors, *Electronic Publishing in Science II, proceedings of joint ICSU Press/UNESCO conference, Paris*. ICSU Press, 2001 (copy at http://arXiv.org/blurb/pg01unesco.html). URL: http://users.ox.ac.uk/~icsuinfo/ginspargfin.htm.
3. P. Ginsparg. Can peer review be better focused?, 2003, Science & Technology Libraries, to appear. URL: http://arXiv.org/blurb/pg02pr.html.
4. Brinkman Outlines Priorities, Challenges for APS in 2002. *APS News*, January 2002. URL: http://www.aps.org/apsnews/0102/010208.html.
5. M. E. J. Newman, Who Is the Best Connected Scientist? A Study of Scientific Coauthorship Networks, Lect. Notes Phys. **650**, 337–370 (2004).
6. R.S. Berry. Is electronic publishing being used in the best interests of science? the scientist's view. In Sir R. Elliot and D. Shaw, editors, *Electronic Publishing in Science II, proceedings of joint ICSU Press/UNESCO conference, Paris*, 2001. URL: http://users.ox.ac.uk/~icsuinfo/berryfin.htm.
7. S. Bachrach, R.S. Berry, M. Blume, T. von Foerster, A. Fowler, P. Ginsparg, S. Heller, N. Kestner, A. Odlyzko, A. Okerson, R. Wigington, and A. Moffat. Who should own scientific papers? *Science*, 281:1459–1460, 1998. URL: http://www.sciencemag.org/cgi/content/full/281/5382/1459.
8. S. Rogers and C. Hurt. How scholarly communication should work in the 21st century. *The Chronicle of Higher Education*, October 18:A56, 1989.
9. P. Ginsparg. First steps towards electronic research communication. *Computers in Physics*, 8(4 (Jul/Aug)), 1994.
10. S. Gass. Transforming scientific communication for the 21st century. *Science and Technology Libraries*, 19(3/4):3–18, 2001.
11. D. Stern. eprint moderator model, 1999, version dated Jan 25, 1999. URL: http://www.library.yale.edu/scilib/modmodexplain.html.

12. R. Kling, L. Spector, and G. McKim. Locally controlled scholarly publishing via the internet: The guild model. *The Journal of Electronic Publishing*, August, 2002. URL: `http://www.press.umich.edu/jep/08-01/kling.html`.

13. J. Willinsky. Copyright contradictions in scholarly publishing. *First Monday*, 7(11 (November)), 2002. URL: `http://firstmonday.org/issues/issue7_11/willinsky/index.html`.

14. J. Kleinberg. Bursty and hierarchical structure in streams. In *Proceedings of the 8th ACM SIGKDD International Conference on Knowledge Discovery and Data Mining*, 2002. URL: `http://www.cs.cornell.edu/home/kleinber/kleinber.html`.

15. H. B. O'Connell. Physicists thriving with paperless publishing. *HEP Lib. Web.*, 6:3, arXiv:physics/0007040, 2002. URL: `http://arXiv.org/physics/0007040`.

16. M. J. Kurtz, G. Eichhorn, A. Accomazzi, C. Grant, M. Demleitner, S. S. Murray, N. Martimbeau, and B. Elwell. The NASA astrophysics data system: Sociology, bibliometrics and impact. 2003. URL: `http://cfa-www.harvard.edu/~kurtz/jasist-submitted.pdf`.

17. P. Ginsparg, P. Houle, T. Joachims, and J.-H. Sul. Mapping subsets of scholarly information. *PNAS, to appear*, 2004. URL: `http://arXiv.org/cs.IR/0312018`.

18. P. Hayes and S. Weinstein. CONSTRUE/TIS: a system for content-based indexing of a database of news stories. In *Annual Conference on Innovative Applications of AI*, 1990.

19. G. Salton and C. Buckley. Term weighting approaches in automatic text retrieval. *Information Processing and Management*, 24(5):513–523, 1988.

20. M. Porter. An algorithm for suffix stripping. *Program (Automated Library and Information Systems)*, 14(3):130–137, July 1980.

21. V. Vapnik. *Statistical Learning Theory*. Wiley, Chichester, GB, 1998.

22. T. Joachims. Text categorization with support vector machines: Learning with many relevant features. In *Proceedings of the European Conference on Machine Learning*, pages 137 – 142, Berlin, 1998. Springer. URL: `http://www-ai.cs.uni-dortmund.de/DOKUMENTE/joachims_98a.ps.gz`.

23. S. Dumais, J. Platt, D. Heckerman, and M. Sahami. Inductive learning algorithms and representations for text categorization. In *Proceedings of ACM-CIKM98*, November 1998.

24. T. Joachims. *Learning to Classify Text Using Support Vector Machines – Methods, Theory, and Algorithms*. Kluwer, 2002.

25. C. Burges. A tutorial on support vector machines for pattern recognition. *Data Mining and Knowledge Discovery*, 2(2):121–167, 1998.

26. N. Cristianini and J. Shawe-Taylor. *An Introduction to Support Vector Machines and Other Kernel-Based Learning Methods*. Cambridge University Press, 2000.

27. I. Guyon, N. Matic, and V. Vapnik. Discovering informative patterns and data cleaning. In *Advances in Knowledge Discovery and Data Mining*, pages 181–203, 1996.

Who Is the Best Connected Scientist?
A Study of Scientific Coauthorship Networks

Mark E.J. Newman

Department of Physics and Center for the Study of Complex Systems, University of
Michigan, Ann Arbor, MI 48109, USA
Santa Fe Institute, 1399 Hyde Park Road, Santa Fe, NM 87501, USA

Abstract. Using data from computer databases of scientific papers in physics, biomedical research, and computer science, we have constructed networks of collaboration
between scientists in each of these disciplines. In these networks two scientists are
considered connected if they have coauthored one or more papers together. We have
studied many statistical properties of our networks, including numbers of papers written by authors, numbers of authors per paper, numbers of collaborators that scientists
have, typical distance through the network from one scientist to another, and a variety
of measures of connectedness within a network, such as closeness and betweenness.
We further argue that simple networks such as these cannot capture the variation in
the strength of collaborative ties and propose a measure of this strength based on the
number of papers coauthored by pairs of scientists, and the number of other scientists
with whom they worked on those papers. Using a selection of our results, we suggest a
variety of possible ways to answer the question "Who is the best connected scientist?"

1 Introduction

A social network is a set of people or groups each of which has connections of
some kind to some or all of the others [1,2]. In the language of social network
analysis, the people or groups are called *actors* and the connections *ties*. Both
actors and ties can be defined in different ways depending on the questions of
interest. An actor might be a single person, a team, or a company. A tie might
be a friendship between two people, a collaboration or common member between
two teams, or a business relationship between companies.

Social network analysis has a history stretching back at least half a century,
and has produced many results concerning social influence, social groupings, inequality, disease propagation, communication of information, and indeed almost
every topic that has interested twentieth century sociology. In the last few years,
it has become the focus of considerable attention in the applied mathematics
and statistical physics communities as well [3–6].

Traditional investigations of social networks have been carried out through
field studies. Typically one looks at a fairly self-contained community such as a
business community [7–9], a school [10,11], a religious or ethnic community [12],
and so forth, and constructs the network of ties by interviewing participants, or
by circulating questionnaires. A study will ask respondents to name those with
whom they have the closest ties, often ranked by subjective closeness, and may

M.E.J. Newman, Who Is the Best Connected Scientist? A Study of Scientific Coauthorship Networks,
Lect. Notes Phys. **650**, 337–370 (2004)
http://www.springerlink.com/

optionally call for additional information about those people or about the nature of the ties.

Studies of this kind have revealed much about the structure of communities, but they suffer from two substantial problems that make them poor sources of data for the kinds of quantitative approaches to network analysis that have been developed in physics and mathematics. First, the data they return are not numerous. Collecting and compiling data from these studies is an arduous process and most data sets contain no more than a few tens or hundreds of actors. It is a rare study that exceeds a thousand actors. This makes the statistical accuracy of many results poor, a particular difficulty for the large-system-size methods adopted in statistical physics. Second, they contain significant and uncontrolled errors as a result of the subjective nature of respondents' replies. What one respondent considers to be a friendship or acquaintance, for example, may be completely different from what another respondent does. In studies of school-children, for instance [10,11,13], it is found that some children will claim friendship with every single one of their hundreds of schoolmates, while others will name only one or two friends. Clearly these respondents are employing different definitions of friendship.

In response to these inadequacies, many researchers have turned instead to other, better documented networks, for which reliable statistics can be collected. Examples include the Internet [14,15], the world wide web [16,17], email networks [18,19], peer-to-peer networks [20,21], power grids [22], telephone call graphs [23], and train routes [24]. These graphs are certainly interesting in their own right, and furthermore might loosely be regarded as social networks, since their structure clearly reflects something about the structure of the society that built them. However, their connection to the "true" social networks discussed here is tenuous at best and so, for our purposes, they cannot offer a great deal of insight.

A more promising source of data is the affiliation network. An affiliation network is a network of actors connected by common membership in groups of some sort, such as clubs, teams, or organizations. Examples studied in the past include women and the social events they attend [25], company directors and the boards of directors on which they sit [7,26], company CEOs and the clubs they frequent [8], and movie actors and the movies in which they appear [22,27]. Data on affiliation networks tend to be more reliable than those on other social networks, since membership of a group can often be determined with a precision not available when considering friendship or other types of acquaintance. Very large networks can be assembled in this way as well, since in many cases group membership can be ascertained from membership lists, making time-consuming interviews or questionnaires unnecessary. A network of movie actors, for example, has been compiled using the resources of the Internet Movie Database,[1] and contains the names of nearly half a million actors—a much better sample on which to perform statistics than most social networks, although it is unclear whether this particular network has any real social interest.

---

[1] http://www.imdb.com/.

In this article we study in detail another affiliation network, one which is a true social network, for which excellent data are available, and which furthermore will be of interest to readers for personal as well as scientific reasons. We study networks in which the actors are scientists and the ties between them are scientific collaborations, as documented in the papers that they write.

2 Coauthorship Networks

Here we construct networks of scientists in which a link between two scientists is established by their coauthorship of one or more scientific papers. These networks are affiliation networks in which actors are linked by their common membership of groups consisting of the authors of a paper. They are more truly social networks than many affiliation networks; it is probably fair to say that most people who have written a paper together are genuinely acquainted with one another, in a way that, for example, movie actors who appeared together in a movie may not be. There are exceptions—some very large collaborations, for example in high-energy physics, will contain coauthors who have never even met—and we discuss these where appropriate. By and large, however, the network reflects genuine professional interaction between scientists, and may be the largest social network ever studied.[2]

The idea of constructing a network of coauthorship is not new. Many readers will be familiar with the concept of the Erdős number, named for Paul Erdős, the Hungarian mathematician, one of the founding fathers of graph theory, among other things [29]. At some point, it became a popular cocktail party pursuit for mathematicians to calculate how far removed they were in terms of publication from Erdős. Those who had published a paper with Erdős were given a Erdős number of 1, those who had published with one of those people but not with Erdős, a number of 2, and so forth. The present author, for example, has an Erdős number of 3, via Robert Ziff and Mark Kac [30–32]. In the jargon of social networks, your Erdős number is the geodesic distance between you and Erdős in the coauthorship network. In recent studies, it has been found that the average Erdős number is about 4.7, and the maximum known finite Erdős number (within mathematics) is 15 [33,34]. These results are probably influenced to some extent by Erdős' prodigious mathematical output: he published at least 1512 papers, more than any other mathematician ever except possibly Leonhard Euler. However, quantitatively similar, if not quite so impressive, results are in most cases found if the network is centered on another mathematician. (On the other hand, fifth-most published mathematician, Lucien Godeaux, produced 644 papers, on 643 of which he was the sole author. He has no finite Erdős number [35]. Clearly sheer size of output is not a sufficient condition for high connectedness.)

---

[2] If one considers the world wide web to be a social network (an issue of some debate [28]), then it certainly dwarfs the networks studied here, with more than six billion pages cataloged by the largest search engines at the time of writing.

There is also a substantial body of work in bibliometrics (a specialty within information science) on extraction of collaboration patterns from publication data [36–41]. However, these studies have not so far attempted to reconstruct actual collaboration networks from bibliographic data, concentrating more on organizational and institutional aspects of collaboration.[3]

In this article, we study networks of scientists using bibliographic data drawn from four publicly available databases of papers. The databases are:

1. Physics E-print Archive:[4] a database of unrefereed preprints in physics, self-submitted by their authors, running from 1991 to the present. This database is subdivided into specialties within physics, such as condensed matter and high-energy physics, and in more recent years has added divisions covering nonlinear sciences, mathematics, computer science, and quantitative biology.
2. Medline:[5] a database of articles on biomedical research published in refereed journals, stretching from 1961 to the present. Entries in the database are updated by the database's maintainers, rather than papers' authors, giving it relatively thorough coverage of its subject area. The inclusion of biomedicine is crucial in a study such as this one. In most countries biomedical research easily dwarfs civilian research on any other topic, in terms of both expenditure and human effort. Any study that omitted it would be leaving out the largest part of current scientific research.
3. SPIRES:[6] a database of preprints and published papers in high-energy physics, both theoretical and experimental, from 1974 to the present. The contents of this database are also professionally maintained. High energy physics is an interesting case socially, having a tradition of much larger experimental collaborations than other disciplines.
4. NCSTRL:[7] a database of preprints in computer science, submitted by participating institutions and stretching back about ten years.

We have constructed networks of collaboration for each of these databases separately and analyzed them using a variety of techniques, some standard, some invented for the purpose.

---

[3] There has been a considerable amount of work on networks of citations between papers, both in information science [36,42,43] and more recently in physics [44]. These networks, though often confused with coauthorship networks, are quite distinct from them; in a citation network the "actors" are papers and the (directed) ties between them are citations of one paper by another. While citation data are plentiful and many results are known, citation networks are not really social networks since the authors of two papers need not be acquainted for one of them to cite the other's work. On the other hand, citation probably does imply a certain congruence in the subject matter of the two papers, which although not a social relationship, may certainly be of interest for other reasons.

[4] http://arxiv.org/
[5] http://www.ncbi.nlm.nih.gov/
[6] http://www.slac.stanford.edu/spires/
[7] http://www.ncstrl.org/

The outline of the article is as follows. In Sect. 3 we discuss some basic statistics, to give a feel for the shape of our networks. Among other things we discuss the typical numbers of papers per author, authors per paper, and number of collaborators of scientists in the various disciplines. In Sect. 4 we look at a variety of measures concerned with paths between scientists in the network. In Sect. 5 we extend our networks to include a measure of the strength of collaborative ties between scientists and examine measures of connectedness in these weighted networks. In Sect. 6 we give our conclusions. This article is an updated and extended version of an earlier two-part report [45,46].

3 Basic Results

For this study, we constructed collaboration networks using data from a five-year period from January 1, 1995 to December 31, 1999, although data for much longer periods were available in some of the databases. There were several reasons for using this fairly short time window. First, older data are less complete than newer for all databases. Second, we wanted to study the same time period for all databases, so as to be able to make valid comparisons between collaboration patterns in different fields. The coverage provided by both the Physics E-print Archive and the NCSTRL database is relatively poor before 1995, and this sets a limit on how far back we can look. Third, the networks change over time, both because people enter and leave the professions they represent and because practices of scientific collaboration and publishing change. In this article we do not address time evolution of the network, although this has been done elsewhere [47,48]. For our purposes, a short window of data is desirable, to ensure that the collaboration network is roughly static during the study.

The raw data for the networks described here are computer files containing lists of papers, including authors' names and possibly other information such as title, abstract, date, journal reference, and so forth. Construction of the collaboration networks is straightforward. The files are parsed to extract author names and as names are found a list is maintained of the ones seen so far—vertices already in the network—so that recurring names can be correctly assigned to extant vertices. An edge is added between each pair of authors on each paper, unless there is an edge between them already. (In other words, we allow only a single edge between any pair of authors, even if they write more than one paper together. In Sect. 5 we will consider more general networks that take repeated coauthorship into account.)

A naive computer program implementing this procedure, in which names were stored in a simple array, would take time $O(pn)$ to run to completion, where p is the total number of papers in the database and n the number of authors. This however turns out to be prohibitively slow for large networks since p and n are of similar size and may be a million or more. Instead therefore, we store the names of the authors in an ordered binary tree, which reduces the running time to $O(p \log n)$, making the calculation tractable, even for the largest databases studied here.

In Table 1 we give a summary of some of the basic results for the networks studied here. We discuss these results in detail in the rest of this section.

3.1 Number of Authors

The size of the databases varies considerably, from over a million authors for Medline to about ten thousand for NCSTRL. In fact, it is difficult to say with precision how many authors there are. One can say how many distinct *names* appear in a database, but the number of names is not necessarily the same as the number of authors. A single author may report their name differently on different papers. For example, F. L. Wright, Francis Wright, and Frank Lloyd Wright could all be the same person. The spelling of a name may also vary because of inconsistencies in transliteration from non-Roman alphabets. Conversely, two authors may have the same name. Grossman and Ion [35] point out that there are two American mathematicians named Norman Lloyd Johnson, who are known to be distinct people and who work in different fields, but between whom computer programs such as ours cannot hope to distinguish. Even additional clues such as home institution or field of specialization cannot reliably be used to distinguish such people, since many scientists have more than one institution or publish in more than one field. The present author, for example, has addresses at the University of Michigan and the Santa Fe Institute, and publishes in statistical physics, sociology, and epidemiology.

In order to control for these biases, we constructed two different versions of each of the collaboration networks studied here, as follows. In the first, we identify each author by his or her surname and first initial only. This method is clearly prone to confusing two people for one, but will rarely fail to identify two names which genuinely refer to the same person. In the second version of each network, we identify authors by surname and all initials. This method can much more reliably distinguish authors from one another, but will also identify one person as two if they give their initials differently on different papers. Indeed this second measure appears to overestimate the number of authors in a database substantially. Networks constructed in these two different fashions therefore give upper and lower bounds on the number of authors, and hence also give bounds on many of the other quantities studied here. In Table 1 we give numbers of authors in each network using both methods, but for many of the other quantities we give only an error estimate based on the separation of the bounds.

3.2 Number of Papers per Author

The average number of papers per author in the various subject areas is in the range of around three to six over the five-year period. The only exception is the SPIRES database, covering high-energy physics, in which the figure is significantly higher at 11.6. One possible explanation for this is that SPIRES is the only database that contains both preprints and published papers. It is possible that the high figure for papers per author reflects duplication of papers

Table 1. Summary of results of the analysis of seven scientific collaboration networks. Numbers in parentheses give an estimate of the error on the least significant figures.

	Medline	Physics E-print Archive				SPIRES	NCSTRL
		complete	astro-ph	cond-mat	hep-th		
total papers	2163923	98502	22029	22016	19085	66652	13169
total authors	1520251	52909	16706	16726	8361	56627	11994
first initial only	1090584	45685	14303	15451	7676	47445	10998
mean papers per author	6.4(6)	5.1(2)	4.8(2)	3.65(7)	4.8(1)	11.6(5)	2.55(5)
mean authors per paper	3.754(2)	2.530(7)	3.35(2)	2.66(1)	1.99(1)	8.96(18)	2.22(1)
collaborators per author	18.1(1.3)	9.7(2)	15.1(3)	5.86(9)	3.87(5)	173(6)	3.59(5)
size of giant component	1395693	44337	14845	13861	5835	49002	6396
first initial only	1019418	39709	12874	13324	5593	43089	6706
as a percentage	92.6(4)%	85.4(8)%	89.4(3)%	84.6(8)%	71.4(8)%	88.7(1.1)%	57.2(1.9)%
2nd largest component	49	18	19	16	24	69	42
clustering coefficient C	0.066(7)	0.43(1)	0.414(6)	0.348(6)	0.327(2)	0.726(8)	0.496(6)
mean distance	4.6(2)	5.9(2)	4.66(7)	6.4(1)	6.91(6)	4.0(1)	9.7(4)
maximum distance	24	20	14	18	19	19	31

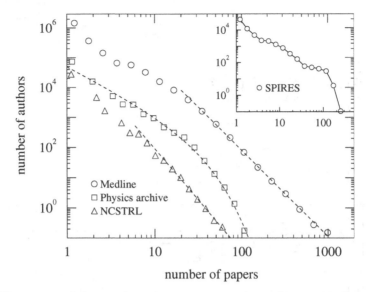

Fig. 1. Histograms of the number of papers written by authors in Medline (circles), the physics archive (squares), and NCSTRL (triangles). The dotted lines are fits to the data as described in the text. Inset: the equivalent histogram for the SPIRES database.

in both preprint and published form. However, the maintainers of the database go to some lengths to avoid this [49], and a more probable explanation is perhaps that publication rates are higher for the large collaborations favored by high-energy physics, since a large group of scientists has more person-hours available for the writing of papers.[8]

In addition to the average numbers of papers per author in each database, it is interesting to look at the distribution p_k of numbers k of papers per author. In 1926, Alfred Lotka showed, using a dataset compiled by hand, that this distribution followed a power law, with exponent approximately -2, a result which is now referred to as Lotka's Law of Scientific Productivity [50]. In other words, in addition to the many authors who publish only a small number of papers, one expects to see a "fat tail" consisting of a small number of authors who publish a very large number of papers. In Fig. 1 we show on logarithmic scales histograms for each of our four databases of the numbers of papers published. (These histograms and all the others shown here were created using the "all initials" versions of the collaboration networks.) For the Medline and NCSTRL databases these histograms follow a power law quite closely, at least in their

---

[8] We can test this hypothesis by dividing the total number of papers over the five-year period in any one of our networks by the total number of authors to get a mean "effort" figure for the number of papers produced by the community per author. From Table 1, this figure is 1.2 for SPIRES, compared to 1.1, 1.4, and 1.9 for NCSTRL, Medline, and the physics archive respectively. Thus the mean effort per person indeed has roughly the same value in high-energy physics as it has in other fields, or is even perhaps a little lower than the average.

tails, with exponents of $-2.86(3)$ and $-3.41(7)$ respectively—somewhat steeper than those found by Lotka, but in reasonable agreement with other more recent studies [51,52,36]. For the physics archive the pure power law is a poor fit. An exponentially truncated power law does much better:

$$p_k = Ck^{-\tau}e^{-k/\kappa}, \tag{1}$$

where τ and κ are constants and C is fixed by the requirement of normalization—see Fig. 1. (The probability p_0 of having zero papers is taken to be zero, since the names of scientists who have not written any papers do not appear in the database.) The exponential cutoff we attribute to the finite time window of five years used in this study which prevents any one author from publishing a very large number of papers. Lotka and subsequent authors who have confirmed his law have not usually used such a window.

It is interesting to speculate why the cutoff appears only in physics and not in computer science or biomedicine. Surely the five-year window limits everyone's ability to publish very large numbers of papers, regardless of their area of specialization? For the case of Medline one possible explanation is suggested by a brief inspection of the names of the most published authors. The top ten, for example, are Suzuki, T., Wang, Y., Suzuki, K., Takahashi, M., Nakamura, T., Tanaka, K., Tanaka, T., Wang, J., Suzuki, Y., and Takahashi, T. The predominance of Japanese names in this list may reflect differences in author attribution practices in Japanese biomedical research, but more probably these are simply common names, and these apparently highly published authors are each several different people who have been conflated in our analysis. (One could perform a partial test of this conjecture by going back to the original articles and checking cases in which authors gave their full first names, rather than just an initial, to see if those names differ in some cases.) Thus it is possible that there is not after all any fat tail in the distribution for the Medline database, only the illusion of one produced by the large number of scientists with commonly occurring names. (This doesn't however explain why the tail appears to follow a power law.) This argument is strengthened by the sheer numbers of papers involved. T. Suzuki published, it appears, 1697 papers, or about one paper a day, including weekends and holidays, every day for the entire five-year course of our study. This seems to be an improbably large output.

Interestingly, no national bias is seen in any of the other databases, and the names that top the list in physics and computer science are not common ones. (For example, the most published authors in the other three databases are Shelah, S. (physics archive),[9] Wolf, G. (SPIRES), and Bestavros, A. (NCSTRL).) Thus it is still unclear why the NCSTRL database should have a power-law tail, though this database is small and it is possible that it does possess a cutoff in the productivity distribution which is just not visible because of the limits of the dataset.

For the SPIRES database, which is shown separately in the inset of the figure, neither pure nor truncated power law fits the data well, the histogram displaying

---

[9] A mathematician, not a physicist.

a significant bump around the 100-paper mark. A possible explanation for this is that a small number of large collaborations published around this number of papers during the time-period studied. Since each author in such a collaboration is then credited with publishing a hundred papers, the statistics in the tail of the distribution can be substantially skewed by such practices.

In the first column of Table 2, we list the most frequent authors in three subject-specific subdivisions of the physics archive: `astro-ph` (astro-physics), `cond-mat` (condensed matter physics), and `hep-th` (high-energy theory). Although there is only space to list the top ten winners in this table, the entire list (and the corresponding lists for the other tables in this article) can be found by the curious reader on the world wide web.[10]

3.3 Numbers of Authors per Paper

Grossman and Ion [35] report that the average number of authors on papers in mathematics has increased steadily over the last sixty years, from a little over 1 to its current value of about 1.5. As Table 1 shows, still higher numbers seem to apply to current studies in the sciences. Purely theoretical papers appear to be typically the work of two scientists, with high-energy theory and computer science showing averages of 1.99 and 2.22 in our calculations. For databases covering experimental or partly experimental subject areas the averages are higher: 3.75 for biomedicine, 3.35 for astrophysics, 2.66 for condensed matter physics. The SPIRES high-energy physics database however shows the most startling results, with an average of 8.96 authors per paper, obviously a result of the presence of papers in the database written by very large collaborations. (Perhaps what is most surprising about this result is actually how small it is. The hundreds strong mega-collaborations of CERN and Fermilab are sufficiently diluted by theoretical and smaller experimental groups, that the number is only 9, and not 90.)

Distributions of numbers of authors per paper are shown in Fig. 2, and appear to have power-law tails with widely varying exponents of $-6.2(3)$ (Medline), $-3.34(5)$ (physics archive), $-4.6(1)$ (NCSTRL), and $-2.18(7)$ (SPIRES). The SPIRES data, which are again shown in a separate inset, display a pronounced peak in the distribution around 200–500 authors. This peak presumably corresponds to the large experimental collaborations which dominate the upper end of this histogram.

The largest number of authors on a single paper was 1681 (in high-energy physics, of course).

3.4 Numbers of Collaborators per Author

The differences between the various disciplines represented in the databases are emphasized still more by the numbers of collaborators that a scientist has, the

---

[10] Complete tables of results for authors in the Physics E-print Archive can be found on the world wide web at `http://www.santafe.edu/~mark/collaboration/`.

Table 2. The authors with the highest numbers of papers, numbers of coauthors, and betweenness, and strongest collaborations in astrophysics, condensed matter physics, and high-energy theory. The figures for betweenness have been divided by 10^6. Full lists of the rankings of all the authors in these databases can be found on the world wide web at http://www.santafe.edu/~mark/collaboration/.

	number of papers	number of co-workers	betweenness	collaboration weight
astro-ph	112 Fabian, A.C.	360 Frontera, F.	2.33 Kouveliotou, C.	16.5 Moskalenko, I.V./Strong, A.W.
	101 van Paradijs, J.	353 Kouveliotou, C.	2.15 van Paradijs, J.	15.0 Hernquist, L./Heyl, J.S.
	81 Frontera, F.	329 van Paradijs, J.	1.80 Filippenko, A.V.	14.0 Mathews, W.G./Brighenti, F.
	80 Hernquist, L.	299 Piro, L.	1.57 Beaulieu, J.P.	13.4 Labini, F.S./Pietronero, L.
	79 Gould, A.	296 Costa, E.	1.52 Nomoto, K.	12.2 Piran, T./Sari, R.
	78 Silk, J.	291 Feroci, M.	1.52 Pian, E.	11.8 Zaldarriaga, M./Seljak, U.
	78 Klis, M.V.D.	284 Pian, E.	1.49 Frontera, F.	11.4 Hernquist, L./Katz, N.
	73 Kouveliotou, C.	284 Hurley, K.	1.35 Silk, J.	11.1 Avila-Reese, V./Firmani, C.
	70 Ghisellini, G.	244 Palazzi, E.	1.33 Kamionkowski, M.	10.9 Dai, Z.G./Lu, T.
	66 Piro, L.	244 Heise, J.	1.28 McMahon, R.G.	10.8 Ostriker, J.P./Cen, R.
cond-mat	116 Parisi, G.	107 Uchida, S.	4.11 MacDonald, A.H.	22.3 Belitz, D./Kirkpatrick, T.R.
	79 Scheffler, M.	103 Ueda, Y.	3.96 Bishop, A.R.	17.0 Shrock, R./Tsai, S.
	75 Das Sarma, S.	96 Revcolevschi, A.	3.36 Das Sarma, S.	15.0 Yukalov, V.I./Yukalova, E.P.
	74 Stanley, H.E.	94 Eisaki, H.	2.96 Tosatti, E.	14.7 Martín-Delgado, M.A./Sierra, G.
	70 MacDonald, A.H.	84 Cheong, S.	2.52 Wang, X.	14.3 Krapivsky, P.L./Ben-Naim, E.
	68 Sornette, D.	83 Isobe, M.	2.38 Revcolevschi, A.	14.1 Beenakker, C.W.J./Brouwer, P.W.
	60 Volovik, G.E.	78 Stanley, H.E.	2.30 Uchida, S.	13.8 Weng, Z.Y./Sheng, D.N.
	56 Beenakker, C.W.J.	76 Shirane, G.	2.21 Sigrist, M.	13.7 Sornette, D./Johansen, A.
	53 Dagotto, E.	76 Scheffler, M.	2.19 Cheong, S.	13.6 Rikvold, P.A./Novotny, M.A.
	50 Helbing, D.	76 Menovsky, A.A.	2.18 Stanley, H.E.	13.0 Scalapino, D.J./White, S.R.
hep-th	78 Odintsov, S.D.	50 Ambjorn, J.	0.98 Odintsov, S.D.	34.0 Lu, H./Pope, C.N.
	73 Lu, H.	44 Ferrara, S.	0.88 Ambjorn, J.	29.0 Odintsov, S.D./Nojiri, S.
	72 Pope, C.N.	43 Vafa, C.	0.88 Kogan, I.I.	18.7 Lee, H.W./Myung, Y.S.
	69 Cvetic, M.	39 Odintsov, S.D.	0.84 Henneaux, M.	18.3 Schweigert, C./Fuchs, J.
	68 Ferrara, S.	39 Kogan, I.I.	0.73 Douglas, M.R.	14.7 Ovrut, B.A./Waldram, D.
	65 Vafa, C.	36 Proeyen, A.V.	0.67 Ferrara, S.	14.7 Kleihaus, B./Kunz, J.
	65 Tseytlin, A.A.	35 Fre, P.	0.63 Vafa, C.	12.9 Mavromatos, N.E./Ellis, J.
	65 Mavromatos, N.E.	35 Ellis, J.	0.60 Khare, A.	12.4 Kachru, S./Silverstein, E.
	63 Witten, E.	35 Douglas, M.R.	0.58 Tseytlin, A.A.	11.7 Kakushadze, Z./Tye, S.H.H.
	54 Townsend, P.K.	34 Lu, H.	0.58 Townsend, P.K.	11.6 Arefeva, I.Y./Volovich, I.V.

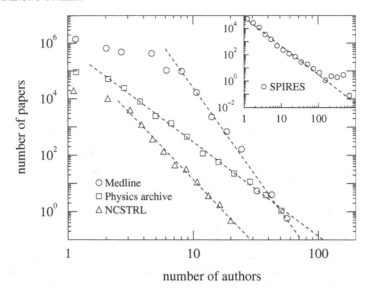

Fig. 2. Histograms of the number of authors on papers in Medline (circles), the physics archive (squares), and NCSTRL (triangles). The dotted lines are the best fit power-law forms. Inset: the equivalent histogram for the SPIRES database, showing a clear peak in the 200 to 500 author range.

total number of people with whom a scientist wrote papers during the five-year period. The average number of collaborators is markedly lower in the purely theoretical disciplines (3.87 in high-energy theory, 3.59 in computer science) than in the wholly or partly experimental ones (18.1 in biomedicine, 15.1 in astrophysics). But the SPIRES high-energy physics database takes the prize once again, with scientists having an impressive 173 collaborators, on average, over a five-year period. This clearly begs the question whether the high-energy coauthorship network can be considered an accurate representation of the social network of the high-energy physics community; it seems unlikely that an author could know 173 colleagues well.

The distributions of numbers of collaborators are shown in Fig. 3. In all cases they appear to have long tails, but only the SPIRES data (inset) fit a power-law distribution well, with a low measured exponent of -1.20. Note also the small peak in the SPIRES data around 700—presumably again a product of the large collaborations.

For the other three databases, the distributions show some curvature. This may, as we have previously suggested, be the signature of an exponential cutoff, produced once again by the finite time window of the study [53]. Redner (personal communication) and Barabási *et al.* [48] have independently suggested alternative explanations based on growth models of networks, although the fundamental causative agent is the same finite time window in these theories also.

Column 2 of Table 2 shows the authors in `astro-ph`, `cond-mat`, and `hep-th` with the largest numbers of collaborators. The winners in this race tend to be

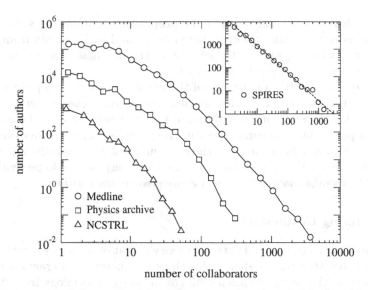

Fig. 3. Histograms of the number of collaborators of authors in Medline (circles), the physics archive (squares), and NCSTRL (triangles). Inset: the equivalent histogram for the SPIRES database, which is well fit by a power law (dotted line).

experimentalists, who conduct research within larger collaborations, although there are exceptions. The high-energy theory database of course contains only theorists, and the smaller numbers of collaborators reflect this.

3.5 Size of the Giant Component

In the theory of random graphs it is known that there is a continuous phase transition with increasing density of edges in a graph at which a *giant component* forms, i.e., a connected subset of vertices whose size scales extensively [54,55]. Well above this transition, in the region where the giant component exists, the giant component usually fills a large portion of the graph, and all other components (i.e., connected subsets of vertices) are small with mean size independent of the size of the network. We see a situation reminiscent of this in all of the graphs studied here: a single large component of connected vertices that fills the majority of the volume of the graph, and a number of much smaller components filling the rest. In Table 1 we show the size of the giant component for each of our databases, both as total number of vertices and as a fraction of system size. In all cases the giant component fills around 80% or 90% of the total volume, except for high-energy theory and computer science, which give smaller figures. A possible explanation of these two anomalies may be that the corresponding databases give poorer coverage of their subjects. The hep-th high-energy database is quite widely used in the field, but overlaps to a large extent with the longer established SPIRES database, and it is possible that some authors neglect it for this reason [49]. The NCSTRL computer science database differs from the others in this study in that the preprints it contains are submitted by participating institutions, of which there are about 160. Preprints from institutions not participating are mostly left out of the database, and its coverage of the subject area is, as a result, incomplete.

The figure of 80–90% for the size of the giant component is a promising one. It indicates that the vast majority of scientists are connected via collaboration, and hence via personal contact, with the rest of their field. Despite the prevalence of journal publishing and conferences in the sciences, person-to-person contact is still of paramount importance in the communication of scientific information, and it is reasonable to suppose that the scientific enterprise would be significantly hindered if scientists were not so well connected to one another.

3.6 Clustering Coefficients

An interesting idea from social network theory is that of *transitivity*, which, along with its sibling structural balance, describes symmetry of interaction amongst trios of actors. "Transitivity" has a different meaning in sociology from its meaning in mathematics and physics, although the two are related. It refers to the extent to which the existence of ties between actors A and B and between actors B and C implies a tie between A and C. The transitivity, or more precisely the fraction of transitive triples, is that fraction of connected triples of vertices which also form "triangles" of interaction. Here a connected triple means an actor who is connected to two others. In the physics literature, this quantity is usually called the clustering coefficient C [22], and can be written[11]

$$C = \frac{3 \times \text{number of triangles on the graph}}{\text{number of connected triples of vertices}}. \tag{2}$$

The factor of three in the numerator compensates for the fact that each complete triangle of three vertices contributes three connected triples, one centered on each of the three vertices, and ensures that $C = 1$ on a completely connected graph. On unipartite random graphs $C = O(n^{-1})$, where n is the number of vertices, and hence goes to zero in the limit of large graph size [22,6]. In social networks it is believed that the clustering coefficient will take a non-zero value even in very large networks, because there is a finite (and probably quite large) probability that two people will be acquainted if they have another acquaintance in common. This is a hypothesis we can test with our collaboration networks. In Table 1 we show values of the clustering coefficient C, calculated from (2), for each of the databases studied, and as we see, the values are indeed large—as large as 0.7 in the case of the SPIRES database and around 0.3 or 0.4 for most of the others.

There are a number of possible explanations for these high values of C. First of all, it may be that they indicate simply that collaborations of three or more people are common in science. Every paper that has three authors clearly contributes a triangle to the numerator of (2) and hence increases the clustering

---

[11] This is one of two slightly different definitions of the clustering coefficient that are in use. See, for instance, [6].

coefficient. This is, in a sense, a "trivial" form of clustering, although it is by no means socially uninteresting.

In fact it turns out that this effect can account for some but not all of the clustering seen in our graphs. One can construct a random graph model of a collaboration network which mimics the trivial clustering effect, and the results indicate that only about a half of the clustering we see is a result of authors collaborating in groups of three or more [56]. The rest of the clustering must have a social explanation, and there are some obvious possibilities:

1. A scientist may collaborate with two colleagues individually, who may then become acquainted with one another through their common collaborator, and so end up collaborating themselves. This is the usual explanation for transitivity in acquaintance networks [1].
2. Three scientists may all revolve in the same circles—read the same journals, attend the same conferences—and, as a result, independently start up separate collaborations in pairs, and so contribute to the value of C, although only the workings of the community, and not any specific person, is responsible for introducing them.
3. As a special case of the previous possibility—and perhaps the most likely case—three scientists may all work at the same institution, and as a result may collaborate with one another in pairs.

Interesting studies could no doubt be made of these processes by combining our network data with data on, for instance, institutional affiliations of scientists. Such studies are, however, perhaps better left to the social scientists who specialize in them.

The clustering coefficient of the Medline database is worthy of brief mention, since its value is far smaller than those for the other databases. One possible explanation of this comes from the unusual social structure of biomedical research, which, unlike the other sciences, has traditionally been organized into laboratories, each with a principal investigator supervising a large number of postdocs, students, and technicians working on different projects. This organization produces a tree-like hierarchy of collaborative ties with fewer interactions within levels of the tree than between them. A tree has no loops in it, and hence no triangles to contribute to the clustering coefficient. Although the biomedicine hierarchy is certainly not a perfect tree, it may be sufficiently tree-like for the difference to show up in the value of C. Another possible explanation comes from the generous tradition of authorship in the biomedical sciences. It is common, for example, for a researcher to be made a coauthor of a paper in return for synthesizing reagents used in an experimental procedure. Such a researcher will in many cases have a less than average likelihood of developing new collaborations with their collaborators' friends, and therefore of increasing the clustering coefficient.

4 Distances and Centrality

The basic statistics of the previous section are certainly of importance, particularly for the construction of network models [22,56–59], but there is much more that we can do with our collaboration networks. In this section, we look at some simple but useful measures of network structure, concentrating on measures having to do with paths between vertices in the network. In Sect. 5 we discuss some shortcomings of these measures, and construct some new and more complex measures that may better reflect true collaboration patterns.

4.1 Shortest Paths

A fundamental concept in graph theory is the *geodesic*, the shortest path of vertices and edges that links two given vertices. There may not be a unique geodesic between two vertices: there may be two or more shortest paths, which may or may not share some vertices. Or there may be no paths between the vertices at all. The geodesic(s) between two vertices s and t can be calculated in time $O(m)$, where m is the number of edges in the graph, using the following algorithm, which is a modified form of the standard breadth-first search [60].

1. Assign vertex s distance zero, to indicate that it is zero steps away from itself, and set $d = 0$.
2. For each vertex i whose assigned distance is d, follow each attached edge to the vertex j at its other end and then do one of the following three things:
   a) If j has not already been assigned a distance, assign it distance $d + 1$. Declare i to be a predecessor of j.
   b) If j has already been assigned distance $d + 1$, then there is no need to do this again, but i is still declared a predecessor of j.
   c) If j has already been assigned a distance less than $d + 1$, do nothing.
3. Set $d \leftarrow d + 1$.
4. Repeat from step (2) until there are no unassigned vertices left.

Now the shortest path (if there is one) between s and t is the path you get by stepping from t to its predecessor, and then to the predecessor of each successive vertex until s is reached. If a vertex has two or more predecessors, then there are two or more shortest paths, each of which must be followed separately if we wish to know all shortest paths between s and t.

In Fig. 4 we show the shortest paths of known collaborations between two of the author's colleagues, Duncan Watts (Columbia) and László Barabási (Notre Dame), both of whom work on networks of various kinds. It is interesting to note that, although the two scientists in question are well acquainted both personally and with one another's work, the shortest path between them does not run entirely through other collaborations in the field. (For example, the connection between the present author and Juan Pedro Garrahan results from our coauthorship of a paper on spin glasses.) Although this may at first sight appear odd, it is probably in fact a good sign. It indicates that workers in the field come from different scientific camps, rather than all descending intellectually from a

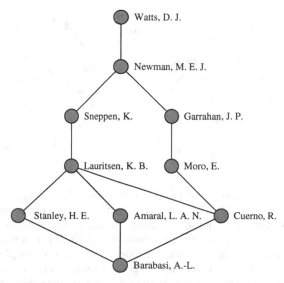

Fig. 4. The geodesics, or shortest paths, in the collaboration network of physicists between Duncan Watts and László Barabási.

single group or institution. This presumably increases the likelihood that those workers will express independent opinions on the open questions of the field, rather than merely spouting slight variations on the same underlying doctrine.

A database that would allow one conveniently and quickly to extract shortest paths between scientists in this way might have some practical use. Kautz *et al.* [61] have constructed a web-based system which does just this for computer scientists, with the idea that such a system might help to create new professional contacts by providing a "referral chain" of intermediate scientists through whom contact may be established.

4.2 Betweenness and Funneling

A quantity of interest in many social network studies is the *betweenness* of an actor i, which is defined as the total number of shortest paths between pairs of actors that pass through i [62]. This quantity is one possible indicator of who the most influential people in the network are. In a network in which information flows entirely or mostly along the shortest paths between actors, those with highest betweenness are the ones who control the flow of information between most others. The vertices with highest betweenness also produce an increase in the geodesic distance between the largest number of pairs of others when removed from the network [1].

Naively, one might think that betweenness would take time of order $O(mn^2)$ to calculate for all vertices, since there are $O(n^2)$ shortest paths to be considered, each of which takes time $O(m)$ to calculate, and until recently the standard network analysis packages such as UCInet and Pajek indeed used $O(mn^2)$

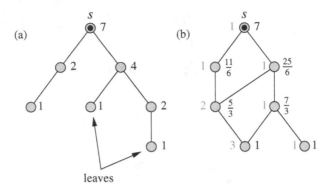

Fig. 5. Calculation of betweenness: (a) When there is only a single shortest path from a source vertex s (top) to all other reachable vertices, those paths necessarily form a tree, which makes the calculation of the contribution to betweenness from this set of paths particularly simple, as described in the text. (b) For cases in which there is more than one shortest path to some vertices, the calculation is more complex. First we must calculate the number of paths from the source s to each other vertex (numbers to left of vertices), and then use these to weight the path counts appropriately and derive the betweenness scores (numbers to right of vertices).

algorithms. Recently however, faster algorithms for betweenness have been discovered by the present author [46] and independently by Brandes [63]. These algorithms can perform the same calculation in time $O(mn)$. Here we describe the algorithm of Newman [46], which is fast enough to allow the exhaustive calculation of betweenness for all vertices in the very large graphs studied here.

We start by performing a breadth-first search to determine the set of shortest paths from some source vertex s to all other vertices that are reachable from s. Consider first the simple case of a network in which there is only a single shortest path from the source vertex to any other. (We will consider other cases in a moment.) The resulting set of paths then forms a tree as shown in Fig. 5a. We can use this tree to calculate betweenness as follows. We find first the "leaves" of the tree, i.e., those nodes such that no shortest paths to other nodes pass through them, and we assign a score of 1 to them—the only path to these vertices is the one that ends there. Then, starting with those vertices that are farthest from the source vertex s on the tree, i.e., lowest in Fig. 5a, we work upwards, assigning a score to each vertex that is 1 plus the sum of the scores on the neighboring vertices immediately below it. When we have gone though all vertices in the tree, the resulting scores are the betweenness counts for the paths from vertex s. (In our calculation we define paths to include the vertices at their ends. Sometimes they are defined to exclude these vertices, in which case the score at each vertex is decreased by 1, except for the source vertex s, which receives a score of zero.) Repeating the process for all possible vertices s and summing the scores, we arrive at the full betweenness scores for shortest paths between all pairs. The breadth-first search and the process of working up through the tree both take worst-case time $O(m)$ and there are n vertices total, so the entire calculation takes time $O(mn)$ as claimed.

This simple case serves to illustrate the basic principle behind the algorithm. In general, however, it is not the case that there is only a single shortest path between any pair of vertices. Most networks have at least some vertex pairs between which there are several geodesic paths of equal length. Figure 5b shows a simple example of a shortest path "tree" for a network with this property. The resulting structure is in fact no longer a tree, and in such cases an extra step is required in the algorithm to correctly calculate the betweenness.

Following Freeman's original definition of betweenness [62], we give multiple shortest paths between a pair of vertices equal weights summing to 1. Note that some of the paths may run through the same vertices for some part of their length, resulting in vertices with greater weight. To calculate correctly what fraction of the paths flow through each vertex in the network, we generalize the breadth-first search part of our algorithm, as follows.

Consider Fig. 5b and suppose we are starting at vertex s. We carry out the following steps:

1. Assign vertex s distance zero, to indicate that it is zero steps from itself, and set $d = 0$. Also assign s a weight $w_s = 1$ (whose purpose will become clear shortly).
2. For each vertex i whose assigned distance is d, follow each attached edge to the vertex j at its other end and then do one of the following three things:
   a) If j has not yet been assigned a distance, assign it distance $d + 1$ and weight $w_j = w_i$.
   b) If j has already been assigned a distance and that distance is equal to $d + 1$, then the vertex's weight is increased by w_i, that is $w_j \leftarrow w_j + w_i$.
   c) If j has already been assigned a distance less than $d + 1$, do nothing.
3. Set $d \leftarrow d + 1$.
4. Repeat from step 2 until there are no vertices that have distance d.

The resulting weights for the example of Fig. 5b are shown to the left of each vertex in the figure.

Physically, the weight on a vertex i represents the number of distinct paths from the source vertex to i. These weights are precisely what we need to calculate our betweennesses, because if two vertices i and j are connected, with j farther than i from the source s, then the fraction of a geodesic path from j through i to s is given by w_i/w_j. Thus, to calculate the contribution to the betweenness from all shortest paths starting at s, we need only carry out the following steps:

1. Find every "leaf" vertex t, i.e., a vertex such that no paths from s to other vertices go though t and assign it a score of $x_t = 1$.
2. Now, starting with the vertices that are farthest from the source vertex s— lower down in a diagram such as Fig. 5b—work up towards s. To each vertex i assign a score $x_i = 1 + \sum_j x_j w_i/w_j$, where the sum is over the neighbors j immediately below vertex i.
3. Repeat from step 2 until vertex s is reached.

The resulting scores are shown to the right of each vertex in Fig. 5b. Now repeating this process for all n source vertices s and summing the resulting scores on the vertices gives us the total betweenness for all vertices in time $O(mn)$.

We have applied this algorithm to our coauthorship networks and in column 3 of Table 2 we show the ten highest betweennesses in the astro-ph, cond-mat, and hep-th subdivisions of the physics archive. While we leave it to the knowledgeable reader to decide whether the scientists named are indeed pivotal figures in their respective fields, we do notice one interesting feature of the results. The betweenness measure gives very clear winners in the competition: the individuals with highest betweenness are well ahead of those with second highest, who are in turn well ahead of those with third highest, and so on. This same phenomenon has been noted in other networks [1,64].

Strogatz has raised an interesting question about social networks which we can address using our betweenness algorithm: are all of your collaborators equally important for your connection to the rest of the world, or do most paths from others to you pass through just a few of your collaborators (S. H. Strogatz, personal communication)? One could certainly imagine that the latter might be true. Collaboration with just one or two senior or famous members of one's field could easily establish short paths to a large part of the collaboration network, and all of those short paths would go through those one or two members. Strogatz calls this effect "funneling." Since our algorithm, as a part of its operation, calculates the vertices through which each geodesic path to a specified actor passes, it is a trivial modification to calculate also how many of those geodesic paths pass through each of the immediate collaborators of that actor, and hence to use it to look for funneling.

Our collaboration networks, it turns out, show strong funneling. For most people, their top few collaborators lie on most of the paths between themselves and the rest of the network. The rest of their collaborators, no matter how numerous, account for only a small number of paths. Consider, for example, the present author. Out of the 44 000 scientists in the giant component of the physics archive collaboration network, 31 000 paths from them to me, about 70%, pass through just two of my collaborators, Chris Henley and Juanpe Garrahan. Another 13 000, most of the remainder, pass through the next four collaborators. The remaining five account for a mere 1% of the total.

To give a more quantitative impression of the funneling effect, we show in Fig. 6 the average fraction of paths that pass through the top 10 collaborators of an author, averaged over all authors in the giant component of the Physics database. The figure shows for example that on average 64% of one's shortest paths to other scientists pass through one's top-ranked collaborator. Another 17% pass through the second-ranked collaborator. The top 10 shown in the figure account for 98% of all paths.

That one's top few acquaintances account for most of one's shortest paths to the rest of the world has been noted before in other contexts. For example, Stanley Milgram, in his famous "small-world" experiment, noted that most of

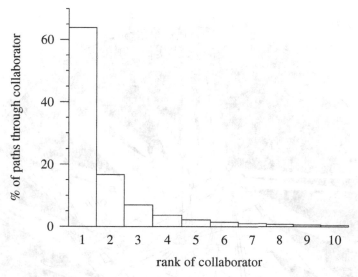

Fig. 6. The average percentage of paths from other scientists to a given scientist that pass through each collaborator of that scientist, ranked in decreasing order. The plot is for the physics archive network, although similar results are found for other networks.

the paths he found to a particular target person in an acquaintance network went through just one or two acquaintances of the target [65]. He called these people "sociometric superstars."

4.3 Average Distances

Breadth-first search allows to us calculate exhaustively the lengths of the shortest paths from every vertex on a graph to every other in time $O(mn)$. We have done this for each of the networks studied here and averaged these distances to find the average distance between any pair of (connected) authors in each of the subject fields studied. These figures are given in the penultimate row of Table 1. As the table shows, these figures are all quite small: they vary from 4.0 for SPIRES to 9.7 for NCSTRL, although this last figure may be artificially inflated by the poor coverage of this database discussed in Sect. 3.5. At any rate, all the figures are very small compared to the number of vertices in the corresponding databases. This "small-world effect," famously discussed by Milgram [65] and by Pool and Kochen [66], is, like the existence of the giant component, probably a good sign for science; it shows that scientific information—discoveries, experimental results, theories—will not have far to travel through the network of scientific acquaintance to reach the ears of those who can benefit by it. Even the *maximum* distances between scientists in these networks, shown in the last row of the table, are not very large, the longest path in any of the networks being just 31 steps long, again in the NCSTRL database.

The explanation of the small-world effect is simple. Consider Fig. 7, which shows all the collaborators of the present author (in all subjects, not just

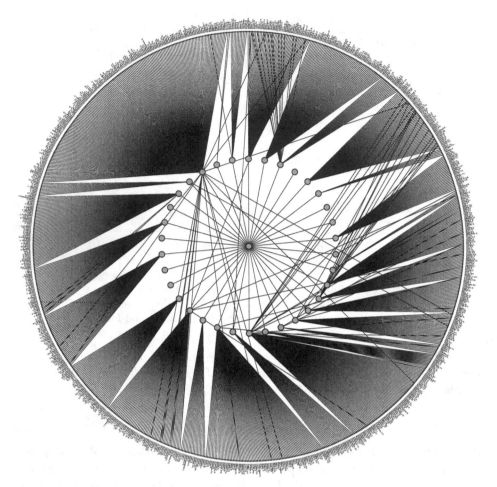

Fig. 7. The point in the center of the figure represents the author of the article you are reading, the first ring his collaborators, and the second ring their collaborators. Collaborative ties between members of the same ring, of which there are many, have been omitted from the figure for clarity.

physics), and all the collaborators of those collaborators—all my first and second neighbors in the collaboration network. As the figure shows, I have 35 first neighbors, but 891 second neighbors. The "radius" of the whole network around me is reached when the number of neighbors within that radius equals the number of scientists in the giant component of the network, and if the increase in numbers of neighbors with distance continues at the impressive rate shown in the figure, it will not take many steps to reach this point.

This simple idea is borne out by theory. In almost all networks, the average distance between pairs of vertices ℓ scales logarithmically with the number of vertices n. In a standard random graph, for instance [54,55], $\ell = \log n / \log z$,

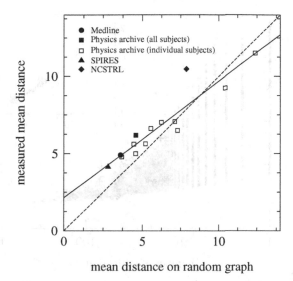

Fig. 8. Average distance between pairs of scientists in the various networks, plotted against average distance on a random graph of the same size and degree distribution. The dotted line shows where the points would fall if measured and predicted results agreed perfectly. The solid line is the best straight-line fit to the data.

where z is the average degree of a vertex, the average number of collaborators in our terminology. In the more general class of random graphs in which the distribution of vertex degrees is arbitrary [67–70], rather than Poissonian as in the standard case, the equivalent expression is [56,71]

$$\ell = \frac{\log(n/z_1)}{\log(z_2/z_1)} + 1, \tag{3}$$

where z_1 and z_2 are the average numbers of first and second neighbors of a vertex. It is widely assumed that this logarithmic behavior extends to most networks, so the small-world effect is not a surprise to those familiar with graph theory. However, it would be nice to demonstrate explicitly the presence of logarithmic scaling in our networks. Figure 8 does this in a crude fashion. In this figure we have plotted the measured value of ℓ, as given in Table 1, against the value given by (3) for each of our four databases, along with separate points for nine of the subject-specific subdivisions of the physics archive. As the figure shows, the correlation between measured and predicted values is quite good. The correlation coefficient is $R^2 = 0.86$, rising to $R^2 = 0.95$ if the NCSTRL database, with its incomplete coverage, is excluded (the diamond in the figure).

Figure 8 needs to be taken with a pinch of salt. Its construction implicitly assumes that the different networks are statistically similar to one another and to the random graphs with the same distributions of vertex degree, an assumption which is almost certainly not correct. Nonetheless, the fact that even with such

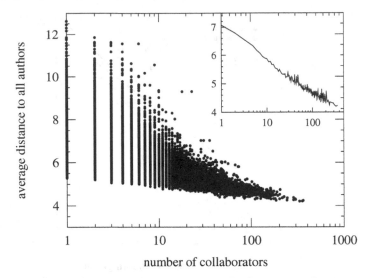

Fig. 9. Scatter plot of the mean distance from each physicist in the giant component of the physics archive network to all others as a function of number of collaborators. Inset: the same data averaged vertically over all authors having the same number of collaborators.

inherent errors the logarithmic behavior is still clearly visible lends at least some credence to its graph theoretical basis.

We can also trivially use our breadth-first search algorithm to calculate the average distance from a single vertex to all other vertices in the giant component. This average is essentially the same as the quantity known as *closeness centrality* to social network analysts.[12] Like betweenness it is also a measure, in some sense, of the centrality of a vertex—authors with low values of this average will, it is assumed, be the first to learn new information, and information originating with them will reach others quicker than information originating with other sources. Average distance is thus a measure of centrality of an actor in terms of their access to information, whereas betweenness is a measure of an actor's control over information flowing between others.

Calculating average distance for many networks returns results which look sensible to the observer. Calculations for the network of collaborations between movie actors, for instance, give small average distances for actors who are famous—ones many of us will have heard of. Interestingly, however, performing the same calculation for our scientific collaboration networks does not give exactly the results we might expect. For example, one finds that the people at the top of the list are always experimentalists. This, you might think, is not such a bad thing: perhaps the experimentalists are better connected people? In a sense, in fact, it turns out that they are. In Fig. 9 we show the average distance from scientists in the physics archive to all others in the giant component

---

[12] Technically, closeness is the reciprocal of the average distance to other vertices [1].

as a function of their number of collaborators. As the figure shows, there is a clear trend towards shorter average distance as the number of collaborators becomes large. This trend is clearer still in the inset, where we show the same data averaged over all authors who have the same number of collaborators. Since experimentalists often work in large groups, it is not surprising to learn that they tend to have shorter average distances to other scientists.

But this brings up an interesting question, one that we touched upon in Sect. 2: while most pairs of people who have written a paper together will know one another reasonably well, there are exceptions. On a high-energy physics paper with 1000 coauthors, for instance, it is unlikely that every one of the 499 500 possible acquaintanceships between pairs of those authors will actually be realized. Our closeness measure does not take into account the tendency for collaborators in large groups not to know one another, or to know one another less well. In the next section we describe a more sophisticated calculation which does do this.

5 Weighted Collaboration Networks

There is more information present in the databases used here than in the simple networks we have constructed from them, which tell us only whether scientists have collaborated or not. In particular, we also know on how many papers each pair of scientists collaborated during the period of the study, and how many other coauthors they had on each of those papers.[13] We can use this information to make an estimate of the strength of collaborative ties.

First of all, it is probably the case, as we pointed out at the end of the previous section, that two scientists whose names appear on a paper together with many other coauthors know one another less well on average than two who were the sole authors of a paper. The extreme case which we discussed of a very large collaboration illustrates this point forcefully, but it applies to smaller collaborations too. Even on a paper with four or five authors, the authors probably know one another less well on average than authors on a paper with fewer. To account for this effect, we weight collaborative ties inversely according to the number of coauthors as follows. Suppose a scientist collaborates on the writing of a paper that has n authors in total, i.e., he or she has $n - 1$ coauthors on that paper. Then we assume that he or she is acquainted with each coauthor $1/(n-1)$ times as well, on average, as if there were only one coauthor. One can imagine this as meaning that the scientist divides his or her time equally between the $n - 1$ coauthors. This is obviously only a rough approximation: in reality a scientist spends more time with some coauthors than with others. However, in the absence of other data, it is the obvious first approximation to make.

---

[13] In fact, the full coauthorship pattern is, like all affiliation networks, most properly represented as a bipartite graph with two kinds of vertices representing scientists and papers, and edges running between scientists and the papers on which their name appears as a coauthor. We have investigated this representation elsewhere [56, 71].

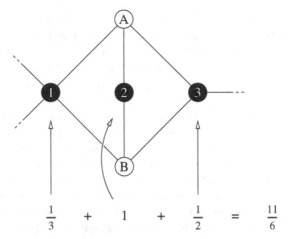

Fig. 10. Authors A and B have coauthored three papers together, labeled 1, 2, and 3, which had respectively four, two, and three authors. The tie between A and B accordingly accrues weight $\frac{1}{3}$, 1, and $\frac{1}{2}$ from the three papers, for a total weight of $\frac{11}{6}$.

Second, authors who have written many papers together will, we assume, know one another better on average than those who have written few papers together. To account for this, we add together the strengths of the ties derived from each of the papers written by a particular pair of individuals. Thus, if δ_i^k is one if scientist i was a coauthor of paper k and zero otherwise, then our weight w_{ij} representing the strength of the collaboration (if any) between scientists i and j is

$$w_{ij} = \sum_k \frac{\delta_i^k \delta_j^k}{n_k - 1}, \tag{4}$$

where n_k is the number of coauthors of paper k and we explicitly exclude from our sums all single-author papers. (They do not contribute to the coauthorship network, and their inclusion in (4) would make w_{ij} ill-defined.) We illustrate this measure for a simple example in Fig. 10.

Note that the equivalent of vertex degree for our weighted network—i.e., the sum of the weights for each of an individual's collaborations—is now just equal to the number of papers they have coauthored with others:

$$\sum_{j(\neq i)} w_{ij} = \sum_k \sum_{j(\neq i)} \frac{\delta_i^k \delta_j^k}{n_k - 1} = \sum_k \delta_i^k. \tag{5}$$

In Fig. 11 we show as an example collaborations between Gerard Barkema (one of the present author's frequent collaborators) and all of his collaborators in the physics archive for the five years of our study. Lines between points represent collaborations, with their thickness proportional to the weights w_{ij} of (4). As the figure shows, Barkema has collaborated closely with myself and with

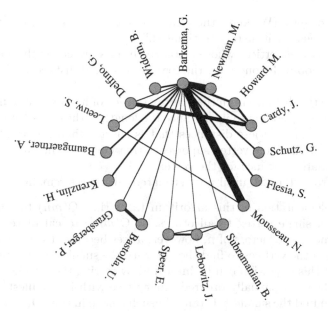

Fig. 11. Gerard Barkema and his collaborators, with lines representing collaborations whose thickness is proportional to our estimate, (4), of the strength of the corresponding tie.

Normand Mousseau, and less closely with a number of others. Also, two of his collaborators, John Cardy and Gesualdo Delfino, have collaborated quite closely with one another.

In the last column of Table 2 we show the pairs of collaborators who have the strongest collaborative ties in three subdivisions of the physics archive.

We have used our weighted collaboration graphs to calculate distances between scientists. In this simple calculation we assumed that the distance between authors is just the inverse of the weight of their collaborative tie. Thus if one pair of authors know one another twice as well as another pair, the distance between them is half as great. Calculating minimum distances between vertices on a weighted graph such as this cannot be done using the breadth-first search algorithm of Sect. 4.1, since the shortest weighted path may not be the shortest in terms of number of steps on the unweighted network. Instead we use Dijkstra's algorithm [72,60], which calculates all distances from a given starting vertex s as follows.

1. Distances from vertex s are stored for each vertex and each distance is labeled either "exact," meaning we have calculated that distance exactly, or "estimated," meaning we have made an estimate of the distance, but that estimate may be wrong. Estimated distances in Dijkstra's algorithm are always upper bounds on the exact distance. We start by assigning an estimated distance of ∞ to all vertices except vertex s to which we assign an estimated

distance of zero. (We know the latter to be exactly correct, but for the moment we consider it merely "estimated.")

2. From the set of vertices whose distances from s are currently marked "estimated," choose the one with the lowest estimated distance, and mark this "exact."

3. Calculate the distance from that vertex to each of its immediate neighbors in the network by adding to its distance the length of the edges leading to those neighbors. Any of these distances that is shorter than a current estimated distance for the same vertex supersedes that current value and becomes the new estimated distance for the vertex.

4. Repeat from step (2), until no "estimated" distances remain.

A naive implementation of this algorithm takes time $O(mn)$ to calculate distances from a single vertex to all others, or $O(mn^2)$ to calculate all pairwise distances. One of the factors of n, however, arises because it takes time $O(n)$ to search through the vertices to find the one with the smallest estimated distance. The speed of this operation can be improved by storing the estimated distances in a binary heap (a partially ordered binary tree with its smallest entry at its root). We can find the smallest distance in such a heap in time $O(1)$, and add and remove entries in time $O(\log n)$. This speeds up the operation of the algorithm to $O(mn \log n)$, making the calculation feasible for the large networks studied here.

It is in theory possible to generalize any of the calculations of Sect. 4 to the weighted collaboration graph using this algorithm and variations on it. For example, we can find shortest paths between specified pairs of scientists, as a way of establishing referrals. We can calculate the weighted equivalent of betweenness by a simple adaption of our algorithm of Sect. 4.2—we use Dijkstra's algorithm to establish the hierarchy of predecessors of vertices and then count paths through vertices exactly as before. We can also study the weighted version of the "funneling" effect using the same algorithm. Here we carry out just one calculation explicitly to demonstrate the idea; we calculate the weighted version of the distance centrality measure of Sect. 4.3, i.e., the average weighted distance from a vertex to all others. In Table 3 we show the winners in this particular popularity contest, along with their numbers of collaborators and papers in the database. Many of the scientists who score highly here do indeed appear to be well connected individuals. For example, number 1 best connected astrophysicist, Martin Rees, is the Astronomer Royal of Great Britain.[14] What is interesting to note however (apart from nonchalantly checking to see if one has made it into the top 10) is that sheer number of collaborators is no longer a necessary prerequisite for being well-connected in this sense (although some of the scientists listed do have a large number of collaborators). The case of D. Youm is particularly startling, since Youm has only two collaborators listed in the database but nonetheless is fifth best connected high-energy theorist (out

---

[14] On being informed of this latest honor, Prof. Rees is reported as replying, "I'm certainly relieved not to be the most disconnected astrophysicist" (H. Muir, *New Scientist,* November 25, 2000, p. 10).

Table 3. The ten best connected individuals in three of the communities studied here, calculated using the weighted distance measure described in the text.

	rank	name	co-workers	papers
astro-ph:	1	Rees, M. J.	31	36
	2	Miralda-Escude, J.	36	34
	3	Fabian, A. C.	156	112
	4	Waxman, E.	15	30
	5	Celotti, A.	119	45
	6	Narayan, R.	65	58
	7	Loeb, A.	33	64
	8	Reynolds, C. S.	45	38
	9	Hernquist, L.	62	80
	10	Gould, A.	76	79
cond-mat:	1	Fisher, M. P. A.	21	35
	2	Balents, L.	24	29
	3	MacDonald, A. H.	64	70
	4	Senthil, T.	9	13
	5	Das Sarma, S.	51	75
	6	Millis, A. J.	43	37
	7	Ioffe, L. B.	16	27
	8	Sachdev, S.	28	44
	9	Lee, P. A.	24	34
	10	Jungwirth, T.	27	17
hep-th:	1	Cvetic, M.	33	69
	2	Behrndt, K.	22	41
	3	Tseytlin, A. A.	22	65
	4	Bergshoeff, E.	21	39
	5	Youm, D.	2	30
	6	Lu, H.	34	73
	7	Klebanov, I. R.	29	47
	8	Townsend, P. K.	31	54
	9	Pope, C. N.	33	72
	10	Larsen, F.	11	27

of eight thousand), because one or both of those collaborators are themselves very well connected, and because their ties to Youm are very strong. Experimentalists no longer dominate the field, although the well-connected among them still score highly.

Note that the number of papers for each of the well-connected scientists listed is high. Having written a large number of papers is, as it rightly should be, always a good way of becoming well connected. Whether you write many papers with many different authors, or many with a few, writing many papers will put you in touch with your peers.

6 Conclusions

In this article we have studied social networks of scientists in which the actors are authors of scientific papers, and a tie between two authors represents coauthorship of one or more papers. Drawing on the lists of authors in four databases of papers in physics, biomedical research, and computer science, we have constructed explicit networks for papers appearing between the beginning of 1995 and the end of 1999. We have cataloged a large number of basic statistics for our networks, including typical numbers of papers per author, authors per paper, and numbers of collaborators per author in the various fields. We also note that the distributions of these quantities roughly follow a power-law form, although there are some deviations which may be due to the finite time window used for the study.

We have also looked at a variety of non-local properties of our networks. We find that typical distances between pairs of authors through the networks are small—the networks form a "small world" in the sense discussed by Milgram—and that they scale logarithmically with total number of authors in a network, in reasonable agreement with the predictions of random graph models. Using a new algorithm for counting the number of shortest paths between vertices on a graph that pass through each other vertex, we have calculated the so-called betweenness measure of centrality on our graphs. We have also shown that for most authors the bulk of the paths between them and other scientists in the network go through just one or two or their collaborators, an effect that Strogatz has dubbed "funneling."

We have suggested a measure of the closeness of collaborative ties that takes account of the number of papers a given pair of scientists have written together, as well as the number of other coauthors with whom they wrote them. Using this measure we have added weightings to our collaboration networks and used the resulting networks to find those scientists who have the shortest average distance to others. Generalization of the betweenness and funneling calculations to these weighted networks is also straightforward.

The calculations presented in this article inevitably represent only a small part of the investigations that could be conducted using large network datasets such as these. We hope, given the high current level of interest in network phenomena, that others will find many further uses for collaboration network data.

Acknowledgements

The author would particularly like to thank Paul Ginsparg for his invaluable help in obtaining the data used for this study. The data used were generously made available by Oleg Khovayko, David Lipman, and Grigoriy Starchenko (Medline), Paul Ginsparg and Geoffrey West (Physics E-print Archive), Heath O'Connell (SPIRES), and Carl Lagoze (NCSTRL). The Physics E-print Archive and NCSTRL are maintained by Cornell University, while Medline and SPIRES are maintained by the National Center for Biotechnology Information and the Stanford Linear Accelerator Center, respectively.

In addition, the author would like to thank Steve Strogatz for suggesting the funneling effect calculation of Sect. 4.2, and Dave Alderson, László Barabási, Sankar Das Sarma, Paul Ginsparg, Rick Grannis, Jon Kleinberg, Laura Landweber, Sidney Redner, Ronald Rousseau, Steve Strogatz, Duncan Watts, Douglas White, and an anonymous referee for many useful comments and suggestions. This work was funded in part by the National Science Foundation under grant number DMS-0234188, by Intel Corporation, and by the Santa Fe Institute.

References

1. S. Wasserman and K. Faust, *Social Network Analysis*. Cambridge University Press, Cambridge (1994).
2. J. Scott, *Social Network Analysis: A Handbook*. Sage, London, 2nd edition (2000).
3. S. H. Strogatz, Exploring complex networks. *Nature* **410**, 268–276 (2001).
4. A.-L. Barabási, *Linked: The New Science of Networks*. Perseus, Cambridge, MA (2002).
5. D. J. Watts, *Six Degrees: The Science of a Connected Age*. Norton, New York (2003).
6. M. E. J. Newman, The structure and function of complex networks. *SIAM Review* **45**, 167–256 (2003).
7. P. Mariolis, Interlocking directorates and control of corporations: The theory of bank control. *Social Science Quarterly* **56**, 425–439 (1975).
8. J. Galaskiewicz and P. V. Marsden, Interorganizational resource networks: Formal patterns of overlap. *Social Science Research* **7**, 89–107 (1978).
9. J. F. Padgett and C. K. Ansell, Robust action and the rise of the Medici, 1400–1434. *Am. J. Sociol.* **98**, 1259–1319 (1993).
10. A. Rapoport and W. J. Horvath, A study of a large sociogram. *Behavioral Science* **6**, 279–291 (1961).
11. T. J. Fararo and M. Sunshine, *A Study of a Biased Friendship Network*. Syracuse University Press, Syracuse (1964).
12. H. R. Bernard, P. D. Killworth, M. J. Evans, C. McCarty, and G. A. Shelley, Studying social relations cross-culturally. *Ethnology* **2**, 155–179 (1988).
13. J. Moody, Race, school integration, and friendship segregation in America. *Am. J. Sociol.* **107**, 679–716 (2001).
14. M. Faloutsos, P. Faloutsos, and C. Faloutsos, On power-law relationships of the internet topology. *Computer Communications Review* **29**, 251–262 (1999).
15. Q. Chen, H. Chang, R. Govindan, S. Jamin, S. J. Shenker, and W. Willinger, The origin of power laws in Internet topologies revisited. In *Proceedings of the 21st Annual Joint Conference of the IEEE Computer and Communications Societies*, IEEE Computer Society (2002).
16. R. Albert, H. Jeong, and A.-L. Barabási, Diameter of the world-wide web. *Nature* **401**, 130–131 (1999).
17. A. Broder, R. Kumar, F. Maghoul, P. Raghavan, S. Rajagopalan, R. Stata, A. Tomkins, and J. Wiener, Graph structure in the web. *Computer Networks* **33**, 309–320 (2000).
18. H. Ebel, L.-I. Mielsch, and S. Bornholdt, Scale-free topology of e-mail networks. *Phys. Rev. E* **66**, 035103 (2002).

19. M. E. J. Newman, S. Forrest, and J. Balthrop, Email networks and the spread of computer viruses. *Phys. Rev. E* **66**, 035101 (2002).
20. L. A. Adamic, R. M. Lukose, A. R. Puniyani, and B. A. Huberman, Search in power-law networks. *Phys. Rev. E* **64**, 046135 (2001).
21. M. Ripeanu, I. Foster, and A. Iamnitchi, Mapping the Gnutella network: Properties of large-scale peer-to-peer systems and implications for system design. *IEEE Internet Computing* **6**, 50–57 (2002).
22. D. J. Watts and S. H. Strogatz, Collective dynamics of 'small-world' networks. *Nature* **393**, 440–442 (1998).
23. J. Abello, A. Buchsbaum, and J. Westbrook, A functional approach to external graph algorithms. In *Proceedings of the 6th European Symposium on Algorithms*, Springer, Berlin (1998).
24. P. Sen, S. Dasgupta, A. Chatterjee, P. A. Sreeram, G. Mukherjee, and S. S. Manna, Small-world properties of the Indian railway network. *Phys. Rev. E* **67**, 036106 (2003).
25. A. Davis, B. B. Gardner, and M. R. Gardner, *Deep South*. University of Chicago Press, Chicago (1941).
26. G. F. Davis and H. R. Greve, Corporate elite networks and governance changes in the 1980s. *Am. J. Sociol.* **103**, 1–37 (1997).
27. L. A. N. Amaral, A. Scala, M. Barthélémy, and H. E. Stanley, Classes of small-world networks. *Proc. Natl. Acad. Sci. USA* **97**, 11149–11152 (2000).
28. B. Wellman, J. Salaff, D. Dimitrova, L. Garton, M. Gulia, and C. Haythornthwaite, Computer networks as social networks. *Annual Review of Sociology* **22**, 213–238 (1996).
29. P. Hoffman, *The Man Who Loved Only Numbers*. Hyperion, New York (1998).
30. P. Erdős and M. Kac, The Gaussian law of errors in the theory of additive number theoretic functions. *Am. J. Math.* **26**, 738–742 (1940).
31. R. M. Ziff, G. E. Uhlenbeck, and M. Kac, The ideal Bose-Einstein gas, revisited. *Phys. Rep.* **32**, 169–248 (1977).
32. M. E. J. Newman and R. M. Ziff, Efficient Monte Carlo algorithm and high-precision results for percolation. *Phys. Rev. Lett.* **85**, 4104–4107 (2000).
33. V. Batagelj and A. Mrvar, Some analyses of Erdős collaboration graph. *Social Networks* **22**, 173–186 (2000).
34. J. W. Grossman, The evolution of the mathematical research collaboration graph. *Congressus Numerantium* **158**, 202–212 (2002).
35. J. W. Grossman and P. D. F. Ion, On a portion of the well-known collaboration graph. *Congressus Numerantium* **108**, 129–131 (1995).
36. L. Egghe and R. Rousseau, *Introduction to Informetrics*. Elsevier, Amsterdam (1990).
37. H. Kretschmer, Coauthorship networks of invisible college and institutionalized communities. *Scientometrics* **30**, 363–369 (1994).
38. O. Persson and M. Beckmann, Locating the network of interacting authors in scientific specialties. *Scientometrics* **33**, 351–366 (1995).
39. G. Melin and O. Persson, Studying research collaboration using co-authorships. *Scientometrics* **36**, 363–377 (1996).
40. Y. Ding, S. Foo, and G. Chowdhury, A bibliometric analysis of collaboration in the field of information retrieval. *Intl. Inform. and Libr. Rev.* **30**, 367–376 (1999).
41. M. Bordens and I. Gómez, Collaboration networks in science. In H. B. Atkins and B. Cronin (eds.), *The Web of Knowledge: A Festschrift in Honor of Eugene Garfield*, Information Today, Medford, NJ (2000).

42. D. J. de S. Price, Networks of scientific papers. *Science* **149**, 510–515 (1965).
43. P. O. Seglen, The skewness of science. *J. Amer. Soc. Inform. Sci.* **43**, 628–638 (1992).
44. S. Redner, How popular is your paper? An empirical study of the citation distribution. *Eur. Phys. J. B* **4**, 131–134 (1998).
45. M. E. J. Newman, Scientific collaboration networks: I. Network construction and fundamental results. *Phys. Rev. E* **64**, 016131 (2001).
46. M. E. J. Newman, Scientific collaboration networks: II. Shortest paths, weighted networks, and centrality. *Phys. Rev. E* **64**, 016132 (2001).
47. M. E. J. Newman, Clustering and preferential attachment in growing networks. *Phys. Rev. E* **64**, 025102 (2001).
48. A.-L. Barabási, H. Jeong, E. Ravasz, Z. Néda, A. Schuberts, and T. Vicsek, Evolution of the social network of scientific collaborations. *Physica A* **311**, 590–614 (2002).
49. H. B. O'Connell, Physicists thriving with paperless publishing. Preprint physics/0007040 (2000).
50. A. J. Lotka, The frequency distribution of scientific production. *J. Wash. Acad. Sci.* **16**, 317–323 (1926).
51. H. Voos, Lotka and information science. *Journal of the American Society for Information Science* (July-August 1974), 270–272 (1974).
52. M. L. Pao, An empirical examination of Lotka's law. *Journal of the American Society for Information Science* (January 1986), 26–33 (1986).
53. M. E. J. Newman, The structure of scientific collaboration networks. *Proc. Natl. Acad. Sci. USA* **98**, 404–409 (2001).
54. P. Erdős and A. Rényi, On the evolution of random graphs. *Publications of the Mathematical Institute of the Hungarian Academy of Sciences* **5**, 17–61 (1960).
55. B. Bollobás, *Random Graphs*. Academic Press, New York, 2nd edition (2001).
56. M. E. J. Newman, S. H. Strogatz, and D. J. Watts, Random graphs with arbitrary degree distributions and their applications. *Phys. Rev. E* **64**, 026118 (2001).
57. A.-L. Barabási and R. Albert, Emergence of scaling in random networks. *Science* **286**, 509–512 (1999).
58. J. M. Kleinberg, Navigation in a small world. *Nature* **406**, 845 (2000).
59. P. L. Krapivsky and S. Redner, Organization of growing random networks. *Phys. Rev. E* **63**, 066123 (2001).
60. T. H. Cormen, C. E. Leiserson, R. L. Rivest, and C. Stein, *Introduction to Algorithms*. MIT Press, Cambridge, MA, 2nd edition (2001).
61. H. Kautz, B. Selman, and M. Shah, ReferralWeb: Combining social networks and collaborative filtering. *Comm. ACM* **40**, 63–65 (1997).
62. L. C. Freeman, A set of measures of centrality based upon betweenness. *Sociometry* **40**, 35–41 (1977).
63. U. Brandes, A faster algorithm for betweenness centrality. *Journal of Mathematical Sociology* **25**, 163–177 (2001).
64. K.-I. Goh, B. Kahng, and D. Kim, Universal behavior of load distribution in scale-free networks. *Phys. Rev. Lett.* **87**, 278701 (2001).
65. S. Milgram, The small world problem. *Psychology Today* **2**, 60–67 (1967).
66. I. de S. Pool and M. Kochen, Contacts and influence. *Social Networks* **1**, 1–48 (1978).
67. B. Bollobás, A probabilistic proof of an asymptotic formula for the number of labelled regular graphs. *European Journal of Combinatorics* **1**, 311–316 (1980).

68. T. Łuczak, Sparse random graphs with a given degree sequence. In A. M. Frieze and T. Łuczak (eds.), *Proceedings of the Symposium on Random Graphs, Poznań 1989*, pp. 165–182, John Wiley, New York (1992).
69. M. Molloy and B. Reed, A critical point for random graphs with a given degree sequence. *Random Structures and Algorithms* **6**, 161–179 (1995).
70. M. Molloy and B. Reed, The size of the giant component of a random graph with a given degree sequence. *Combinatorics, Probability and Computing* **7**, 295–305 (1998).
71. M. E. J. Newman, D. J. Watts, and S. H. Strogatz, Random graph models of social networks. *Proc. Natl. Acad. Sci. USA* **99**, 2566–2572 (2002).
72. R. K. Ahuja, T. L. Magnanti, and J. B. Orlin, *Network Flows: Theory, Algorithms, and Applications*. Prentice Hall, Upper Saddle River, NJ (1993).

Information Dynamics in the Networked World

Bernardo A. Huberman and Lada A. Adamic

HP Labs, 1501 Page Mill Road, Palo Alto CA 94304, USA

Abstract. We review three studies of information flow in social networks that help reveal their underlying social structure, how information spreads among them and why small world experiments work.

1 Introduction

The problem of information flows in social organizations is relevant to issues of productivity, innovation and the sorting out of useful ideas from the general chatter of a community. How information spreads determines the speed by which individuals can act and plan their future activities. Moreover, information flows take place within social networks whose nature is sometimes difficult to establish. This is because the network itself is sometimes different from what one would infer from the formal structure of the group or organization.

The advent of email as the predominant means of communication in the information society now offers a unique opportunity to observe the flow of information along both formal and informal channels. Not surprisingly, email has been established as an indicator of collaboration and knowledge exchange [1–5]. Email is also a good medium for social network research because it provides plentiful data on personal communication in an electronic form. This volume of data enables the discovery of shared interests and relationships where none were previously known [6].

In this chapter we will review three studies that utilized networks exposed by email communication. In all three studies, the networks analyzed were derived from email messages sent through the Hewlett Packard Labs email server over the period of several months in 2002 and 2003. The first study, by Tyler et al. [4], develops an automated method applying a betweenness centrality algorithm to rapidly identify communities, both formal and informal, within the network. This approach also enables the identification of leadership roles within the communities. The automated analysis was complemented by a qualitative evaluation of the results in the field.

The second study, by Wu et al. [7] analyzes email patterns to model information flow in social groups, taking into account the observation that an item relevant to one person is more likely to be of interest to individuals in the same social circle than those outside of it. This is due to the fact that the similarity of node attributes in social networks decreases as a function of the graph dis-

B.A. Huberman and L.A. Adamic, Information Dynamics in the Networked World, Lect. Notes Phys.
650, 371–398 (2004)
http://www.springerlink.com/

tance. An epidemic model on a scale-free network with this property has a finite threshold, implying that the spread of information is limited. These predictions were tested by measuring the spread of messages in an organization and also by numerical experiments that take into consideration the organizational distance among individuals.

Since social structure affects the flow of information, knowledge of the communities that exist within a network can also be used for navigating the networks when searching for individuals or resources. The study by Adamic and Adar [8], does just this, by simulating Milgram's small world experiment on the HP Labs email network. The small world experiment has been carried out a number of times over the past several decades, each time demonstrating that individuals passing messages to their friends and acquaintances can form a short chain between two people separated by geography, profession, and race. While the existence of these chains has been established, how people are able to navigate without knowing the complete social networks has remained an open question. Recently, models have been proposed to explain the phenomenon, and the work of Adamic and Adar is a first study to test the validity of these models on a social network.

2 Email as Spectroscopy

Communities of practice are the informal networks of collaboration that naturally grow and coalesce within and outside organizations. Any institution that provides opportunities for communication among its members is eventually threaded by communities of people who have similar goals and a shared understanding of their activities [9]. These communities have been the subject of much research as a way to uncover the reality of how people find information and execute their tasks. (for example, see [10–12], or for a survey see [13]).

These informal networks coexist with the formal structure of the organization and serve many purposes, such as resolving the conflicting goals of the institution to which they belong, solving problems in more efficient ways [14], and furthering the interests of their members. Despite their lack of official recognition, informal networks can provide effective ways of learning, and with the proper incentives actually enhance the productivity of the formal organization [15–17].

Recently, there has been an increased amount of work on identifying communities from online interactions (a brief overview of this work can be found in [1]). Some of this work finds that online relationships do indeed reflect actual social relationships, thus adding effectively to the "social capital" of a community. Ducheneaut and Bellotti [18] conducted in-depth field studies of email behavior, and found that membership in email communities is quite fluid and depends on organizational context. Mailing lists and personal web pages also serve as proxies for social relationships [19], and the communities identified from these online proxies resemble the actual social communities of the represented individuals. Because of the demonstrated value of communities of practice, a fast, accurate method of identifying them is desirable.

Classical practice is to gather data from interviews, surveys, or other field-work and to construct links and communities by manual inspection (see [20,21] or an Internet-centric approach in [22]). These methods are accurate but time-consuming and labor-intensive, prohibitively so in the context of a very large organization. Alani et al. [23] recently introduced a semi-automated utility that uses a simple algorithm to identify nearest neighbors to one individual within a university department.

The method of Tyler et al. [4] uses email data to construct a network of cor-respondences, and then discovers the communities by partitioning this network. It was applied to a set of over one million email messages collected over a period of roughly two months at HP Labs in Palo Alto, an organization of approxi-mately 400 people. The only pieces of information used from each email are the names of the sender and receiver (i.e., the "to:" and "from:" fields), enabling the processing of a large number of emails while minimizing privacy concerns.

The method was able to identify small communities within the organiza-tion, and the leaders for those communities, in a matter of hours, running on a standard Linux desktop PC. This experiment was followed by a qualitative evaluation of the experimental results in the "field", which consisted of sixteen face-to-face interviews with individuals in HP Labs. The interviews validated the results obtained by the automated process, and provided interesting perspectives on the communities identified. We describe the results in more detail below.

2.1 Identifying Communities

It is straightforward to construct a graph based on email data, in which vertices represent people and edges are added between people who exchanged at least a threshold number of email messages. Next, one can identify communities: subsets of related vertices, with many edges connecting vertices of the same subset, but few edges lying between subsets [24].

The method of Wilkinson and Huberman [25], related to the algorithm of Girvan and Newman [24], partitions a graph into discrete communities of nodes and is based on the idea of betweenness centrality, or betweenness, first proposed by Freeman [26]. The betweenness of an edge is defined as the number of all-pair shortest paths that traverse it. This property distinguishes inter-community edges, which link many vertices in different communities and have high between-ness, from intra-community edges, whose betweenness is low.

To illustrate the community discovery process, consider the small graph shown in Fig. 1. This graph consists of two well-defined communities: the four vertices denoted by squares, including vertex A, and the nine denoted by circles, including vertex B. Edge AB has the highest betweenness, because all paths between any circle and square must pass through it. If one were to remove it, the squares and circles would be split into two separate communities. The al-gorithm of Wilkinson et al. repeatedly identifies inter-community edges of large betweenness such as AB and removes them, until the graph is resolved into many separate communities.

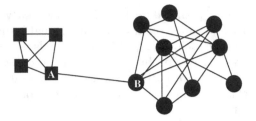

Fig. 1. An example graph with edge AB having high betweenness.

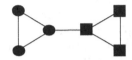

Fig. 2. The smallest possible graph of two viable communities.

Because the removal of an edge strongly affects the betweenness of many others, the values were repeatedly updated with the fast algorithm of Brandes [27, 28,24]. The procedure stops removing edges when it cannot further meaningfully subdivide communities. Figure 2 shows the smallest possible component that can be subdivided into two viable subcommunities. It has 6 nodes, consisting of two triangles linked by one edge. A component with fewer than 6 nodes cannot be subdivided further.

Components of size ≥ 6, for example the group of size nine in Fig. 1, can also constitute single cohesive communities. Figure 3 shows how the algorithm determines when to stop subdividing a community. The edge XY has the highest betweenness, but removing it would separate a single node, which does not constitute a viable community. In general, the single edge connecting a leaf vertex (such as X in Fig. 3) to the rest of a graph of N vertices has a betweenness of $N-1$, because it contains the shortest path from X to all $N-1$ other vertices. The stopping criterion for components of size ≥ 6 is therefore that the highest betweenness of any edge in the component be equal to or less than $N-1$.

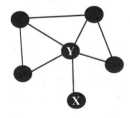

Fig. 3. An example graph of one community that does not contain distinct subcommunities.

2.2 Multiple Community Structures

As mentioned above, the removal of any one edge affects the betweenness of all the other edges, particularly in large, real-world graphs such as the email graph. Early in the process, there are many inter-community edges which have high betweenness and the choice of which to remove, while arbitrary, dictates which edges will be removed later. For example, a node belonging to two communities can be placed in one or the other by the algorithm, depending on the order in which edges are removed. One can take advantage of this arbitrariness to repeatedly partition the graph into many different "structures" or sets of communities. These sets are then compared and aggregated into a final list of communities.

Wilkinson and Huberman [25] introduced randomness into the algorithm by calculating the shortest paths from a random subset as opposed to all the nodes. The algorithm cycles randomly through at least m centers (where m is some cutoff) until the betweenness of at least one edge exceeds the threshold betweenness of a "leaf" vertex. The edge whose betweenness is highest at that point is removed, and the procedure is repeated until the graph has been separated into communities. The modified algorithm may occasionally remove an intra-community edge, but such errors are unimportant when a large number of structures is aggregated.

Applying this modified process n times yields n community structures imposed on the graph. One can then compare the different structures and identify communities. For example, after imposing 50 structures on a graph, one might find: a community of people A, B, C, and D in 25 of the 50 structures; a community of people A, B, C, D, and E in another 20; and one of people A, B, C, D, E and F in the remaining 5. This result is reported in the following way: A(50) B(50) C(50) D(50) E(25) F(5) which signifies that A, B, C, and D form a well-defined community, E is related to this community, but also to some other(s), and F is only slightly, possibly erroneously, related to it. For details of the aggregation procedure, please see [25].

The entire process of determining community structure within the graph is displayed below.

- For i iterations, repeat {
    1. Identify disjoint components of the graph.
    2. For each component, check to see if component is a community.
        - If so, remove it from the graph and output it.
        - If not, remove edges of highest betweenness, using the modified Brandes algorithm for large components, and the normal algorithm for small ones. Continue removing edges until the community splits in two.
    3. Repeat step 2 until all vertices have been removed from the graph in communities. }
- Aggregate the i structures into a final list of communities.

2.3 Results

The algorithm was applied to email data from the HP Labs mail server from the period November 25, 2002 to February 18, 2003, with 185,773 emails exchanged between the 485 HP Labs employees. For simplicity, emails that had an external origin or destination were omitted. Messages sent to a list of more than 10 recipients were likewise removed, as these emails were often lab-wide announcements (rather than personal communication), which were not useful in identifying communities of practice.

A graph was constructed from this data by placing edges between any two individuals that had exchanged at least 30 emails in total, and at least 5 in both directions. The threshold eliminated infrequent or one-way communication, and eliminated some individuals from the graph who either sent very few emails or used other email systems. The resulting graph consisted of 367 nodes, connected by 1110 edges.

There was one giant connected component of 343 nodes and six smaller components ranging in size from 2 to 8. The modified Brandes algorithm detected 60 additional distinct communities within the giant component. The largest community consisted of 57 individuals. The mean community size was 8.4, with standard deviation 5.3. A comparison of these communities with information from the HP corporate directory revealed that 49 of the 66 communities consisted of individuals entirely within one lab or organizational unit. The remaining 17 contained individuals from two or more organizations within the company.

2.4 Identifying Leadership Roles

In addition to identifying formal and informal work communities, it is also possible to draw inferences about the leadership of an organization from its communication data. One method is to visualize the above graph of the HP Labs email network with a standard force-directed spring algorithm [29], shown in Fig. 4. This spring layout of the email network does not use any information about the actual organization structure, and yet high level managers (the reddest nodes are at the top of the hierarchy) are placed close to the center of the graph. The trend is quantified in Table 1, which lists the average hierarchy depth (levels from the lab director) as a function of the position in the layout from the center.

Note that there is a group of 6 nodes in the upper right portion of the graph that are quite removed from the center, but are relatively high in the organizational hierarchy. This is the university relations group that reports directly to the head of HP Labs, but has no other groups reporting to it. Hence the layout algorithm correctly places them on the periphery of the graph, since their function, that of managing HP's relationship with universities, while important, is not at the core of day-to-day activities of the labs.

Evaluating communication networks with this technique could provide information about leadership in communities about which little is known. Sparrow proposed this approach for analyzing criminal networks [30], noting that "Euclidean centrality is probably the closest to the reality" of the current criminal

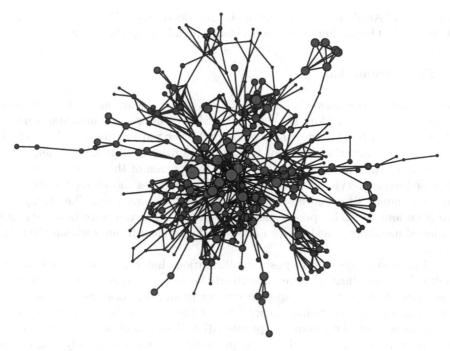

Fig. 4. The giant connected component of the HP Labs email network. The redness and size of a vertex indicates an individual's closeness to the top of the lab hierarchy (red-close to top, blue-far from top, black-no data available).

Table 1. Average hierarchy depth by distance from center in layout

distance from center	number of vertices	average depth in hierarchy
< 0.1	14	2.6
0.1 to 0.2	32	3.0
0.2 to 0.3	56	3.2
0.3 to 0.4	66	4.0
0.4 to 0.5	56	4.0
0.5 to 0.6	45	4.2
0.6 to 0.7	42	4.0
0.7 to 0.8	12	3.9
0.8 to 0.9	13	3.8

network analysis techniques. More recently, Krebs applied centrality measures and graphing techniques [31] to the terrorist networks uncovered in the 9/11 aftermath. He found that the average shortest path was unusually long for such a small network, and concluded that the operation had traded efficiency for secrecy - individuals in one part of the network did not know those in other parts of the network. If one cell had been compromised, the rest of the network would remain relatively unaffected. Several social network centrality measures pointed to Mohamed Atta's leadership role in the attacks of Sept. 11. The role was also confirmed by Osama bin Laden in a video tape following the attacks.

2.5 Field Evaluation

The HP Labs social network, being much less covert, could readily be compared to the structure of the formal organization. Nevertheless, the informal communities identified by the algorithm could not be verified in this way. Tyler et al. decided to validate the results of their algorithm by conducting a brief, informal field study. Sixteen individuals chosen from seven of the sixty communities identified were interviewed informally. The communities chosen represented various community sizes and levels of departmental homogeneity. They ranged in size from four to twelve people, and three out of the seven were heterogeneous (included members of at least two different departmental units within the company).

All sixteen subjects gave positive affirmation that the community reflected reality. More specifically, eleven described the group as reflecting their department, four described it as a specific project group, and one said it was a discussion group on a particular topic. Nine of the sixteen (56.25%) said nobody was missing from the group, six people (37.5%) said one person was missing, and one person (6.25%) said two people were missing. Conversely, ten of the sixteen (62.5%) said that everybody in the group deserved to be there, whereas the remaining six (37.5%) said that one person in the group was misclassified.

The interviews confirmed that most of the communities identified were based on organization structure. However, the communities also tended to include people who were de facto department members, but who did not technically appear in the department's organization chart, such as interns or people whose directory information had changed during the two months of the study. Finally, the algorithm seemed to succeed in dividing departmental groups whose work is distinct, but lumped together groups whose projects overlap.

Heterogeneous, cross-department communities are of particular interest because they cannot be deduced from the formal organization. The interviews revealed that most of them represented groups formed around specific projects, and in one case, a discussion forum. For example, one community contained three people from different labs coordinating on one project: a technology transfer project manager, a researcher who was the original designer of a piece of PC hardware, and an engineer redesigning the hardware for a specific printer.

2.6 Discussion

The power of this method for identifying communities and leadership is in its automation. It does an effective job of uncovering communities of practice with nothing more than email log ("to:" and "from:") data. The betweenness centrality measures can be further augmented to incorporate weights on the edges, representing, for example, the frequency of communication along a link [32].

Because the method of Wilkinson et al. [25] needs to re-run the Brandes algorithm every time an edge is removed, the algorithm has a running time of $O(n^3)$. Even faster algorithms, that can identify communities in $O(n^2)$ [33] and $O(n)$ [34] time have since been developed. The simplicity and speed of these new algorithms means that they can be applied to organizations of thousands to hundreds of thousands and produce results efficiently.

Communities identified in this automated way lack the richness in contextual description provided by ethnographic approaches. They do not reveal the nature or character of the identified communities, the relative importance of one community to another, or the subtle inter-personal dynamics within the communities. These kinds of details can only be uncovered with much more data- or labor-intensive techniques. However, in cases where an organization is very large, widely dispersed, or incompletely defined (informal), this method provides an suitable alternative or compliment to the more traditional, labor-intensive approaches.

3 Information Flow in Social Groups

In the previous section we saw that individuals tend to organize both formally and informally into groups based on their common activities and interests. In this section we examine how this structure in the interaction network affects the way information spreads. This is not unlike the transmission of an infectious agent among individuals, where the pattern of contacts determines how far a disease spreads. Thus one would expect that epidemic models on graphs are relevant to the study of information flow in organizations.

As we will show shortly (see, for example Fig. 8), e-mail networks can form scale free graphs. This is of particular interest, since recent work on epidemic propagation on scale free networks found that the threshold for an epidemic is zero, implying that a finite fraction of the graph becomes infected for arbitrarily low transmission probabilities [35–37]. The presence of additional network structure was found to further influence the spread of disease on scale-free graphs [38–40].

There are, however, differences between information flows and the spread of viruses. While viruses tend to be indiscriminate, infecting any susceptible individual, information is selective and passed by its host only to individuals the host thinks would be interested in it. The information any individual is interested in depends strongly on their characteristics. Furthermore, individuals with similar characteristics tend to associate with one another, a phenomenon known as homophily [41–43]. Conversely, individuals many steps removed in a social network on average tend not to have as much in common, as shown in a study [19] of a network of Stanford student homepages and illustrated in Fig. 5.

Wu et al. [7] introduced an epidemic model with decay in the transmission probability of a particular piece of information as a function of the distance between the originating source and the current potential target. This epidemic model on a scale-free network has a finite threshold, implying that the spread

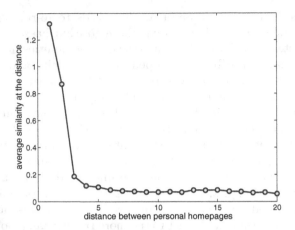

Fig. 5. Average similarity of Stanford student homepages as a function of the number of hyperlinks separating them.

of information is limited. The predictions were further tested by observing the prevalence of messages in an organization and also by numerical experiments that take into consideration the organizational distance among individuals.

Consider the problem of information transmission in a power-law network of interacting individuals, where the degree distribution is given by

$$p_k = Ck^{-\alpha}e^{-k/\kappa}, \tag{1}$$

where $\alpha > 1$, there is an exponential cutoff at κ and C is determined by the normalization condition. A real world graph will at the very least have a cutoff at the maximum degree $k = N$, where N is the number of nodes, and many networks show a cutoff at values much smaller than N. For their analysis, Wu et al. [7] made use of generating functions, whose application to graphs with arbitrary degree distributions is discussed in [44]. The generating function of the distribution is

$$G_0(x) = \sum_{k=1}^{\infty} p_k x^k = \frac{Li_\alpha(xe^{-k/\kappa})}{Li_\alpha(e^{-1/\kappa})}. \tag{2}$$

where $Li_n(x)$ is the nth polylogarithm of x.

Following the analysis in [45] for the SIR (susceptible, infected, removed) model, one can estimate the probability $p_l^{(1)}$ that the first person in the community who has received a piece of information will transmit it to l of their neighbors. Using the binomial distribution, we find

$$p_l^{(1)} = \sum_{k=l}^{\infty} p_k \binom{k}{l} T^l (1 - T)^{k-l}, \tag{3}$$

where the superscript "(1)" refers to first neighbors, those who received the information directly from the initial source. The *transmissiblity* T is the average total probability that the information will be transmitted across an edge in the network from a infective individual to a susceptible neighbor. T is derived in [45] as a function of r_{ij}, the rate of contacts between the two nodes, and τ_i, the time a node remains infective. If we assume to a first approximation that r_{ij} and τ_i are iid randomly distributed according to the distributions $P(r)$ and $P(\tau)$, then the item will propagate as if all transmission probabilities are equal to a constant T.

$$T = \langle T_{ij} \rangle = 1 - \int_0^\infty dr d\tau P(r) P(\tau) e^{-r\tau} \tag{4}$$

The generating function for $p_m^{(1)}$ is given by

$$G^{(1)}(x) = \sum_{l=0}^\infty \sum_{k=l}^\infty p_k \binom{k}{l} T^l (1-T)^{k-l} x^l \tag{5}$$

$$= G_0(1 + (x-1)T) = G_0(x; T). \tag{6}$$

Suppose the transmissibility decays as a power of the distance from the initial source. We choose this weakest form of decay as the results that are obtained from it will also be valid for stronger functional forms. Then the probability that an mth neighbor will transmit the information to a person with whom he has contact is given by

$$T^{(m)} = (m+1)^{-\beta} T, \tag{7}$$

where $\beta > 0$ is the decay constant. $T^{(m)} = T$ at the originating node ($m = 0$) and decays to zero as $m \to \infty$.

The generating function for the transmission probability to 2nd neighbors can be written as

$$G^{(2)}(x) = \sum_k p_k^{(1)} [G_1^{(1)}(x)]^k = G^{(1)}(G_1^{(1)}(x)), \tag{8}$$

where

$$G_1^{(1)}(x) = G_1(x; 2^{-\beta} T) = G_1(1 + (x-1)2^{-\beta} T) \tag{9}$$

and

$$G_1(x) = \frac{\sum_k k p_k x^k}{x \sum_k k p_k} = \frac{G_0'(x)}{G_0'(1)} \tag{10}$$

is the generating function of the degree distribution of a vertex reached by following a randomly chosen edge, not counting the edge itself [44]. Similarly, if we define $G^{(m)}(x)$ to be the generating function for the number of mth neighbors affected, then we have

$$G^{(m+1)}(x) = G^{(m)}(G_1^{(m)}(x)) \quad \text{for } m \geq 1, \tag{11}$$

where

$$G_1^{(m)}(x) = G_1(x; (m+1)^{-\beta}T) = G_1(1 + (x-1)(m+1)^{-\beta}T). \qquad (12)$$

Or, more explicitly,

$$G^{(m+1)}(x) = G^{(1)}(G_1^{(1)}(G_1^{(2)}(\cdots G_1^{(m)}(x)))). \qquad (13)$$

The average number z_{m+1} of $(m+1)$th neighbors is

$$z_{m+1} = G^{(m+1)'}(1) = G_1^{(m)'}(1)G^{(m)'}(1) = G_1^{(m)'}(1)z_m. \qquad (14)$$

So the condition that the size of the outbreak (the number of affected individuals) remains finite is given by

$$\frac{z_{m+1}}{z_m} = G_1^{(m)'}(1) < 1, \qquad (15)$$

or

$$(m+1)^{-\beta}TG_1'(1) < 1. \qquad (16)$$

Note that $G_1'(1)$ does not diverge when $\alpha < 3$ due to the presence of a cutoff at κ. For any given T, the left hand side of the inequality above goes to zero when $m \to \infty$, so the condition is eventually satisfied for large m. Therefore the average total size

$$\langle s \rangle = \sum_{m=1}^{\infty} z_m \qquad (17)$$

is always finite if the transmissibility decays with distance.

To compare this result with previous results on disease spread on scale-free networks, we take as an example a network made up of 10^6 vertices. We can define an epidemic to be an outbreak affecting more than 1% or 10^4 vertices. Thus for fixed α, κ and β, we can define T_c as the transmissibility above which $\langle s \rangle$ would be made to exceed 10^4.

Figure 6 shows the numerical results of the variation of T_c as a function of α. When $\beta = 0$ (there is no decay in transmission probability), $\kappa = \infty$, and $\alpha < 3$, T_c is zero and epidemics encompassing more than 10^4 vertices occur for arbitrarily small T, as was found in [36]. Keeping β at zero and adding a cutoff at $\kappa = 100$ produces a non-zero critical transmissibility T_c, as was found in [45]. For $\alpha = 2$, a typical value for real-world networks, T_c is still very near zero, meaning that for most values of T, epidemics do occur. However, when we impose a decay in transmissibility by setting β to 1, T_c rises substantially. For example, T_c jumps to 0.54 at $\alpha = 2$ and rises rapidly to 1 as α increases further, implying that the information may not spread over the network.

In order to validate empirically that the spread of information within a network of people is limited, and hence distinct from the spread of a virus, a sample

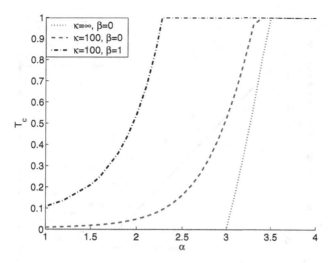

Fig. 6. T_c as a function of α. The three different curves, from bottom to top are: 1) no decay in transmission probability, no exponential cutoff in the degree distribution ($\kappa = \infty, \beta = 0$). 2) $\kappa = 100, \beta = 0$, 3) $\kappa = 100, \beta = 1$.

from the mail clients of 40 individuals (30 within HP Labs, and 10 from other areas of HP, other research labs, and universities) was gathered. Each volunteer executed a program that identified URLs and attachments in the messages in their mailboxes, as well as the time the messages were received. This data was cryptographically hashed to protect the privacy of the users. By analyzing the message content and headers, the data was restricted to include only messages which had been forwarded at least one time, thereby eliminating most postings to mailing lists and more closely approximating true inter-personal information spreading behavior. The median number of messages in a mailbox in the sample was 2200, indicating that many users keep a substantial portion of their email correspondence. Although some messages may have been lost when users deleted them, it was assumed that a majority of messages containing useful information had been retained.

Figure 7 shows a histogram of how many users had received each of the 3401 attachments and 6370 URLs. The distribution shows that only a small fraction (5% of attachments and 10% of URLs) reached more than 1 recipient. Very few (41 URLs and 6 attachments) reached more than 5 individuals, a number which, in a sample of 40, starts to resemble an outbreak. In follow-up discussions with the study subjects, the content and significance of most of these messages was identified. 14 of the URLs were advertisements attached to the bottom of an email by free email services such as Yahoo and MSN. These are in a sense viral, because the sender is sending them involuntarily. It is this viral strategy that was responsible for the rapid buildup of the Hotmail free email service user base. 10 URLs pointed to internal HP project or personal pages, 3 URLs were for external commercial or personal sites, and the remaining 14 could not be identified.

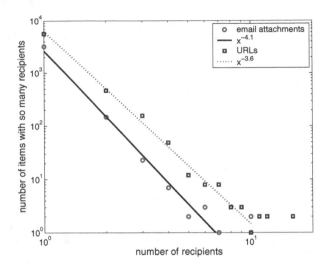

Fig. 7. Number of people receiving URLs and attachments

The next portion of the analysis analyzed the effect of decay in the transmission probability on the email graph at HP Labs. The graph was constructed from recorded logs of all incoming and outgoing messages over a period of 3 months. The graph has a nearly power-law out degree distribution, shown in Fig. 8, including both internal and external nodes. Because all of the outgoing and incoming contacts were recorded for internal nodes, their in and out degrees were higher than for the external nodes for which we could only record the email they sent to and received from HP Labs. A graph with the internal and external nodes mixed (as in [46]) was used to specifically demonstrate the effect of a decay on the spread of email in a power-law graph.

The spread of a piece of information was simulated by selecting a random initial sender to infect and following the email log containing 120,000 entries involving over 7,000 recipients in the course of a week. Every time an infective individual (one willing to transmit a particular piece of information) was recorded as sending an email to someone else, they had a constant probability p of infecting the recipient. Hence individuals who email more often have a higher probability of infecting. It is also assumed that an individual remains infective for a period of 24 hours.

Next a decay was introduced in the one-time transmission probability p_{ij} as $p\, d_{ij}^{-1.75}$, where d_{ij} is the distance in the organizational hierarchy between individuals i and j. The exponent roughly corresponds to the decay in similarity between homepages shown in Fig. 5. Here $r_{ij} = p_{ij}\, f_{ij}$, where f_{ij} is the frequency of communication between the two individuals, obtained from the email logs. The decay represents the fact that individuals closer together in the organizational hierarchy share more common interests. Individuals have a distance of one to their immediate superiors and subordinates and to those they share a superior

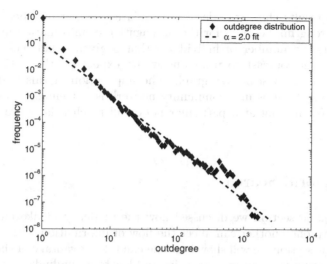

Fig. 8. Outdegree distribution for all senders (224,514 in total) sending email to or from the HP Labs email server over the course of 3 months. The outdegree of a node is the number of correspondents the node sent email to.

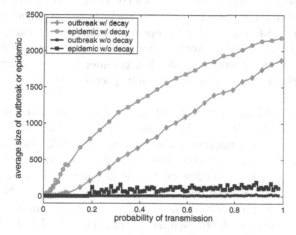

Fig. 9. Average outbreak and epidemic size as a function of the transmission probability p. The total number of potential recipients is 7119.

with. The distance between someone within HP labs and someone outside of HP labs was set to the maximum hierarchical distance of 8.

Figure 9 shows the variation in the average outbreak size, and the average epidemic size (chosen to be any outbreak affecting more than 30 individuals). Without decay, the epidemic threshold falls below $p = 0.01$. With decay, the threshold is set back to $p = 0.20$ and the outbreak epidemic size is limited to about 50 individuals, even for $p = 1$.

As these results show, the decay of similarity among members of a social group has strong implications for the propagation of information among them. In particular, the number of individuals that a given email message reaches is very small, in contrast to what one would expect on the basis of a virus epidemic model on a scale free graph. The implication of this finding is that merely discovering hubs in a community network is not enough to ensure that information originating at a particular node will reach a large fraction of the community.

4 Small World Search

In the preceding section we discussed how the tendency of like individuals to associate with one another can affect the flow of information within an organization. In this section we will show how one can take advantage of the very same network structure to navigate social ties and locate individuals.

The observation that any two people in the world are most likely linked by a short chain of acquaintances, known as the "small world" phenomenon has been the focus of much research over the last forty years [47–50]. In the 1960's and 70's, participants in small world experiments successfully found paths from Nebraska to Boston and from Los Angeles to New York. In an experiment in 2001 and 2002, 60,000 individuals were able to repeat the experiment using email to form chains with just four links on average across different contents [51]. The small world phenomenon is currently exploited by commercial networking services such as LinkedIn, Friendster, and Spoke[1] to help people network, for both business and social purposes.

The existence of short paths is not particularly surprising in and of itself. Although many social ties are "local" meaning that they are formed through one's work or place of residence, Watts and Strogatz [52] showed that it takes only a few "random" links between people of different professions or location to create short paths in a social network and make the world "small". In addition, Pool and Kochen [53] have estimated that an average person has between 500 and 1,500 acquaintances. Ignoring for the moment overlap in one's circle of friends, one would have $1,000^2$ or $1,000,000$ friends of friends, and $1,000^3$ or one billion friends-of-friends-of-friends. This means that it would take only 2 intermediaries to reach a number of people on the order of the population of the entire United States.

Although the existence of short paths is not surprising, it is another question altogether how people are able to select among hundreds of acquaintances the correct person to form the next link in the chain. Killworth and Barnard [50] performed the "reverse" experiment to measure how many acquaintances a typical person would use as a first step in a small world experiment. Presented with

---

[1] http://www.linkedin.com/,
http://www.friendster.com/,
http://www.spokesoftware.com/

1,267 random targets, the subjects chose about 210 different acquaintances on average, based overwhelmingly on geographic proximity and similarity of profession to the targets.

Recently, mathematical models have been proposed to explain why people are able to find short paths. The model of Watts, Dodds, and Newman [54] assumes that individuals belong to groups that are embedded hierarchically into larger groups. For example an individual might belong to a research lab, that is part of an academic department at a university, that is in a school consisting of several departments, that is part of a university, that is one of the academic institutions in the same country, etc. The probability that two individuals have a social tie to one another is proportional to $\exp^{-\alpha h}$, where h is the height of their lowest common branching point in the hierarchy.

The decay in linking probability means that two people in the same research laboratory are more likely to know one another than two people who are in different departments at a university. The model assumes a number of separate hierarchies corresponding to characteristics such as geographic location or profession. In reality, the hierarchies may be intertwined, for example professors at a university living within a short distance of the university campus, but for simplicity, the model treats them separately.

In numerical experiments, artificial social networks were constructed and a simple greedy algorithm was performed where the next step in the chain was selected to be the neighbor of the current node with the smallest distance along any dimension. At each step in the chain there is a fixed probability, called the attrition rate, that the node will not pass the message further. The numerical results showed that for a range of the parameter α and number of attribute dimensions, the networks are "searchable", meaning that a minimum fraction of search paths find their target.

Kleinberg [55,56] posed a related question: in the absence of attrition, when does the length of the chains scale in the same way as the average shortest path. Unlike the study of Watts. et al., there is no attrition - all chains run until completion, but need to scale as the actual shortest path in the network does. In the case of a small world network, the average shortest path scales as $\ln(N)$, where N is the number of nodes. Kleinberg proved that a simple greedy strategy based on geography could achieve chain lengths bounded by $(\ln N)^2$ under the following conditions: nodes are situated on an m-dimensional lattice with connections to their $2m$ closest neighbors and additional connections are placed between any two nodes with probability $p \sim r^{-m}$, where r is the distance between them. Since in the real world our locations are specified primarily by two dimensions, longitude and latitude, the probability is inversely proportional to the square of the distance. A person should be four times as likely to know someone living a block away, than someone two city blocks away. However, Kleinberg also proved that if the probabilities of acquaintance do not follow this relationship, nodes would not be able to use a simple greedy strategy to find the target in polylogarithmic time.

The models of both Watts et al. and Kleinberg show that the probability of acquaintance needs to be related to the proximity between individuals' attributes in order for simple search strategies using only local information to be effective. Below we describe experiments empirically testing the assumptions and predictions of the proposed two models.

4.1 Method

In order to test the above hypothesis, Adamic and Adar [8] applied search algorithms to email networks derived from the email logs at HP Labs already described in Sect. 2. A social contact was defined to be someone with whom an individual had exchanged at least 6 emails each way over the period of approximately 3 months. The bidirectionality of the email correspondence guaranteed that a conversation had gone on between the two individuals and hence that they are familiar with one another.

Imposing this constraint yielded a network of 436 individuals with a median number of 10 acquaintances and a mean of 13. The degree distribution, shown in Fig. 10, is highly skewed with an exponential tail. This is in contrast to the raw power-law email degree distribution, used in Sect. 3 and shown in Fig. 8, pertaining to both internal and external nodes and possessing no threshold in email volume. A scale free distribution in the raw network arises because there are many external nodes emailing just one individual inside the organization, and there are also some individuals inside the organization sending out announcements to many people and hence having a very high degree. However, once we impose a higher cost for maintaining a social contact (that is, emailing that contact at least six times and receiving at least as many replies), then there are few individuals with many contacts.

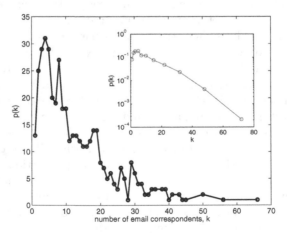

Fig. 10. Degree distribution in the HP Labs email network. Two individuals are linked if they exchanged at least 6 emails in either direction. The inset shows the same distribution, but on a semilog scale, to illustrate the exponential tail of the distribution

4.2 Simulating Milgram's Experiment on an Email Network

The resulting network, consisting of regular email patterns between HP Labs employees, had 3.1 edges separating any two individuals on average, and a median of 3. Simulations were performed on the network to determine whether members of the network would be able to use a simple greedy algorithm to locate a target. In this simple algorithm, each individual can use knowledge only of their own email contacts, but not their contacts' contacts, to forward the message.

Three different strategies were tested, at each step passing the message to the contact who is either

- best connected
- closest to the target in the organizational hierarchy
- sitting in closest physical proximity to the target

The first strategy selects the individual who is more likely to know the target by virtue of the fact that he/she knows so many people. It has been shown [57], that this is an effective strategy in power-law networks with exponents close to 2 (the case of the unfiltered HP Labs email network), but that it performs poorly in graphs with a Poisson degree distribution that has an exponential tail. Since the distribution of contacts in the filtered HP network was not power-law, the high degree strategy was not expected to perform well, and this was verified through simulation. The median number of steps required to find a randomly chosen target from a random starting point was 17, compared to the three steps in the average shortest path. Even worse, the average number of steps is 40. This discrepancy between the mean and the median is a reflection of the skewness of the distribution: a few well connected individuals and their contacts are easy to find, but some individuals who do not have many links and are not connected to highly connected individuals are difficult to locate using this strategy.

The second strategy consisted of passing the message to the contact closest to the target in the organizational hierarchy. The strategy relies on the observation, illustrated in Figs. 11 and 13 that individuals closer together in the organizational hierarchy are more likely to email with one another. Figure 12 illustrates such a search, labelling nodes by their hierarchical distance (h-distance) from the target. The h-distance is computed as follows: a node has distance one to their manager and to everyone they share a manager with. Distances are then recursively assigned, so that each node has h-distance 2 to their first neighbor's neighbors, and h-distance 3 to their second neighbor's neighbors, etc. A simple greedy strategy using information about the organizational hierarchy worked extremely well. The median number of steps was only 4, close to the median shortest path of 3. With the exception of one individual, whose manager was not located on site, and who was consequently difficult to locate, the mean number of steps was 4.7, meaning that not only are people typically easy to find, but nearly everybody can be found in a reasonable number of steps.

In the original experiment by Milgram the completed chains were divided between those that reached the target through his professional contacts and those that reached him through his hometown. On average those that relied on

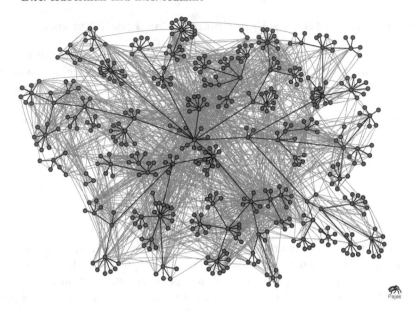

Fig. 11. Email communications within HP Labs (gray lines) mapped onto the organizational hierarchy (black lines). Note that email communication tends to "cling" to the formal organizational chart.

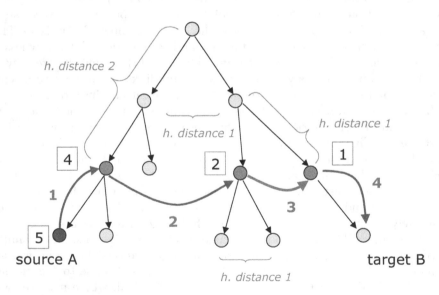

Fig. 12. Example illustrating a search path using information about the target's position in the organizational hierarchy to direct a message. Numbers in the square give the h-distance from the target.

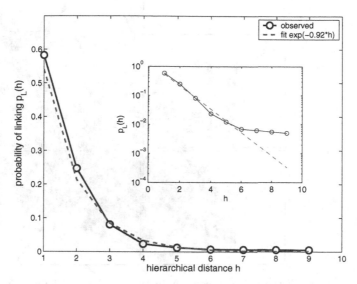

Fig. 13. Probability of linking as a function of the separation in the organizational hierarchy. The exponential parameter $\alpha = 0.92$, in the searchable range according to the model of Watts et al.[54]

geography took 1.5 steps longer to reach the target, a difference found to be statistically significant. In the words of Travers and Milgram [48], the following seemed to occur: "Chains which converge on the target principally by using geographic information reach his hometown or the surrounding areas readily, but once there often circulate before entering the target's circle of acquaintances. There is no available information to narrow the field of potential contacts which an individual might have within the town."

Performing the small world experiment on the HP email network using geography produced a similar result, in that geography could be used to find most individuals, but was slower, taking a median number of 7 steps, and a mean of 12. Figure 14 shows the email correspondence mapped onto the physical layout of the buildings. Individuals' locations are given by their building, the floor of the building, and the nearest building post (for example "H15") to their cubicle. The distance between two cubicles was approximated by the "street" distance between their posts (for example "A3" and "C10" would be $(C - A) \times 25' + (10 - 3) \times 25' = 2 \times 25' + 7 \times 25' = 225$ feet apart). Adding the x and y directions separately reflects the interior topology of the buildings where one navigates perpendicular hallways and cannot traverse diagonally. If individuals are located on different floors or in different buildings, the distance between buildings and the length of the stairway are factored in.

The general tendency of individuals in close physical proximity to correspond holds: over 87% percent of the 4000 email links are between individuals on the same floor, and overall individuals closer together are more likely to correspond. Still, individuals maintain disproportionately many far-flung contacts while not

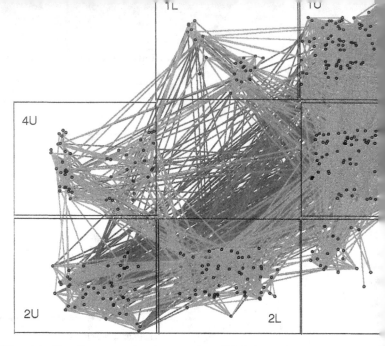

Fig. 14. Email communications within HP Labs mapped onto appro... location based on the nearest post number and building given for each box represents a different floor in a building. The lines are color code... physical distance between the correspondents: red for nearby individu... away contacts.

getting to know some of their close-by neighbors. The relationship ability of acquaintance and cubicle distance r between two indiv... in Fig. 15, is well-fitted by a $1/r$ curve. However, Kleinberg has s... optimum relationship in two dimensional space is $1/r^2$ - a stro... probability of acquaintance than the $1/r$ observed.

In the case of HP Labs, the geometry may not be quite two... because it is complicated by the particular layout of the buildin... optimum relationship may lie between $1/r$ and $1/r^2$. In any case $1/r$ probability of linking shows a tendency consistent with Milg... tions about the original small world experiment. At HP Labs, be... constraints, re-organizations, and personal preferences, employees... be removed from some of the co-workers they interact with. This hi... strategy relying solely on geography, because one might get physic... to the target, but still need a number of steps to find an individual... with them.

Figure 16 shows a histogram of chain lengths resulting from... each of the three strategies. It shows the clear advantage of usin...

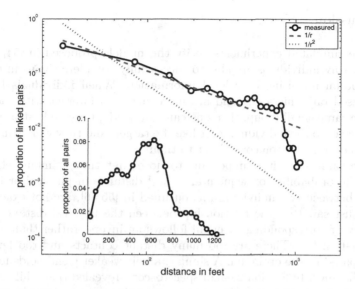

Fig. 15. Probability of two individuals corresponding by email as a function of the distance between their cubicles. The inset shows how many people in total sit at a given distance from one another.

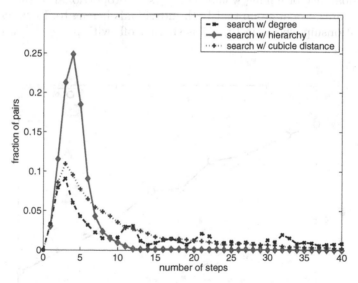

Fig. 16. Results of search experiments utilizing either knowledge of the target's position in the organizational hierarchy or the physical location of their cubicle.

position in organizational hierarchy as opposed to his/her cubicle location to pass a message through one's email contact. It also shows that both searches using information about the target outperform a search relying solely on the connectivity of one's contacts.

4.3 Discussion

The above simulated experiments verify the models proposed in [54] and [55] to explain why individuals are able to successfully complete chains in the small world experiments using only local information. When individuals belong to groups based on a hierarchy and are more likely to interact with individuals within the same small group, then one can safely adopt a greedy strategy - pass the message onto the individual most like the target, and they will be more likely to know the target or someone closer to them.

At the same time it is important to note that the optimum relationship between the probability of acquaintance and distance in physical or hierarchical space between two individuals, as outlined in [55,56], are not exactly satisfied. We just saw that the relationship between the physical distance and the probability of corresponding by email follows an inverse rather than an inverse square relationship. There are too many distant contacts and too few nearby ones compared to the optimum. A similar, albeit weaker trend holds for organizational distance. In Section 2 email spectroscopy revealed that while collaborations mostly occurred within the same organizational unit, they also frequently bridged different parts of the organization or broke up a single organizational unit into noninteracting subgroups. The optimum relationship derived in [56] for the probability of linking would be inversely proportional to the size of the smallest organizational group that both individuals belong to. However, the observed relationship, shown in Fig. 17 is slightly off, with $p \sim g^{-3/4}$, g being the group size.

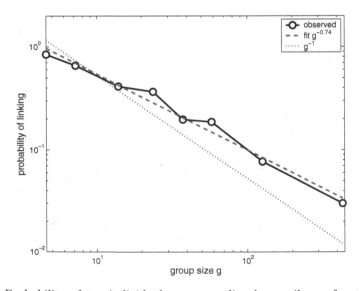

Fig. 17. Probability of two individuals corresponding by email as a function of the size of the smallest organizational unit they both belong to. The optimum relationship derived in [56] is $p \sim g^{-1}$, g being the group size. The observed relationship is $p \sim g^{-3/4}$.

Overall, the results of the email study are consistent with the model of Watts et al. [54]. This model does not require the search to find near optimum paths, but simply determines when a network is "searchable", meaning that fraction of messages reach the target given a rate of attrition. The relationship found between separation in the hierarchy and probability of correspondence, shown in Fig. 13, is well within the searchable regime identified in the model.

The study of Adamic and Adar is a first step, validating these models on a small scale. The email study gives a concrete way of observing how the small world chains can be constructed. Using a very simple greedy strategy, individuals across an organization could reach each other through a short chain of coworkers. It is quite likely that similar relationships between acquaintance and proximity (geographical or professional) hold true in general, and therefore that small world experiments succeed on a grander scale for the very same reasons.

5 Conclusion

In this chapter we reviewed three studies of information flow in social networks. The first developed a method of analyzing email communication automatically to expose communities of practice and their leaders. The second showed that the tendency of individuals to associate according to common interests influences the way that information spreads throughout a social group. It spreads quickly among individuals to whom it is relevant, but unlike a virus, is unable to infect a population indiscriminately. The third study showed why small world experiments work - how individuals are able to take advantage of the structure of social networks to find short chains of acquaintances. All three studies relied on email communication to expose the underlying social structure, which previously may have been difficult and labor-intensive to obtain. We expect that these findings are also valid with other means of social communication, such as verbal exchanges, telephony and instant messenger systems.

Acknowledgements

We would like to thank Eytan Adar, Joshua Tyler, and TJ Giuli for their comments and suggestions.

References

1. B. Wellman, "Computer networks as social networks," *Science*, vol. 293, pp. 2031–34, 2002.
2. S. Whittaker and C. Sidner, "Email overload: exploring personal information management of email," in *Proceedings of CHI'96 Conference on Computer Human Interaction*, (New York), pp. 276–283, Logos Verlag, 21996.
3. R. Guimerà, L. Danon, A. Díaz-Guilera, F. Giralt, and A. Arenas, "Self-similar community structure in organizations," *Physical Review E*, vol. 65, p. 065103, 2003.

4. J. R. Tyler, D. M. Wilkinson, and B. A. Huberman, "Email as spectroscopy: Automated discovery of community structure within organizations," in *Proceedings of the International Conference on Communities and Technologies*, (Netherlands), Kluwer Academic Publishers, 2003.
5. J.-P. Eckmann, E. Moses, and D. Sergi, "Dialog in e-mail traffic." http://xyz.lanl.gov/abs/cond-mat/0304433", 2003.
6. M. F. Schwartz and D. C. M. Wood, "Discovering shared interests among people using graph analysis," *Communications of the ACM*, vol. 36, no. 8, pp. 78–89, 1993.
7. F. Wu, B. A. Huberman, L. A. Adamic, and J. Tyler, "Information flow in social groups," *Physica A*, 2004. (to appear).
8. L. A. Adamic and E. Adar, "How to search a social network." submitted for publication, http://www.hpl.hp.com/shl/papers/socsearch, 2002.
9. W. G. Ouchi, "Markets, bureaucracies, and clans," *Administrative Science Quarterly*, vol. 25, pp. 129–141, 1980.
10. P. Blau and W. Scott, *Formal organizations. A comparative approach*. Lndn. Routledge & Kegan Paul, 1963.
11. R. S. Burt, "Models of network structure," *Annual Review of Sociology*, vol. 6, pp. 79–141, 1980.
12. S. Wasserman and K. Faust, *Social network analysis*. Cambridge: Cambridge University Press, 1994.
13. W. R. Scott, *Organizations: Rational, Natural, and Open Systems*. Englewood Cliffs, NJ: Prentice-Hall, 1992.
14. B. Huberman and T. Hogg, "Communities of practice: Performance and evolution," *Computational and Mathematical Organization Theory*, vol. 1, pp. 73–92, 1995.
15. M. Crozier, *The Bureaucratic Phenomenon*. Chicago: University of Chicago Press, 1964.
16. D. Crane, *Invisible Colleges: Diffusion of Knowledge in Scientific Communities*. Chicago: University of Chicago Press, 1972.
17. J. Lave and E. Wenger, *Situated Learning: Legitimate Peripheral Participation*. Cambridge University Press, 1991.
18. N. Ducheneaut and V. Bellotti, "A study of email work processes in three organizations." to appear in the Journal of CSCW.
19. L. A. Adamic and E. Adar, "Friends and neighbors on the web," *Social Networks*, vol. 25, no. 3, pp. 211–230, 2003.
20. T. Allen, *Managing the Flow of Technology*. Cambrige, MA: MIT Press, 1984.
21. P. Hinds and S. Kiesler, "Communication across boundaries: Work, structure, and use of communication technologies in a large organization," *Organization Science*, vol. 6, no. 4, pp. 373–393, 1995.
22. L. Garton, C. Haythornwaite, and B. Wellman, "Studying on-line social networks," *Journal of Computer Mediated Communication*, vol. 3, no. 1, 1997.
23. H. Alani, K. O'Hara, and N. Shadbolt, "Ontocopi: Methods and tools for identifying communities of practice, intelligent information processing conference," in *IFIP World Computer Congress (WCC)*, 2002.
24. M. Girvan and M. Newman, "Community structure in social and biological networks," *Proc. Natl. Acad. Sci. USA*, vol. 99, pp. 8271–8276, 2002.
25. D. Wilkinson and B. A. Huberman, "A method for finding communities of related genes." *Proc. Natl. Acad. Sci. USA*, Feb. 2, 2004, 10.1073/pnas.0307740100.
26. L. Freeman, "A set of measures of centrality based on betweenness," *Sociometry*, vol. 40, pp. 35–41, 1977.

27. U. Brandes, "A faster algorithm for betweenness centrality," *Journal of Mathematical Sociology*, vol. 25, no. 2, pp. 163–177, 2001.
28. M. Newman, "Who is the best connected scientist? a study of scientific coauthorship networks," *Phys. Rev. E*, vol. 64, p. 016131, 2001.
29. T. M. J. Fruchterman and E. M. Reingold, "Graph drawing by force-directed placement," *Software - Practice and Experience*, vol. 21, no. 11, pp. 1129–1164, 1991.
30. M. K. Sparrow, "The application of network analysis to criminal intelligence: An assessment of the prospects," *Social Networks*, vol. 13, pp. 251–274, 1991.
31. V. E. Krebs, "Uncloaking terrorist networks," *First Monday*, vol. 7, April 2002.
32. M. E. J. Newman, "Scientific collaboration networks. ii. shortest paths, weighted networks, and centrality," *Phys. Rev. E*, vol. 64, p. 016132, 2001.
33. M. E. J. Newman, "Fast algorithm for detecting community structure in networks." http://aps.arxiv.org/abs/cond-mat/0309508/, 2003.
34. F. Wu and B. A. Huberman, "Discovering communities in linear time: a physics approach," *Europhysics Letters*, 2004. to appear.
35. Z. Dezso and A.-L. Barabasi, "Halting viruses in scale-free networks," *Phys. Rev. E*, vol. 65, p. 055103, 2002.
36. R. Pastor-Satorras and A. Vespignani, "Epidemic spreading in scale-free networks," *Phys. Rev. Lett.*, vol. 86, no. 14, pp. 3200–3203, 2001.
37. M. E. J. Newman, S. F., and J. Balthrop, "Email networks and the spread of computer viruses," *Phys. Rev. E*, vol. 66, p. 035101, 2002.
38. V. M. Eguiluz and K. Klemm, "Epidemic threshold in structured scale-free networks," *Phys. Rev. Lett.*, vol. 89, p. 108701, 2002.
39. A. Vazquez, M. Boguna, Y. Moreno, R. Pastor-Satorras, and A. Vespignani, "Topology and correlations in structured scale-free networks," *Physical Review E*, vol. 67, p. 046111, 2003.
40. M. E. J. Newman, "Assortative mixing in networks," *Phys. Rev. Lett*, vol. 89, p. 208701, 2002.
41. P. Lazarsfeld and R.K.Merton, "Friendship as a social process: A substantive and methodological analysis," in *Freedom and Control in Modern Society* (M. Berger, T. Abel, and C. Page, eds.), New York: Van Nostrand, 1954.
42. J. Touhey, "Situated identities, attitude similarity, and interpersonal attraction," *Sociometry*, vol. 37, pp. 363–374, 1974.
43. S. Feld, "The focused organization of social ties," *American Journal of Sociology*, vol. 86, pp. 1015–1035, 1981.
44. M. E. J. Newman, S. H. Strogatz, and D. J. Watts, "Random graphs with arbitrary degree distribution and their applications," *Phys. Rev. E*, vol. 64, p. 026118, 2001.
45. M. Newman, "The spread of epidemic disease on networks," *Phys. Rev. E*, vol. 66, p. 016128, 2002.
46. H. Ebel, L.-I. Mielsch, and S. Bornholdt, "Scale-free topology of e-mail networks," *Phys. Rev. E*, vol. 66, p. 035103, 2002.
47. S. Milgram, "The small-world problem," *Psychology Today*, vol. 1, pp. 62–67, 1967.
48. J. Traver and S. Milgram, "An experimental study of the small world problem," *Sociometry*, vol. 32, pp. 425–443, 1969.
49. C. C. Lundberg, "Patterns of acquaintanceship in society and complex organization: A comparative study of the small world problem," *Pacific Sociological Review*, vol. 18, pp. 206–222, 1975.
50. P. Killworth and H. Bernard, "Reverse small world experiment," *Social Networks*, vol. 1, pp. 159–192, 1978.
51. P. S. Dodds, R. M., and D. J. Watts, "An experimental study of search in global social networks," *Science*, vol. 301, pp. 827–829, 2003.

52. D. J. Watts and S. H. Strogatz, "Collective dynamics of small-world networks," *Nature*, vol. 393, pp. 440–442, 1998.
53. I. Pool and M. Kochen, "Contacts and influence," *Social Networks*, vol. 1, pp. 5–51, 1978.
54. D. J. Watts, P. S. Dodds, and M. E. J. Newman, "Identity and search in social networks," *Science*, vol. 296, pp. 1302–1305, 2002.
55. J. Kleinberg, "Navigation in a small world," *Nature*, vol. 406, 2000.
56. J. Kleinberg, "Small-world phenomena and the dynamics of information," *Advances in Neural Information Processing Systems (NIPS)*, vol. 14, 2001.
57. L. A. Adamic, R. M. Lukose, A. R. Puniyani, and B. A. Huberman, "Search in power-law networks," *Phys. Rev. E*, vol. 64, p. 046135, 2001.

Emergence of Complexity in Financial Networks

Guido Caldarelli[1], Stefano Battiston[2], Diego Garlaschelli[3], and
Michele Catanzaro[1]

[1] INFM UdR Roma1 Dipartimento di Fisica, Università La Sapienza, P.le Moro 5, 00185 Roma, Italy.
[2] Laboratoire de Physique, Statistique ENS, 24 rue Lhomond, 75005 Paris France.
[3] INFM UdR Siena and Dipartimento di Fisica, Università di Siena, Via Roma 56, 53100 Siena Italy.

Abstract. We present here a brief summary of the various possible applications of network theory in the field of finance. Since we want to characterize different systems by means of simple and universal features, graph theory could represent a rather powerful methodology. In the following we report our activity in three different subfields, namely the board and director networks, the networks formed by prices correlations and the stock ownership networks. In most of the cases these three kind of networks display scale-free properties making them interesting in their own. Nevertheless, we want to stress here that the main utility of this methodology is to provide new measures of the real data sets in order to validate the different models.

1 Introduction

The study of topological properties of networks has recently received great attention [1]. In particular it has been shown that many natural systems display an unexpected amount of correlation [2] with respect to traditional models [3]. Graphs are mathematical objects formed by vertices connected by edges. An important characterization of a graph is given by the degree of vertices, that is the number of edges per vertex. In an Erdős-Rényi random graph the degree distribution has a poissonian form, whereas in many cases of interest ranging from the WWW [4,5] to the Internet [6,7] to social networks [8] the degree is power law distributed. The scale-free behavior can be reproduced by two classes of models. The growth models where new sites enter and choose a site to be linked with through the "rich gets richer" rule of preferential attachment [9]. Interestingly, there are also other ways to reproduce such scale invariance by means of static models where "good gets richer" [10]). Growth plays a major role in technological networks like Internet and WWW, while in financial networks the size of the system is basically unchanged, even if the system is subject to a process of evolution in which some old nodes are replaced by new ones and some links are rewired.

Like the distribution of the connectivity degree of vertices, assortativity and clustering are among the most interesting quantities to consider for classifying and describing these complex networks. The average degree $k_{nn}(k)$ of the neighboring site of a site of degree k has been introduced in [7] in order to measure assortativity. k_{nn} increases if nodes are correlated by degree (assortative net-

G. Caldarelli, S. Battiston, D. Garlaschelli, and M. Catanzaro, Emergence of Complexity in Financial Networks, Lect. Notes Phys. **650**, 399–423 (2004)
http://www.springerlink.com/

works). It decreases if they are anti-correlated (disassortative networks). The tendency for nodes in a social network to form connections preferentially to others similar to them [11] has been proposed as the key ingredient for the formation of communities in networks [12,13]. It is possible to distinguish the technological networks, where instead, the behavior is rather degree-disassortative, so that vertices tend to be linked to others different from them. Despite the relative simplicity of such behavior, few models [14–16] of network growth are able to reproduce the formation of communities and no one explains the difference between social and technological networks.

The clustering coefficient c for a given site i gives the probability that two nearest neighbors of vertex i are also neighbors each other [2]. The average clustering coefficient, $c(k)$ for sites whose degree is k measures the tendency to form cliques where each nearest neighbor of a node (with degree k) is connected to each other. In real-world scale-free networks this usually decreases with a power-law $c(k) \propto k^{\psi}$ because hubs tend to play the role of connections between separate clusters in the graph, i.e. clusters that have few other interconnections than the ones passing through the hub. Then the high degree node tends to have low clustering coefficient.

We focus in this paper on three different kinds of complex networks with relevance in finance: the network of boards and directors of the largest corporations, the network of stock price correlation and the network of shareholders in the stock market. For each network we report the more relevant topological properties and we present models of network formation accounting for some of the observed properties.

2 The Board and Director Networks

We start with the network of boards and directors, a complex network in finance which is also a social network.

This can be represented as a bipartite graph where two classes of nodes are present (boards and directors) and an edge is always drawn between nodes belonging to different classes. In particular, a director is linked to a company if he serves on the board of directors of the company. Such a graph can be projected onto the board network and onto the director network, where two boards (directors) are connected by a weighted edge, the weight being proportional to the number of directors (boards) they share. In fact, it occurs very often that two boards share some of their directors, and in this case they are said to be *interlocked*.

An example of board network is shown in Fig. 1: nodes represent boards of directors, two boards are connected by an edge if they are interlocked. The network represents the boards 1 degree of separation (in term of edges) away from the board of Chase Manhattan Bank.

Boards of directors of corporations make decisions about the long-term strategy, such decisions having considerable impact on the economic performance of the corporation and collectively on the economy of a country.

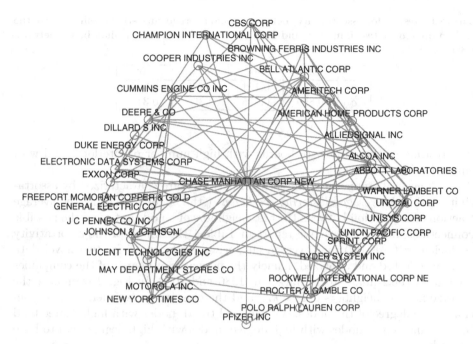

Fig. 1. The network of boards 1 degree of separation away from Chase Manhattan Bank's board.

Because large corporations' boards are organized in a networks leading the economy of a country, some issues are particularly relevant about these networks: what the topological properties are and if they are similar in different countries. What these topological properties mean about the corporate directorate elite as a leading class. What mechanism of network formation can explain the observed features. What effect the network structure has on the process of decision making.

2.1 Topological Properties of Board and Director Networks

A number of recent works have been devoted to the study of the topological properties of the board and director networks. Davis et al. [17] have studied the network of the boards of Fortune 1000 in 1999 and have shown that both the director network and the board network have Small World properties.

Newman et al. [18] have applied on the same data set a random graph model showing that using the generating function method, it is possible to reproduce very accurately the degree distribution of the director network. On the contrary, their model fails in predicting the degree distribution of the board network. In fact the director network turns out to be *assortative* as observed commonly in social networks, meaning that directors with high (low) degree tend to be connected to directors with high (low) degree. As a consequence even if the random graph model predicts the right degree distribution for the director network it

Table 1. Results for assortativity coefficient r and average clustering coefficient $\bar{c}$ in the Italian director network in 1986 and 2002 (D86, D02)) and the Italian board network (B86, B02)

	D86	D02	B86	B02
r	0.131	0.121	0.250	0.322
$\bar{c}$	0.899	0.915	0.356	0.318

underestimates the number of boards with high number of interlocks and with small number of interlocks.

As a general empirical finding, social networks are characterized by assortativity and high average *clustering coefficient* (the latter measuring the average fraction of connection between the first neighbors of a node out of all the possible connections among them). We report in Table 1 the values of the assortativity coefficient r (see [19]) and average clustering coefficient $\bar{c}$ for two novel data sets we collected and analyzed, namely the networks of boards of the companies quoted on the Italian stock Market in 1986 and 2002. In Fig. 2 we report the average nearest neighbors degree k_{nn} and the clustering coefficient c as a function of the degree of the nodes. As a general trend, nodes with high degree tend to be connected to nodes with high degree, nodes with high degree tend to have low values of c.

Newman et al. [20] have recently argued that the presence of groups or communities in a social network is able to produce alone both assortativity and clustering. They develop a model in which nodes belong to one or more groups and have probability p to be connected to another node of the same group. Instead they are never connected to nodes of groups they do not belong to. If groups have heterogeneous size, than nodes who belong to a small group tend to have low degree and are connected to others in the same group, who also have low degree.

This model explains about 40% of the observed assortativity in the Fortune 1000 network. This means that some additional sociological mechanism is at work, probably the fact that new board members are more likely to be recruited among those who are already connected to some of the current board member.

Some of us have recently proposed a new model [21] to reproduce assortativity in social networks. This is a network growth model and is a generalization of the Barabási-Albert preferential attachment model [9], which is known to produce a scale free network with no clustering and no assortativity. The generalization of that model proposed by Catanzaro et al. allows for growth by addition of new links between old nodes. In detail, at every step of growth:

1. with probability p a new node is wired to an existing one with the Barabási-Albert preferential attachment rule ('rich gets richer').

$$p\frac{k_i}{\sum_{j=1,N} k_j}. \tag{1}$$

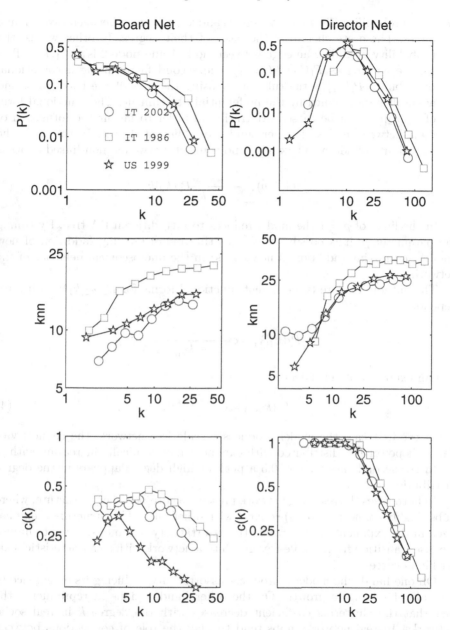

Fig. 2. Comparison of some properties of board networks (**left**) and director networks (**right**) in Fortune 1000 and MIB.

Top: degree distribution. **Middle**: average neighbors degree k_{nn} as a function of the degree of the nodes. **Bottom**: clustering coefficient $c(k)$ as function of the degree of the nodes

2. with probability $(1-p)$ a new edge is added (if absent) between two existing nodes. These are chosen on the basis of their degree. In other words, the probability of adding an edge between node 1 and node 2 is $\tilde{P}(k_1, k_2)$. This can be written as $P_1(k_1)P_2(k_2|k_1)$, the second factor being a conditional probability. $P_1(k_1)$ is the rule for choosing the first of the two nodes, and again it is determined by the preferential attachment. The functional form of $P_2(k_2|k_1)$ can be chosen arbitrarily, in order to favour the formation of links between nodes of either similar or different degree. In this way, the probability of adding a new edge and connecting two old non-linked nodes is

$$(1-p)\frac{k_i}{\sum_{j=1,N} k_j}P_2(k_2|k_1) \tag{2}$$

In the limit of $p=1$ the model reduces to a traditional BA tree. By tuning the parameter p, it is possible to weight the role of growing (addition of new nodes) and mixing (addition of new edges) in the microscopical behavior of the network.

The authors explore two different functional forms for $P_2(k_2|k_1)$: an inverse dependence

$$P_2(k_2|k_1) \propto \frac{1}{|k_1 - k_2| + 1} \tag{3}$$

and an exponential dependence

$$P_2(k_2|k_1) \propto e^{-|k_1-k_2|}. \tag{4}$$

In the first case the model produces a scale free network: the connectivity degree is power law distributed with exponent monotonically increasing with p.

In the second case for $p < 0.5$ a peak at high degree appears in the degree distribution.

In both cases the resulting network presents a core-periphery structure, where hubs (highly connected nodes) connect with other hubs. This structure is emphasized in the exponential case, where the assortativity becomes so large to induce a phase transition from a scale-free graph to a network with a characteristic scale for high degrees.

On one hand, the model reproduces assortativity, reducing its emergence to the role of mixing in growth. On the other hand, it fails in reproducing the fact that the clustering coefficient decreases with the degree k in real social networks. In real networks hubs tend to play the role of connections between separate clusters in the graph, with few links between each other (apart from the ones attached to the hub). Therefore these nodes tend to have low clustering coefficient. In this model, on the other hand, all the hubs are aggregated together. Thus, while producing an assortative network it cannot reproduce a network with $c(k)$ decreasing with k. See [21] for more details.

As a possible extension of such model we suggest that the choice of $P_1(k_1)$ could be modified as follows: when adding a new link between old nodes, the first one of these could be chosen at random, instead that with a preferential attachment rule. With such a choice, high-degree nodes would not be privileged, and the formation of a cluster of hubs might be avoided. Such an attachment rule in the director network could be justified with the assumption that degree is related to prestige as suggested in the sociological literature [17]. It is quite reasonable to assume that individuals would like to be connected to others with similar or higher prestige and to loose connection with individuals with lower prestige. As a result individuals would tend to be connected to others with similar prestige.

As a conclusion, a general model for network formation reproducing all the statistical features observed in social networks is still object of search. We reported some common features of boards and directors networks for US and Italy and we discussed some recent models producing networks with assortativity and high clustering coefficient.

2.2 The Impact of Network Structure on the Decision Making Process

Topological properties reveal interesting features of the boards-directors economical system. For example, it should be noted that both projections of the bipartite graph display a giant connected component containing around 90 % of the nodes. Moreover, while positive assortativity suggests a prestige related dynamics (see above), high clustering coefficient and small world property describe a social system where common "friendships" and close proximity to each other are typical. As well, it is interesting to observe that the Italian network, for which we have two time snapshots, seems to be stationary, as suggested by the substantial invariance of the shape of the degree distribution in the years. A deeper investigation is required to explore the differences (if any) between Italian and US market.

Some recent works have focussed on the influence of the structure of the interlock network on the decisions made by boards. There are essentially two kinds of decisions a board is faced with. *Local* decisions regard topics specific to the board, such as the appointment of a vice president, for which boards can be assumed not to influence each other. Battiston et al. [22] have shown the role of subsets of well connected directors on decisions of this type.

By contrast, *global* decisions concern topics of general interest to the economy such as whether to increase or decrease investments in development or in advertisement, which depend on the belief in economical growth or recession. In these cases, decisions previously made in some boards might influence other boards, through the presence of shared directors.

In a recent model, Battiston et al. [23] investigate the conditions under which a large majority of boards making a same decision can emerge in the network. In their model board directors are engaged in a decision making dynamics based on "herd behavior" and boards influence each other through shared directors.

They find that imitation of colleagues and opinion bias due to the interlock do not trigger an avalanche of identical decisions over the board network, whereas the information about interlocked boards decisions does. There is no need to invoke global public information, nor external driving forces. This model provides a simple endogenous mechanism to explain the fact that boards of the largest corporations of a country can, in the span of a few months, take the same decisions about general topics.

3 Network of Price Correlations

The case of study is given by a network whose vertices are a fixed number of stocks continuously traded at the New York Stock Exchange (NYSE) and the edges are obtained by considering the return cross-correlations. The network is a Minimal Spanning Tree (MST) connecting all the stocks. Spanning trees are subgraphs of a graph which connect all the vertices of the original graph without forming any loop.

It is interesting to note that when the stock correlations are described through this method it is very easy to validate the models of portfolio dynamics. Specifically we show that a simple model of uncorrelated Gaussian return time series and the widespread one-factor model do not reproduce the topological quantities of interest. This last model is the starting point of the Capital Asset Pricing Model [24] that is one of the most widely known models.

3.1 The MST Formation

The topological characterization of the correlation based MST of real data has been already studied in [25]. But in our approach we use a smaller number N of stocks, and number of time records T bigger than N. Our choice is motivated by the request that the correlation matrix be positive definite. When the number of variables is larger than the number of time records the covariance matrix is only positive semi-definite [26]. Moreover, the application of the random matrix theory to the spectral properties of the correlation matrix can be applied only when $T/N > 1$.

We consider the daily price return $r_i(t)$ of asset i on day t. Given a portfolio composed of N assets traded simultaneously in a time period of T trading days, we extract the $N \times N$ correlation matrix. From each correlation coefficient $\rho_{i,j}$ we computed a metric distance $d_{i,j} = \sqrt{2(1 - \rho_{i,j})}$ between asset i and j through this relation [27,28]. The distance matrix is then used to determine the MST connecting all the assets. We start with the two nearest sites (with respect to the distance we just defined) and we connect them by a link. We then consider the second shorter distance and we connect the sites involved. This last link may happen to connect two new sites or it may link one of the previous two sites with a new one. From the third shortest distance on, however, the link corresponding to that distance may form a loop together with the previous links. We establish a new link only if this *does not* form any loop with previous links. Iterating this procedure of establishing links among vertices we obtain the minimal spanning tree of the network. The method of constructing the MST linking N objects is known in multivariate analysis as the nearest neighbor single linkage cluster algorithm [26].

3.2 The Data and the One Factor Model

The data set used here consists of daily closure prices for 1071 stocks traded at the NYSE and continuously present in the 12-year period 1987-1998 (3030 trading days). It is worth noting that the ratio $T/N \simeq 2.83$ is significantly larger than one. With our choice the correlation matrix is positive definite and the theoretical results of the random matrix theory are valid. Figure 3 shows the

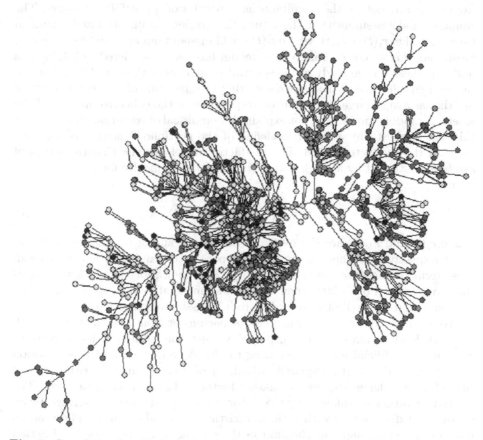

Fig. 3. Correlation based minimal spanning tree of real data from daily stock returns of 1071 stocks for the 12-year period 1987-1998 (3030 trading days). The node color is based on Standard Industrial Classification system. The correspondence is: red for mining - cyan for construction - yellow for manufacturing - green for transportation, communications, electric, gas and sanitary services - magenta for wholesale trade - black for retail trade - purple for finance, insurance and real estate - orange for service industries - light blue for public administration.

graph of the MST for the real data. The spatial coordinates have no meaning here, the graph is just represented in a planar layout using the Pajek software [29]. Different colors of the nodes correspond to the main industrial sector of each firm according to the Standard Industrial Classification system [32] for the main industry sector of each firm and the correspondence is reported in the figure caption. Regions corresponding to different sectors are clearly seen. Examples are clusters of stocks belonging to the financial sector (purple), to the transportation, communications, electric gas and sanitary services sector (green) and to the mining sector (red). The mining sector stocks are observed to belong to two subsectors one containing oil companies (located on the right side of the figure) and one containing gold companies (left side of the figure).

The empirical MST of real data can be compared with the results obtained from simple models of the simultaneous dynamics of a portfolio of assets. The simplest model assumes that the return time series are uncorrelated Gaussian time series, i.e. $r_i(t) = \epsilon_i(t)$, where $\epsilon_i(t)$ are Gaussian random variables with zero mean and unit variance. This type of model has been considered in [33,34] as a null hypothesis in the study of the spectral properties of the correlation matrix. In the cited references it has been shown that the spectrum of the real correlation matrix has a very large eigenvalue corresponding to the collective motion of the assets. A random model does not explain this empirical observation and therefore this fact clarifies why a better modeling of the portfolio dynamics is obtained by using the one-factor model. The one-factor model assumes that the return of assets is controlled by a single factor (or index). Specifically for any asset i we have

$$r_i(t) = \alpha_i + \beta_i r_M(t) + \epsilon_i(t), \tag{5}$$

where $r_i(t)$ and $r_M(t)$ are the return of the asset i and of the market factor at day t respectively, α_i and β_i are two real parameters and $\epsilon_i(t)$ is a zero mean noise term characterized by a variance $\sigma_{\epsilon_i}^2$. Our choice for the market factor is the Standard & Poor's 500 index and we assume that $\epsilon_i = \sigma_{\epsilon_i} w$, where w is a random variable distributed according to a Gaussian distribution.

We estimate the model parameters for each asset from real time series with ordinary least squares method [24] and we use the estimated parameters to generate an artificial market according to (5). A consequence of this equation is that the variance (the squared volatility) of asset i can be written as the sum of a term depending on the market factor and an idiosyncratic term. The fraction of variance explained by the factor r_M is approximately described by an exponential distribution with a characteristic scale of about 0.16. The random model can be considered as the limit of the one factor model when the fraction of variance explained by the factor goes to zero.

3.3 Results of the Models

In the MST obtained with the random model few nodes have a degree larger than few units. This implies that the MST is composed by long files of nodes. These

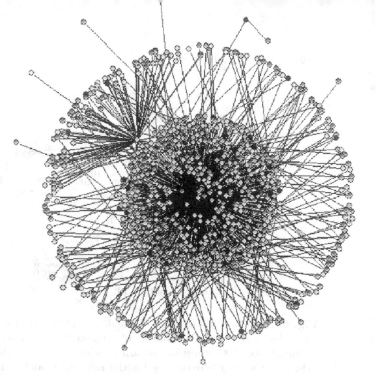

Fig. 4. Correlation based minimal spanning tree of a numerical simulation of factor model. The color codes are those used in figure 3.

files join at nodes of connectivity equal to few units. The MST obtained w one-factor model is very different from the one obtained with the random In Fig. 4 we show the MST obtained in a typical realization of the one model performed with the control parameters obtained as described abo evident that the structure of sectors of Fig. 3 is not present in Fig. 4. In f MST of the one-factor model has a star-like structure with a central noc largest fraction of nodes link directly to the central node and a smaller fra composed by the next-nearest neighbors. Very few nodes are found at a d of three links from the central node. The central node corresponds to C Electric and the second most connected node is Coca Cola. It is worth that these two stocks are the two most highly connected nodes in the re also.

In order to characterize quantitatively the structure of the MST we m of two topological quantities. The first one is the distribution of the degr random graph this quantity is distributed according to a binomial distr which for large networks tends to a Poisson distribution.

The second topological quantity is frequently used for oriented grap any vertex i in the tree we count the total number of vertices a in the

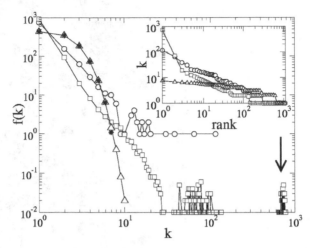

Fig. 5. Frequency distribution of the degree of the MST of real data (circle). We also show the mean degree distribution of random (triangle) and one-factor (square) model averaged over 100 numerical realizations of the MST. The stars are the theoretical values of the degree frequency for the random model in mean field limit. The inset shows the corresponding rank plot of the degree in the three cases.

subtree whose root is i. This quantity is called drainage basin area in oriented graphs of river networks [35], whereas it is usually referred as the in-degree component in graph theory. To calculate the in-degree component in a correlation based MST, we orient the MST according to the number of steps each node is far from the most connected node (sink). When more than one sink is present in the MST a preferential one is randomly chosen among them.

We report in Fig. 5 the frequency distribution for the degree k for the real data and for the average over 100 realizations of the random model and of the one factor model. The degree distribution for the MST of the real data shows a power law behavior with exponent -2.6 for one decade followed by a set of isolated points with high degree. A power law behavior with a similar exponent has been observed in [25] and in another recent study [36]. The highest degree $k_{max} = 115$ is observed for the General Electric, one of the most capitalized company in the NYSE. As pointed out in a previous work [30], some important companies clearly emerge for its high degree value indicating that they act as a reference for other companies. The random model displays an approximately exponential decay of the degree distribution. The value of the maximum degree is small, $k_{max} = 7.34 \pm 0.92$, showing that no asset plays a central role in the MST. The correlation based MST of the random model can be considered as the MST of a set of N points randomly distributed in an Euclidean space with $d = T$ dimension [31]. The N points have independent identically Gaussian distributed coordinates $\mathbf{r_i} = (r_i(1), r_i(2), ..., r_i(T))$ with $i = 1, 2, ..., N$. It has been shown that the distribution of degree of the random MST in Euclidean space converges to a specific distribution in the mean field limit $d \to \infty$ [37]. The numerical

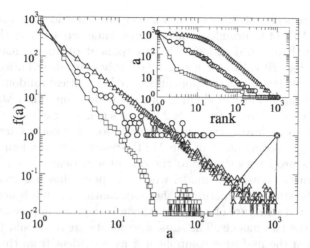

Fig. 6. Frequency distribution of the in-degree component a of the MST of real data (circle). We also show the mean in-degree component distribution of random (triangle) and one-factor (square) model averaged over 100 numerical realizations of the MST. The inset shows the corresponding rank plot of the in-degree component for the three cases.

values of the degree frequency obtained from this mean field limit are shown as a star in Fig. 5 for $k = 1, .., 7$. The agreement of theoretical values with the numerical simulations is very good showing that the mean field limit is already a good approximation for our T parameter.

The MST obtained from the one factor model is characterized by a rapidly power-law decaying degree distribution and by an asset with a very high value of the degree, which is indicated by an arrow in Fig. 5. The value of the maximum degree is $k_{max} = 718 \pm 29$. The corresponding asset is the center of the star like structure of Fig. 4. The region with highest value of the degree contains information about the stocks that act as reference for a large set of other stocks. To get more insight in the structure of this high k region we show a rank plot of the degree both for real data and for the considered models in the inset of Fig. 5. For the real market it is evident the presence of a region of power law extending for more than one decade. On the other hand, for the random model many nodes have a similar value of the degree which is ranging for less than an order of magnitude. This is due to the fact that there is no hierarchy in the random model. The rank plot of the degree of the MST for the one factor model has not a scale free behavior. Indeed, there is a single highly connected node (the center) and a rapidly decaying degree as a function of the rank. This fact corresponds to the simple one-center hierarchy of the MST of the one-factor model.

A discrepancy between real data and models is also observed in the frequency distribution of the in-degree component a defined above. Figure 6 shows the frequency distribution of the in-degree component for real and surrogate data. The inset of Fig. 6 shows the rank plot of the same data. In all three cases the in-degree component distribution has a power law behavior. This is particularly clear for the MST of the random uncorrelated time series where the power law lasts for more than two decades with an exponent of approximately -1.6. It is known that for critical random trees the probability distribution of tree size decays as a power-law with an exponent $3/2$ [38]. A critical random tree is a tree in which the mean number of children of each node is one. In a MST the mean degree is exactly equal to $2n/(n-1) \simeq 2$. Hence when we orient the MST from the root to the leaves we have a tree with one child for each node. Our result shows that the in-degree component of the MST arising from random uncorrelated time series has properties similar to the one of a critical random tree. This is not the case for the one-factor model where the power law has greater absolute slope due to the star-like structure of the tree. Neither models is actually able to catch the oriented structure of real data whose in-degree component distribution is in between the two models. The same arguments are also valid for the region of high values of the in-degree component a as is evident from the rank plot in the inset.

3.4 Comparison Between Data and Models

These results show that the topology of the MST for the real and for the considered artificial markets is different for both nodes with high degree and nodes with low degree. If we define the importance of a node as its degree (or its in-degree component), from our analysis emerges that the real market has a hierarchical distribution of importance of the nodes whereas the considered models are not able to catch such a hierarchical complexity. Specifically, in the random model the fluctuations select randomly few nodes and assign them small values of degree. Thus the MST of the random model is essentially non hierarchical. On the other hand the MST of the one factor model shows a simple one-center hierarchy. The MST of real market shows a more structured hierarchy of the importance of the stocks which is not captured by the considered models. The topology of stock return correlation based MST shows large scale correlation properties characteristic of complex networks in the native as well as in an oriented form. Such properties cannot be reproduced at all, even as a first approximation, by simple models as a random model or the widespread one-factor model.

4 The Stock Investment Network

We now consider a rather different financial network formed by the companies traded in a stock market and by the corresponding shareholders. A directed link is drawn from the vertex representing a company to the vertex representing a shareholder of the company itself. While this investment relationship graph has in principle a bipartite nature (vertices can be assigned to two classes, companies and investors), it happens frequently that some shareholders of a certain company are themselves companies the shares of which are traded in the market.

Therefore the resulting network is in general a directed one where a significant fraction of listed companies are owners of other listed companies, with no well-defined bipartite structure. It is possible to consider the subnetwork restricted to the owners which are listed companies themselves (hereafter the *restricted network*). This yields the structure reported in Figs. 7, 8, 9 for different markets, providing a description of the interconnections among stocks.

Whenever considering the whole investment relationships we will instead refer to the *extended network* or simply the network. We define the *portfolio diversification* as the in-degree k_i of the investor i, corresponding to the number of different assets in its portfolio. Vertices with zero in-degree are listed companies holding no shares of other stocks. The out-degree of a vertex is the number of

Fig. 7. Shareholder network for the Italian case.

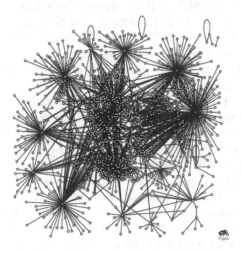

Fig. 8. Shareholder network for the NYSE case.

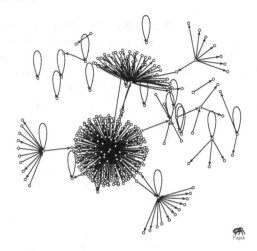

Fig. 9. Shareholder network for the Nasdaq case.

shareholders of the corresponding asset. Since the data are obviously restricted to a limited number of investors per each asset, the out-degree of a company is a biased quantity and we cannot deal with its statistical description. We note that a weight can be assigned to each link, defined as the fraction s_{ij} of the shares outstanding of asset j held by i multiplied by the market capitalization c_j of the asset j. We define *portfolio volume* the quantity $v_i = \sum_j s_{ij} c_j$ representing the total wealth in the portfolio of i.

4.1 Data Analysis

The data in our analysis report the shareholders of all stocks traded in the New York Stock Exchange (NYSE), in the National Association of Security Dealers Automated Quotations (NASDAQ), both in the year 2002, and in the Italian stock market (MIB) in the year 2002. The corresponding number M of assets in the markets is 2053, 3063 and 240 respectively.

On both the extended and the restricted nets, we consider the statistical distribution $P'(k)$ of the number of vertices with in-degree greater than or equal to k (see Fig. 10a). In all the extended nets the distribution always displays a power-law tail of the form $P'(k) \propto (k)^{1-\gamma}$. The corresponding probability density is $P(k) \propto (k)^{-\gamma}$, where the values of the exponent γ are given by $\gamma_{nys} = 2.37$, $\gamma_{nas} = 2.22$, $\gamma_{mib} = 2.97$. Market investments are therefore characterized by a scale-free topology which resembles that displayed by many other complex networks. By contrast, the restricted nets display no power-law behaviour (see Fig. 10b). This means that the relevant contribution to the scale-free nature of market investments comes from the investors outside the market.

It is also possible to compute the cumulative distribution $\theta'(v)$ describing the number of investors with portfolio volume greater than or equal to v. As shown in Fig. 11a, the tail of the distribution displays again a power-law behaviour

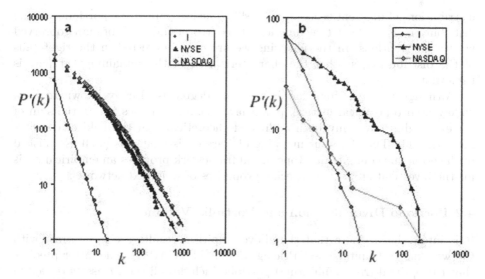

Fig. 10. Integrated degree distribution for the extended networks (**a** on the left) and the reduced one (**b** on the right).

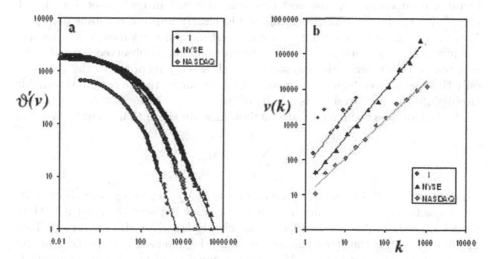

Fig. 11. Empirical properties of the portfolio volume. a) Cumulative volume distributions computed on the extended nets. b) Scaling of v against k in real networks (data points) and the power-law trend expected from the model (lines).

$\theta'(v) \propto v^{1-\alpha}$. The corresponding probability density is $\theta(v) \propto v^{-\alpha}$, with $\alpha_{nys} = 1.95, \alpha_{nas} = 2.09, \alpha_{mib} = 2.24$. It is interesting to observe that v represents the invested wealth. Therefore the observed power-law tails can be considered a generalization to z market investment context of the well-known Pareto tails. Such tails appear in the right hand side of the curves representing the wealth distributions of different economies [39–42]. The left part of the portfolio volume distribution also reflects the (functionally controversial) form of many observed wealth distributions. In the following we are only interested in the right tails of the distribution, so that the characterization of the remaining left part is irrelevant.

Although the scale-free character of the degree is already known to be a widely used topological feature, power-law distributions describing the sum of vertex weights have only been addressed theoretically in the field of complex networks [43]. Therefore our mapping of Pareto distributions (well established in the economic context) in a topological framework provides an empirical basis for the investigation of these specific properties of weighted networks.

4.2 Portfolio Diversification and Portfolio Volume

It is interesting to note that the above empirical results are in contrast with the well-known Capital Asset Pricing Model [24] (CAPM). The latter predicts that the optimal (risk-minimizing) portfolio includes all the M assets traded in the market and it is such that the amount of wealth invested in each asset i is proportional to the market capitalization c_i of i. In other words, if i decides to invest a total volume v_i, the best choice is to invest in each asset j a capital $v_i c_j / \sum_{k=1}^{M} c_k$. In our framework this would clearly imply a constant in-degree $k_i = M$ for each agent i, even if the total invested volume v varies greatly among the investors. One could suspect that the (very different) observed form of $P(k)$ is a biased result, since the investors of each asset j recorded in the data are only those who invested in j an amount of money larger than a certain threshold. However, since this threshold is usually of the form λc_j (for example, $\lambda = 0.0005$ in the Italian data), the recorded shareholdings are those that obey:

$$v_i c_j / \sum_{k=1}^{M} c_k > \lambda c_j,$$

which is satisfied if the volume v_i is such that $v_i > \lambda \sum_k c_k$ independently of c_j. In other words, the shareholdings of an investor behaving according to the CAPM model are either all observed (if v_i is sufficiently large) or all unobserved. This means that, taking the above bias into account, the observed $P(k)$ would again be peaked at the single value $k = M$. These arguments, although quite simplified, suggest that the non-trivial form of the in-degree distribution can genuinely witness the deviation of the investors behaviour from the ideal scenario explored by the CAPM. The above analysis provides therefore additional evidence of the inadequacy of the model, whose predictions are already commonly thought to be unrealistic (for instance by analysing price trends, see the preceding section).

The ideal scenario at the basis of the CAPM model is that all agents are equally informed about the market and process this information in the same way. Clearly, both hypotheses are unrealistic and result in a series of afore-mentioned predictions which differ from the empirical findings. In the following section we try to relax the above hypotheses into a stochastic model allowing

the agents to have heterogeneous availability of information and to make different choices even when equally informed. One basic idea of the model is that,
since the investors in our data are large long-term investors (and not short-term
speculators), their choices depend on the detailed (often private) information
that they need to gather concerning, for instance, the budgets and management
strategies of companies. For this information to be acquired, investors have to
face significant costs. Clearly, diversifying a portfolio by adding the asset j in it
is convenient when the cost of acquiring information about j is smaller than the
expected profit (in terms of risk reduction) associated to j. As a consequence,
large-volume portfolios (corresponding to holders who can face large information
acquisition costs) are therefore likely to display a large diversification as well.
The second ingredient of the model is to allow for two equally wealthy agents
to make different choices (due for instance to different preferred investment sectors), even if assets with better expected long-term performance are statistically
more likely to be chosen.

4.3 Fitness Model Driven by Pareto's Law

The above simple ideas (heterogeneous choices of the investors and information
acquisition costs) can be directly implemented by generalizing a recent model
[10] of network formation to the directed case. We assume that the probability
f_{ij} that the shareholder i invests in the asset j is a function of two quantities,
namely the total volume v_i that i decides to invest and a second quantity y_j
(which can also be a vector) characterizing j (such as its price history, expected
trend, etc.). By assuming the simplest separable form

$$f_{ij} = g(v_i)h(y_j)$$

we can directly express the expected in-degree k of an investor with volume v as

$$k(v) = g(v)h_{tot}$$

where h_{tot} is the total value of $h(y)$ computed over all assets. If the above expression can be inverted to yield $v(k)$, it is possible to compute the corresponding
in-degree distribution:

$$P(k) = \theta[v(k)]\frac{d}{dk}v(k)$$

which clearly depends on the volume distribution $\theta(v)$. The corresponding expressions for $k^{out}(v)$ and $P(k^{out})$ can be easily obtained, however (as we mentioned above) our information concerning k^{out} is incomplete and we cannot therefore use it to validate the model. Similarly, any hypothesis on the form of $h(y)$
would yield results that cannot be tested on the data. For this reason, we avoid
any rigorous definition of y and deal only with the quantities related to v.

   We therefore proceed by suggesting an explicit form for the quantities $\theta(v)$
and $g(v)$. The volume distribution $\theta(v)$ is chosen to display a power-law tail as
the observed one. We therefore set $\theta(v) \propto v^{-\alpha}$. The form of the attachment

probability $g(v)$ can instead be chosen by analogy with the traditional *preferential attachment* [9] mechanism. In the latter scheme, the attachment probability is chosen as an increasing function of the pre-existing vertex degree. While the common choice in the models is the linear one [1], its functional form can be measured on real networks [1] and is found to be proportional to k^β. The case $\beta = 1$ is the *linear* preferential attachment case, while $\beta > 1$ and $\beta < 1$ are the *superlinear* and *sublinear* cases respectively. In the system under consideration, the analogous choice would be $g(v) \propto v^\beta$. Substituting $\theta(v) \propto v^{-\alpha}$ and $g(v) \propto v^\beta$ in the expression for $P(k)$ yields (for large k)

$$P(k) \propto (k)^{-\gamma} \qquad \gamma = (\alpha + \beta - 1)/\beta$$

We therefore recover, as in the original model [10], that the scale-free degree distribution can be obtained by letting the connection probability f_{ij} depend on a power-law distributed quantity v.

An independent test of the model can be performed by noting that the above hypotheses also result in the following expectation for $k(v)$:

$$k(v) \propto v^\beta$$

where β, once the values of γ and α are fixed to the empirical ones, is constrained to the value $\beta = (1 - \alpha)/(1 - \gamma)$. In Fig. 11b we superimpose the above prediction to the points obtained from the data for each network. Indeed, the model expectations are rigorously verified by the data, except for the low k region in MIB. A possible explanation for this anomaly is that, as we checked, these points correspond to those investors holding a very large fraction (about 50%) of the shares of an asset, whose portfolio has therefore a large volume even if its diversification is small. Clearly, these investors are the effective controllers of a company. While in both US markets the fraction of links in the network corresponding to such a large weight is of the order of 10^{-4} (so that their effect is irrelevant on the plot of Fig. 11b), in MIB it equals the extraordinarily larger value 0.13. This determines the peak at small k superimposed to the power-law trend in the Italian market, and singles out an important difference between MIB and the US markets. This suggests that the proposed mechanism fits well the investors' behaviour, apart from that of the effective holders of a company.

Interestingly, in all cases $\beta > 1$, corresponding to a *superlinear* attachment mechanism. It is however worth noting that, while the traditional preferential attachment rule yields scale-free topologies only in the linear case [1], here we observe power-law degree distributions in the nonlinear case as well. This is a remarkable result, since in order to obtain the empirical forms of $P(k)$ the exponent β does not need to be fine tuned, and the results are therefore more robust under modification of the model hypotheses.

4.4 Further Topological Features of the Shareholder Networks

Besides proposing a model for the portfolio diversification, we report the study of other topological features of the shareholder networks allowing to characterize the different markets under study.

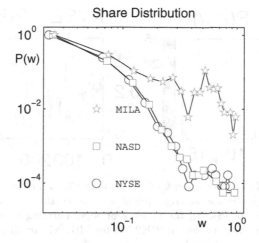

Fig. 12. Distribution of the fraction w of shares of a stock owned by a holder in the three shareholding networks. The distribution is computed over all stocks and holders in each data set.

Differently from social networks, shareholding networks present low or absent clusterization. Only 1%. of nodes have non-zero clustering coefficient in NYSE, versus 12%. in MIB. In Nasdaq there is no clustering at all. In fact, US markets consist of few stars with hundreds of leaves, the centers typically being investors non quoted on the market.

Shareholding relationships, represented as edges departing from the owned stock and pointing to the shareholder, are characterized by the following quantities: the in-degree or portfolio diversification k^{stock} of the stock (if this is also a shareholder of other stocks), the in-degree k of the shareholder, the market capitalization C^s of stock, the market capitalization C^h of shareholder (if this is also a quoted company), the percentage w_{ij} of shares of the stock i owned by the shareholder j.

The distributions of the values w of the percentage of shares of single stocks owned by single holders are shown in Fig. 12. The scale is linear-log. NYSE and NASD have similar fast decaying distributions. MIB displays instead a bump for values of shares just above 50%, meaning that in significative fraction of cases the company is completely controlled a single holder. This is another obvious difference between the US markets and the Italian market. Correlation between amount of share and portfolio size will make this difference more clear.

The distributions $P(k, w)$ of the number of shareholding relationships involving a shareholder with portfolio size k and amount of shares w are shown in Fig. 13 for NYSE and MIB SN. In US markets largest shares are held by holders of any in-degree. In MIB largest shares are held by holders with low in-degree, meaning that holders that have total control of a company tend to own few stocks. In particular the picture shows an obvious cluster of relationships in which the holder owns more than 40% of the shares but has a low portfolio

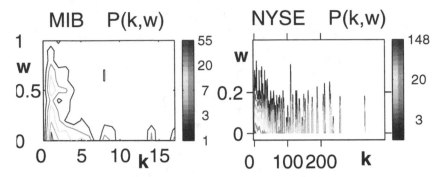

Fig. 13. Horizontal axis: in-degree k of the node at the end of an edge (the portfolio size of the holder), vertical axis: edge weight w (the fraction of shares owned by the holder). Contour plot of P(k,w), the number of edges entering a node with in-degree k and weight w. **left**: distribution of P(K2,W) for MIB. **right**: distribution of P(k,w) for NYSE. Logarithmic color scale.

diversification. The cluster involves 24% of the shareholders. These holders are typically (95%) non quoted companies. This means that there is a consistent fraction of shareholders which control a quoted company and are not themselves quoted on the market.

In all three markets there is a trend for companies to own companies with smaller market capitalization (data not shown). We don't find any correlation between market capitalization and amount of owned shares. Values of market capitalization were available only for quoted companies and not for all holders.

4.5 Effective Control Indexes

So far we did consider the weights of the networks under study in our analysis, but we didn't take into account the relative importance of a shareholder of a stock with respect to the other shareholders of that same stock. It is clear that the concentration of the ownerships plays a crucial role in financial strategy. We thus compute two indexes to capture the fact that a 10 % shareholder holds much more control if the other shareholders hold 1 % each, than if they hold 10 % each. This information is not contained in the amount of share alone nor in the W distribution over all nodes. We define the following quantities.

$$SI = \frac{\left(\sum_{i \in holders} w_i\right)^2}{\sum_{i \in holders} w_i^2} \tag{6}$$

where w_{ij} is, as defined above, the percentage of shares of the stock i owned by the shareholder j. SI (Stock Index) gives the effective number of holders controlling the stock. SI is close to 1 when there is a dominating holder. SI is equal to N when there are N equally important holders. For each holder j and each stock i we compute:

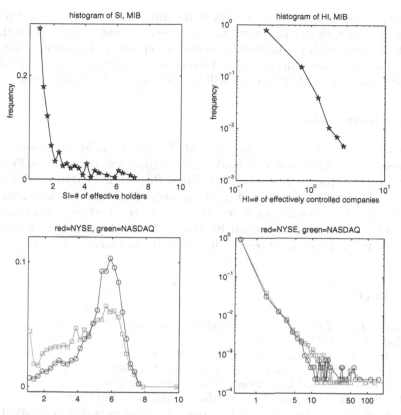

Fig. 14. Stock index and Holder index distribution in the shareholding networks. Top row: MIB. Bottom row: NASDAQ and NYSE. Left: SI distribution. Right: HI distribution. SI measures the effective number of holders of a stock. HI measures the equivalent number of stocks controlled by a holder.

$$h = \frac{w_{ij}^2}{(\sum_{k \in holders} w_{ik})^2} \qquad (7)$$

Then for each holder we sum the above quantity for each of the stocks in his portfolio.

$$HI = \frac{\sum_{i \in stocks\ owned\ by\ j} w_{ij}^2}{(\sum_{k \in holders\ of\ stock\ i} w_{ik})^2} \qquad (8)$$

HI (Holder Index) gives the effective number of stocks controlled by a holder.

We report the distributions of SI and HI in Fig. 14. The results show that in MIB the concentration of power among holders is distributed in a very different way from US markets. In MIB there most companies are controlled by a single holder. In the US markets most companies are controlled by about 6 holders.

Finally, in order to investigate whether holders controlling effectively several stocks tend to have special topological properties in the network, we have searched for possible correlation between the indexes defined above and properties as in-degree, cluster coefficient and betweenness centrality. We find that SI tend to decrease with the value of betweenness centrality. This means that companies that are more central tend to be owned more evenly by several holders. A further analysis in this direction will be the subject of a future work.

Acknowledgements

We would like to thank G. Bonanno, M. Castri, F. Lillo, R. Mantegna, L. Pietronero, E. Sciubba and V. Servedio. The authors acknowledge the European Commission support within the FET Open Project IST-2001-33555 COSIN. Some of the graphes represented in the figures of this paper have been done with the Pajek shareware software [29]. Data of the Fortune 1000 companies in 1999 has been kindly provided by Gerald Davis [17]. The other data sets presented in the paper have been collected by the authors.

References

1. Albert, R. and Barabási, A.-L. Statistical mechanics of complex networks. Rev. Mod. Phys. 74, 47-97 (2002).
2. D. J. Watts and Strogatz, Nature **393**, 440 (1998).
3. P. Erdős, A. Rényi, Bull. Inst. Int. Stat. **38**, 343 (1961).
4. B.A. Huberman and L.A. Adamic *Nature* **399**, 130 (1999).
5. R. Albert, H. Jeong, and A. L. Barabási Nature **401**, 130 (1999).
6. G. Caldarelli, R. Marchetti and L.Pietronero, Europhysics Letters **52**, 386 (2000).
7. R. Pastor-Satorras, A. Vazquez and A. Vespignani, Phys. Rev. Lett. **87**, 258701 (2001).
8. M. E. J. Newman, D. J. Watts, and S. H. Strogatz, Proc. Natl. Acad. Sci. USA **99**, 2566 (2002).
9. A. L. Barabási and R. Albert, Science **286**, 509 (1999)
10. G. Caldarelli, A. Capocci, P. De Los Rios and M.A. Muñoz, Phys. Rev. Lett. 89, 278701 (2002)
11. M.E.J. Newman, M. Girvan *to appear in Proceedings of the XVIII Sitges Conference on Statistical Mechanic* **99**, 12583 (2003).
12. M.Boguna, R. Pastor-Satorras, A. Vespignani *ArXiv:cond-mat/0301149)*.
13. M.E.J. Newman *Phys. Rev.* E 67, 026126 (2003).
14. G. Bianconi and A.-L. Barabási *Europhysics Letters* **54**, 436 (2001).
15. G. Caldarelli, A. Capocci, P. De Los Rios and M.A. Muñoz, *Physical Review Letters* 89, 258702 (2002).
16. D.S. Callaway, J.E. Hopcroft, J.M. Kleinberg, M.E.J. Newman and S.H. Strogatz *Phys.Rev.E* **64**, 041902 (2001).
17. Davis, G.F., Yoo, M., Baker, W.E., The small world of the American corporate elite, 1982-2001, *Strategic Organization* 1: 301-326 (2003).
18. M. E. J. Newman, S. H. Strogatz, and D. J. Watts, Random graphs with arbitrary degree distributions and their applications, Phys. Rev. E 64, 026118 (2001).
19. M. E. J. Newman, Assortative mixing in networks, Phys. Rev. Lett. 89, 208701 (2002).

20. M. E. J. Newman and Juyong Park, Why social networks are different from other types of networks, Phys. Rev. E, in press.
21. M. Catanzaro, G. Caldarelli, L. Pietronero, Assortative model for social networks, cond-mat 0308073 v1
22. Battiston, S., Bonabeau, E., Weisbuch G., Decision making dynamics in corporate boards, *Physica A*, 322, 567 (2003).
23. Battiston, S., Weisbuch G., Bonabeau, E., Decision spread in the corporate board network, submitted.
24. Y. J. Campbell, A. W. Lo, A. C. Mackinlay *The Econometrics of Financial Markets*, (Princeton University Press, Princeton,1997) and references therein.
25. N Vandewalle, F Brisbois and X Tordoir Quantitative Finance **1**, 372 (2001).
26. K. V. Mardia, J. T. Kent and J. M. Bibby *Multivariate Analisys*, (CA: Academic, San Diego, 1979).
27. J. C. Gower, Biometrika **53**, 325 (1966).
28. R. N. Mantegna, Eur. Phys. J. B **11**, 193 (1999).
29. V. Batagelj, A. Mrvar: Pajek – Program for Large Network Analysis. Connections, 21(1998)2, 47-57. Home page for downloads: http://vlado.fmf.uni-lj.si/pub/networks/pajek/
30. G. Bonanno, F. Lillo and R. N. Mantegna, Quantitative Finance **1**, 96 (2001).
31. R. N. Mantegna and H. E. Stanley *An introduction to econophysics: correlations and complexity in finance* (Cambridge University press, Cambridge, 2000).
32. The Standard Industrial Classification system can be found at http://www.osha.gov/oshstats/naics-manual.html
33. L. Laloux, P. Cizeau, J. P. Bouchaud and M. Potters, Phys. Rev. Letters **83**, 1467 (1999).
34. V. Plerou, P. Gopikrishnan, B. Rosenow, L. A. Nunes Amaral and H. E. Stanley, Phys. Rev. Lett. **83**, 1471 (1999).
35. I. Rodriguez-Iturbe and A. Rinaldo, *Fractal River Basins*, (Cambridge University Press, Cambridge, 1997).
36. J.-P. Onnela, A. Chackraborti, K. Kaski, J. Kertész ArXiv:cond-mat/0303579 and ArXiv:cond-mat/0302546
37. M. D. Penrose, The Annals of Probability **24**, 1903 (1996).
38. T.E. Harris, *The Theory of Branching Processes* (Dover, New York, 1989).
39. V. Pareto *Cours d'Économie Politique* (Macmillan, London, 1897). Reprinted in *Oeuvres Complétes* (Droz, Geneva, 1965).
40. W.W. Badger, *Mathematical models as a tool for the social science* (Gordon and Breach, New York, 1980).
41. C. Dagum, & M. Zenga, (eds.) *Income and Wealth Distribution, Inequality and Poverty* (Springer-Verlag, Berlin, 1990).
42. J. J. Persky, Pareto's law. Journal of Economic Perspectives **6**, 181-192 (1992).
43. Yook, S. H., Jeong, H., Barabási, A.-L. and Tu, Y. Weighted evolving networks. Phys. Rev. Lett. 86, 5835-5838 (2001).
44. H. Markovitz, *Portfolio Selection: Efficient Diversification of Investments* (Wiley, New York, 1959).

Topology, Hierarchy, and Correlations
in Internet Graphs

Romualdo Pastor-Satorras[1], Alexei Vázquez[2], and Alessandro Vespignani[3]

[1] Department de Física i Enginyeria Nuclear, Universitat Politècnica de Catalunya,
Campus Nord, 08034 Barcelona, Spain
[2] Department of Physics, University of Notre Dame, Notre Dame, IN 46556, USA
[3] Laboratoire de Physique Théorique, Bâtiment 210 Université de Paris-Sud 91405
ORSAY Cedex, France

Abstract. We present a statistical analysis of different metrics characterizing the
topological properties of Internet maps, collected at two different resolution scales: the
router and the autonomous system level. The metrics we consider allow us to confirm
the presence of scale-free signatures in several statistical distributions, as well as to
show in a quantitative way the hierarchical nature of the Internet. Our findings are
relevant for the development of more accurate Internet topology generators, which
should include, along with the properties of the degree distribution, the hierarchical
signatures reviewed in the present work.

1 Introduction

The relentless growth of the Internet goes along with a wide range of internet-
working problems related to routing protocols, resource allowances, and physical
connectivity plans. The study and optimization of algorithms and policies related
to such problems heavily rely on theoretical analysis and simulations that use
model abstractions of the actual structure of the Internet. On the other hand, in
order to extract the maximum benefit from these studies, it is necessary to work
with reliable Internet topology generators. The basic priority at this respect is
to best define the topology to use for the network being simulated. This implies
the characterization of how routers, hosts, and physical links interconnect with
each other in shaping the actual Internet.

In the last years, several research groups have started to deploy technologies
and infrastructures devoted to obtain a more detailed picture of the Internet.
These studies, aimed at tracking and visualizing the Internet large scale topology
and/or performance, are leading to Internet mapping projects at different resolution scales. These projects typically collect data on Internet elements (routers,
domains) and the connections among them (physical links, peer connections), in
order to create a graph-like [1] representation of large parts of the Internet in
which the vertices represent those elements and the edges represent the respective
connections. Mapping projects focus essentially on two levels of topological representation. First, by inferring router adjacencies it has been possible to measure
the Internet router (IR) level topology. The second measured topology works at
the autonomous system (AS) level and the connectivity obtained from AS routing path information. Although these two representations are related, it is clear

R. Pastor-Satorras, A. Vázquez, and A. Vespignani, Topology, Hierarchy, and Correlations in Internet
Graphs, Lect. Notes Phys. **650**, 425–440 (2004)
http://www.springerlink.com/   © Springer-Verlag Berlin Heidelberg 2004

that they describe the Internet at quite different length scales. In fact, each AS usually groups a large number of routers, often geographically scattered, and therefore the AS maps can be considered in some sense a coarse-grained version of the IR maps.

Internet maps exhibit an extremely large degree of heterogeneity and the use of statistical tools becomes mandatory to provide a proper mathematical characterization of this system [2]. Statistical analysis of the Internet maps have pointed out, to the surprise of many researchers, a very complex connectivity pattern with fluctuations extending over several orders of magnitude [3]. In particular, a power-law behavior has been observed in metrics and statistical distributions of Internet maps at different levels [3–11]. This evidence makes the Internet an example of the so-called *scale-free* (SF) networks [12,13] and uncovers a peculiar structure that cannot be satisfactorily modeled with traditional topology generators. Previous Internet topology generators, based in the classical Erdös and Rényi random graph model [14,15] or in hierarchical models, yielded an exponentially bounded connectivity pattern, with very small fluctuations and in clear disagreement with the recent empirical findings. A theoretical framework for the origin of scale-free graphs has been put forward by Barabási and Albert [12] by devising a novel class of dynamical growing networks. At the same time, several Internet topology generators yielding power-law distributions have been subsequently proposed [16–18].

Data gathering projects [19–23] are progressively making available larger AS and IR level maps which are susceptible of more accurate statistical analysis and raise new and challenging questions about the Internet topology. For instance, statistical distributions show deviations from the pure power-law behavior and it is important to understand to which extent the Internet can be considered a scale-free graph. The way these scaling anomalies—usually signaled by the presence of cut-offs in the corresponding statistical distributions—are related to the Internet finite size and physical constraints is a capital issue in the characterization of the Internet and in the understanding of the dynamics underlying its growth. A further important issue concerns the fact that the Internet is organized on different hierarchical levels, with a set of backbone links carrying the traffic between local area providers. This structure is reflected in a hierarchical arrangement of administrative domains and in a different usage of links and degree of vertices. The interplay between the scale-free nature and the hierarchical properties of the Internet is still unclear, and it is an important task to find metrics that can exploit and characterize the hierarchical features on the AS and IR levels. Finally, although one would expect Internet AS and IR level maps to exhibit similar scale-free behavior, the different resolution in of maps might lead to a noticeable diversity of metrics properties.

In this paper we present a detailed statistical analysis of large AS and IR level maps [19,21,22]. We study the scale-free properties of these maps, focusing on the degree and betweenness distributions. While scale-free properties are confirmed for maps at both levels, IR level maps show also the presence of an exponential cut-off, that can be related to constraints acting on the physical degree and load of routers. Power-law distributions with a cut-off are a general feature of scale-free phenomena in real finite systems and we discuss their origin in the context of growing networks [12,13]. At the AS level we confirm the presence of a strong scale-free character for the large-scale degree and betweenness distributions.

Furthermore, we propose two metrics based on the clustering and the degree correlation functions, that appear to highlight the hierarchical properties of Internet maps. In particular, these metrics clearly distinguish between the AS and IR levels, which show a very different behavior at this respect. While IR level maps appear to possess a quite weak hierarchical structure, AS maps fully exploit the hierarchy of domains around which the Internet revolves. The differences between the two levels might be very important in the developing of reliable Internet topology generators. Indeed, the testing of Internet protocols working at different levels might need of topology generators accounting for the different properties observed. Hierarchical features are also important to scrutinize theoretical models proposing new dynamical growth mechanisms for the Internet as a whole.

2 Internet Maps

The Internet is usually portrayed as an undirected graph [1]. Depending on the meaning assigned to the vertices and edges of the associated graph, we can obtain different levels of representation, each one corresponding to a different degree of coarse-graining with respect to the physical Internet [2].

Internet Router (IR) level: In the IR level maps, vertices represent the routers, while edges represent the physical connections among them. In general, all mapping efforts at the IR level are based on computing router adjacencies from *traceroute* sequences sent to a list of networks in the Internet. The traceroute command performed from a single source provides a spanning tree from that source to every other (reachable) vertex in the network. By merging the information obtained from different sources it is possible to construct IR level maps of different portions of the Internet. In order to detect all the various cross-links, however, a large number of source probes is needed. In addition, the instability of paths between routers and other technical problems—such as multiple alias interfaces—make the mapping a very difficult task [24]. These difficulties have been diversely tackled by different Internet mapping projects, such as the Lucent project at Bell Labs [23], the Cooperative Association for Internet Data Analysis (CAIDA) [20], and the SCAN project at the Information Sciences Institute [22].

Autonomous System (AS) level: Nowadays the Internet can be partitioned into autonomously administered domains which vary in size, geographical extent, and function. Each domain may exercise traffic restrictions or preferences, and handle internal traffic according to particular autonomous policies. This fact has stimulated the separation of the inter-domain routing from the intra-domain routing, and the introduction of the Autonomous Systems Number (ASN). Each AS refers to one single administrative domain of the Internet. Within each AS, an Interior Gateway Protocol is used for routing purposes. Between ASs, an Exterior Gateway Protocol provides the inter-domain routing system. The Border

Gateway Protocol (BGP) is the most widely used inter-domain protocol. In particular, it assigns a 16-bit ASN to identify, and refer to, each AS. In the AS level graphs each vertex represents an AS, while each edge between two vertices represents the existence of a BGP peer connection among the corresponding ASs. It is important to stress that each AS groups many routers together, and the traffic carried by a link is the aggregation of all the individual end-host flows between the corresponding ASs. The AS map can be constructed by looking at the BGP routing tables. In fact, the BGP routing tables of each AS contain a spanning tree from that vertex to every other (reachable) AS. We can then try to reconstruct the complete AS map by merging the connectivity information coming from a certain fraction of these spanning trees. This method has been actually used by the National Laboratory for Applied Network Research (NLANR) [19], using the BGP routing tables collected at the Oregon route server, that gathers BGP-related information since 1997. Enriched maps can be obtained from some other public sources, such as Looking Glass sites and the Reseaux IP Européens (RIPE) [9], getting about 40% of new AS-AS connections.

These graph representations do not model individual hosts, too numerous, and neglect edge properties such as bandwidth, actual data load, or geographical distance. For these reasons, the graph-like representation must be considered as an overlay of the basic topological structure: the skeleton of the Internet. Moreover, the data collected for the two levels are different, and both representations may be incomplete or partial to different degrees. In particular, measurements may not capture all the vertices present in the actual network and, more often, they do not include all the edges among vertices. It is not our purpose here to argue about the reliability of the different maps. However, the conclusions we shall present in this paper seem rather stable in time for the different maps [6]. Hopefully, this fact means that, despite the different degrees of completeness, the present maps represent a fairly good statistical sampling of the Internet as a whole.

In order to perform our analysis, we shall use the IR map collected during October/November 1999 by the SCAN project with the Mercator software as representative of the Internet router level. This map is composed by 228263 vertices and 320149 edges. At the autonomous system level we consider the AS map collected at Oregon route server, dated May 25, 2001, made up by 11174 vertices and 23409 edges.

3 Average Properties

We start by discussing some metrics commonly used in graph theory: the vertex degree k_i, the minimum path distance between pairs of vertices d_{ij}, and the clustering coefficient c_i. The degree k_i of a vertex is defined as the number of edges incident to that vertex, $i.e.$ the number of connections of that vertex with other vertices in the network. If vertices i and j are connected we will say that they are nearest neighbors. The minimum path distance d_{ij} between a pair of vertices i and j is defined as the minimum number of vertices traversed by a path that goes from one vertex to the other. The clustering coefficient c_i [25] of the vertex i is defined as the probability that two vertices connected to i are also connected to each other. Numerically it can be computed as the ratio between the number of edges e_i in the sub-graph identified by the nearest neighbors of i and its maximum possible value $k_i(k_i - 1)/2$, corresponding to a complete sub-graph, i.e. $c_i = 2e_i/k_i(k_i - 1)$. The clustering coefficient c_i takes values of order $\mathcal{O}(1)$ for ordered networks. On the other hand, for random graphs [14,15], which are constructed by connecting vertices at random with a fixed probability p, the clustering coefficient is of order $\mathcal{O}(N^{-1})$.

The average degree of the AS and IR maps is $\langle k \rangle_{AS} = 4.2$ and $\langle k \rangle_{IR} = 2.8$, respectively. This values are of order $\mathcal{O}(1)$, and therefore both maps can be considered as *sparse* graphs. Despite the small average degree, however, the average minimum path distance is also very small, compared to the size of the maps, i.e. $\langle d \rangle_{AS} = 3.6$ and $\langle d \rangle_{IR} = 9.5$. That is, the Internet exhibits what is known as the "small-world" effect [25]: in average one can go from one vertex to any other in the system passing through a very small number of intermediate vertices. This observation precludes the possibility that the Internet is shaped as a regular two-dimensional grid, since in this case its characteristic distance would scale with the number of nodes as $\langle d \rangle \sim N^{1/2}$; with the present Internet size, such scaling would imply that information packets would pass through 10^2–10^3 more nodes, depleting all communication capabilities of this network. Since the network is nevertheless sparse, this necessarily implies that there are some shortcuts which connect different regional networks, strongly decreasing the value of $\langle d \rangle$ [25].

The small world evidence is strengthened by the empirical finding of clustering coefficients of value $\langle c \rangle_{AS} = 0.30$ and $\langle c \rangle_{IR} = 0.03$, orders of magnitude larger than the corresponding value for a Erdös and Rényi random graph of the same size, $\mathcal{O}(N^{-1})$. As discussed above, this fact implies that neighbors of the same vertex are very likely on their turn connected among themselves. The high clustering coefficient of the Internet maps is probably due to geographical constraints. In Internet graphs, all edges are equivalent. Yet, the physical connections are characterized by a real space length. The larger is this length, the higher the cost of installation and maintenance of the physical line, favoring therefore the preferential connection between nearby vertices. It is therefore likely that vertices within the same geographical region will have a large number of connections among them, increasing in this way the clustering coefficient.

The differences observed in the metrics' average values are consistent with the fact that the AS map is a coarse-grained representation of the IR map. The IR level map is, for instance, sparser, and its average minimum path distance is larger. The IR map has a small average degree, because routers have a finite capacity and, therefore, can have a limited number of connections. On the contrary, ASs can have in principle any number of connections, since they represent the aggregation of a large number of routers. This implies that AS maps have a greater number of vertices with a high number of connections (hubs), providing the large-distance shortcuts needed to produce a small average minimum path distance.

4 Scale-Free Properties

The analysis of the average clustering presented in the previous section makes clear that the Internet does not resemble a star-shaped architecture with just a few gigantic hubs and a multitude of singly connected vertices, nor a random graph structure. On the other hand, the small average path distance rules out as well the possibility of a regular grid architecture. These evidences suggest a peculiar topology that will be clearly identified by looking at the detailed statistical distributions. In particular, Faloutsos *et al.* [3] pointed out for the first time that the connectivity properties of the Internet AS maps are characterized by a probability distribution that a vertex has k edges (*i.e.* it is connected to other k vertices) with the form $p(k) \sim k^{-\gamma}$, where $\gamma \simeq 2.1$ is a characteristic exponent. Noticeably, exponent values $2 < \gamma \leq 3$ imply a well defined $\langle k \rangle$, while $\langle k^2 \rangle$ is diverging in the case of networks with infinite size. This behavior signals the presence of *scale-free* degree properties; *i.e.* there is no characteristic degree above which the probability is decaying exponentially to zero. In other words, there is a statistically significant probability that a vertex has a very large number of connections compared to the average degree $\langle k \rangle$. In addition, the implicit divergence of $\langle k^2 \rangle$ is signalling the extreme heterogeneity of the connectivity pattern, since it implies that statistical fluctuations are unbounded. The work of Faloutsos *et al.* was followed by different studies of AS maps [6,9, 26], and IR maps [4,10]. Here, we will review the analysis of scale-free properties in recent AS and IR level maps.

We start by considering the integrated degree distribution $P(k) = \sum_{k'=k}^{\infty} p(k')$. In the case of a pure power-law probability distribution $p(k) \sim k^{-\gamma}$, we expect the functional behavior $P(k) \sim k^{1-\gamma}$. In Fig. 1 we show the integrated degree distribution for the AS map. In this case a clear power law decay with exponent $\gamma = 2.1 \pm 0.1$ is observed, as it has been already reported elsewhere [3,6,26]. The reported distribution is also stable in time as found by analyzing different time snapshot of the AS level maps obtained by the NLANR [6]. The heavy tailed nature of the degree distribution has important consequences in the dynamics of processes taking place on top of this networks. Indeed, recent studies about network resilience to removal of vertices [27] and virus spreading [28] have shown that the relevant parameter is the ratio $\kappa = \langle k^2 \rangle / \langle k \rangle$ between the first two moments of the degree distribution. If $\kappa \gg 1$ then the network manifests some properties that are not observed for networks with exponentially decaying degree distributions. For instance, we can randomly remove practically all the vertices in the network and a giant connected component [15] will still exist. In the AS map we observe a wide degree distribution with a factor κ, which, being determined effectively by the tail of the distribution, takes on a quite large value compared with $\langle k \rangle$.

The degree distribution of the IR level map has a long tail with power-law behavior that is, however, smoothed by a clear exponential cut-off. The existence of a power-law tendency for small connectivities is better seen for the probability distribution $p(k)$, as shown in Fig. 2. A power law fit of the form $p(k) \sim k^{-\gamma}$ for $k \leq 100$ yields the exponent $\gamma = 2.1 \pm 0.1$, in agreement

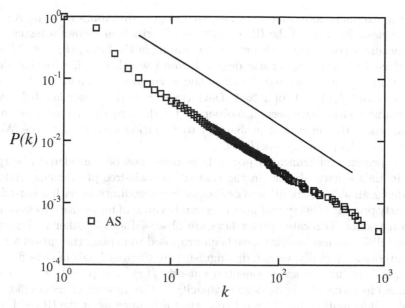

Fig. 1. Integrated degree distribution $P(k)$ for the AS map. The solid line corresponds to a power law decay $P(k) \sim k^{1-\gamma}$ with exponent $\gamma = 2.1$.

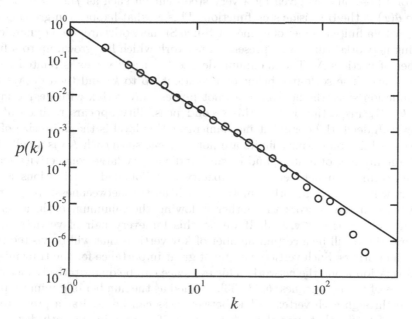

Fig. 2. Degree distribution $p(k)$ for the IR map. The solid line is a power law decay $p(k) \sim k^{-\gamma}$ with $\gamma = 2.1$.

with the exponent found for the integrated degree distribution in the AS map. Nevertheless, for $k \gg 50$ the IR map degree distribution follows a faster decay. This picture is consistent with a finite size scaling of the form $p(k) = k^{-\gamma} f(k/k_c)$ [29]. Here k_c is a characteristic degree beyond which the distribution decays faster than a power law, and $f(x)$ has the asymptotic behavior $f(x) = \text{const.}$ for $x \ll 1$ and $f(x) \ll 1$ for $x \gg 1$. Deviations from the power law behavior at large connectivities have been also observed for the larger maps reported in [10]. In that work, the integrated probability distribution was fitted to the Weibull distribution $P(k) = a \exp[-(k/k_c)^{\beta}]$.

The presence of truncated power laws must not be considered a surprise, since it finds a natural place in the context of scale-free phenomena. Actually, bounded scale-free distributions (*i.e.* power-law distributions with a cut-off) are implicitly present in every real world system because of finite-size effects or physical constraints. Truncated power laws are observed also in other real networks [30] and different mechanisms have been proposed to explain the cut-off for large connectivities. Actually, we can distinguish two different kinds of cut-offs in real networks. The first is an exponential cut-off, $f(x) = \exp(-x)$, which can be explained in terms of a finite degree capacity of the network elements [30] or incomplete information [31]. This is likely what is happening at the IR level, where the finite capacity constraint (maximum number of router interfaces) is, in our opinion, the dominant mechanism affecting the tail of the degree distribution. A second possibility is given by a very steep cut-off such as $f(x) = \theta(k_c - x)$, where $\theta(x)$ is the Heaviside step function. This is what happens in growing networks with a finite number of elements. Since SF networks are often dynamically growing networks, this case represents a network which has grown up to a finite number of vertices N. The maximum degree k_c of any vertex is related to the network age. The scale-free behavior is evident up to k_c and then decays as a step function since the network does not possess any vertex with degree larger than k_c. By inspecting Fig. 1, this second possibility appears realized at the AS level. Indeed, the dominant mechanism at this level is the finite size of the network, while connectivity limits are not present, since each AS is a collection of a large number of routers, and it can handle a very large connectivity load.

The connection between finite capacity and bounded distributions is also evident when inspecting other metrics, such as the betweenness. To go from one vertex in the network to another following the minimum path, a certain sequence of vertices is visited. If we do this for every pair of vertices in the network, there will be a certain number of key vertices that will be visited more often than others. Such vertices will be of great importance for the transmission of information along the network. This evidence can be quantitatively measured by means of the betweenness b_i [32,33], defined as the number of minimum paths that go through each vertex i. The betweenness can be used as a proxy to the amount of traffic that goes through a vertex, if the minimum path distance is considered as the metric defining the optimal path between pairs of vertices. In this sense, it has been recently studied for AS maps under the name of "load" [34].

Considering the betweenness as a static estimate of the amount of traffic that a vertex supports, then, if routers have a bounded capacity, the betweenness

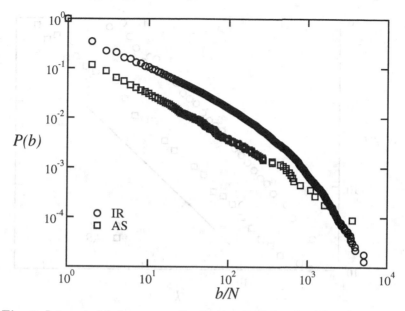

Fig. 3. Integrated betweenness distribution $P(b)$ for the AS and IR maps.

distribution should also be bounded at large values. On the contrary, this effect should be absent for the AS maps. In order to check this possibility, we analyze the probability distribution $p(b)$ that a vertex has betweenness b. The integrated betweenness distribution $P(b) = \sum_{b'=b}^{\infty} p(b')$ for the AS and IR maps is shown in Fig. 3. The AS distribution is well fitted by a power law $P(b) \sim b^{1-\gamma_b}$ with an exponent $\gamma_b = 1.9 \pm 0.1$. In the case of the IR map, on the other hand, the betweenness distribution follows a truncated power law, in analogy to what is observed for the connectivity distribution.

The similar behavior shown by the degree and betweenness distribution suggests the possibility of a relationship between their respective power law exponents. Indeed, the betweenness and the degree can be simply related if one assumes that the average number of shortest paths b_k passing over a vertex of degree k scales as

$$b_k \sim k^{\beta}. \tag{1}$$

By inserting this expression into the integrated betweenness distribution, we obtain

$$P(k) \sim k^{\beta(1-\gamma_b)}. \tag{2}$$

Since we have that $P(k) \sim k^{1-\gamma}$, we obtain the scaling relation

$$\beta = \frac{\gamma - 1}{\gamma_b - 1}. \tag{3}$$

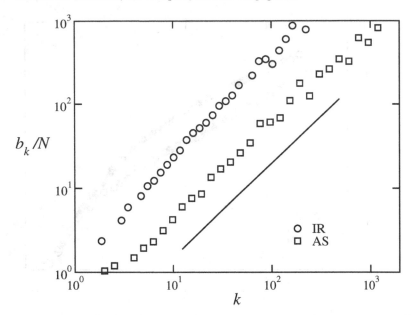

Fig. 4. Betweenness b_k as a function of the vertex degree k. The solid line is a power law $b_k \sim k^\beta$, with $\beta = 1.1$.

In Fig. 4 we report the direct measurement of the average betweenness of a vertex of degree k. We obtain a power-law behavior with an exponent $\beta \sim 1.1$, in agreement with the scaling relation equation (3), for the independently measured values of γ and γ_b.

It is worth to stress that while the power law truncation is an expected feature of finite systems, the scale-free regime is the important signature of an emergent cooperative behavior in the Internet dynamical evolution. This dynamics plays therefore a central role in the understanding and modeling of the Internet. In this perspective, the developing of a statistical mechanics approach to complex networks [12] is providing a new dynamical framework where the distinctive statistical regularities of the Internet can be understood in term of the basic processes ruling the appearance or disappearance of vertices and edges.

5 Hierarchy and Correlations

The topological metrics analyzed so far give us a distinction between the AS and IR maps with respect to the large degree and betweenness properties. The difference becomes, however, more evident if we consider properties related with the existence of hierarchy and correlations. The primary known structural difference in the Internet is the distinction between *stub* and *transit* domains (or ASs) [2]. Transit ASs correspond to large backbones, providing national or international connectivity, or to regional providers serving large metropolitan areas. Stub ASs, on the other hand, correspond to campus networks and local area networks. The purpose of transit ASs is to provide connectivity to stubs, minimizing the necessity of direct stub-stub connections. For this reason, transit ASs are well interconnected among them, and link stub ASs in their geographical neighborhood. The primary characteristic of this domain hierarchy is that traffic paths between vertices in the same domain stay entirely within that domain. For instance, stub ASs handle all traffic that originates and terminates inside the AS boundaries, while a routing path between two vertices in different stub ASs goes generally through one or more transit ASs. This traffic division can be schematically represented as a hierarchical structure, roughly partitioned into international connections, national connection, regional networks, and local area networks. Vertices providing access to international connections or national backbones are of course on top level of this hierarchy, since they make possible the communication between regional and local area networks.

A heuristic way to quantify the hierarchical nature of the Internet levers on the concept of *backbones*, links that carry traffic for a large fraction of all the possible source-destination pairs. In other words, the traffic is not evenly spread on all the links, but is canalized through the more central links and hubs in the Internet, which are more used than others. This feature can be measured by studying the traffic (or load) carried by the different edges and vertices. The actual load, however, is not a topological quantity, and cannot be directly computed from Internet maps. Nevertheless, we can use as a proxy for its value the number of shortest paths among source-destinations pairs that go through each vertex and edge, a definition that corresponds to the vertex and edge betweenness introduced in the previous section. The vertex betweenness distribution shown in Fig. 3 reveals that the load is not uniformly distributed, but has a tendency to concentrate on a few vertices, while a large fraction of peripheral vertices have a small betweenness value. It is possible to identify those vertices that concentrate most of the load by analyzing the average betweenness b_k of the vertices with degree k, see Fig 4. In this plot we can check that the betweenness is an increasing function of the degree for both the AS and IR maps, which indicates that the vertices carrying the largest load are indeed those with the largest degree, which, together with the interconnections among them, form a well-defined backbone concentrating most of the traffic carried by the network.

This hierarchical structure will introduce some correlations in the network and it is an important issue to understand how these features manifest at the topological level. In order to quantify the presence of hierarchies in Internet maps we introduce two metrics based on the clustering coefficient and the nearest neighbor average degree [6].

The clustering coefficient is defined as the probability that two neighbors of a given vertex are also connected to each other. In Sect. 3 we have shown that the clustering coefficient for the AS and IR maps is orders of magnitude larger than the one expected for a random graph and, therefore, that they are far from being random. Further information can be extracted if one computes the clustering coefficient as a function of the vertex degree [6,35]. In particular, the average clustering coefficient c_k of vertices with degree k is defined as

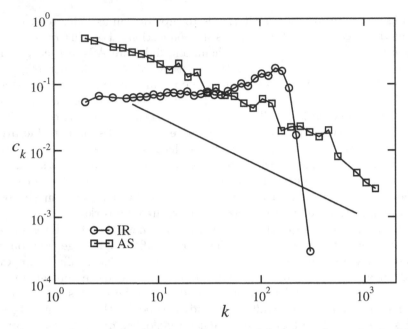

Fig. 5. Average clustering coefficient as a function of the vertex degree for the AS and IR maps. The solid line is given by a power law $c_k \sim k^{-0.8}$.

$$c_k = \frac{1}{Np(k)} \sum_i c_i \, \delta_{k_i,k}, \tag{4}$$

where $c(i)$ is the clustering coefficient of vertex i, $\delta_{k_i,k}$ is the Kronecker symbol and the sum runs over all possible vertices in the graph. In Fig. 5 we plot the average clustering coefficient c_k for vertices with degree k. In the case of the AS maps this quantity follows a behavior that can be approximated by a power law decay with an exponent around 0.8. For the IR map, however, except for a sharp drop for large values of k, attributable to low statistics, it takes an almost constant value. This implies that in the AS level vertices with a small number of connections have larger local clustering coefficients than those with a large degree. This behavior is consistent with the picture described above of highly clustered regional networks sparsely interconnected by national backbones and international connections. The regional clusters of ASs are probably formed by a large number of vertices with small degree but large clustering coefficients. Moreover, they should also contain vertices with large connectivities that are connected with the other regional clusters. These large degree vertices will be on their turn connected to vertices in different clusters which are not interconnected and, therefore, will have a small local clustering coefficient. On the contrary, in the IR level map these correlations are absent. Somehow the domain hierarchy does not produce any signature at the single router scale, where the geographic constraints and connectivity bounds probably play a more important role.

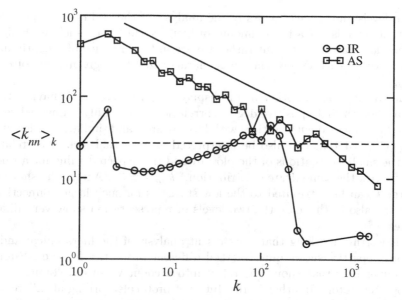

Fig. 6. Nearest neighbors average degree for the AS and IR maps. The solid line is given by the power law decay $\langle k_{nn}\rangle_k \sim k^{-0.55}$. The horizontal dashed line marks the value in the absence of correlations, $\langle k_{nn}\rangle_k^0 = \langle k^2\rangle / \langle k\rangle = 26.9$, computed for the IR map.

These observations for the clustering coefficient are supported by another metric related with the correlations between vertex connectivities. These correlations are quantified by the conditional probability $p_c(q\,|\,k)$ that, given a vertex with degree k, it is connected to a vertex with degree q. With the available data, a direct plot of $p_c(q\,|\,k)$ results very noisy and difficult to interpret [36]. Thus in [6] we suggested to measure instead the nearest neighbors average degree of the vertices of degree k,

$$\langle k_{nn}\rangle_k = \sum_q q\, p_c(q\,|\,k), \tag{5}$$

and to plot it as a function of the degree k. If there are no degree correlations (*i.e.* for a random network), then $p_c^0(q\,|\,k) = q\,p(q)/\langle k\rangle$, where $p(q)$ is the degree distribution, and we obtain $\langle k_{nn}\rangle_k^0 = \langle k^2\rangle / \langle k\rangle$, which is independent of k. The corresponding plots for the AS and IR maps are shown in Fig. 6. For the AS map we observe a power-law decay for more than two decades, with a characteristic exponent 0.55. This decay is the signature of the presence of negative degree correlations, a property also known in physics and social sciences as *dissasortative mixing* [37]; *i.e.* high degree vertices have a statistical majority of vertices with low degree, while the opposite holds for low degree vertices. This property is another clear signature of the structural organization of the Internet at the AS level. Vertices connectivity properties are arranged in a hierarchy of levels, in which vertices at the top levels are more interconnected with vertices at the bottom levels and vice-versa. As in the analysis of the c_k function, we can observe that there is not a finite amount of hierarchical levels (such as the binary classification stub-transit), but rather we are in the presence of a continuum of levels, in which each degree class k is characterized by a given value of c_k and $\langle k_{nn} \rangle_k$.

The IR map, on the other hand, displays a quite different behavior, that is compatible with slight positive degree correlations (*i.e.* highly connected vertices tend to be connected with vertices with large degree, and similarly for low degree nodes [37]), but that can also be interpreted as an actual lack of correlations, given the small fluctuations of the plot around the expected value for a random network with the same degree distribution, $\langle k_{nn} \rangle_k^0 \simeq 30$. Again, the sharp drop for large k can be attributed to the low statistics for such large connectivities. Therefore, also in this case the two levels of representation show very different features.

It is worth remarking that the present analysis of the hierarchical and correlation properties shows quite marked different properties for the AS and IR maps, differences that should be taken into account when developing Internet topology generators. In other words, Internet protocols working at different representation levels must be thought as working on different topologies. Topology generators as well must include these differences, depending on the level at which we intend to model the Internet topology.

6 Conclusions

The increasing availability of larger Internet maps and the proliferation of growing network models with scale-free features have recently stimulated a more detailed statistical analysis aimed at the identification of distinctive metrics and features for the Internet topology. At this respect, we have presented a detailed statistical analysis of several metrics on Internet maps collected at the router and autonomous system levels. The present analysis confirms the presence of a power-law (scale-free) behavior for the degree distribution, as well as for the betweenness distribution, that can be associated to a measure of the load of the vertices in the maps. The exponential cut-offs observed in the IR maps, associated to the limited capacity of the routers, are absent in the AS level, which conglomerate a large number of routers and are thus able to bear a larger load. The analysis of the clustering coefficient and the nearest neighbors average degree shows in a quantitative way the presence of strong correlations in the Internet degree at the AS level, correlations that can be related to the hierarchical distribution of this network. These correlations, on the other hand, seem to be of different sign and much smaller (or even nonexistent) at the IR level. The correlation properties clearly indicate the presence of strong differences between the IR and AS levels of representation. These evidences represent a step forward in the characterization of the Internet topology, and will be helpful for scrutinizing more thoroughly the actual validity of the network models proposed so far, and as ingredient in the elaboration of new and more realistic Internet topology generators [36].

Acknowledgements

This work has been partially supported by the European Commission - Fet Open project COSIN IST-2001-33555. R.P.-S. acknowledges financial support from the Ministerio de Ciencia y Tecnología (Spain). We thank T. Erlebach for the help in the data collection process.

References

1. B. Bollobás, *Modern Graph Theory*, (Springer-Verlag, New York, 1998).
2. R. Pastor-Satorras and A. Vespignani, *Evolution and Structure of the Internet: A Statistical Physics Approach*, (Cambridge University Press, Cambridge, 2004).
3. M. Faloutsos, P. Faloutsos, and C. Faloutsos, ACM SIGCOMM '99, Comput. Commun. Rev. **29**, 251 (1999).
4. R. Govindan and H. Tangmunarunkit, Proc. of IEEE Infocom 2000, Tel Aviv, Israel
5. G. Caldarelli, R. Marchetti, and L. Pietronero, Europhys. Lett. **52**, 386 (2000).
6. R. Pastor-Satorras, A. Vázquez, and A. Vespignani, Phys. Rev. Lett. **87**, 258701 (2001); A. Vázquez, R. Pastor-Satorras, and A. Vespignani, Phys. Rev. E **65**, 066130 (2002).
7. W. Willinger, R. Govindan, S. Jamin, V. Paxson, and S. Shenker, Proc. Natl. Acad. Sci USA **99** 2573, (2002)
8. H. Tangmunarunkit, R. Govindan, S. Jamin, S. Shenker, and W. Willinger, Comput. Commun. Rev. **32**, 76 (2002)
9. Q. Chen, H. Chang, R. Govindan, S. Jamin, S. J. Shenker, and W. Willinger, Proceedings of IEEE Infocom 2002, New York, USA.
10. A. Broido and K. C. Claffy, San Diego Proceedings of SPIE International symposium on Convergence of IT and Communication. Denver, CO. 2001
11. D. Vukadinovic, P. Huang and T. Erlebach in Innovative Internet Computing Systems (I2CS 2002) Lecture Notes in Computer Science, Springer-Verlag, Berlin, 2002.
12. R. Albert and A.-L. Barabási, Rev. Mod. Phys. **74**, 47 (2002).
13. S. N. Dorogovtsev and J. F. F. Mendes, *Evolution of networks: From biological nets to the Internet and WWW*, (Oxford University Press, Oxford, 2003).
14. P. Erdös and P. Rényi, Publ. Math. Inst. Hung. Acad. Sci. **5**, 17 (1960).
15. B. Bollobás, *Random graphs* (Academic Press, London, 1095).
16. A. Medina, I. Matta, and J. Byers, Comput. Commun. Rev. **30**, 2 (2000).
17. A. Medina and I. Matta, *BRITE: a flexible generator of Internet topologies"*, Tech. Rep. BU-CS-TR-2000-005, Boston University, 2000.
18. C. Jin, Q. Chen, and S. Jamin, *"INET: Internet topology generators"*, Tech. Rep. CSE-TR-433-00, EECS Dept., University of Michigan, 2000.
19. The National Laboratory for Applied Network Research (NLANR), sponsored by the National Science Foundation, provides Internet routing related information based on BGP data (see http://moat.nlanr.net/).

20. The Cooperative Association for Internet Data Analysis (CAIDA), located at the San Diego Supercomputer Center, provides measurements of Internet traffic metrics (see http://www.caida.org/home/).

21. Topology project, Electric Engineering and Computer Science Department, University of Michigan (http://topology.eecs.umich.edu/).

22. Mapping the Internet within the SCAN project at the Information Sciences Institute (http://www.isi.edu/div7/scan/).

23. B. Cheswick and H. Burch, Internet mapping project at Lucent Bell Labs (http://www.cs.bell-labs.com/who/ches/map/).

24. H. Burch and B. Cheswick, IEEE computer, **32**, 97 (1999).

25. D. J. Watts and S. H. Strogatz, Nature **393**, 440 (1998).

26. H. Chou, (http://xxx.lanl.gov/abs/cs.NI/0012019).

27. R. A. Albert, H. Jeong, and A.-L. Barabási, Nature **406**, 378 (2000); D. S. Callaway, M. E. J. Newman, S. H. Strogatz, and D. J. Watts, Phys. Rev. Lett. **85**, 5468 (2000); R. Cohen, K. Erez, D. ben-Avraham, and S. Havlin, Phys. Rev. Lett. **86**, 3682 (2001).

28. R. Pastor-Satorras and A. Vespignani, Phys. Rev. Lett. **86**, 3200 (2001).

29. S. N. Dorogovtsev and J. F. F. Mendes, Advances in Physics **51**, 1079 (2002).

30. L. A. N. Amaral, A. Scala, M. Barthélémy, and H. E. Stanley, Proc. Natl. Acad. Sci. USA **97**, 11149 (2000).

31. S. Mossa, M. Barthélémy, H. E. Stanley, and L. A. N. Amaral, Phys. Rev. Lett. **88**, 138701 (2002).

32. M. E. J. Newman, Phys. Rev. E **64**, 016131-1 (2001); **64**, 016132-1 (2001)

33. U. Brandes, Journal of Math. Sociology, **25**, 35 (2001).

34. K.-I. Goh, B. Kahng, and D. Kim, Phys. Rev. Lett. **87**, 278201 (2001).

35. E. Ravasz and A. Barabási, Phys. Rev. E **67**, 026112 (2003).

36. K.-I. Goh, B. Kahng, and D. Kim, Phys. Rev. Lett. **88**, 108701 (2002).

37. M. E. J. Newman, Phys. Rev. Lett. **89**, 208701 (2002).

Part IV

Biological Networks

Characteristics of Biological Networks

Albert-László Barabási[1], Zoltán N. Oltvai[2], and Stefan Wuchty[1]

[1] Department of Physics, University of Notre Dame, Notre Dame, IN 46556, USA
[2] Department of Pathology, Northwestern University, Chicago, IL 60611, USA

Abstract. Network principles describe uniformly systems as diverse as the cell or the Internet. The emergence of these networks is driven by self-organizing processes that are governed by simple but generic laws. While unraveling the complex and interwoven systems of different interacting units, it has become clear that the topology of networks of different origin share the same characteristics on the large scale. In biological systems, networks appear in many different disguises ranging from protein interactions to metabolic networks. In this paper, we survey the most prominent characteristics of biological networks focusing on the emergence of scale-free architecture and hierarchical arrangement of functional modules. Finally, we present empirical evidence that cohesive parts of the protein interaction network have a significantly higher tendency to be evolutionary conserved.

1 Introduction

Integrating the information collected about the world requires breaking the studied systems into comprehensible small parts and understanding the ways these parts interact with each other. In many cases, the mutual relationships between the components are best described as complex networks which offer us a new way to categorize systems of very different origin in a single framework [1,2]. This approach has uncovered unexpected similarities between the organization of various complex systems, indicating that the networks describing real systems are governed by generic organization principles and mechanisms.

During the last decade, genomics has produced an incredible quantity of molecular interaction data, contributing to maps of specific cellular networks. The emerging fields of transcriptomics and proteomics have the potential to join the already extensive data sources provided by the genome wide analysis of gene expression at the mRNA and protein level [3–5]. Indeed, extensive protein-protein interaction maps generated for a variety of organisms including viruses [6,7], prokaryotes, like *H.pylori* [8] and eukaryotes, like *S. cerevisiae* [9–15], *C.elegans* [16,17] and *D. melanogaster* [18] triggered a series of investigations which aim on the topological properties of the underlying networks [15,19,20]. Beyond the current focus on uncovering the structure of genomes, proteomes and interactomes of various organisms, some of the most extensive datasets are the metabolic maps [21,22], laying the foundation of an increasing number of studies focusing on the architecture of the metabolism [23–25].

A.-L. Barabási, Z.N. Oltvai, and S. Wuchty, Characteristics of Biological Networks, Lect. Notes Phys.
650, 443–457 (2004)
http://www.springerlink.com/   © Springer-Verlag Berlin Heidelberg 2004

Understanding the driving forces which shape networks with diverse constituents and conceiving why these networks share similar topological features allows biology to combine the numerous details about molecular interactions into a single framework, offering means to address the structure of the cell as a whole.

2 Basic Network Features

The most elementary networks measure is a nodes degree (or connectivity), reflecting the number of neighbors k a particular node has. Basically, various network topologies are mainly classified by the connectivity distributions $P(k)$ of their nodes.

As in most networks, there exist multiple paths between any two nodes i and j. A useful distance measure is the length of the shortest path, l_{ij}. The mean path length is defined as

$$\langle l \rangle = \frac{2}{N(N-1)} \sum_{i<j} l_{ij} \,, \tag{1}$$

offering a measure of the networks navigability. A network which can be traversed by a relatively small number of steps is often called a 'small world' network: l depends logarithmically on the network's actual size, $l \sim \log N$. First illustrated with social networks, two randomly chosen individuals can be connected by only six intermediate acquaintances [26] although the actual worldwide web of acquaintances consists of billions of nodes.

Nodes in many real systems exhibit a tendency to cluster, a network property which can be quantified using the clustering coefficient [27]. This topological measure reflects the extent to which the neighbors of a particular node are connected to each other. For example, in a friendship network C tells the degree to which friends of a particular person are friends with each other as well. Formally, the clustering coefficient of node i is defined as

$$C_i = \frac{2n_i}{k_i(k_i - 1)}, \tag{2}$$

where n_i denotes the number of links connecting the k_i neighbors of node i to each other. Accordingly, we can define the average clustering coefficient as

$$\langle C \rangle = \frac{1}{N} \sum_{i=1}^{N} C_i. \tag{3}$$

An additional important measure of the networks structure is the function $C(k)$, defined as the average clustering coefficient of all nodes with k links. If $C(k)$ is independent of k the network is either homogeneous or it is dominated by numerous small tightly linked clusters. In contrast, if $C(k)$ follows $C(k) \sim k^{-\beta}$, the network has a hierarchical architecture meaning that sparsely connected nodes are parts of highly cohesive areas [28–31]. In such hierarchical networks, communication between the different highly clustered neighborhoods are maintained by a few hubs.

As we will see below, the degree distribution $P(k)$ and the k dependence of $C(k)$ can have generic features, allowing us to classify various networks. Parameters such as the average degree $\langle k \rangle$, average path length $\langle l \rangle$ and average clustering coefficient $\langle C \rangle$ characterize the unique properties of the particular network under consideration.

3 Network Models

The main role of network models is to explain the emergence and behavior of some of the most important network characteristics. As we will see, models play a crucial role in shaping our understanding of complex networks.

3.1 Random Networks

While graph theory initially focused on regular graphs, since the 1950's large networks with no apparent design principles were described as random graphs [32], proposed as the simplest and most straightforward realization of a complex network. According to the Erdős-Rényi (ER) model of random graphs [33], we start with N nodes and connect each pair of nodes with probability p, creating a graph with approximately $pN(N-1)/2$ randomly distributed links (first column in Fig. 1). The ER graph has a Poisson-like degree distribution indicating that most nodes have approximately the same number of links, $k \approx \langle k \rangle$ (first column in Fig. 2) and exhibits the small-world property, seen as a proportionality between the mean path length and the logarithm of the number of nodes, $L \sim \log N$.

The growing interest in complex systems prompted many scientists to ask a simple question: Are real networks behind diverse complex systems, like the cell, fundamentally random?

3.2 Scale-Free Networks

A highly nontrivial development in our understanding of complex networks was the discovery that for most large networks, such as the world wide web [34], metabolic and protein interaction networks [15,23,25], the degree distribution follows a power-law

$$P(k) \sim k^{-\gamma}. \tag{4}$$

These networks are called scale-free, as a power-law does not support the existence of a characteristic scale. Two mechanisms, absent from the classical random

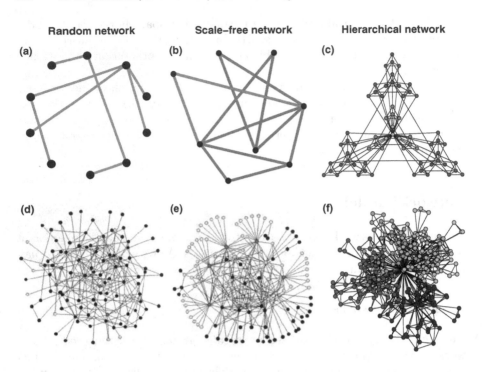

Fig. 1. (a) The random network model is constructed by laying down N nodes and connecting each pair of nodes with probability p. The figure shows a particular realization of such a network for $N = 10$ and $p = 0.2$. (b) The scale-free model assumes that the network constantly grows by the addition of new nodes. The figure shows the network at time t (nodes connected by green links) and after the addition of a new node at time $t+1$ (red links). With the introduction of new nodes, already highly connected ones are more favored to be connected to the new one than less connected nodes. This procedure is called *preferential attachment*. (c) The iterative construction of a hierarchical network starts from a fully connected cluster of four nodes (blue), which is replicated three times. Subsequently, the peripheral nodes of each replica (green) are connected to the central node of the original module. Repeating the replication and the connection step with the 16-node module (red) leads to a 64-nodes network which provides scale-free topology and is built by nested modules. (d) The random network is rather homogeneous, i.e. most nodes have approximately the same number of links. (e) In contrast, a scale-free network is extremely inhomogeneous: while the majority of nodes has one or two links, a few nodes have a large number of links preserving the systems integrity. To show this, five nodes with the highest number of links are colored red, and their first neighbors are colored green. While in the random network only 27% of the nodes are reached by the five most connected nodes, in the scale-free network more than 60% are, demonstrating the key role hubs play in the scale-free network. Note, that both networks contain the same number of nodes and links. (f) A hierarchical network still preserves its scale-free organization and displays inherent modularity of nodes. The nodes affiliation to a certain module is indicated by different colors. However, the underlying networks structure clearly indicates blurred boundaries of its modules.

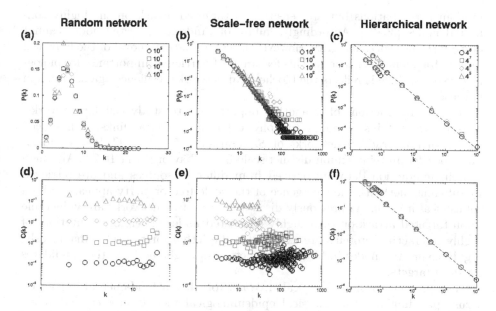

Fig. 2. (a) For the random graph, the degree distribution, $P(k)$, which gives the probability that a randomly selected node has exactly k edges, follows a Poisson distribution which is strongly peaked at the average degree $\langle k \rangle$ and decays exponentially for large k. (b), (c) $P(k)$'s of a scale-free and a hierarchical network do not have a peak and decay as a power-law, $P(k) \sim k^{-\gamma}$. (d), (e) For both the random and the scale-free network, the $C(k)$ function, which denotes the mean clustering coefficient for nodes with exactly k links, is independent of k. (f) In contrast, $C(k)$ of a hierarchical network depends on k, decaying as $C(k) \sim k^{-1}$. Insets correspond to the number of the underlying networks.

network model, are responsible for the emergence of this power-law degree distribution [34,35]. First, most networks grow through the addition of new nodes, that link to nodes already present in the system. Second, in most real networks there is a higher probability to link to a node with a large number of connections, a property called preferential attachment. The scale-free model introduced by Barabási and Albert (BA) (second column in Figs. 1 and 2) incorporates these features. Starting from a small graph, at each time step a node with m links is added to the network, connecting to a previously present node i with probability

$$\Pi_i = k_i / \sum_j k_j, \qquad \qquad \cdot (5)$$

where k_i is the degree of node i. The network generated by this growth process will be scale-free with degree exponent $\gamma = 3$. In a scale-free network the probability that a node is highly connected ($k \gg \langle k \rangle$) is statistically more significant than in a random graph. Thus, the networks properties are governed by a relatively small number of highly connected nodes or hubs. An important consequence of the hubs is that scale-free networks exhibit high tolerance to random perturbations but are sensitive to targeted attack on the highly connected nodes [36–38]. Accordingly, failure of randomly selected nodes cannot destroy the networks integrity. However, the systematic removal of the hubs will rapidly fragment the network. This feature is of particular importance for biological systems, since it reflects the biochemical networks resilience against random mutations.

Protein interactions offer another opportunity to study cellular networks, considering proteins as nodes and physical interactions as links. It has been shown that interaction networks of *S. cerevisiae*, *H.pylori*, *C.elegans* and *D. melanogaster* proteins exhibit distinct scale-free behavior [15,17,18,20]. Although protein interaction data are derived from different sources and are retrieved by different methods, the emergence of the scale-free property appears to be a robust feature [39]. As previously discussed, scale-free networks are vulnerable upon targeted attack on their highly connected nodes. Therefore, mutations of highly interacting proteins are expected to be lethal for the cell. Analogously, highly connected nodes in biochemical networks might be potential candidates for drug targets.

The presence of hubs in a scale-free network has a fundamental impact on virus spreading as well. Classical epidemiological models predict that infectious diseases with transmission probability under an epidemic threshold will inevitably die out. However, in scale-free networks the epidemic threshold is reduced to zero [40]. Thus, as some social and sexual networks are known to exhibit a scale-free topology [41], even extremely weakly infectious viruses can spread and prevail, making random immunization ineffective.

3.3 Hierarchical Networks

Many real networks are expected to be fundamentally modular, meaning that the network can be seamlessly partitioned into a collection of modules. Each module is expected to perform an identifiable task, separable from the function of other modules [42–45]. Therefore, we must reconcile the scale-free property with the networks potential modularity. Numerical simulations indicate that neither the random nor the scale-free network model are modular.

In order to account for the coexistence of modularity, local clustering and scale-free topology in real systems, we have to assume that clusters combine in an iterative manner, generating a hierarchical network [29,46]. Such networks emerge from an iterative duplication and integration of clustered nodes, a process which in principle can be repeated indefinitely. Our starting point is a small cluster of four densely linked nodes. Next, we generate three replicas of this hypothetical module and connect the three external nodes of the replicated clusters to the central node of the old cluster, obtaining a large 16-node module. Subsequently, we again generate three replicas of this 16-node module, and connect the 9 peripheral nodes of each replica to the central node of the old module, obtaining a new module of 64 nodes (third column of Fig. 1).

The hierarchical network model seamlessly integrates a scale-free topology with an inherent modular structure by generating a network that has a power

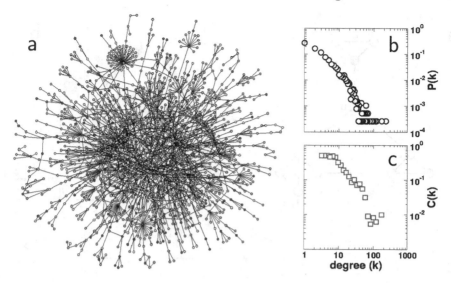

Fig. 3. Map and statistical characteristics of the protein-protein interaction network of
S. cerevisiae [15]. **a:** The color code of nodes refers to the phenotypic effect the deletion
of the respective protein has on the organism (red: lethal, green: viable, orange: slowed
growth, yellow:unknown). **b:** Connectivity distribution of the proteins degree k follows
approximately a power-law $P(k) \sim k^{-\gamma}$ indicating the scale-free nature of the proteome
c: The clustering coefficient measuring the cliquishness of the networks nodes exhibits
a power-law dependence on the nodes degree $C(k) \sim k^{-\beta}$, indicating the presence of
modules in the network.

law degree distribution with degree exponent $\gamma = 1 + \ln 4/\ln 3 = 2.26$. Yet,
the most important signature of this hierarchical modularity is the fact that
the clustering coefficient, $C(k)$, scales as k^{-1} (third column of Fig. 2). Note,
that for the network generated by the ER and BA models $C(k)$ is independent
of k.

In protein-protein interaction networks (Fig. 3a), the power-law dependence
of the degree distribution $P(k) \sim k^{-\gamma}$ indicates its scale-free topology (Fig. 3b)
[15]. In addition, we observe that the clustering coefficient of this network $C(k)$
scales as $C(k) \sim k^{-\beta}$, indicating the presence of modules in the network (Fig. 3c)
[39,47,48]. A cell's metabolism consists of a set of biochemical reactions shaping
a network featuring chemical substances (the metabolites) as nodes and their
participation in these reactions as links. Similar to protein-protein interaction
networks, metabolic networks have scale-free topology since their degree distributions feature a power-law [15]. Furthermore, the clustering coefficient varies
with the inverse degree, $C(k) \sim k^{-1}$, indicating an inherent hierarchical modularity [28]. In order to discern the discrete modules, for each pair of nodes, we
define the topological overlap [28]

$$O_T = \frac{J_n(i,j)}{\min(k_i, k_j)}, \tag{6}$$

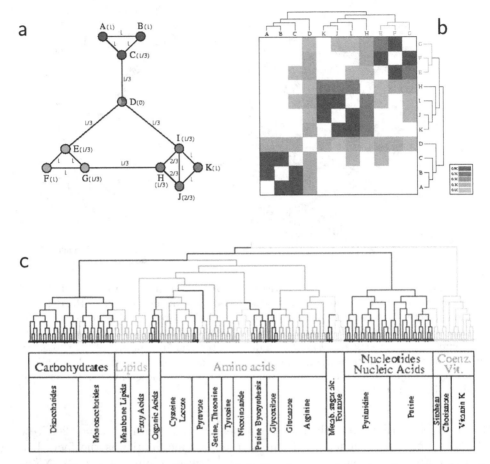

Fig. 4. Uncovering the underlying modularity of a complex network. **a:** Topological overlap illustrated on a hypothetical network. On each link, we indicate the topological overlap for the connected nodes, and in parentheses next to each node we indicate the node's clustering coefficient. **b:** The topological overlap matrix corresponds to the small network shown in a. The rows and columns of the matrix were reordered by the application of an average-linkage clustering method to its elements, allowing us to identify those nodes which have a high topological overlap. The associated tree reflects the three distinct modules built into the model of a, as well as the fact that the *EFG* and *HIJK* modules are closer to each other in a topological sense than to the *ABC* module. **c:** Hierarchies of topological modules in the *E. coli* metabolism. The branches of the tree obtained by average-linkage clustering of the topological overlap of metabolites [28] are color-coded to reflect the predominant biochemical classification of their substrates. The biochemical classes represent carbohydrate metabolism (blue); nucleotide and nucleic acids metabolism (red); protein, peptide and amino acid metabolism (green); lipid metabolism (cyan); aromatic compound metabolism (dark pink); monocarbon metabolism (yellow) and coenzym metabolism(light orange) [21].

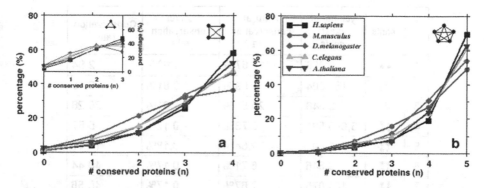

Fig. 5. Relationship between the topology of a protein interaction network and the evolutionary conservation of individual proteins. The panel shows the detailed conservation rates of fully connected 3- (inset a), 4- (a) and 5- (b) node motifs. For example, (b) indicates that in human less than 1% of the pentagon motifs found originally in yeast have fully disappeared (i.e., none of their components have an ortholog), and only 1.5% of motifs have a single ortholog component (n = 1), while for more than 69% of the motifs each of the five proteins have been conserved (n = 5).

where $J_n(i,j)$ denotes the number of nodes to which both i and j are linked (plus 1 if there is a direct link between i and j) and $\min(k_i, k_j)$ is the smaller of the k_i and k_j degrees. This topological measure, which scales from 0 to 1, reflects the extent to which two metabolites i and j interact with the same substrates. If the latter are part of larger metabolic modules they appear to have a higher topological overlap with their neighbors (Fig. 4a,b). The application of average-linkage clustering to the obtained overlap matrix has been used to uncover the topological modules present in the metabolism (Fig. 4c). The organization of clusters identified a hierarchy of nested topological modules of increasing sizes and decreasing interconnectedness. The hierarchical tree offers a breakdown of the metabolism into several large modules which are further partitioned into smaller but more integrated submodules, reflecting a certain degree of inherent self-similarity. Some of these modules have been found to be in excellent agreement with the known functional classification of metabolites [49].

Other approaches to discern modules in metabolic networks focused on the appearance of edges in mutual shortest paths in the network [50,51]. The most frequent edges were identified and removed in an iterative manner, uncovering again the underlying functional modules. Modularity does not, however, imply clear-cut subnetworks which are linked in well-defined ways. In fact, the boundaries of modules are often considerably blurred, triggered by highly connected nodes which interconnect modules.

3.4 Network Motifs

Many biological functions are carried out by the integrated activity of highly interacting cellular components referred to as functional modules. Motifs, con-

	Motifs	Number of yeast motifs	Natural conservation rate	Random conservation rate	Conservation ratio
1	••	9,266	13.67%	4.63%	2.94
2	⁘	167,304	4.99%	0.81%	6.15
3	⁘	3,846	20.51%	1.01%	20.28
4	⁙	3,649,591	0.73%	0.12%	5.87
5	⁙	1,763,891	2.64%	0.18%	14.67
6	⁙	9,646	6.71%	0.17%	40.44
7	⁙	164,075	7.67%	0.17%	45.56
8	⁙	12,423	18.68%	0.12%	157.89
9	⁙	2,339	32.53%	0.08%	422.78
10	⁘	25,749	14.77%	0.05%	279.71
11	⁙	1,433	47.24%	0.02%	2,256.67

Fig. 6. The evolutionary conservation of motif constituents. The third column denotes the number of motifs of a given kind found in the yeast protein interaction network of 3,174 proteins, obtained by counting all subgraphs of two to five nodes. We identified 678 proteins that have an ortholog in each of the five studied higher eukaryotes, and identified all motifs for which each component belongs to this evolutionary conserved protein subset. The natural conservation rate shows what fraction of the original yeast motifs are evolutionary fully conserved. For example, we find that 47% of the fully connected pentagons (#11) have each of their five proteins conserved in each of the five higher eukaryotes. The random conservation rate therefore denotes the fraction of motifs which are found to be fully conserved for the random ortholog distribution. The last column denotes the ratio between the natural and the random conservation ratios, indicating that all motifs are highly conserved, some (for example #11) having a natural conservation rate 2,256 times higher than expected in the absence of correlations between protein conservation rate and the topology of a given motif.

sidered as topologically distinct interaction patterns within complex networks, may represent the simplest building blocks of such modules [45,52]. Owing to their small size, motifs can be explicitly identified and enumerated in various cellular networks, but their biological significance, if any, remains undetermined. A well-known signature of the conservation of specific cellular functions is the evolutionary conservation of orthologous proteins that are responsible for selected functions. We call proteins of different species orthologous if they have diverged in two lineages from a common ancestor. If there is indeed an evolutionary pressure to maintain specific motifs, their components should be evolutionary conserved. An analysis aiming on the evolutionary conservation of motifs determined from a web of yeast protein interactions [53] uncovers a trend toward the preferential retention of highly cohesive motifs [54]. Figure 5 shows that larger fully connected motifs are widely conserved as a whole, meaning that all the motif's constituents have an ortholog in higher eukaryotes such as *H.sapiens*, *M.musculus*, *D.melanogaster*, *C.elegans* and *A.thaliana* [55].

These results indicate that the orthologs are not randomly distributed in the yeast protein interaction network, but are the building blocks of cohesive motifs, which tend to be evolutionary conserved. If the same number of orthologs were randomly placed on the yeast protein interaction network, mimicking the absence of any correlations between the network topology and the ortholog position, the motif conservation observed above should disappear. Indeed, motifs under such random ortholog distribution display a trend opposite to that observed for the original non-random system: The larger the motif, the smaller is the likelihood that each of its components is conserved (Table in Fig. 6). The influence of the global network topology on the retention rate of specific local motifs is best quantified by calculating the ratio between the real and the random conservation rate. Interestingly, for each motif this conservation ratio is bigger than one and increases significantly for larger motifs (table in Fig. 6). These exceptionally high conservation rates strongly suggest that participation within motifs significantly influences the evolutionary conservation of their specific components.

4 Conclusions

The power-law degree distribution, the quantitative signature of a scale-free network, has emerged as one of the few universal laws characterizing cellular networks. Of even greater immediate importance is the intriguing possibility of using the insights provided by the scale-free models as a framework to facilitate the analysis of biological networks at a higher level of abstraction. Such approaches could reveal salient features of biological phenomena missed by non-network based approaches. The appearance of hierarchical modularity in biological networks supports the assumption that evolution acts on many levels. The accumulation of local changes, affecting the small highly integrated modules, slowly impacts the larger, less integrated modules as well. Thus, evolution might act in self-similar fashion, copying and reusing existing modules to further increase the organism's complexity. This conclusion is further supported by the lately uncovered tendency that the constituents of highly cohesive interaction motifs are preferentially conserved, as orthologs in higher eukaryotes. Obviously, the topological placement of proteins is a crucial factor for their tendency to be evolutionary conserved suggesting that proteins organized in a cohesive subpart of the interaction network might be duplicated and subsequently altered as a whole in order to develop new biological functions. Especially in the face of eukaryotic evolution, this perception might be suitable to describe the explosion of complexity in the development of the single-celled *S. cerevisiae* toward the multicellular *H. sapiens*.

The universal applicability of these network concepts to the different layers of biological informations - genome, transcriptome, proteome and metabolome – allows us to draw a picture of the complex relationships between these components. In a simple complexity pyramid, various molecular components - genes,

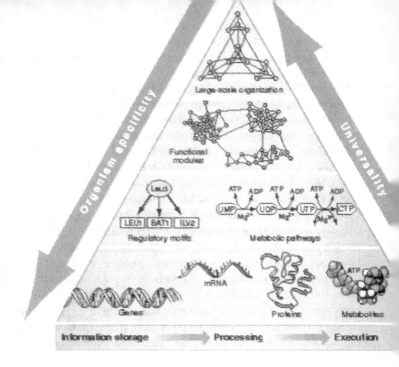

Fig. 7. From the *particular* to the *universal*: The bottom (level 1) of the schematic representation of the cell's functional organization: genome, proteome and metabolom. Insights into the cell's organization can be consider the components to be linked by functional relationships, such motifs and metabolic pathways (level 2). In turn, they are the building erational modules (level 3) which are nested and considerably blurre scale-free hierarchical architecture (level 4). Although the individual c unique, the topological properties of biological networks share astoundi This suggests that universal organizing principles apply to all kinds o works [56].

RNAs, proteins and metabolites – are linked by functional relatic rise to recurrent patterns such as metabolic pathways and gene motifs. On a succeeding level of abstraction, motifs and pathways a integrated to form functional modules which are responsible for di functions [42]. These modules are nested, considerably blurred a scale-free, hierarchical architecture (Fig. 7).

The recent progress in biological networks analysis [57] has su covered the skeleton and organization of networks, offering impo about the assembly and functionality of components and subnetwo we will need to go several steps further addressing the dynamic as ous cellular networks. Especially, the analysis of fluxes and fluctuat links in metabolic and regulatory pathways will play a major role, significantly influencing potential biotechnological applications.

References

1. Réka Albert and Albert-László Barabási. Statistical mechanics of complex networks. *Rev. Mod. Phys.*, 74:47–97, 2002.
2. S. N. Dorogovtsev and J. F. F. Mendes. Evolution of Random Networks. *Adv. Phys.*, 51:1079–1187, 2002.
3. A. Pandey and M. Mann. Proteomics to study genes and genomes. *Nature*, 405:837 – 846, 2000.
4. H. Caron, B. van Schaik, M. van der Mee, F. Baas, G. Riggins, P. van Sluis, M.-C. Hermus, R. van Asperen, K. Boon, P. A. Voute, S. Heisterkamp, A. van Kampen, and R. Versteeg. The Human Transcriptome Map: Clustering of Highly Expressed Genes in Chromosomal Domains. *Science*, 291:1289–1292, 2001.
5. C.B. Burge. Chipping away at the transcriptome. *Nature Genet.*, 27:232–234, 2001.
6. M. Flajolet, G. Rotondo, L. Daviet, F. Bergametti, G. Inchauspe, P. Tiollais, C. Transy, and P. Legrain. A genomic approach to the hepatitis c virus. *Gene*, 242:369–379, 2000.
7. S. McGraith, T. Holtzman, B. Moss, and S. Fields. Genome-wide analysis of vaccinia virus protein-protein interactions. *Proc. Natl. Acad. Sci. USA*, 97:4879–4884, 2000.
8. J.-C. Rain, L. Selig, H. DeReuse, V. Battaglia, C. Reverdy, S. Simon, G. Lenzen, F. Petel, J. Wojcik, V. Schächter, et al. The protein-protein interaction map of Helicobacter pylori. *Nature*, 409:211–215, 2001.
9. T. Ito, K. Tashiro, S. Muta, R. Ozawa, T. Chiba, M. Nishizawa, K. Yamamoto, S. Kuhara, and Y. Sakaki. Towards a protein-protein interaction map of the budding yeast: A comprehensive system to examine two-hybrid interactions in all possible combinations between the yeast proteins. *Proc. Nat. Acad. Sci. USA*, 97(3):1143–1147, 2000.
10. T. Ito, T. Chiba, R. Ozawa, M. Yoshida, M. Hattori, and Y. Sakaki. A comprehensive two-hybrid analysis to explore the yeast protein interactome. *Proc. Nat. Acad. Sci. USA*, 98(8):4569–4574, 2001.
11. B. Schwikowski, P. Uetz, and S. Fields. A network of protein-protein interactions in yeast. *Nature Biotechn.*, 18:1257–1261, 2000.
12. P. Uetz, L. Giot, G. Cagney, T.A. Mansfield, R.S Judson, J.R. Knight, D. Lockshorn, V. Narayan, M. Srinivasan, P. Pochart, et al. A comprehensive analysis of protein-protein interactions of Saccharomyces cerevisiae. *Nature*, 403:623–627, 2000.
13. A.C. Gavin, M. Bösche, R. Krause, P. Grandi, M. Marzioch, A. Bauer, J. Schultz, J.M. Rick, A.-M. Michon, C.-M. Cruciat, et al. Functional organization of the yeast proteome by systematic analysis of protein complexes. *Nature*, 415:141–147, 2002.
14. Y. Ho, A. Gruhler, A. Heilbut, G.D. Bader, L. Moore, S.-L. Adams, A. Millar, P. Taylor, K. Bennett, K. Boutillier, et al. Systematic identification of protein complexes in Saccharomyces cerevisiae by mass spectrometry. *Nature*, 415:180 – 183, 2002.

15. H. Jeong, S.P. Mason, A.-L. Barabási, and Z.N. Oltvai. Lethality and centrality in protein networks. *Nature*, 411:41–42, 2001.

16. A.J.M. Walhout, R. Sordella, X.W. Lu, J.L. Hartley, G.F. Temple, M.A. Brasch, N. Thierry-Mieg, and M. Vidal. Protein interaction mapping in C. elegans using proteins involved in vulval development. *Science*, 287:116–122, 2000.

17. S. Li, C.M. Armstrong, N. Bertin, H. Ge, S. Milstein, M. Boxem, P.-O. Vidalain, J.-D.J. Han, A. Chesneau, T. Ha, et al. A map of the interactome network of the metazoan C. elegans. *Science*, 303:540–543, 2004.

18. L. Giot, J.S. Bader, C. Brouwer, A. Chaudhuri, B. Kuang, Y. Li, Y.L. Hao, C.E. Ooi, B. Godwin, E. Vitols, et al. A protein interaction map of Drosophila melanogaster. *Science*, 302:1727–1736, 2004.

19. S. Wuchty. Interaction and Domain Networks of Yeast. *Proteomics*, 2:1715–1723, 2002.

20. A. Wagner. The Yeast Protein Interaction Network Evolves Rapidly and Contains Few Redundant Duplicate Genes. *Mol. Biol. Evol*, 18(7):1283–1292, 2001.

21. R. Overbeek, N. Larsen, G.D. Pusch, M. D'Souza, E. Selkov Jr, N. Kyrpides, M. Fonstein, N. Maltsev, and E. Selkov. WIT: integrated system for high-throughput genome sequence analysis and metabolic reconstruction. *Nucleic Acids Res.*, 28:123–125, 2000.

22. P. D. Karp, M. Riley, M. Saier, I.T. Paulsen, S.M. Paley, and A. Pellegrini-Toole. The EcoCyc and MetaCyc databases. *Nucl. Acids Res.*, 28:56–59, 2000.

23. H. Jeong, B. Tombor, R. Albert, Z.N. Oltvai, and A.-L. Barabási. The large-scale organization of metabolic networks. *Nature*, 407:651–654, 2000.

24. D.A. Fell and A. Wagner. The small world of metabolism. *Nature Biotech.*, 189:1121–1122, 2000.

25. A. Wagner and D.A. Fell. The small world inside large metabolic networks. *Proc. R. Soc. Lon. B*, 268:1803–1810, 2001.

26. S. Milgram. The Small-World Problem. *Psychology Today*, 2:60–67, 1967.

27. D. J. Watts and S. H. Strogatz. Collective dynamics of small-world networks. *Nature*, 393:440–442, 1998.

28. E. Ravasz, A.L. Somera, D.A. Mongru, Z.N. Oltvai, and A.-L. Barabási. Hierarchical organization of modularity in metabolic networks. *Science*, 297:1551–1555, 2002.

29. E. Ravasz and A.-L. Barabási. Hierarchical Organization in Complex Networks. *Phys. Rev. E*, 67:026122, 2002.

30. S. N. Dorogovtsev, A. V. Goltsev, and J. F. F. Mendes. Pseudofractal Scale-free Web. *Phys. Rev. E*, 65:066122, 2002.

31. S. Jung, S. Kim, and B. Kahng. A Geometric Fractal Growth Model for Scale Free Networks. *Phys. Rev. E*, 65:056101, 2002.

32. B. Bollobás. *Random Graphs*. Academic Press, London, 1985.

33. P. Erdös and A. Rényi. On the evolution of random graphs. *Publ. Math. Inst. Hung. Acad. Sci.*, 5:17–61, 1960.

34. A.-L. Barabási and R. Albert. Emergence of scaling in random networks. *Science*, 286:509–512, 1999.

35. A.-L. Barabási, R. Albert, and H. Jeong. Mean-field theory for scale-free random networks. *Physica A*, 272:173–187, 1999.

36. R. Albert, H. Jeong, and A.-L. Barabási. Attack and error tolerance of complex networks. *Nature*, 406:378, 2000.

37. R. Cohen, K. Erez, D. ben Avraham, and S. Havlin. Resilience of the internet to random breakdowns. *Phys. Rev. Lett.*, 85(21):4626–4628, 2000.

38. R. Cohen, K. Erez, D. ben Avraham, and S. Havlin. Breakdown of the internet under intentional attack. *Phys. Rev. Lett.*, 86(16):3682–3685, 2001.

39. S. H. Yook, Z.N. Oltvai, and A.-L. Barabaśi. Functional and topological characterization of protein interaction networks, 2004. Proteomics, in press.

40. R. Pastor-Satorras and A. Vespignani. Epidemic spreading in scale-free networks. *Phys. Rev. Lett.*, 86:3200–3203, 2001.

41. F. Liljeros, C.R. Edling, L.A.N. Amaral, and Y. Aberg. The web of human sexual contacts. *Nature*, 411:907–908, 2001.

42. L. H. Hartwell, J. J. Hopfield, S. Leibler, and A. W. Murray. From molecular to modular cell biology. *Nature*, 402:C47–C52, 1999.

43. Y.I. Wolf, G. Karev, and E.V. Koonin. Scale-free networks in biology: new insights into the fundamentals of evolution? *Bioessays*, 24:105–109, 2002.

44. D.A. Lauffenburger. Cell signaling pathways as control modules: Complexity for simplicity. *Proc. Natl. Acad. Sci. USA*, 97:5031–5033, 2000.

45. S.S. Shen-Orr, R. Milo, S. Mangan, and U. Alon. Network motifs in the transcriptional regulation network of E.coli. *Nature Genet.*, 31:64 – 68, 2002.

46. A.-L. Barabási, E. Ravasz, and T. Vicsek. Deterministic scale-free networks. *Physica A*, 299:559–564, 2001.

47. A.W. Rives and T. Galitski. Modular organisation of cellular networks. *Proc. Natl. Acad. Sci. U.S.A.*, 100:1128–1133, 2003.

48. V. Spirin and L. Mirny. Protein complexes and functional modules in molecular networks. *Proc. Natl. Acad. Sci. USA*, 100:12123–12128, 2003.

49. S.Y. Gerdes, M.D. Scholle, J.W. Campbell, G. Balázsi, E. Ravasz, M.D. Daugherty, A.L. Somera, N.C. Kyripides, I. Anderson, M.S. Gelfand, et al. Experimental determination and system level analysis of essential genes in Escherichia coli mg1655. *J. Bact.*, 185(19):5673–5684, 2003.

50. P. Holme, M. Huss, and H. Jeong. Subnetwork hierarchies in biochemical pathways. *Bioinformatics*, 19(4):532–538, 2003.

51. M. Girvan and M.E.J. Newman. Community structure in social and biological networks. *Proc. Natl. Acad. Sci. USA*, 99:7821–7826, 2002.

52. R. Milo, S.S. Shen-Orr, S. Itzkovitz, N. Kashtan, D. Chklovskii, and U. Alon. Network motifs: Simple building locks of complex networks. *Science*, 298:824–827, 2002.

53. I. Xenarios, L. Salwinski, X.J. Duan, P. Higney, S.-M. Kim, and David Eisenberg. Dip, the database of interacting proteins: a research tool for studying cellular networks of protein interactions. *Nucl. Acids Res.*, 30:303–305, 2002.

54. S. Wuchty, Z.N. Oltvai, and A.-L. Barabási. Evolutionary conservation of motif constituents in the yeast protein interaction network. *Nature Genetics*, 35:176–179, 2003.

55. M. Remm, C.E.V. Storm, and E.L. Sonnhammer. Automatic clustering of orthologs and in-paralogs from pairwise species comparisons. *J. Mol. Biol.*, 314:1041–1052, 2001.

56. Z.N. Oltvai and A.-L. Barabási. Life's Complexity Pyramid. *Science*, 298:763–764, 2002.

57. A.-L. Barabási and Z.N. Oltvai. Network biology: Understanding the cells's functional organization. *Nature Rev. Genetics*, 5:101–113, 2004.

Boolean Modeling
of Genetic Regulatory Networks

Réka Albert

Department of Physics, Pennsylvania State University, University Park, PA 16802, USA

Abstract. Biological systems form complex networks of interaction on several scales, ranging from the molecular to the ecosystem level. On the subcellular scale, interaction between genes and gene products (mRNAs, proteins) forms the basis of essential processes like signal transduction, cell metabolism or embryonic development. Recent experimental advances helped uncover the qualitative structure of many gene control networks, creating a surge of interest in the quantitative description of gene regulation. We give a brief description of the main frameworks and methods used in modeling gene regulatory networks, then focus on a recent model of the segment polarity genes of the fruit fly Drosophila melanogaster.

The basis of this model is the known interactions between the products of the segment polarity genes, and the network topology these interactions form. The interactions between mRNAs and proteins are described as logical (Boolean) functions. The success in reproducing both wild type and mutant gene expression patterns suggests that the kinetic details of the interactions are not essential as long as the network of interactions is unperturbed. The model predicts the gene patterns for cases that were not yet studied experimentally, and implies a remarkable robustness toward changes in internal parameters, initial conditions and even some mutations.

The success of this approach also suggests a wide applicability of real-topology-based Boolean modeling for gene regulatory networks. In cases when the information about the system is incomplete, Boolean modeling can verify the sufficiency of interactions and can propose ways to complete the network. After a coherent picture is obtained, more realistic kinetic models can be used to gain additional insights into the functioning of the system.

1 Introduction

1.1 Complex Networks in Genetic Regulation

Recent remarkable progress in molecular biology has led to a complete map of the genomes of many organisms, and the identification and classification of the proteins is well under way. The next major challenge is to determine all the interactions between genes, proteins and other cellular components and to integrate this knowledge into a system-level understanding. It is now widely recognized that the networks of interaction and regulation between cellular entities are highly complex, and their understanding needs a concerted effort between experiment, modeling and theory.

R. Albert, Boolean Modeling of Genetic Regulatory Networks, Lect. Notes Phys. **650**, 459–481 (2004)
http://www.springerlink.com/
© Springer-Verlag Berlin Heidelberg 2004

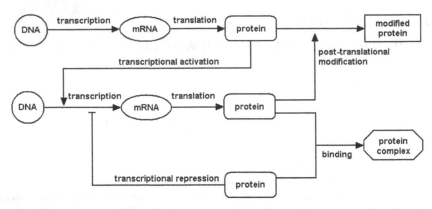

Fig. 1. Genes regulate each other's activity through regulatory networks. Gene transcription into mRNA is influenced by transcription factors, themselves products of other genes. In addition, post-translational modifications lead to proteins with modified properties.

Genes and gene products interact and form networks on several levels [1]. On the genomic level, a class of proteins called transcription factors can activate or inhibit the transcription of genes into mRNAs. Since these transcription factors are themselves products of genes, the ultimate effect is genes regulating each other's expression by forming so called gene regulatory networks. Proteins can participate in diverse chemical reactions that lead to modified proteins with different functions than the originals. Several proteins can also bind to form protein complexes with new roles. Proteins that are members of a class called enzymes catalyze the biochemical reactions forming the base of cellular processes (e.g. metabolism). In many cases the different-level interactions are integrated, for example in signal transduction networks the presence of an external signal (a chemical attractor or a hormone) triggers a cascade of interactions that can involve both biochemical reactions and regulation.

The focus of this article is gene regulatory networks whose components are genes, mRNAs and proteins, and the interactions include transcription, translation, transcriptional regulation and posttranslational reactions (see Fig. 1). We can realize from this description that gene regulatory networks cannot be completely described by a standard graph of nodes and edges. Specifically, the nodes have distinct identities as they correspond to diverse cellular components, and the edges can have two different signatures (signs) corresponding to activation and inhibition.

Gene regulatory networks play a crucial role during development, the process in which a unicellular egg gives rise to an adult [2]. Each cell in a developing embryo has the same DNA, but at no time in their life cycle are all of their genes expressed, i.e. transcribed into mRNA and synthesizing protein. The basis of cell differentiation is differential gene expression, and this is accomplished by interactions between genes, i.e. gene regulatory networks [3].

1.2 Modeling Gene Regulatory Networks

When trying to understand the role and functioning of a gene regulatory network, the first step is to assemble the components of the network and the interactions between them. This structural information needs to be complemented with information or hypotheses regarding the kinetics of the interactions. Since development is a dynamic process in which the expression of genes can constantly change, gene network models need to have a dynamical aspect, i.e. they need to define a state variable for each component, and study how this state changes by the interactions in the network. This state variable can correspond to the concentration of mRNAs and proteins, or it can be a binary value corresponding to the qualitative statement that a gene is expressed or not.

A complete gene regulatory network model incorporates experimental knowledge about the components and their interactions as well as the initial state of these components, and leads to the known final state or dynamical behavior of the network. Validated models then are able to investigate cases that cannot be explored experimentally, for example changes in the initial state, in the components or in the interactions, and they can lead to predictions and insights into the functioning of the system.

1.3 Pioneering Work in Modeling Gene Regulatory Networks

The experimental advances in the mapping of gene regulatory networks are fairly recent, but modeling general aspects of gene regulatory networks dates back to the end of 1960s thanks to the pioneering work of Stuart Kauffman and René Thomas.

In the absence of experimental results, Stuart Kauffman considered an idealized representation of a typical (random) gene network [4,5]. He assumed that genes are equivalent, and their interactions form a directed graph in which each gene receives inputs from a fixed number K of randomly selected neighbors. The state of genes is described by binary (ON/OFF) variables, and the dynamic behavior of each variable, that is, whether it will be ON or OFF at next moment, is governed by a Boolean function. In general, a Boolean or logical function is written as a statement acting on the inputs using the logical operators "and", "or" and "not" and its output is 1(0) if the statement is true (false). In a Random Boolean Network (RBN) the functions governing the state of each node are randomly selected from the 2^{2^K} possible K-input Boolean functions, and kept fixed afterward. Kauffman studied the dynamics of these RBNs, focusing on the attractors (usually cycles) in the state space of the whole network. He discovered the existence of a phase transition in an RBN of size N depending on the value of the parameter K. For $K > 2$ there are around N/e possible cycles whose length scales exponentially with N, however, for $K = 2$ both the number and length of the limit cycles is only $\sqrt{N}$. Kauffman proposed to identify the number of attractors of a gene regulatory network with the number of possible cell types, and noted that the number of cell types seems to increase approximately with the square root of the number of genes per cell, suggesting that gene regulatory networks are in the ordered regime, or on the edge between order and chaos. The RBN models spawned a lot of research in the physics literature, see e.g. [6–9].

While Kauffman proposed a dynamic view on randomly connected gene regulatory networks, René Thomas developed a detailed logical description of the mechanisms governing transcriptional regulation, including the effects of DNA domains such as promoters, initiators, terminators, and the concepts of genetic dominance and recessivity [10]. This formalism was later refined to include multilevel variables and used to study feedback loops, i.e. circular chains of interaction. These loops can be classified into two categories based on the number of negative (inhibitory) interactions in the loop: if this number is even, the loop is positive, and if the number of negative interactions is odd, the loop is a negative feedback loop. Thomas found that a positive feedback loop is a necessary condition for the existence of multiple steady states, while a negative feedback loop with two or more elements is a necessary condition for stable limit cycles [11]. Biologically this means that cell differentiation is based on positive feedback loops, and homeostasis (stability to small perturbations) is based on negative feedback loops. The logical framework introduced by René Thomas was successfully applied to various gene regulatory networks playing a role in the flower morphogenesis of the wall cress *Arabidopsis thaliana* [12] and the development of the fruit fly *Drosophila melanogaster* [13,14].

1.4 Current Models

Broadly speaking, the modeling approaches to gene regulatory networks can be divided into two main groups. In the 'discrete-state' approach each network node (mRNA or protein) is assumed to have a small number of discrete states and the regulatory interactions between nodes are described by logical functions similar to those used in programming. Typically time is also quantized, and the network model that describes how gene products interact to determine the state at the next time gives rise to a discrete dynamical system [12–18].

A more detailed level of description is used in the 'continuous-state' approach, in which the levels of mRNAs and proteins are assumed to be continuous functions of time, and their evolution is modeled by differential equations with mass-action kinetics or other rate laws for the production and decay of all components [19–21]. In order to compare with usually ON/OFF type experimental gene expression profiles, the continuous concentrations are transformed into binary variables using thresholds.

In this paper I shall focus on a recent model of the segment polarity genes developed in collaboration with Hans Othmer [18]. In this model we concentrate on the products (mRNAs, proteins) of segment polarity genes. We reconstruct the network of interaction between these components from experimental data, and assume that these interactions can be expressed as Boolean functions. We find that the dynamic behavior of this model always leads to steady states, and these steady states are in very good agreement with the experimental data on the gene expression pattern of wild type and mutant *Drosophila* embryos. In addition, the model leads to insights into the functioning of this network, the most important being that the network topology is a main source of robustness.

2 The Segment Polarity Gene Network

The genes involved in embryonic pattern formation in the fruit fly *Drosophila melanogaster*, as well as the majority of the interactions between them, are known (for recent reviews see [22–24]). As in other arthropods, the body of the fruit fly is composed of segments, and determination of the adult cell types in these segments is controlled by about 40 genes organized in a hierarchical cascade of gene families [25]. These gene families are expressed in consecutive stages of embryonic development and have a spatial expression pattern that is successively more precisely-defined (see Fig. 2). The genes at one step initiate or

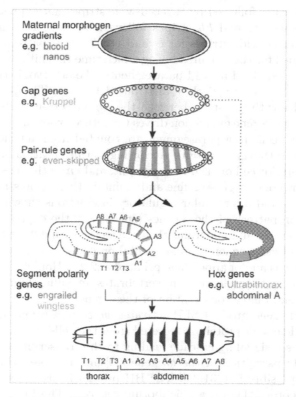

Fig. 2. The segmentation of the fruit fly embryo is governed by a hierarchy of gene families, starting from maternal genes. Each of these genes encode for transcription factors, and are responsible for the initiation of the genes in the next family. While the genes in the first three steps are transient, the segment polarity genes maintain a stable pattern for three hours. Reproduced with permission from [23].

modulate the expression of those involved in the next step of the cascade. While most of these genes act only transiently, the segment polarity genes are expressed throughout the life of the fly, and their periodic spatial pattern is maintained for at least 3 hours of embryonic development.

The best characterized segment polarity genes include *engrailed* (*en*), *wingless* (*wg*), *hedgehog* (*hh*), *patched* (*ptc*), *cubitus interruptus* (*ci*), *smoothened* (*smo*) and *sloppy paired* (*slp*)[1]. The segment polarity genes encode for diverse proteins including the transcription factors Engrailed (EN), Sloppy Paired (SLP), and Cubitus Interruptus (CI), the secreted proteins Wingless (WG) and Hedgehog (HH), and the transmembrane proteins Patched (PTC) and Smoothened (SMO)[2].

2.1 Wild Type Patterns of the Segment Polarity Genes

The segment polarity genes are activated by the pair-rule genes at about 3 hours after fertilization. The initial state of the segment polarity genes includes two-cell-wide SLP stripes followed by two-cell-wide stripes not expressing SLP [26], single-cell-wide *wg*, *en* and *hh* stripes followed by three cells not expressing them, and three-cell-wide stripes for *ci* and *ptc* [2]. This pattern is maintained almost unmodified for three hours, during which time the initially homogeneous-looking embryo is divided into 14 parasegments (the embryonic counterparts of the adult segments) by regularly - distributed furrows. The position of these furrows coincides with the space between the *wg* and *en* -expressing cells, thus the periodicity of the gene expression drives the future external appearance of the embryo [25]. The cells in a parasegment are counted from anterior (toward the head) to posterior (toward the tail). According to this notation, *wg* is expressed in the most posterior cell of each parasegment, and *en* in the most anterior cell.

The segment polarity genes refine and maintain their expression through the network of intra- and intercellular regulatory interactions shown in Fig. 3. The stable expression pattern of these genes (specifically the expression of *wingless* and *engrailed*) defines and maintains the borders between different parasegments and contributes to subsequent developmental processes, including the formation of denticle patterns and of appendage primordia [2,25]. Homologs of the segment polarity genes have been identified in vertebrates, including humans, which suggests strong evolutionary conservation of these genes.

The pair-rule gene product SLP activates *wg* transcription and represses *en* transcription. The WG protein is secreted from the cells that synthesize it [25, 27] and initiates a signaling cascade leading to the transcription of *en* [28]. EN promotes the transcription of the *hh* gene [29] and represses the transcription of *ci* [30] and possibly *ptc* [31,32]. The HH protein is also secreted, and binds to the HH receptor PTC on a neighboring cell [22]. The intracellular domain

---

[1] Many of these genes were named for the phenotypic changes their mutations cause, e.g. a *wingless* mutant fruit fly does not have wings.

[2] These notations follow the convention that names of genes and mRNAs are italicized, while names of proteins are capitalized.

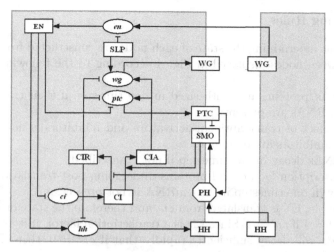

Fig. 3. The network of interactions between the segment polarity genes. The shape of the nodes indicates whether the corresponding substances are mRNAs (ellipses), proteins (rectangles) or protein complexes (octagons). The edges of the network signify either biochemical reactions (e.g. translation) or regulatory interactions (e.g. transcriptional activation). The edges are distinguished by their signatures, i.e. whether they are activating ($\rightarrow$) or inhibiting ($\dashv$). Terminating arrows ($\rightarrow$) indicate translation, post-translational modifications (in the case of CI), transcriptional activation or the promotion of a post-translational modification reaction (e.g., SMO determining the activation of CI). Terminating segments ($\dashv$) indicate transcriptional inhibition or in the case of SMO, the inhibition of the post-translational modification reaction CI$\rightarrow$CIR.

of PTC forms a complex with SMO [33] in which SMO is inactivated by a post-translational conformation change [34]. Binding of HH to PTC removes the inhibition of SMO, and activates a pathway that results in the modification of CI [34]. The CI protein can be converted into one of two transcription factors, depending on the activity of SMO. When SMO is inactive, CI is cleaved to form CIR, a transcriptional repressor that represses wg, ptc [35] and hh transcription [36,37]. When SMO is active, CI is converted to a transcriptional activator, CIA, that promotes the transcription of wg and ptc [35,37–39].

3 Description of the Model

In the model, each mRNA or protein is represented by a node of a network, and the interactions between them are encoded as directed edges (see Fig. 3). The state of each node is 1 or 0, according as the corresponding substance is present or not. The states of the nodes can change in time. We choose a time interval that is larger or equal to the duration of all transcription and translation processes, and we use this interval as the length of a unit timestep. The next state of node i is determined by a Boolean function of its state and the states of those nodes that that have edges incident on it.

3.1 Updating Rules

The functions determining the state of each node are constructed from the inter-actions between nodes displayed in Fig. 3 according to the following rules (see also Fig. 4)

(i) mRNAs/proteins are synthesized in one timestep if their transcriptional activators/mRNAs are present;

(ii) the effect of transcriptional activators and inhibitors is never additive, but rather, inhibitors are dominant;

(iii) mRNAs decay in one timestep if not transcribed;

(iv) transcription factors and proteins undergoing post-translational modifi-cation decay in one timestep if their mRNA is not present.

For example, EN is translated from en, and therefore the state of EN at time $t+1$, $EN^{t+1} = 1$ if $en^t = 1$. Since EN is a transcription factor, it is assumed that its expression will decay sufficiently rapidly that if $en^t = 0$, then $EN^{t+1} = 0$. These two assumptions mean that

$$EN^{t+1} = en^t. \tag{1}$$

Table 1 gives an overview of the Boolean functions for each node. In each case, subscripts signify spatial position (i.e. cell number) and superscripts signify time[3].

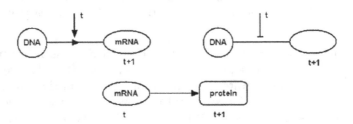

Fig. 4. Assumptions for the kinetics of the interactions. We assume that the timescale for turning ON or OFF is the same. Transcription requires the presence of activators and the absence of inhibitors; translation requires the presence of the mRNA.

3.2 Representing the State of the System

Expression of the segment polarity gene occurs in stripes that encircle the em-bryo, and therefore we treated the two-dimensional pattern as one-dimensional. We considered a line of 12 cells corresponding to three parasegment primor-dia (i.e. the spatial regions that will become the parasegments), and imposed

---

[3] In coding these rules we have used an equivalent description of a function giving the output of every possible combination of inputs.

Table 1. The Boolean functions used in the model. The functions are based on the known interactions between mRNAs and proteins shown in Fig. 3, and on the temporal assumptions listed above. In general the updating rule gives the expression of a node at time $t + 1$ as a function of the expression of its effector nodes at time t. However, there are three exceptions: we assume that the expression of SLP does not change, and that the activation of SMO and the binding of PTC to HH are instantaneous.

Node	Boolean updating function
SLP_i	$SLP_i^{t+1} = SLP_i^t = \begin{cases} 0 \text{ if } i\%4 = 1 \text{ or } i\%4 = 2 \\ 1 \text{ if } i\%4 = 3 \text{ or } i\%4 = 0 \end{cases}$
wg_i	$wg_i^{t+1} = (CIA_i^t \text{ and } SLP_i^t \text{ and not } CIR_i^t)$
	or $[wg_i^t \text{ and } (CIA_i^t \text{ or } SLP_i^t) \text{ and not } CIR_i^t]$
WG_i	$WG_i^{t+1} = wg_i^t$
en_i	$en_i^{t+1} = (WG_{i-1}^t \text{ or } WG_{i+1}^t) \text{ and not } SLP_i^t$
EN_i	$EN_i^{t+1} = en_i^t$
hh_i	$hh_i^{t+1} = EN_i^t \text{ and not } CIR_i^t$
HH_i	$HH_i^{t+1} = hh_i^t$
ptc_i	$ptc_i^{t+1} = CIA_i^t \text{ and not } EN_i^t \text{ and not } CIR_i^t$
PTC_i	$PTC_i^{t+1} = ptc_i^t \text{ or } (PTC_i^t \text{ and not } HH_{i-1}^t \text{ and not } HH_{i+1}^t)$
PH_i	$PH_i^t = PTC_i^t \text{ and } (HH_{i-1}^t \text{ or } HH_{i+1}^t)$
SMO_i	$SMO_i^t = \text{not } PTC_i^t \text{ or } HH_{i-1}^t \text{ or } HH_{i+1}^t$
ci_i	$ci_i^{t+1} = \text{not } EN_i^t$
CI_i	$CI_i^{t+1} = ci_i^t$
CIA_i	$CIA_i^{t+1} = CI_i^t \text{ and } (SMO_i^t \text{ or } hh_{i-1}^t \text{ or } hh_{i+1}^t)$
CIR_i	$CIR_i^{t+1} = CI_i^t \text{ and not } SMO_i^t \text{ and not } hh_{i\pm1}^t$

periodic boundary conditions on the ends. We used four cells per parasegment primordium because when expression of the segment polarity genes begins, a given gene is expressed in every fourth cell. The state of the system includes a 12-cell wide 1-dimensional periodic pattern for each node in the network which we represented as a series of black/gray boxes corresponding to cells in which the given node is ON/OFF (see Fig. 5). To make the periodicity of the pattern clear, we separated the patterns corresponding to distinct parasegments by short white spaces.

4 Modeling the Wild Type Segment Polarity Genes

The first step in validating the model is testing whether it captures the wild type behavior of the system. Therefore we started from the known initial pattern of the segment polarity genes and updated their states according to the rules presented on Table 1, checking whether they become stationary.

The initial state of each parasegment primordium includes SLP present in the last (posterior) two cells, wg present in the last cell, en and hh expressed in the first (anterior) cell, and ci and ptc expressed in the posterior three cells [25,26,29,31,32]. Since the proteins are translated after the mRNAs are transcribed, we assumed that the proteins are not expressed in the initial state. The

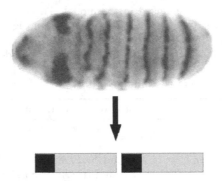

Fig. 5. Because the expression pattern of the segment polarity genes is symmetrical, we represent it with a one-dimensional pattern corresponding to the anterior-posterior axis. We identify the state of each node in the network with a periodic succession of black/ gray squares corresponding to cells that express/do not express the given node. Each segment of four cells corresponds to a parasegment primordium.

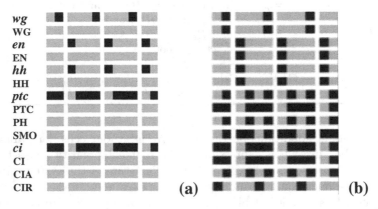

Fig. 6. Wild-type expression patterns of the segment polarity genes. Here and hereafter left corresponds to anterior and right to posterior in each parasegment. Horizontal rows correspond to the pattern of individual nodes - specified at the left side of the row - over two full and two partial parasegments. Each parasegment is assumed to be four cells wide. A black (gray) box denotes a node that is ON (OFF). (a) The experimentally-observed initial state. *en, wg* and *hh* are expressed in every fourth cell, while the broad *ptc* and *ci* stripes are complementary to *en*. (b) The steady state given by the model when initialized with the pattern in (a). This pattern is in excellent agreement with the observed gene expression patterns. After [18].

one-dimensional representation of the mRNA and protein patterns is shown in Fig. 6a.

We iterated the dynamical system defined by the rules in Table 1 starting from the initial state described above. We found that after only 6 time steps, the expression pattern stabilizes in a time-invariant spatial pattern (see Fig. 6b) that coincides with the experimentally observed stable expression of the segment polarity genes. Indeed, wg and WG are expressed in the most posterior cell of each parasegment [40], while en, EN, hh and HH are expressed in the most anterior cell of each parasegment [29,40], ptc is expressed in two cells, one on each side of the en-expressing cells [25,31]. SMO is present in a broad region ranging from the wg-expressing to the en-expressing cells [41]. ci is expressed almost ubiquitously, with the exception of the cells expressing en [30,38]. CIA is expressed in the neighbors of the HH-expressing cells, while CIR is expressed far from the HH-expressing cells [35].

Thus the model [18] demonstrates that the interaction between the segment polarity genes is able to maintain their expression after initialization. The success in reproducing the stable expression pattern of these genes is a strong indication that the kinetic details of the interactions do not matter, just their signature and the regulatory network they form. This conclusion is in agreement with the results of the continuous-state model of von Dassow *et al.* [20].

5 The Functional Topology of the Segment Polarity Network

The success of our model demonstrates that the topology of the regulatory network has a determining role in its dynamics. Nevertheless, knowledge of the topology alone is not enough to determine what will happen in the network. Moreover, the presence of dual interaction signatures precludes us from using standard graph theoretical tools to analyze this network.

To obtain a better insight into the connection between topology and dynamics, we proposed the construction of an expanded graph that reflects the function of the network. The first step of this expansion is adding complementary pseudo-nodes corresponding to every node whose negated state enters the Boolean rules on Table 1. The second step is to introduce composite pseudo-nodes for nodes whose states are terms of a conjunction in these rules (see Table 2). Consider the transcription of the hh gene. Figure 3 shows that hh has two incoming edges, one from EN and one from CIR, and Table 1 shows that transcription of the hh gene requires both the presence of the EN protein and the absence of the CIR protein. We introduce complementary pseudo-node, $\overline{\text{CIR}}$, that is expressed whenever CIR is not, and connect it to CIR with a symmetrical edge. Then we add the composite pseudo-node $\overline{\text{ECR}}$, and we draw two directed edges starting from EN and $\overline{\text{CIR}}$ and ending in $\overline{\text{ECR}}$, to represent the dependence of $\overline{\text{ECR}}$ on the expression of EN and $\overline{\text{CIR}}$ (see Fig. 7). Now hh receives inputs only from $\overline{\text{ECR}}$.

Figure 8 shows the nodes and edges corresponding to the mRNAs and proteins in the second cell of the parasegment together with the pseudo-nodes these mRNAs and proteins interact with, both cell-autonomously, and in the neighboring cells. Although the introduction of the pseudo-nodes increases the number of nodes in the network, it eliminates the distinction between edges based on their signatures; all directed edges in Fig. 8 now signify activation. However, there are differences in the way multiple activating edges are taken into account: multiple

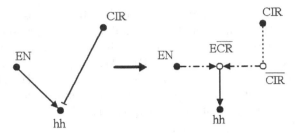

Fig. 7. Illustration of the network expansion process used to construct the functional topology. To express the logical rule governing the transcription of *hh* graphically, we introduce the complementary node $\overline{\text{CIR}}$ and the composite node $\overline{\text{ECR}}$. The expanded network contains real nodes (filled circles) and pseudo-nodes (open circles), an inter-dependence relation between CIR and $\overline{\text{CIR}}$ (dotted line), edges corresponding to the activation of $\overline{\text{ECR}}$ (dash-dotted lines) and a single edge expressing the activation of *hh* transcription. After [18].

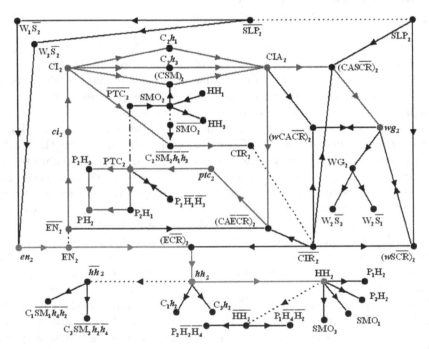

Fig. 8. Functional topology of the network affecting the second cell of the parasegment. Pseudo-nodes with multiple indexes correspond to intercellular interactions and either receive some of their inputs from the neighboring cells, or contribute to the expression of the nodes in the neighboring cells (not shown). Symmetrical edges between nodes and their complementaries are drawn with dotted lines, double arrows denote a pair of oppositely directed edges. The colored edges illustrate two antagonistic activating clusters starting from *en* (purple) and $\overline{\text{EN}}$ (red).

Table 2. Definition of the symbols for pseudo-nodes used in Fig. 8. The state of each composite node is determined from the logical function giving its relation to the state of its "parent" nodes.

Symbol of pseudo-node	Relation to parent node(s)
Complementary nodes	
$\overline{EN}$	not EN
$\overline{hh}$	not hh
$\overline{HH}$	not HH
$\overline{PTC}$	not PTC
$\overline{SLP}$	not SLP
$\overline{SMO}$	not SMO
Composite nodes corresponding to a single cell	
$(CA\overline{ECR})_2$	CIA_2 and $\overline{EN_2}$ and $\overline{CIR_2}$
$(CA\overline{SCR})_2$	CIA_2 and SLP_2 and $\overline{CIR_2}$
$(CSM)_2$	CI_2 and SMO_2
$(\overline{ECR})_2$	$\overline{EN_2}$ and $\overline{CIR_2}$
$(wCA\overline{CR})_2$	wg_2 and CIA_2 and $\overline{CIR_2}$
$(wS\overline{CR})_2$	wg_2 and SLP_2 and $\overline{CIR_2}$
Composite nodes corresponding to intercellular interactions	
$C_i h_j$	CI_i and hh_j
$C_i\overline{SM_i h_j h_k}$	CI_i and $\overline{SMO_i}$ and $\overline{hh_j}$ and $\overline{hh_k}$
$P_i\overline{H_j H_k}$	PTC_i and $\overline{HH_j}$ and $\overline{HH_k}$
$P_i H_j$	PTC_i and HH_j
$W_i S_j$	WG_i and SLP_j

edges ending in composite pseudo-nodes are added by the operator "and", while multiple edges ending in real nodes are cumulated by the operator "or".

Figure 8 illustrates the heterogeneous functional topology of the segment polarity network. The majority of nodes have few edges, but there are key nodes with a large number of incoming or outgoing edges. For example, $\overline{CIR}$ has 5 outgoing edges, while HH has 4. The important role of HH in the network is reflected in the fact that it affects the future expression of 4 other proteins (CIA, CIR, PTC and PH) in the neighboring cells, for a total of 8 nodes. Other nodes such as CIA have several incoming edges, indicating that they can be activated in many ways. A single node, SLP, has only outgoing edges because it is constitutively present; all others have both incoming and outgoing edges (the apparent exceptions interact with nodes in the neighboring cells).

The functional network of Fig. 8 gives insight into the time-evolution of the expression of the segment polarity genes. For example, we can determine the cluster of nodes that can be activated by the expression of a given node (see colored nodes in Fig. 8). The absence of EN (or conversely the presence of $\overline{EN}$) gives the largest activated cluster, containing ci, CI, CIA, ptc, PTC, PH, wg, and WG. A separate activated cluster starts with the presence of en, and contains EN, hh and HH. These activating clusters indicate that the cells expressing en and hh never express wg, ptc or ci. This separation into anterior and posterior compartments expressing different genes is well-known, in fact, it is the basis for calling these genes "segment polarity genes" [2].

While the majority of the activating effects propagate outside the cell, there are three cases in which an activation can return to its source. In other words, three short positive feedback loops exist in the network of Fig. 8. The first two loops connect wg_2 with $(wCA\overline{CR})_2$ or $(wS\overline{CR})_2$ and the third connects PTC_2 with $P_2\overline{H_1H_3}$. These loops ensure the maintenance of wg and PTC if all the conditions for the expression of the pseudo-node in the cycle are met. The successful activation of the wg cycle can induce the stable expression of en and hh in those neighboring cells where neither SLP nor CIR is expressed, and stable expression of PTC leads to stable CIR expression two cells removed from en expression.

6 Gene Mutations

An important method for inferring gene interactions experimentally is to silence selected genes by mutations. These null mutant genes are not able to synthesize protein, and if that protein is a transcription factor, the effects of the mutation propagate through the system (see Fig. 9). Our model is able to simulate the effect of null mutations by setting the state of the transcript to OFF and not updating it during the evolution of the system.

Our results indicate that if any of en, wg or hh are blocked, the steady state is a pattern with no en, wg, ptc or hh, as in Fig. 10a. We can see from Fig. 8 that each of these mutations disrupts intercellular signaling, causing ubiquitous expression of CIR, which in turn leads to ubiquitous repression of transcription. This result is in excellent agreement with all experimental observations regarding en, wg and hh mutant embryos [29,31,42–44].

If the ptc gene is blocked, we obtain a pattern with broad wg, en and hh stripes (see Fig. 10b). Indeed, the network in Fig. 8 shows that if ptc is deactivated, $\overline{PTC}$ will be ubiquitous, causing all CI to be transformed into CIA, which leads to two-cells-wide wg and en/hh expression. This pattern agrees with the experimental results on ptc mutants [29,43–45]. Moreover, our results are in agreement with all experimental observations of double mutants as well [29,40, 42,46].

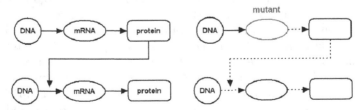

Fig. 9. Gene mutations that disrupt transcription factors propagate through the system, affecting multiple nodes.

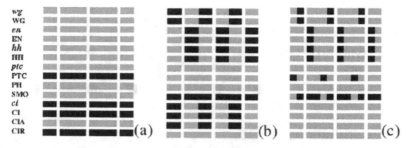

Fig. 10. Segment polarity gene expression patterns predicted for gene mutations. (a) Pattern with no segmentation. This pattern arises if any of *wg*, *en* or *hh* is kept OFF in the model and is independent of the initial state of the other genes. (b) Broad expression pattern. The stripes of *en*, *wg* and *hh* are two-cells-wide, while the *ci* stripe narrows and CIR is not expressed. This state arises when *ptc* is kept OFF, regardless of the initial state of other genes. (c) Almost normal pattern obtained for *ci* mutants and wild type initial conditions.

If all the other genes are initiated normally, we find that the effect of a *ci* deletion does not affect the *en*, *wg* and *hh* patterns (see Fig. 10c). Indeed, Fig. 8 shows that the deactivation of *ci* leads to the disappearance of CIA and CIR, but wild-type *wg* can still be maintained by SLP[39,44]. In conclusion, the model is in agreement with every observed gene pattern in mutants (see Fig. 11) and provides predictions for genes whose expression was not studied experimentally.

7 Determination of the Steady States and Their Domains of Attraction

The fact that the model reproduces the results of numerous experiments remarkably well suggests that the structure of the model is essentially correct, and warrants exploration of problems that have not been studied experimentally. For example, we can determine the complete set of stable steady state patterns of segment polarity gene expression, and estimate the domain of attraction of these states. The former can be done analytically by noting that these are fixed points of the discrete dynamical system, and so $x_i^{t+1} = x_i^t$, where x corresponds to any node in the network. Thus a steady state is the solution of the system of equations obtained from Table 1 by simply removing the time indices (see [18]). We obtain 10 solutions that correspond to four distinct patterns (see Fig. 12) and their slight variations.

The first steady state is the pattern with no segmentation first presented in Fig. 10a. The second corresponds to the wild-type pattern first shown in Fig. 6b. The third steady state has two-cells-wide *en* and *wg* stripes like the *ptc* mutant (see Fig. 12c). In the fourth distinct steady state *wg* is expressed in the anterior neighbor of its wild-type position, while the *en*/*hh* stripe is displaced posteriorly (see Fig. 12d). This expression pattern corresponds to an embryo

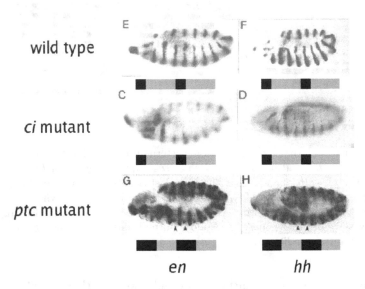

Fig. 11. Comparison between the experimental results (embryo pictures showing the expression pattern of *en* and *hh*, after [29]) and predictions of the model (black and gray patterns) for two gene mutations. The model indicates that *ci* mutation can preserve the pattern of *en* and *hh*, while *ptc* mutation doubles their expression, in agreement with experiments.

with no parasegmental grooves, since the end of the *wg* stripe does not meet the beginning of the *en* stripe.

While each of the steady states can be obtained starting from suitable nearby states, the number of initial conditions leading to a chosen stable pattern, i.e., its domain of attraction, can be very different. Consider first the number of initial states that lead to the wild-type steady state. If we fix all nodes but one in their wild-type pattern, there are $2^4 = 16$ distinct initial patterns corresponding to the four cells of the parasegment. We do this for each of the 14 variable nodes in turn (we do not change the expression of SLP) and find that the number of initial patterns leading to the wild-type steady state is 3 for *wg* or WG variation, 4 for *en*, EN, *hh* or HH variation, 8 for *ptc*, PTC, CI or CIA variation, and 16 for PH, SMO, *ci* or CIR variation. When the initial pattern of all 14 nodes can vary, there are $3^2 \cdot 4^4 \cdot 8^4 \cdot 16^4 \sim 6 \times 10^{11}$ prepatterns that lead to the wild type steady state, which is a fraction of 8×10^{-6} of the total number of initial states $N_{st} = 16^{14}$.

We find that the network is very robust with respect to missing initial expression of nodes. We have determined that the minimal prepatterning that leads to wild-type stable expression is as follows.

- *wg* **is wild type,**
- *en* **and** *hh* **are not expressed,**
- *ptc* **is expressed in the third cell of the parasegment primordium,**
- *ci* **and the proteins are not expressed.**

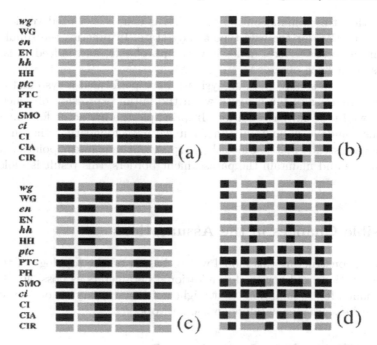

Fig. 12. Various stable patterns of the segment polarity genes. (a) Steady state with no segmentation. (b) Wild-type type expression pattern. (c) Steady state with broad *en*, *wg* and *hh* domains. (d) Ectopic pattern with displaced *wg*, *en* and *hh* stripes.

In summary, it is enough to initiate the expression of two genes in two cells per parasegment primordium, and the interactions between the segment polarity genes will initiate the others. This result suggests a remarkable error-correcting ability for the segment polarity gene network.

Note that the minimal prepattern contains the wild-type expression of *wg*. If *wg* is not expressed initially we find that the final pattern is like Fig. 12a, regardless of the initial pattern of the other nodes. Consequently, a fraction of at least $1/16^{th}$ of the initial states leads to the pattern of Fig. 12a. This finding suggests that *wg* has a special role in the functioning of the segment polarity network, and has to be activated at a specific time and specific cells in order to obtain wild-type gene expression.

In the other limit, broader than wild-type initial expression of any node except PH, SMO, *ci* and CIR leads to the pattern with broad stripes as in Fig. 12c. This pattern is obtained in the vast majority of prepatterns, comprising about 90% of the total number of initial state and its features were frequently observed in overexpression experiments [43,44,47].

The minimal prepattern needed for the ectopic pattern with displaced *wg* and *en* stripes (Fig. 10c) is *wg* expression in the third cell of the parasegment primordium (the same as its steady pattern), and *ptc* expression in the last cell of the parasegment primordium, where the wild-type stripe of *wg* would normally be. Note that this minimal initial condition is simply a shifted version of the minimal condition for the wild-type steady state. In practice the simultaneous ectopic initiation of several nodes is very improbable, and indeed, this steady state has never been observed.

In addition to steady state analysis we have performed a systematic analysis of the dynamics of the network when the initial expression of genes differs from the wild-type initial condition. In principle the attractor for some initial conditions could be periodic in time, but we have found that the only stable attractors are steady states. Since the purpose of the segment polarity network is to stabilize and maintain the parasegment borders, this result is biologically realistic.

8 Possible Changes in the Assumptions

Our goal in constructing this model was to base it on the topology of the regulatory interactions and have as few additional parameters as possible. However, there remain a few assumptions that might not be necessary or, on the contrary, could reveal essential constraints on the network.

8.1 Equal Timescale for Synthesis and Decay

We assumed that the expression of mRNAs/proteins decays in one time step if their transcriptional activators/mRNAs are switched off. This conjecture is probably too severe, as the decay time of proteins is usually longer than the time their synthesis takes. Therefore we studied a variant of the model in which the expression of a protein is maintained for at least two steps [18]. We find that this variant leads to exactly the same steady states as the original model, and these states have approximately the same basins of attraction[4]. The two-step model reaches the wild-type steady state shown in Fig. 6b if it is started from the initial pattern of Fig. 6a, and leads to the same states for gene mutations. The only change is in the intermediate states visited en route to the final state: both the wild-type and the broad type pattern stabilizes on average 30% faster using the two-step assumption. On the other hand, the pattern with no segmentation is reached at a slightly lower rate than in the original model. In conclusion, the two-step assumption provides a more realistic modeling of the decay of the proteins without changing the conclusions of the model.

8.2 Assumptions for WG and PTC

The model contains two exceptions to the one-step decay rule through the assumptions of persistence of existent wg and PTC expression. The stability of these nodes has a major role in stabilizing the expression of the segment polarity genes, reflected in the existence of the cycles in the functional topology

---

[4] Note that in this case limit cycles are possible.

of the network (Fig. 8) and in the fact that the steady states are completely determined by the pattern of wg and PTC. It is therefore important to check what happens if wg and PTC decay in one step as other mRNAs and proteins.

If we assume that

$$wg_i^{t+1} = CIA_i^t \text{ and } SLP_i^t \text{ and not } CIR_i^t, \tag{2}$$

it is still possible to arrive at the wild-type steady state, but only for much more restricted initial states. Furthermore, the resilience of the network to mutations in ci is destroyed, and all initial states lead to the steady state with no segmentation. Since it is observed experimentally that ci null mutants still display almost normal segmentation [44], we can conclude that the stability of wg is required for the functioning of the segment polarity genes. This suggests a special role for SLP as the main activator of wg, and underlies the need for its stability.

If we do not assume the maintenance of initial PTC expression, the pattern of PTC will follow that of its transcript and split into two stripes. This will cause the complete disappearance of CIR and the only steady state will be the pattern with broad stripes as in Fig. 10b. Thus the persistence of PTC is a major requirement for the function of the segment polarity network, and suggests that the protein has special structural properties.

8.3 Four-Cell-Wide Parasegments

During the three hours of stable segment polarity gene patterning the parasegment is enlarged due to two rounds of divisions [2]. While the wg stripe remains a single cell wide, the en stripe widens to three cells. The maintaining of this en requires WG transport, and, indeed, wingless protein is seen to diffuse over a distance of 2-3 cell diameters [48]. In order to determine if our model is able to describe the segment polarity gene patterns in later stages, we applied it to the transition between a four- and eight-cell-wide parasegment. We started with the wild-type pattern of Fig. 6a and assumed that each cell divides into two identical cells, with the same genes expressed in each of the two. We also assumed that WG and HH can be transported through the nearest neighbors of the cells expressing their mRNAs. The model leads to the steady state represented in Fig. 13, with a single cell wide wg stripe, three cell wide en and hh stripes, and two ptc stripes flanking the en domain. This steady state agrees perfectly with the wild-type pattern observed in 8hr old embryos [25,30,31,41,38].

8.4 Stable SLP

Throughout our analysis we assumed that the expression of SLP does not change. To test whether this assumption is necessary, we have studied the effects of inactivated and overexpressed SLP. We obtain seven final states for inactivated SLP, but none of them corresponds to the wild-type pattern. The closest state, obtained when we start from wild-type initial conditions, has a wild-type wg and ptc pattern, but en and hh are expressed on both sides of the wg stripe and CIR

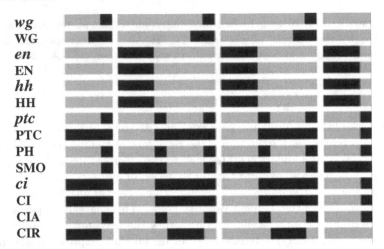

Fig. 13. Stable expression pattern of the segment polarity genes after a round of cell division, as obtained from our model. We assume that at this stage WG and HH can be transported through the neighboring cells. This pattern is in good agreement with experimental observations of 8hr old embryos. After [18].

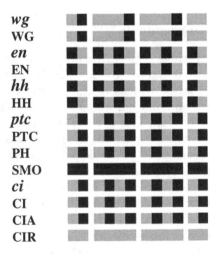

Fig. 14. The pattern obtained from our model when we start from wild-type initial conditions, but SLP is not functional. Note that *en* is expressed on both sides of the *wg* stripe. After [18].

is not absent (Fig. 14). At this point this state is a theoretical prediction that can be verified by conditional SLP mutants (i.e., mutants that have normal pair-rule activity, but no segment polarity activity). We also find that ubiquitously expressed SLP leads to the state with no segmentation presented in Fig. 10b. This finding is in agreement with experimental results [26]. Based on these results, and the important role of SLP in maintaining *wg* transcription, we conclude that the SLP protein plays a vital role in this network.

9 Conclusions

The model demonstrates that the topology of the regulatory network plays a determining role in the dynamics and the stability of the segment polarity genes. The success in reproducing wild type and mutant gene patterns indicates that the kinetic details of the interactions do not matter, as long as their net effect is preserved. Our simulations also suggest a remarkable robustness and error correcting ability of the segment polarity gene network. We found that a large fraction of initiation delays can be rescued, and the network can compensate even for some gene mutations . The model also gives numerous predictions that can be tested experimentally. First, we concluded that the *wingless* gene plays a key role in the system, and it is imperative that it be initiated at the right time in the right pattern. However, non-initiation of *engrailed* and *hedgehog* can be rescued by the activity of the network. Experiments with conditional mutants defective in initiation could verify these predictions. Second, we found that the state of the segment polarity genes can evolve into a pattern with displaced stripes if initiated in a certain way. While this ectopic initiation is difficult, it should be possible. Finally, we concluded that the stable expression of SLP is a crucial requirement; this could be tested by the isolation of the segment polarity- and pair-rule roles of SLP.

The two-step model represents a step toward modeling the transition from the initial state to the final steady state of the segment polarity network. A more realistic model would assume different time intervals (expressed in number of steps) for the decay for mRNAs and proteins. While this extension would involve unknown parameters, the condition of reaching the same steady states as the original model would provide constraints on the variability of the decay rates. Another direction where the model could be extended is to consider a two-dimensional array of cells. It is known experimentally that the stripes of segment polarity genes are not initiated as straight lines, but have jagged borders [2]. During the functioning of the segment polarity network these stripes straighten, and the parasegment borders become sharp. A two-dimensional simulation of our model could lead to important insights into this process.

The model presented here is part of a larger family of models using a logical approach to gene regulatory networks [13–17,50]. As illustrated by these models, this approach enables the integration of qualitative observations on gene interactions into a coherent picture, while adding a minimum of additional kinetic hypotheses. The analysis of a Boolean model is more tractable than that for a model based on differential equations, which inevitably has numerous unknown parameters, and a Boolean model facilitates a more systematic study of the possible steady states and their basins of attraction[5]. We envision realistic

---

[5] Note that Boolean logic can be extended to so-called polynomial logic applicable to multi-level variables, see [49].

topology-based Boolean modeling as an important first step in understanding the interplay between the topology and dynamics of gene regulatory networks and testing the completeness of available topological information. While the segment polarity gene network was successfully modeled by a simple synchronous binary Boolean model, other networks require more detailed models incorporating asynchronous updating and/or multi-level variables [13,14]. Of course there are undoubtedly systems, such as metabolic networks, for which a Boolean approach might not be an appropriate first level of analysis.

Acknowledgements

The original work presented here was done in collaboration with Hans G. Othmer and supported in part by NIH Grant #GM 29123.

References

1. B. Alberts et al., Molecular Biology of the Cell, 4th edn. (2003).
2. L. Wolpert, R. Beddington, J. Brockes, T. Jessell, P. Lawrence and E. Meyerowitz, Principles of Development, (Current Biology Ltd., London 1998).
3. E.H. Davidson et al., Science 295, 1669 (2002).
4. S. A. Kauffman, J. Theor. Biol. 22, 437 (1969).
5. S. A. Kauffman, The origins of Order, (Oxford University Press, New York, 1993).
6. B. Derrida and Y. Pomeau, Europhys. Lett. 1, 45 (1986).
7. G. Weisbuch and D. Stauffer, J. Phys. (Paris) 48 11 (1987).
8. B. Luque and R. V. Solé, Phys. Rev. E 55, 257 (1997).
9. R. Albert and A.-L. Barabási, Phys. Rev. Lett. 84, 5660 (2000).
10. R. Thomas, J. Theor. Biol. 42, 563 (1973).
11. R. Thomas and R. D'Ari, Biological Feedback (CRC Press, Boca Raton, Ann Arbor, Boston, 1990).
12. L. Mendoza, D. Thieffry and E. R. Alvarez-Buylla, Bioinformatics 15, 593 (1999).
13. L. Sánchez and D. Thieffry, J. Theor. Biol. 211, 115 (2001).
14. A. Ghysen and R. Thomas, BioEssays 25, 802 (2003).
15. J. W. Bodnar, J. Theor. Biol. 188, 391 (1997).
16. J. W. Bodnar and M. K. Bradley, Cell Biochem. and Biophys. 34, 153 (2001).
17. C.-H. Yuh, H. Bolouri and E. H. Davidson, Development 128, 617 (2001).
18. R. Albert and H. G. Othmer, J. Theor. Biol 223, 1 (2003).
19. J. Reinitz and D. H. Sharp, Mechanisms of Development 49, 133 (1995).
20. G. von Dassow, E. Meir., E. M. Munro and G. M. Odell, Nature 406, 188 (2000).
21. V. V. Gursky, J. Reinitz and A. M. Samsonov, Chaos 11, 132 (2001).
22. P. W. Ingham and A. P. McMahon, Genes Dev. 15, 3059 (2001).
23. B. Sanson, EMBO Reports 2, 1083 (2001).
24. V. Hatini and S. DiNardo, Trends in Genetics 17, 574 (2001).
25. J. E. Hooper and M. P. Scott, The Molecular Genetic Basis of Positional Information in Insect Segments. In: Early Embryonic Development of Animals, ed. by W. Hennig (Springer, Berlin 1992) pp. 1-49.
26. K. M. Cadigan, U. Grossniklaus and W. J. Gehring, Genes Dev. 8, 899 (1994).
27. S. Pfeiffer and J.-P. Vincent, Cell & Dev. Biol. 10, 303 (1999).

28. K. M. Cadigan and R. Nusse, Genes Dev. **11**, 3286 (1997).
29. T. Tabata, S. Eaton and T. B. Kornberg, Genes Dev. **6**, 2635 (1992).
30. S. Eaton and T. B. Kornberg, Genes. Dev. **4**, 1068 (1990).
31. A. Hidalgo and P. Ingham, Development **110**, 291-301 (1990).
32. A. M. Taylor, Y. Nakano, J. Mohler and P. W. Ingham, Mechanisms of Development **42**, 89 (1993).
33. M. van den Heuvel and P. W. Ingham, Nature **382**, 547 (1996).
34. P. W. Ingham, EMBO J. **17**, 3505 (1998).
35. P. Aza-Blanc and T. B. Kornberg, Trends in Genetics **15**, 458 (1999).
36. J. T. Ohlmeyer and D. Kalderon, Nature **396**, 749 (1998).
37. N. Méthot and K. Basler, Cell **96**, 819 (1999).
38. C. Alexandre, A. Jacinto and P. W. Ingham, Genes Dev. **10**, 2003 (1996).
39. T. von Ohlen and J. E. Hooper, Mechanisms of Development **68**, 149 (1997).
40. P. W. Ingham, A. M. Taylor and Y. Nakano, Nature **353**, 184 (1991).
41. J. Alcedo, Y. Zou and M. Noll, Molecular Cell **6**, 457 (2000).
42. S. DiNardo, E. Sher, J. Heemskerk-Jongens, J. A. Kassis and P. H. O'Farrell, Nature **332**, 45 (1988).
43. C. Schwartz, J. Locke, C. Nishida and T. B. Kornberg, Development **121**, 1625 (1995).
44. A. Gallet, C. Angelats, S. Kerridge and P. P. Thérond, Development **127**, 5509 (2000).
45. A. Martinez-Arias, N. Baker and P. W. Ingham, Development **103**, 157 (1988).
46. A. Bejsovec and E. Wieschaus, Development **119** 501 (1993).
47. J. Heemskerk, S. DiNardo, R. Kostriken and P. H. O'Farrell, Nature **352**, 404 (1991).
48. A. Bejsovec and A. Martinez-Arias, Development **113**, 471 (1991).
49. R. Laubenbacher and B. Stigler, Polynomial models for biochemical networks (preprint, 2003).
50. I. Shmulevich, E. Dougherty and W. Zhang, Proc. IEEE **90**, 1778 (2002).

Theoretical Neuroanatomy: Analyzing the Structure, Dynamics, and Function of Neuronal Networks

Anil K. Seth and Gerald M. Edelman

The Neurosciences Institute, 10640 John Jay Hopkins Drive, San Diego, CA 92121 USA

Abstract. The mammalian brain is an extraordinary object: its networks give rise to our conscious experiences as well as to the generation of adaptive behavior for the organism within its environment. Progress in understanding the structure, dynamics and function of the brain faces many challenges. Biological neural networks change over time, their detailed structure is difficult to elucidate, and they are highly heterogeneous both in their neuronal units and synaptic connections. In facing these challenges, graph-theoretic and information-theoretic approaches have yielded a number of useful insights and promise many more.

1 Introduction

The human cerebral cortex contains a network consisting of approximately 30 billion neurons and about 1×10^{15} connections. The state of each neuron varies in complex ways over time, and neurons are diverse in their intrinsic properties and in the number of connections they make. Moreover, the structure of the brain continuously changes as a result of ongoing interactions among neural substrates, neural activity, and the embodied action of the organism within an environment [1–3].

The aim of this chapter is to describe some recent developments in understanding structural, dynamical and functional aspects of neural networks from the perspective of network theory. We focus our analysis on vertebrate cortical networks (the cortex is the heavily folded outer layer of the brain). Given this focus, our coverage is not intended to be comprehensive, and we present a mixture of previous results and new modeling data.

The structure of a neural network is reflected by the set of synaptic connections among neurons at a given time. These synapses have different strengths (weights) which may vary over time as a result of plasticity processes. Dynamics here refers to the neural firing patterns that a neural network supports. This is the level of effective connectivity in the brain [4], and there is now substantial evidence that cognitive and perceptual states are closely related to dynamical patterns of neural activity [5–7]. Finally, we use the term 'function' to refer to the role of these firing patterns within the context of a larger system [8], in this case the generation of adaptive behavior for the organism within its environment.

Structural and neuroanatomical analyses are most directly related to current network theory. Representing neural networks as directed graphs allows the ap-

A.K. Seth and G.M. Edelman, Theoretical Neuroanatomy: Analyzing the Structure, Dynamics, and Function of Neuronal Networks, Lect. Notes Phys. **650**, 483–511 (2004)
http://www.springerlink.com/

plication of a broad range of analytical tools from graph theory and statistical cluster analysis [9–14]. At present, much less is known about the detailed connectivity of the vertebrate cortex in comparison to artificial networks such as the internet (see, for example, [15]). However, the topology of biological neural networks can be described at many different spatial scales, and graph-theoretic analysis at the level of interconnected brain regions has revealed distinctive features of structural organization including so-called 'small-world' characteristics [16].

Network structure is particularly important inasmuch as it provides a substrate for dynamic processes [17,18]. Most dynamical data recorded from neural tissue consist of sequences of discrete action potentials — or 'spikes' — with varying inter-spike-intervals, and there is increasing evidence that precise spike timing is significant for normal neural operations [19,20]. Statistical information theory provides a very general means of characterizing these dynamics, and while there is a growing literature concerning the measurement of the information content of single neural spike trains [21–24], we focus here on the dynamics generated by *networks* of neuronal elements. We describe several information-theoretic measures appropriate for this task, including entropy, 'integration' and 'neural complexity' [25,14]. Theoretical models show that networks optimized to generate each kind of dynamics possess distinctive structural motifs [14]. Of particular interest is neural complexity [25], which is maximized for networks that show an even balance between dynamical segregation and dynamical integration. Networks that generate high complexity contain dense local clusters that are linked by reciprocal bridges. Intriguingly, cortical connection matrices have very similar structural properties, suggesting that they may be near-optimal for generating complex dynamics [14].

To be useful, a network-theoretic approach to neuroscience must take *function* into account. Since biological neural networks are embodied in organisms which are themselves embedded within environments, the most general function of a biological neural network is to generate action for the organism that is adaptive within the current environment. It has been suggested [14] that adaptation to rich sensory environments and motor demands may require complex neural dynamics. In this chapter we describe results that support this hypothesis, based on a model of target fixation by a simulated agent that is controlled by an artificial neural network [26].

While neuroscience has its specific explanatory targets and sources of data, the problem of understanding interactions among network structure, dynamics, and function is a very general one. Some aspects of the analyses presented in this chapter will prove useful for elucidating these interactions in network systems of many different kinds.

2 Structure

The anatomical or structural connectivity of a neuronal system is determined by the network of connections linking its elements at a given time. This network can be described at a variety of spatial scales, from synaptic connections among individual neurons, to fiber bundles linking local neuronal groups, to the massively parallel pathways connecting distributed areas of the brain.

In the vertebrate (and especially the human) cortex, global connectivity patterns at the level of individual neurons remain largely unknown [27]. Most recent analyses of brain connectivity have made use of datasets such as those describing connectivity at the level of segregated brain regions in the macaque monkey visual cortex [28,29] and the cat cortex [30,31]. While these datasets have proven very useful, they are far from complete. Recent efforts to combine results across many anatomical studies are improving this situation and have resulted in the development of online databases with increasingly detailed connectivity information [32]. For example, as of December 5, 2003 the CoCoMac database (accessible at www.cocomac.org) contained details of 33,850 connections among 6,466 distinct sites in the macaque brain [33]. Future analyses using this database, and others like it, may reveal many presently unknown features of neuroanatomical organization.

2.1 Graph Theoretic Analysis

Neural networks can be described as directed graphs (digraphs) G_{NK} with N vertices (nodes) and K edges. The connection matrix $\mathbf{C_{ij}}(G)$ of G contains elements c_{ij} holding the connection strength between node j (source) and node i (target). If no information is available about relative connection strengths, as is often the case, then $\mathbf{C_{ij}}(G)$ is written as a binary matrix, with entries 1 (connection present) and 0 (connection not present). Paths within G_{NK} refer to any ordered sequence of distinct nodes linking a source j to a target i with the condition that no node is visited more than once, unless $i = j$, in which case the path is a cycle.

Given G_{NK} and the corresponding $\mathbf{C_{ij}}(G)$ matrix, many standard tools of graph theory can be applied to yield insights into the structure of a neural network system [9–14].

The average connectivity of a digraph is the total number of connections present divided by the total number of connections possible. For the human cerebral cortex, this value appears to be very small, perhaps as low as 9.7 x 10^{-7} [34]. However, most synapses in the brain are between neurons that share the same local neighborhood, so that locally defined average connectivity values are much higher. For example, Nicoll and Blakemore estimate that neurons located within 300μm of each other in rat visual cortex are connected with a probability of 0.09 [35]. Average connectivity also increases at larger spatial scales; at the level of cortical areas, connection densities may be as high as 0.36 [28].

Average connectivity by itself does not give much insight into the structural organization of the brain. More useful are local measures such as the 'matching index' which gives the proportion of connections shared by two nodes i and j, normalized by the total number of connections belonging to both nodes [36]. A high matching index between two nodes suggests a possible functional overlap. For example, primate cortical areas FST and MSTd have a matching index of

0.71 [12], and electrophysiological studies indicate that cells in these areas do indeed have similar response properties [37].[1]

Another useful local measure is the cluster index f_{clust}, which is defined for a node i as the number of connections present among i and its immediate neighbors divided by the maximal number of connections possible among this subset [16]. The (global) cluster index for a graph G is the mean of f_{clust} for all nodes. A high global cluster index indicates a 'cliquish' graph in which neighboring nodes connect mainly among each other. Since most synaptic connections are made locally, neural networks may be expected to have high cluster indices, and indeed the macaque visual cortex has a cluster index $f_{clust} = 0.57$, much higher than the value for a randomly connected network with equivalent N and K ($f_{clust} \simeq 0.36$) [12].

The cluster index f_{clust} is often measured in conjunction with the 'characteristic path length' of a graph, l_{path}, which is a measure of the mean separation between any two nodes [16]. Formally, $l_{path}(G)$ is the global average of the distance matrix $\mathbf{D}_{ij}(G)$ whose elements d_{ij} hold the shortest path between nodes j and i. Highly clustered networks usually have high values for l_{path}, whereas random networks have low f_{clust} and low l_{path}. In a widely cited paper, Watts & Strogatz [16] identified a class of networks which combine high clustering with short characteristic path lengths; these 'small-world' networks consist mainly of local connections with a small proportion of randomly rewired edges. Small-world characteristics are present in networks of many different kinds [16,39], and analysis of cortical connection matrices also reveals high cluster indices in combination with comparatively short path lengths [14,12]; for example, the macaque visual cortex has $f_{clust} = 0.57$ (see above) and $l_{path} = 1.64$ (l_{path} for an equivalent random network $\simeq 1.60$).

'Scale-free' networks have the property that the probability $P(k)$ that a node connects to k other nodes follows a power law, i.e. $P(k) = k^{-\gamma}$, where typically $2 < \gamma < 3$. Such power-law degree distributions contrast with randomly connected networks which have Poissonian distributions with a characteristic average degree [40]. As with small-world networks, scale-free characteristics have been found in a wide variety of networks, including the internet, the world-wide web, and biological systems such as metabolic reaction networks [41,42]. As yet, however, no data have been found to indicate the existence of scale-free networks in the brain [12], although of course this does not rule out their discovery in the future.

It has been suggested that a distinctive feature of neurobiological structure and dynamics is *reentry*; the ongoing recurrent exchange of signals along massively parallel reciprocal pathways among neural areas [43–47]. The possibility of reentrant activity in a network can be estimated to a first approximation by the fraction of reciprocal connectivity among areas, i.e. the relative abundance

---

[1] Areas FST (floor of the superior temporal sulcus) and MSTd (dorsomedial superior temporal cortex) are visual areas whose cells show selective responses to the correlations in visual input that result from movement through a visual environment (optic flow) [38,37].

of cycles of length two. This quantity, designated as f_{recip}, is high for most cortical connection matrices [14]. For example, the macaque visual cortex has $f_{recip} = 0.77$, a fraction much higher than the expected value of 0.3 for equivalent random networks [28,12].

A more general measure of reentry is provided by the quantity $p_{cyc}(q)$, which captures the relative abundance of cycles of length q in a network (note that $p_{cyc}(2)$ is equivalent to f_{recip}) [14]. Both macaque visual cortex and cat cortex have values for $p_{cyc}(q)$, for cycles of lengths up to at least $q = 5$, that are significantly above those obtained for equivalent random networks [14].

2.2 Optimal Wiring

None of the measures presented so far take into account the physical separations among neurons. Yet biological neural networks exist within the three-dimensional confines of the skull. Moreover, neural material is metabolically expensive, both to develop and to sustain in operation. These considerations have led some researchers to propose that neuroanatomical structure is arranged in such a way as to minimize the total wiring length among neural structures [48-51].

While some evidence exists in support of these ideas from studies of invertebrate nervous systems [49], analyses of cortical connection matrices are less convincing. Both the macaque visual cortex and the cat cortex have minimal wiring lengths that are significantly shorter than those of equivalent random networks; however there exist many other networks with equivalent N and K that have even shorter wiring lengths [14]. In any case, it seems improbable that the sole selection pressure during brain evolution was minimization of wiring length. More likely, anatomical arrangements evolved primarily to support dynamical activity patterns that contributed to the generation of adaptive behavior by the organism, with metabolic and developmental constraints playing a secondary role.

2.3 Statistical Analysis

Statistical techniques complement graph-theoretic analyses by detecting consistent patterns in complex data sets, usually by some form of dimensionality reduction. A constellation of techniques is currently available; here we briefly mention only one: non-metric multidimensional scaling (NMDS) [52,53,12]. NMDS rearranges objects in a low dimensional space (usually 2D or 3D) so that the rank-ordering of the original (high-dimensional) distances among them are best preserved. Proximity in a NMDS diagram indicates that two nodes are strongly interconnected and/or share a relatively large proportion of their connections.

Young et al. [54] applied NMDS to the connection matrix of the macaque visual system. They found that the resulting 2D configuration showed a clear separation between two distinct groups of nodes. This separation corresponds to the commonly accepted distinction between 'dorsal' and 'ventral' streams of visual processing [55].[2] Alternative methods, which give results that are largely consistent with the NMDS analysis, are reviewed by Hilgetag et al. [56,12].

3 Dynamics

The networks of the brain support the exchange of signals among neurons. There is now substantial evidence that the resulting activity patterns are closely related to our cognitive and perceptual states [5–7]. A central challenge for a network-theoretic approach to neuroscience is to relate structural descriptions to dynamical patterns of activity at different spatial and temporal scales. In this section we focus on the application of statistical information theory as a means of characterizing neural network dynamics [25,14].

3.1 Information-Theoretic Analysis

The dynamical – or effective – connectivity of a neuronal system consists of the pattern of temporal correlations, or deviations from statistical independence, in the activities of neuronal elements that are generated by their interactions [4]. It has been suggested that cortical networks exhibit a balance between two main principles of dynamical organization, segregation and integration [25,14,57, 58]. Cortical networks contain many kinds of specialized neuronal units that are anatomically segregated from each other; for example, cells in different regions of visual cortex are specialized to respond to color, orientation, motion and so forth [59,60]. At the same time, in order to support globally coherent cognitive and perceptual states, the activity of these segregated elements has to be integrated across space and over time.

Global Dynamical Measures. A useful description of the effective connectivity of a neural system is the joint probability distribution function of the activities of its neuronal elements. Assuming that this function is Gaussian, this is equivalent to the covariance matrix of the system. Importantly, the covariance matrix captures the *total* effect of all (anatomical) connections within a system on deviations from statistical independence of the activities of a pair of elements, not just the effect of any direct anatomical connection linking them [61].

Covariance matrices can be numerically estimated by direct observation of system activity over time. Alternatively, if linear dynamics are assumed, the covariance matrix of a system X can be derived analytically from the anatomical connectivity matrix $\mathbf{C_{ij}}(X)$:

$$\mathbf{COV}(X) = \mathbf{Q}^T * \mathbf{Q}, \qquad \mathbf{Q} = [1 - \mathbf{C_{ij}}(X)]^{-1}, \tag{1}$$

where ^{T} indicates matrix transpose [25].

---

[2] The dorsal visual stream is also called the 'where' stream and is usually associated with object location, whereas the ventral stream - the 'what' stream - is usually associated with object identification.

Once the covariance matrix is obtained, a number of global dynamical measures can be calculated. Three such measures, which are based on the foundations of statistical information theory, are entropy $H(X)$, 'integration' $I(X)$ and 'neural complexity' $C_n(X)$ [25,14].

For a system X, the entropy $H(X)$ measures the system's overall degree of statistical independence. Assuming stationarity, $H(X)$ can be calculated using the standard formula

$$H(X) = 0.5\ln((2\pi e)^N |\mathbf{COV}(X)|), \tag{2}$$

where $\mathbf{COV}(X)$ is the $N \times N$ covariance matrix of X, and $|.|$ denotes the matrix determinant [25,62].

The integration $I(X)$ measures the system's overall *deviation* from statistical independence. All elements in a highly integrated neural system are tightly coupled in their activity. With x_i denoting the i'th element, $I(X)$ can be calculated using

$$I(X) = \sum_{i=1}^{N} H(x_i) - H(X). \tag{3}$$

Perhaps the most interesting global dynamical measure is neural complexity $C_n(X)$, which measures the extent to which a system balances dynamical segregation and dynamical integration [25]. The component parts of a neurally complex system are differentiated; however, as larger and larger subsets of elements are considered they become increasingly integrated (see Fig. 1). The term 'complexity' may be considered appropriate for such systems since they are intermediate between the two relatively simple extremes of disorder (maximal statistical independence) and order (maximal statistical dependence).

The neural complexity $C_n(X)$ of a system X is calculated by summing the average Mutual Information (MI) between subsets of different sizes, for all possible bipartitions of the system:

$$C_n(X) = \sum_{k=1}^{n_t/2} \langle MI(X_j^k; X - X_j^k) \rangle, \tag{4}$$

where n_t is the total number of ways of bipartitioning X, X_j^k is the j'th bipartition of size k, and $\langle . \rangle$ is the average across index j. The MI between two subsets (A and B) measures the uncertainty about A that is accounted for by the state of B, and is defined as $MI(A; B) = H(A) + H(B) - H(AB)$ [63] (see Fig. 1).

Another, closely related measure of complexity expresses the portion of entropy that is accounted for by interactions among all the elements of a system [58]:

$$C(X) = H(X) - \sum_{k=1}^{N} H(x_i | X - x_i). \tag{5}$$

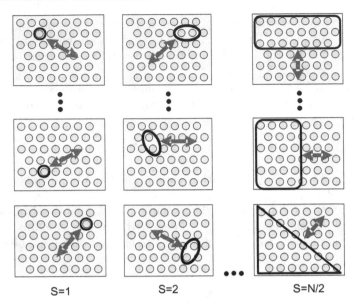

Fig. 1. Neural complexity is defined as the ensemble average MI between subsets of a given size and their complement, summed over all subset sizes (adapted from Fig. 2 in [64]). Small circles represent neuronal elements and bold arrows indicate MI between subsets and the remainder of the system. Shown in the figure are a selection of subsets of size 1 (S=1), size 2 (S=2), and size $N/2$ for system size N (S=$N/2$). A neurally complex system is one in which small subsets of the system show high statistical independence, but large subsets show low statistical independence.

This measure takes on high values if single elements are highly informative about the system to which they belong, without being overly alike. It has the advantage of being easier to compute than the first formulation $C_n(X)$, and all results described in this chapter use this formulation.

Dynamical Cluster Analysis. In Sect. 2.3 we described how statistical techniques such as NMDS could be used to identify clusters of highly interconnected nodes from high dimensional connectivity data. Cluster analysis can also performed at the level of network dynamics, where a dynamical cluster corresponds to a strongly interactive subset of elements. Note that dynamically identified clusters need not correspond to structurally identified clusters, although in many cases they may do so.

A dynamical cluster is characterized by a high level of statistical dependence among its elements and, at the same time, a low level of statistical dependence with elements outside the cluster. Clusters can be identified by calculating the cluster index $Cl(X_i{}^k)$, which is defined as follows [65]:

$$Cl(X_i{}^k) = I(X_i{}^k)/MI((X_i{}^k); X - (X_i{}^k)), \tag{6}$$

where $X_i{}^k$ denotes the i'th subset of size k, $I(X_i{}^k)$ indicates its integration, and $MI((X_i{}^k); X - (X_i{}^k))$ denotes its mutual information with the rest of the system.

Note that Tononi et al. [65] describe this cluster index as a measure of *functional* clustering rather than *dynamical* clustering. However, as we remarked in the introduction, function here refers to the role of a particular pattern of dynamics in the context of the behavior of the organism.

Graph Selection. To explore the relationship between network structure and network dynamics, Sporns et al. [14] applied an evolutionary search procedure to look for distinct anatomical motifs associated with different measures of effective connectivity. This procedure uses a global dynamical measure as a fitness (or cost) function, and it is implemented as follows. First, an initial population of U random graphs is created, each with N nodes and K edges with fixed identical positive weights w_{ij}. The 'fitness' of each graph is then assessed by deriving the corresponding covariance matrix and calculating the global dynamical measure, either $H(X)$, $I(X)$ or $C(X)$. After all members of U have been assessed, the graph with highest fitness is selected and all others are discarded. The selected graph is then replicated $U - 1$ times to create the next generation. Each replication introduces a small amount of variation by randomly rewiring r edges of the selected graph. After a sufficient number of generations (these authors typically used 3,000 generations, a population size $U = 10$ and a rewiring rate $r = 1$) the members of U should have near-optimal structures for generating the particular dynamical measure used as the selection criterion.[3]

Sporns et al. selected separately for graphs with high entropy $H(X)$, high integration $I(X)$, and high neural complexity $C(X)$. In each case they found that the resulting graphs ($N = 32, K = 256$) had distinctive structural features, as revealed both by simple visual inspection and by analysis using the graph-theoretic measures described in Sect. 2.1. Graphs optimized for high entropy contained mostly reciprocal connections without any apparent local clustering; they had high f_{recip}, low f_{clust}, as well as a short diameter $(diam_G)$ and a short characteristic path length (l_{path}). Graphs optimized for integration, by contrast, were highly clustered (high f_{clust}), had low values for f_{recip}, large diameters and large characteristic path lengths. Visual inspection of these graphs revealed a very large central cluster loosely connected to an outlying mesh of nodes. Graphs optimized for neural complexity were the most similar to the 'small-world' class of networks. These graphs usually are comprised of a number of dense groups of nodes linked by a relatively small number of reciprocal bridges; they had high values for f_{recip} and f_{clust}, as well as low values for l_{path} and $diam_G$.

---

[3] Sporns et al. [14] applied three neurobiologically motivated constraints during graph selection. First, the number of incoming edges to each node was kept constant (8). Second, all graphs were required to be 'strongly connected' such that at least one path exists between all pairs $\{i, j\}$ of nodes. Third, small self-inhibitory weights w_{ii} were applied to each graph such that the total variance – given by the sum of the diagonal terms in $\mathbf{COV}(X)$ – remained constant.

3.2 Dynamical Properties of Cortical Networks

To relate these theoretical observations to empirical data, Sporns et al. [14] analyzed the complexity of the connection matrices of the macaque visual cortex and the cat cortex. They assumed linear dynamics, equal connection strengths, and used equation (1) to derive covariance matrices corresponding to the respective connection matrices. They found that both connection matrices gave rise to effective connectivity with high neural complexity $C(X)$ as compared to random networks with equivalent N and K. Indeed, the matrices seemed to be near-optimal for generating highly complex dynamics. Random rewiring of edges, in virtually all cases, led to a reduction in the complexity of the corresponding effective connectivity.

To identify the sets of anatomical areas responsible for these complex dynamics, these authors applied dynamical cluster analysis (see Sect. 3.1) to the macaque visual cortex connection matrix. This analysis revealed an hierarchical organization which separated into two distinct streams corresponding to the 'dorsal' and 'ventral' visual processing streams [55], together with a small subset of areas which were strongly interactive with both streams (these authors suggest that these areas are strong candidates for mediating inter-stream interactions). Overall, this dynamical analysis was highly consistent with the structural cluster analyses of Young et al. [54] and Hilgetag et al. [56].

Another application of dynamical cluster analysis compared PET[4] data obtained from normal and schizophrenic subjects performing a set of cognitive tasks [65]. This study found significant differences in cluster profiles between these two groups, despite the absence of differences in overall levels of activity. Future application of this analysis using imaging methods with higher temporal resolution (for example magnetoencephalography, see [67]) may be extremely revealing.

3.3 Matching Complexity and Degeneracy

While entropy, integration and neural complexity provide measures of the intrinsic dynamics of a neural system, it is also of interest to characterize the dynamics of neural systems as they interact with a surrounding environment. Measures of 'matching complexity' [68] and 'degeneracy' [69,70] fulfil this role respectively for systems connected to an input or to an output.

The matching complexity between system X and input S is defined as the total complexity when the input is present $C(X)^T$ minus the intrinsic complexity $C(X)^I$ and the complexity $C(X)^S$ that is directly attributable to the input S [68]:

$$M(X; S) = C(X)^T - C(X)^I - C(X)^S. \tag{7}$$

A high level of matching complexity indicates that there is strong match between the statistical structure of the system and that of the input. A low value indicates statistical 'novelty'.

---

[4] Positron Emission Tomography: A brain imaging technique based on measurement of metabolic activity [66].

Degeneracy refers to the ability of elements (or sets of elements) that are structurally different to perform the same function or yield the same output [69, 70]. Degeneracy can be expressed as the mutual information between the various subsets of a system and the system's output [69]:

$$D(X; O) = MI^P(X, O) - \sum MI^P(x_i; O | X - x_i; O), \tag{8}$$

where $MI^P(x_i; O | X - x_i; O)$ refers to the conditional mutual information between each element and O, given the mutual information between the rest of the system and O. Degeneracy is high for systems in which many different elements affect the output in a similar ways but at the same time can have independent effects. This property contributes to the robustness of a system [70].

Sporns et al. [14] extended their graph selection method (see Sect. 3.1) to select for graphs that exhibited high matching complexity with respect to an input, and, separately, high degeneracy with respect to an output. In the former case, a subset of 8 nodes was connected to an 8-node sensory sheet S that had a particular input pattern described by a covariance matrix **COVs**. In the latter case, a subset of nodes was chosen as a representation of output, and selection was carried out based on the global measure of degeneracy with respect to a particular output pattern.

In all cases, selection for matching or degeneracy resulted in graphs with high neural complexity $C(X)$ [14]. This suggests that high $C(X)$ may reflect not only an intrinsic balance between dynamical integration and segregation, but may also correspond to the ability of a network rapidly to distribute input signals and robustly generate output signals.

4 Function

As we remarked in the introduction, the most general function of a biological neural network is to generate adaptive behavior for the organism within its environment [71]. The analysis of graphs with high degeneracy and high matching complexity begins to address this issue of function. However, in the studies described above, there is no *behavior* as such: networks are coupled either to a static input or a static output, and there is no sense in which the output at a given time affects the input at a subsequent time.

In this section, we tackle these concerns by describing a relatively simple model of target fixation by a simulated head/eye system that is controlled by an artificial neural network. This model explicitly involves behavior, and its analysis provides support for the hypothesis that complex neural dynamics facilitate adaptation to rich sensory environments and motor demands [14]. An extended analysis of this model is given in Seth & Edelman [26].

As in the studies described above, the present model makes use of the technique of evolutionary graph selection. However, instead of using a global dynamical measure as a fitness function, in this case the fitness of a network is

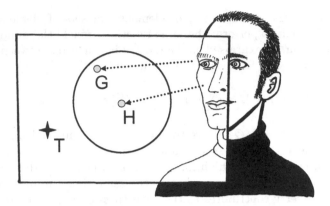

Fig. 2. Simulation model of target fixation, depicting projections onto a plane ($100u$ x $100u$, where u denotes an arbitrary spatial unit), of head direction H and gaze direction G. Also shown are target position T, and maximum offset of G from H (circle, radius $35u$). The position of G is calculated as a vector sum of H and an eye direction (E, not shown).

determined by its ability to support adaptive behavior when employed as a control mechanism for a simulated head/eye system engaged in a target fixation task. To explore how network structure, dynamics, and function interact, we select networks for their target fixation ability in a variety of conditions that differ in the (qualitative) complexity of the environment and the phenotype (the head/eye system). Networks which evolved in different conditions are then analyzed in terms of their behavioral properties, their structural properties, and their dynamical properties.

4.1 Methods

Model Outline. This section describes basic properties of the model; specifics are provided in Appendix A. The simulated environment is a simple planar area, 100x100 arbitrary units u, within which a target (T) can appear. The simulated head/eye system, or phenotype, is represented by the projection onto the plane of a head direction (H) and a gaze direction (G), which is the combination of the head direction (H) and an eye direction (E) relative to the head (Fig. 2).

The velocities of H and E in the x, y plane, and thus the position of G, are controlled by a neural network ($N = 32$, $K = 256$). The edges of this network have real-valued weights, and certain nodes are specified as sensory inputs and others as motor outputs (see below). The remainder are 'interneurons' mediating the transformation of input signals into output signals. Evolutionary algorithms are used to specify the connectivity and weight distribution of networks so that they give rise to adaptive behavior, which in this case consists of maximizing the time for which G and T are aligned (i.e. keeping the target fixated) while simultaneously minimizing the offset between H and G (i.e. keeping the head

Table 1. Phenotype parameters in the four conditions (E_S, E_T, E_H, and E_C). V_{maxH}, V_{maxE}: maximum velocity of head and eye, A_H, A_E: motor gain of head and eye, m_H, m_E: momentum of head and eye, *lag*: the time-lag between head motor node output and head movement.

Condition	V_{maxH}	V_{maxE}	A_H	A_E	m_H	m_E	*lag*
E_S	7.0	7.0	1.0	1.0	0.0	0.0	0
E_T	7.0	7.0	1.0	1.0	0.0	0.0	0
E_H	5.0	10.0	0.33	1.0	0.75	0.05	10
E_C	5.0	10.0	0.33	1.0	0.75	0.05	10

and eye aligned). A constraint on the system is that G must remain within $35u$ of H, as indicated by the circle in Fig. 2.

Environment/Phenotype Conditions. We specify four conditions which are distinguished by properties of the target and the head/eye phenotype. The phenotype is defined by seven parameters: V_{maxE} specifies the maximum velocity of H, V_{maxH} specifies the maximum velocity of E with respect to H; motor gain parameters A_H and A_E specify scaling factors relating motor node output to H and E velocity; momentum parameters m_H and m_E specify the inertial resistance of H and E, and *lag* specifies a time-lag between head motor node output and head movement. These parameters constrain the movements of H, E and G as described in Appendix A (which also describes how T is updated).

The values taken by the parameters in the four conditions are shown in Table 1. In condition E_S (for 'simple') both the head and eye have the same maximum velocity, identical motor gains, and zero momentum; the target is stationary. Condition E_T (for 'tracking') presents a more complex environment in which the target may occasionally jump to a different random location and/or drift at a slow speed in a random direction; the phenotype is the same in condition E_S. Condition E_H (for 'head') keeps the simple environment of condition E_S but introduces a more complex phenotype in which E can move twice as fast as H, has a higher motor gain, a much lower momentum, and in which there is a non-zero time-lag. Finally, condition E_C (for 'complex') combines the properties of both conditions E_T and E_H and is therefore the richest of the four.

Network Implementation. Each behavioral trial begins with the head and eye aligned and pointing to the center of the plane. The target position is initialized at a randomly selected location within $20u$ of this point. All trials last for 600 time steps, each of which involves updating the state of the network controller (X) as well as the positions of E, G, H, and, in conditions E_T and E_C only, T. The network is updated using, for all nodes j,

$$s_j(t) = f\left(s_{in}(j,t) + \sum_{i=1}^{K} \mathbf{C_{ij}}(X)s_i(t-1)\right), \qquad s_j(0) = 0.0, \qquad (9)$$

where $s_j(t)$ is the output of node j at time t, .$s_{in}(j, t)$ denotes the sensory input to node j at time t (if any), f is a sigmoid function with input range ± 10.0 and output range ± 1.0, and $\mathbf{C_{ij}}(X)$ is the connection matrix of X.

Six nodes are specified as 'sensory' inputs, two responding to the x, y displacement of G from T (with a maximum range of $\pm 50.0u$), and four delivering 'proprioceptive' information, two of which reflect the displacement of H from the center of the plane (range $\pm 50.0u$), and two of which reflect the displacement of G from H (range $\pm 35.0u$). In all cases, input values are linearly scaled to the range ± 1.0. Four nodes are specified as output nodes, two influencing H velocity in the x, y plane, and two influencing the velocity of E. Appendix A describes how the activities of these nodes, along with the values in Table 1, are used to update the positions of E, G, and H.

Network Structure. Evolutionary algorithms (EAs) were used to specify the connectivity and weight distribution of networks so that they supported target fixation behavior.[5] We ran a total of 40 separate EAs: 10 replications of the selection process for each of the four environment/phenotype conditions.

Each EA evolved a population of 64 networks over 2000 generations, with each network initialized (generation 0) by randomly allocating K connections (each connection strength assigned randomly in the range ± 1.0), subject to the constraint that each node had 8 incoming connections. Each generation involved evaluating the fitness of each network, as described below, and then using stochastic rank-based selection to replace low-fitness networks with mutated versions of high-fitness networks. Each mutation of a network involved randomly rewiring 1 connection (preserving in-degree) and also modifying the strength of each connection (probability 0.05 per connection) in the range ± 0.1.

The fitness of a network was calculated as the mean of four separate behavioral trials, with the fitness of each trial (ϕ) given by

$$\phi = \frac{t_f}{t_{tot}} + c_0(1.0 - c_1\bar{d}), \tag{10}$$

where t_f denotes the number of time steps for which the target was fixated (within a tolerance of $3u$), t_{tot} the total number of time steps in the trial (600), and $\bar{d}$ the mean offset for the trial between H and G. The constants c_0 and c_1 were selected in order to balance the fitness contributions due to target fixation and those due to minimizing the offset between H and G ($c_0 = \frac{1}{4}, c_1 = \frac{1}{35}$).

---

[5] We used a distributed evolutionary algorithm [72]: Each EA was initialized by arranging the population on an 8x8 toroidal grid and evaluating the fitness of each network. Each subsequent generation involved 64 repeats of the following: A random grid position was chosen determining a 3x3 sub-grid. Stochastic rank-based selection was then used to select a weak member of this sub-grid for replacement, and a strong member as the 'parent'. A mutated copy of the parent then replaced the weak member and was evaluated (the parent was also re-evaluated with probability 0.75).

Table 2. Summary of target fixation performance. Shown are mean percentage of each behavioral trial for which T was fixated (within a tolerance of $3u$), and mean displacement of G from H, for each network type. Average values were calculated from 10 behavioral trials of each evolved network in the corresponding environment/phenotype condition. Standard errors are given in parentheses.

Network type	fixation %	$|G-H|$ (u)
S-network	98.81 (0.01)	0.28 (0.07)
T-network	92.43 (0.51)	0.92 (0.08)
H-network	96.15 (0.79)	1.94 (0.19)
C-network	84.76 (1.23)	15.41 (0.50)

4.2 Results

For convenience, networks which evolved in condition E_S will be referred to as S-networks, with the same nomenclature for conditions E_T (T-networks), E_H (H-networks) and E_C (C-networks).

Behavioral Analysis. The target fixation behavior of evolved networks was assessed by measuring the percentage of each behavioral trial for which the target was fixated (i.e. the fraction of time steps for which the displacement between H and G was less than $3u$). Table 2 shows that average fixation performance was very high for S-networks, T-networks, and H-networks, and slightly lower for C-networks.

Figure 3 shows representative examples of successful fixation behavior for S-networks and C-networks in the corresponding conditions (E_S and E_C respectively). Inspection of the trajectories of the head and the eye indicates that the behavioral dynamics for C-networks were qualitatively more complex than for S-networks. A non-trivial coordination of head and eye is needed in E_C conditions in order to maintain consistent fixation of the target. Table 2 affirms

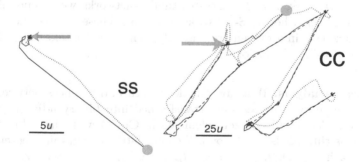

Fig. 3. Typical successful fixation behavior for S-networks in condition E_S (left) and for C-networks in condition E_C (right), showing trajectories of G (solid line), H (dotted line) and T (dashed line, E_C only), gray circle indicates initial positions of G and H, gray arrows indicate initial positions of T. Successful target tracking by C-networks is indicated by the overlap of the trajectories of G and T.

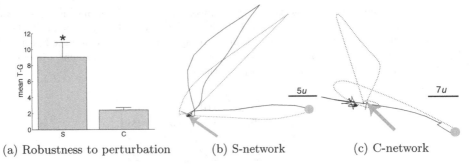

(a) Robustness to perturbation   (b) S-network   (c) C-network

Fig. 4. Robustness of S-networks and C-networks to a novel perturbation: (a) Mean displacement of G from T during enforced H movement, calculated over the 20 time steps of enforced movement and the 10 time steps following for 10 S-networks and 10 C-networks. Asterisk indicates statistical significance ($p < 0.001$, two-tailed t test). (b,c) trajectories generated by a typical (b) S-network and (c) C-network during this operation (solid line G, dotted line H, gray circles indicates initial positions of G and H, gray arrows indicate positions of T). For the S-network, fixation is lost (both H and G are displaced from the target), whereas for the C-network, fixation is maintained (G remains close to the target even though H is displaced).

that the average displacements between H and G were considerably larger in for C-networks than for all other network types, while target fixation was achieved reliably in all cases.

To better compare the behavioral properties of S-networks and C-networks, they were reevaluated in a novel condition involving unexpected perturbations. As in condition E_S, the environment was initialized with a stationary target T. An evolved S-network or C-network was then introduced and allowed to fixate. After 100 time steps, a head velocity (V_{hx}, V_{hy}) of between 1.0 and 3.0 (u per time step) was induced for 20 time steps in a random direction. Importantly, neither network type had been selected to respond adaptively to this perturbation. Nevertheless, Fig. 4 illustrates that C-networks were generally able to maintain fixation, whereas S-networks were not. These results show that network optimization in a rich environmental/phenotypic context can facilitate the emergence of robust behavior.

Structural Analysis. To assay reliable structural differences between network types, connectivity matrices were transformed into binary adjacency matrices $A_{ij}(X)$ by replacing all non-zero elements in $C_{ij}(X)$ with the value 1 (we also tried various thresholds, see below). We calculated four graph-theoretic quantities from each $A_{ij}(X)$: f_{recip}, $diam_G$, l_{path}, and f_{clust} (see Sect. 2.1).

Table 3 shows the results of applying these metrics to the fittest networks from each EA in each condition. Values were also calculated for an additional 64 random networks (R-networks) with equivalent N and K (and a per-node indegree of 8). The table indicates that there are no significant differences between network types (including random) in any of the metrics.

Table 3. Structural analysis of network types showing values for f_{recip}, $diam_G$, l_{path} and f_{clust}. Each entry in the table shows mean and standard error (in parentheses) calculated from 10 networks of each type (64 for R-networks).

Network type	f_{recip}	$diam_G$	l_{path}	f_{clust}
S-network	0.27 (0.028)	3.1 (0.32)	1.82 (0.01)	0.25 (0.001)
T-network	0.26 (0.006)	3.2 (0.42)	1.83 (0.02)	0.25 (0.001)
H-network	0.26 (0.019)	3.3 (0.67)	1.83 (0.03)	0.25 (0.001)
C-network	0.27 (0.023)	3.0 (0.01)	1.82 (0.01)	0.24 (0.001)
R-network	0.27 (0.042)	3.0 (0.01)	1.82 (0.01)	0.25 (0.001)

Why are no differences observed? It may be that the transformation of $\mathbf{C}_{ij}(X)$ into $\mathbf{A}_{ij}(X)$ overemphasizes the importance of weak connections. We tested this possibility by recalculating values after thresholding adjacency matrices to include only relatively strong connections. We tested a variety of thresholds, but in no case did we observe differences in structural measures among conditions. This suggests that specific connection strengths may be critical for the behavioral properties of the networks. Importantly, this is *not* to say that network anatomy is irrelevant to the behavioral differences among the network types; rather, it emphasizes the need for graph-theoretic measures which can be usefully applied to weighted networks as well as to binary networks [73,74].

4.3 Dynamical Analysis

Intrinsic Dynamics. As a first approach to describing the dynamical properties of evolved networks, we treated the evolved $\mathbf{C}_{ij}(X)$ matrices as in Sect. 3.1, i.e., assuming linear dynamics and activation by Gaussian noise, we applied equation (1) to derive the corresponding covariance matrices. An important property of this method is that $\mathbf{COV}(X)$, and therefore also the resulting values of $C(X)$, $I(X)$, and $H(X)$, are *independent of behavioral context*. This method is therefore best described as characterizing the *intrinsic* dynamical properties of a network.

For each EA, the fittest member of the final generation was used to generate a covariance matrix, which in turn was used to calculate the intrinsic neural complexity $C(X)$, integration $I(X)$, and entropy $H(X)$ of the network dynamics. Values were also calculated for 64 R-networks.

Figure 5 shows mean and standard deviations of intrinsic $C(X)$, $I(X)$, and $-H(X)$, for each network type. In contrast to the structural analysis (see above), clear differences are evident: Random R-networks scored lowest on all three dynamical measures, and C-networks scored highest. Differences between C-networks and all other networks were significant at the $p < 0.01$ level by two-tailed t test for $C(X)$, $I(X)$, and $H(X)$. These observations indicate that adaptation to comparatively complex environmental/phenotypic conditions (E_C) endows C-networks with above random neural complexity and integration, and below random entropy. Furthermore, only networks with these dynamical characteristics displayed the behavioral robustness and flexibility as described above.

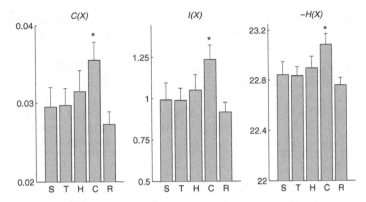

Fig. 5. Mean and standard error intrinsic $C(X)$, $I(X)$, and $-H(X)$ for S-networks (S), T-networks (T), H-networks (H), C-networks (C), and R-networks (R). Asterisks indicate that the value for C-networks is significantly higher than for all other network types (p < 0.01, two-tailed t tests). Values of $C(X)$, $I(X)$ and $-H(X)$ for S-networks, T-networks, and H-networks are significantly lower than the corresponding values for C-networks, and significantly higher than the corresponding values for R-networks (p < 0.01, two-tailed t tests). There are no significant differences in $C(X)$, $I(X)$ or $-H(X)$ among S-networks, T-networks, and H-networks. All distributions are normal (p < 0.05, Bera-Jarques test).

Figure 5 also shows that S-networks, H-networks, and T-networks scored at intermediate levels on all measures. Values of $C(X)$, $I(X)$ and $-H(X)$ for these networks are significantly lower than the corresponding values for C-networks, and significantly higher than the corresponding values for R-networks (p < 0.01, two-tailed t tests); there are no significant differences in $C(X)$, $I(X)$ or $-H(X)$ among S-networks, T-networks, and H-networks. Changes in intrinsic dynamics, therefore, depend both on properties of the environment (condition E_T) and on properties of the phenotype (condition E_H).

While these results show that neurally complex network dynamics can accompany adaptive behavior in rich environmental/phenotypic conditions, intrinsic neural complexity is not unique in this respect, since similar patterns of results are apparent for both intrinsic integration $I(X)$ and entropy $H(X)$.

Interactive Dynamics. An alternative approach to characterizing network dynamics is to derive covariance matrices directly from recordings of node activities during behavior, i.e. while the network is actively transforming input signals into output signals. We define the resulting dynamics as 'interactive' since they are relative to a particular behavior, environment, and phenotype. One important property of interactive dynamics, as compared to intrinsic dynamics, is that they enable comparison of dynamics generated by the same network in different behavioral regimes or in different environments/phenotypes.

Each behavioral trial yields a (N x 600) matrix **F** which contains the individual activity records of each node for all 600 time steps. This matrix can be used to generate a covariance matrix according to:

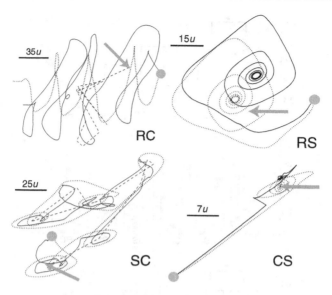

Fig. 6. Representative behavior for various network-environment/phenotype combinations, showing trajectories of G (solid line), H (dotted line) and T (dashed line), gray circles indicate initial positions of G and H, gray arrows indicate initial positions of T. RC: R-network in condition E_C, RS: R-network in condition E_S, SC: S-network in condition E_C, CS: C-network in condition E_S. Trajectories of a C-network in condition E_C and an S-network in condition E_S are shown in Fig. 3.

$$\mathbf{COV}(X) = cov(\dot{\mathbf{F}}), \tag{11}$$

where $\dot{\mathbf{F}}$ is the first derivative (with respect to time) of the activity matrix and $cov()$ is a standard covariance function. This approach was chosen because rates of change of node activities are more likely to reflect interactions between the network and its phenotype and environment, than are absolute activity levels. $\mathbf{COV}(X)$ can then be used to calculate corresponding values of interactive $C(X)$, $I(X)$ and $H(X)$ in just the same way as for the calculation of intrinsic dynamics. Note that calculation of interactive dynamics does not require equation (1) and so does not assume linear system dynamics.

We compared interactive dynamics for C-networks, S-networks, and random R-networks evaluated in both E_C and E_S conditions, recording both network activities and behavioral trajectories. Figure 6 shows representative trajectories from the various combinations (trajectories of a C-network in condition E_C and an S-network in condition E_S are shown in Fig. 3). R-networks in E_C and E_S conditions never achieve fixation, and their behavior is highly variable. S-networks in E_C conditions are unable to achieve or maintain fixation despite a general tendency to track towards the target, and C-networks in E_S conditions rapidly achieve fixation despite some persistent oscillation in the head direction. Notice that in all except the last of these cases, the behavioral dynamics are rich, even though the behavior itself is not adaptive.

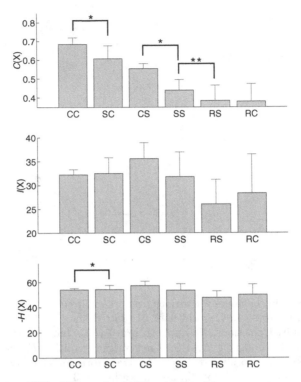

Fig. 7. Interactive $C(X)$, $I(X)$, and $-H(X)$ calculated from covariance matrices derived from recorded neural activity during behavior. CC: C-network in condition E_C, SC: S-network in condition E_C, CS: C-network in condition E_S, SS: S-network in condition E_S, RS: R-network in condition E_S, RC: R-network in condition E_C. Each column shows mean and standard error calculated from 10 repetitions of each combination of network and condition. There is a significant correlation between the rank ordering of combination (from 'CC' to 'RC') and mean $C(X)$ ($r = 0.97$, p < 0.01; Spearman's rank correlation). No significant correlations exist for $I(X)$ or $-H(X)$ (p > 0.1). Asterisks indicate statistically significant differences between combinations ((*) p < 0.01, (**) p < 0.05, two-tailed t tests; only adjacent combinations were tested). All distributions are normal (p < 0.05, Bera-Jarques test).

Figure 7 shows the interactive dynamics generated in each combination of network and condition. The top row shows a clear gradient in interactive $C(X)$ proceeding from C-networks in condition E_C ('CC'; high), to R-networks in conditions E_S and E_C ('RS', 'RC'; low). Intermediate values of $C(X)$ were observed for simple networks in rich environmental/phenotypic conditions ('SC'), and for complex networks in simple conditions ('CS'). C-networks in condition E_C ('CC') have significantly higher $C(X)$ than all other combinations. The remaining significant differences among combinations are also consistent with the steady gradient in interactive $C(X)$.

Strikingly, only $C(X)$ shows a pattern of values reflecting rich adaptive behavior. Neither $I(X)$ (Fig. 7, middle row) nor $H(X)$ (bottom row) show any such sensitivity. This is in contrast to the intrinsic analysis, in which all three dynamical measures behaved in a similar way (Fig. 5).

It is also notable that R-networks evoke low interactive $C(X)$ despite displaying the rich behavioral patterns shown in Fig. 6. The contrast between R-networks and C-networks, together with the steady gradient in interactive $C(X)$ across combinations (Fig. 7, top row), suggests that $C(X)$ is indeed selectively sensitive to the dynamics of rich *adaptive* behavior. Figure 7 also suggests that high interactive $C(X)$ depends on a combination of environmental and phenotypic properties. The intermediate values of $C(X)$ in Fig. 7 ('SC' and 'CS') are consistent with the intermediate values of intrinsic $C(X)$ associated with rich environments (condition E_T) and complex phenotypes (condition E_H) shown in Fig. 5.

4.4 Summary

In this model, evolutionary algorithms were used to generate neural networks able to support target fixation in environment/phenotype combinations of qualitatively different levels of complexity. Not surprisingly for a selectional system, we found that those networks which evolved under rich environmental/phenotypic conditions exhibited (qualitatively) more complex behavior than networks which evolved in comparatively simple conditions (Fig. 3). When compared in a condition involving a novel perturbation, networks which evolved in rich conditions showed greater robustness than networks which evolved in simple conditions (Fig. 4).

This robustness was reflected by significantly higher neural complexity $C(X)$ for networks in evolved in rich conditions, than for networks which evolved in relatively simple conditions, or for equivalent random networks (figs. 5 and 7). This was true for both intrinsic and interactive methods of calculating dynamics, where the former are derived analytically from network connectivity, and the latter are computed from observed network activity during behavior. However, while intrinsic dynamics did not differentiate between neural complexity $C(X)$, integration $I(X)$ or entropy $H(X)$ (Fig. 5), interactive dynamics revealed that only $C(X)$ consistently associated with adaptive behavior in rich environmental/phenotypic conditions (Fig. 7).

Both types of dynamical analysis indicate that the magnitude of neural complexity depended on a combination of environmental, phenotypic, and mechanistic properties. Networks evolved in conditions of intermediate richness (E_T and E_H) generated intermediate values of intrinsic $C(X)$ (Fig. 5). Networks evolved in rich environmental/phenotypic conditions, and tested in simple conditions, and *vice-versa*, generated intermediate values of interactive $C(X)$ (Fig. 7). Taken together these results show that, in the present target fixation task, neural complexity is selectively sensitive to the dynamics of adaptive behavior in rich environmental and phenotypic conditions.

5 General Discussion

This chapter has surveyed a network-theoretic approach to neuroscience at the levels of structure, dynamics and function. While the coverage has not been comprehensive, we have described some key results and techniques that can be applied not just to vertebrate cortical networks but to the analysis of complex network systems of many different kinds. For example, the new results described in the previous section show the importance of drawing a sharp distinction between the dynamical patterns a network may support, and the functional applications that these patterns may serve.

5.1 Structure

Structure provides the most basic level of analysis of cortical networks. Many interesting features are revealed by the application of graph-theoretic and statistical analytical tools, most notably the presence of small-world characteristics as well as the prevalence of reentrant connectivity [14].

However, current structural analysis has certain important limitations. Detailed structural information is hard to obtain, especially at the microscopic level of neuron-to-neuron connectivity, especially with regard to the human brain. Moreover, this fine structure is continually changing as a result of a host of activity-dependent plasticity processes. Also, most methods of structural analysis assume unweighted, binary networks (although see [73–75]) and also assume that all nodes are essentially identical. Neither assumption is remotely satisfied in the case of the brain, and it is certain that both neuronal diversity and differences among synaptic strengths are essential for normal brain function.

Nonetheless, the rapid development of novel graph-theoretic methods [41,18], together with the accelerating accumulation of detailed neuroscientific data sets [32,33], promises that future structural analyses will reveal further important features of the structural organization of neural systems.

5.2 Dynamics

Anatomical structure gives rise to neural dynamics. The tools of statistical information theory are well suited to the analysis of these dynamics. It bears emphasizing that the global dynamical measures described in this chapter apply equally well to dynamics generated by both binary and weighted networks; in this sense, at least, dynamical analysis may be more generally applicable than structural analysis.

However, some assumptions have to be made in the application of these measures. For example, the analytical derivation of a covariance matrix from a connectivity matrix requires that linear dynamics be assumed. At least in some cases, linear and non-linear systems behave similarly with respect to effective connectivity [58], and it has also been argued that large-scale dynamics of interacting brain areas are accurately represented by linear systems [76,77]. Also, one may drop assumptions of linear dynamics by deriving covariances directly from recorded activity (see Sect. 4.3).

Calculation of entropy from covariance further assumes that network activity can be described as a stationary Gaussian process. This condition is by definition satisfied for calculation of intrinsic dynamics (see Sect. 3.1), but may not be strictly satisfied in the calculation of interactive dynamics from network activity during behavior. Indeed, it is a fact that many natural processes are not well modeled as stationary. Information-theoretic treatment of non-stationary signals has been widely discussed [78,79,62], and as long as deviations from stationarity are not too extreme, such techniques remain well placed to provide insight into the structure of dynamical interactions that a network produces.

More generally, the present methods are cross-sectional: Covariance matrices are derived from observations of network activity at successive time slices, so that correlations over time exhibited in the activity profiles of *single nodes* are overlooked (by definition, the analytical calculation of intrinsic dynamics assumes that there are no such correlations). A contrasting approach, which focuses on these correlations, is provided by a growing literature concerned with measuring the information content of single spike trains [21–24]. Integrating these two approaches stands out as an important challenge for theoretical neuroscience.

Notably absent from this chapter has been any mention of neural synchrony. Some of the most obvious features of human brain dynamics are the prominent oscillations in different frequency bands known as the delta (3-5Hz), alpha (8-13Hz), beta (10-20Hz) and gamma (35-80Hz) rhythms. Different functional roles have been proposed (but not proven) for these rhythms; for example, delta oscillations have been associated with the maintenance of items in short-term memory [80], and alpha oscillations are strongest during sleep and relaxed wakefulness [81], and may represent an 'idling state' of the brain.

Gamma oscillations have a controversial interpretation: it has been suggested that gamma oscillations serve to 'bind' together disparate neural processes into globally integrated activity patterns [45,82,83]. In support of this idea, detailed computer simulations of visual cortical areas have shown that reentrant interactions can synchronize the activity of anatomically segregated neural areas, leading to coherent perceptual performance and behavior in visually complex environments [44,47]. Computer simulations have also suggested that small-world architectures may be particularly suitable for facilitating neural synchrony [84].

5.3 Function

Linking dynamic patterns to functional roles is perhaps the least visited of the tasks facing a science of networks. However, it is at least as important as understanding structural and dynamical features by themselves. Many structural and dynamical features of brains have been selected by evolution precisely because of the adaptive functions they provide for the organism. Functional criteria can also be described for networks of many other kinds, for example power grids (the transmission of power from source to consumer) and telephone networks (the maintenance of uninterrupted service for clients).

As for the brain, we remarked above that neural synchrony may play an essential role in binding disparate neural processes to a common purpose. More generally, the results described in this chapter support the view that the brain is a selectional system [43], in which complex neural dynamics may facilitate adaptation to rich sensory environments and motor demands by providing a special kind of flexibility in the balance between dynamical integration and dynamical segregation. It has even been suggested that neurally complex dynamics in the thalamocortical system constitute the neural correlates of conscious states in humans and other animals [64,46,85]. Further empirical research and theoretical analysis of the interactions between dynamics and function are therefore likely to be valuable for neuroscience in the broader context of modeling the situated organism [86].

5.4 Summary

While important insights can be, and have been gained by analyses at the levels of structure, dynamics, and function separately, it must be recognized that in biological systems these levels are in complex and continuous interaction. Structure is continually changing as a result of activity-dependent plasticity processes. These structural changes evoke dynamical changes which shape the behavior of the organism, and behavior itself determines the correlations in the sensory signals that impinge on these networks, triggering further dynamical changes and structural alterations. Finding a language in which to articulate these complex couplings is the major challenge for a network-theoretic approach to neuroscience. Success in this task will advance not just neuroscience, but also our understanding of many other network systems in which interactions among structure, dynamics, and function are important.

Acknowledgements

This research was supported by the Neurosciences Research Foundation. Special thanks to Bruno van Swinderen for Fig. 2, and to George Reeke and our anonymous reviewers for helpful comments.
Address correspondence to Anil K. Seth (seth@nsi.edu).

Appendix A: Implementation Details

Motor node outputs $(s_{hx}, s_{hy}, s_{ex}, s_{ey})$ are used to update E, H and G according to:

$$V_{hx}(t) = m_H V_{hx}(t-1) + A_H(1-m_h)s_{hx}(t-lag), \tag{12}$$

$$V_{hy}(t) = m_H V_{hy}(t-1) + A_H(1-m_h)s_{hy}(t-lag), \tag{13}$$

$$V_{ex}(t) = m_E V_{ex}(t-1) + A_E(1-m_E)s_{ex}(t), \tag{14}$$

$$V_{ey}(t) = m_E V_{ey}(t-1) + A_E(1-m_E)s_{ey}(t), \tag{15}$$

where $V_{hx}(t), V_{hy}(t)$ represent the velocity of H in the x,y directions respectively at time t, and $V_{ex}(t), V_{ey}(t)$ represent the velocity of E. In all cases, $V(t)$ is bounded by the corresponding value of V_{max}. These values, as well as those of A_H, A_E, m_H, m_E, and *lag*, are specified for each condition in the text. The positions of E, H and G in the x, y plane are then updated using:

$$H(t) = H(t-1) + V_{hx}(t) + V_{hy}(t), \tag{16}$$

$$E(t) = E(t-1) + V_{ex}(t) + V_{ey}(t), \tag{17}$$

$$G(t) = H(t) + E(t), \tag{18}$$

where H(t) and G(t) represent the positions of H and G on the x, y plane at time t, and E(t) represents the position of the eye relative to H. If the distance between G and H exceeds $35.0u$ then G is not updated at that time step. In conditions E_C and E_T the target position (T) is also updated at each time step, alternating between 50 time steps of drift and 50 time steps without drift. Each period of drift is in a random direction at a random speed in the range 0.5 to 1.5 (u per time step). During the intervening periods T is stationary except for occasional jumps (with a probability of 0.025 per time step), each to a randomly chosen location within a radius of $25u$ to $31u$ of its previous location. The target cannot leave the $100^2 u^2$ area: If drift is leading it out of bounds the appropriate velocity component is reversed at the boundary.

References

1. G.M. Edelman. *The remembered present.* Basic Books, Inc., New York, NY, 1989.
2. A. Clark. *Being there: Putting brain, body, and world together again.* MIT Press, Cambridge, MA, 1997.
3. D. Buonomano and M. Merzenich. Cortical plasticity: From synapses to maps. *Annual Review of Neuroscience,* 21:149–186, 1998.
4. K. Friston. Functional and effective connectivity in neuroimaging: A synthesis. *Human Brain Mapping,* 2:56–78, 1994.
5. S.L. Bressler. Large-scale cortical networks and cognition. *Brain Research Reviews,* 20:288–304, 1995.
6. R.S.J. Frackowiak, K.J. Friston, C.D. Frith, R.J. Dolan, and J.C. Mazziotta. *Human brain function.* Academic Press, San Diego, CA, 1997.
7. F. Varela, J.-P Lachaux, E. Rodriguez, and J. Martiniere. The brainweb: Phase synchronization and large-scale integration. *Nature Reviews Neuroscience,* 2:229–239, 2001.
8. R. Cummins. Functional analysis. *Journal of Philosophy,* 72:741–764, 1975.
9. F. Harary. *Graph theory.* Addison-Wesley, Reading, MA, 1969.
10. B. Bollobás. *Random graphs.* Academic Press, London, 1985.
11. B. Jouve, P. Rosentiehl, and M. Imbert. A mathematical approach to the connectivity between the cortical visual areas of the macaque monkey. *Cerebral Cortex,* 8:28–39, 1998.
12. C.C. Hilgetag, R. Kötter, K.E. Stephan, and O. Sporns. Computational methods for the analysis of brain connectivity. In G.A. Ascoli, editor, *Computational neuroanatomy: Principles and methods,* pages 295–331. Humana Press, Totowa, NJ, 2002.
13. O. Sporns. Graph theory methods for the analysis of neural connectivity patterns. In R. Kötter, editor, *Neuroscience Databases. A Practical Guide,* pages 169–183. Kluwer Publishers, Boston, MA, 2002.

14. O. Sporns, G. Tononi, and G.M. Edelman. Theoretical neuroanatomy: Relating anatomical and functional connectivity in graphs and cortical connection matrices. *Cerebral Cortex*, 10:127–141, 2000.
15. R. Albert, H. Jeong, and A.-L. Barábasi. Diameter of the world wide web. *Nature*, 401:130–131, 1999.
16. D.J. Watts and S.H. Strogatz. Collective dynamics of 'small world' networks. *Nature*, 393:440–442, 1998.
17. S.H. Strogatz. Exploring complex networks. *Nature*, 410:268–276, 2001.
18. M.E.J. Newman. The structure and function of complex networks. *SIAM Review*, 45(2):167–256, 2003.
19. G.Q. Bi and M.M. Poo. Synaptic modifications in cultured hippocampal neurons: dependence on spike timing, synaptic strength, and postsynaptic cell type. *Journal of Neuroscience*, 18:10464–10472, 1998.
20. Y.X. Fu, K. Djupsund, H. Gao B. Hayden, K. Shen, and Y. Dan. Temporal specificity in the cortical plasticity of visual space representation. *Science*, 296:1999–2003, 2002.
21. W. Bialek, I. Nemenman, and N. Tishby. Predictability, complexity, and learning. *Neural Computation*, 13:2409–2463, 2001.
22. M. Costa, A.L. Goldberger, and C.K. Peng. Multiscale entropy analysis of complex physiological time series. *Physical Review Letters*, 89:681–682, 2002.
23. L. Paninski. Estimation of entropy and mutual information. *Neural Computation*, 15:1191–1253, 2003.
24. G.N. Reeke and A.D. Coop. Estimating the temporal interval entropy of neuronal discharge. *Neural Computation*, in press.
25. G. Tononi, O. Sporns, and G.M. Edelman. A measure for brain complexity: Relating functional segregation and integration in the nervous system. *Proceedings of the National Academy of Science (USA)*, 91:5033–5037, 1994.
26. A.K. Seth and G.M. Edelman. Environment and behavior influence the complexity of evolved neural networks. *Adaptive Behavior*, in press.
27. F. Crick and E. Jones. Backwardness of human neuroanatomy. *Nature*, 361:109–110, 1993.
28. D.J. Felleman and D.C. Van Essen. Distributed hierarchical processing in the primate cerebral cortex. *Cerebral Cortex*, 1:1–47, 1991.
29. M.P. Young. The organization of neural systems in the primate cerebral cortex. *Philosophical Transactions of the Royal Society of London: Series B*, 252:13–18, 1993.
30. J.W. Scannell, C. Blakemore, and M.P. Young. Analysis of connectivity in the cat cerebral cortex. *Journal of Neuroscience*, 15, 1995.
31. J.W. Scannell, G.A.P.C. Burns, C.C. Hilgetag, M.A. O'Neil, and M.P. Young. The connectional organization of the cortico-thalamic system of the cat. *Cerebral Cortex*, 9:277–299, 1999.
32. R. Kötter. Neuroscience databases: Tools for exploring brain structure-function relationships. *Philosophical Transactions of the Royal Society of London: Series B*, 356:1111–1120, 2001.
33. K.E. Stephan, L. Kamper, A. Bokzurt, G.A.P.C. Burns, M.P. Young, and R. Kötter. Advances in database methodology for the collation of connectivity data on the macaque brain (CoCoMac). *Philosophical Transactions of the Royal Society of London: Series B*, 356:1159–1186, 2001.
34. J.M.J. Murre and D.P.F. Sturdy. The connectivity of the brain: Multi-level quantitative analysis. *Biological Cybernetics*, 73:529–545, 1995.

35. A. Nicoll and C. Blakemore. Patterns of local connectivity in the neocortex. *Neural Computation*, 5:665–680, 1993.

36. C.C. Hilgetag. *Mathematical approaches to the analysis of neural connectivity in the mammalian brain.* PhD thesis, Faculty of Medicine, University of Newcastle upon Tyne, 1999.

37. L. Lagae, D.K. Xiao, S. Raiquel, H. Maes, and G.A. Orban. Position invariance of optic flow component selectivity differentiates monkey MST and FST cells from MT cells. *Invest. Ophthalmol. Vis. Sci.*, 32:823, 1991.

38. J.J. Gibson. *The ecological approach to visual perception.* Houghton-Mifflin, Boston, 1979.

39. D.J. Watts. *Small worlds.* Princeton University Press, Princeton, NJ, 1999.

40. P. Erdös and A. Rényi. On random graphs. *Publicationes Mathematicae*, 6:290–297, 1959.

41. R. Albert and A.-L. Barábasi. Statistical mechanics of complex networks. *Reviews of Modern Physics*, 74:47–97, 2002.

42. H. Jeong, B. Tombor, R. Albert, Z. Oltvai, and A.-L. Barábasi. The large-scale organization of metabolic networks. *Nature*, 407:651–654, 2000.

43. G.M. Edelman. *Neural Darwinism.* Basic Books, New York, 1987.

44. G. Tononi, O. Sporns, and G.M. Edelman. Reentry and the problem of integrating multiple cortical areas: Simulation of dynamic integration in the visual system. *Cerebral Cortex*, 2(4):31–35, 1992.

45. G.M. Edelman. Selection and reentrant signaling in higher brain function. *Neuron*, 10:115–125, 1993.

46. G.M. Edelman and G. Tononi. *A universe of consciousness: How matter becomes imagination.* Basic Books, New York, 2000.

47. A.K. Seth, J.L. McKinstry, G.M. Edelman, and J.L. Krichmar. Visual binding through reentrant connectivity and dynamic synchronization in a brain-based device. *Cerebral Cortex*, in press.

48. G. Mitchison. Neuronal branching patterns and the economy of cortical wiring. *Proceedings of the Royal Society of London: Series B. Biological Sciences.*, 245:151–158, 1991.

49. C. Cherniak. Component placement optimization in the brain. *Journal of Neuroscience*, 14:2418–2427, 1994.

50. C. Cherniak. Optimal-wiring models of neuroanatomy. In G.A. Ascoli, editor, *Computational neuroanatomy: Principles and methods*, pages 71–83. Humana Press, Totowa, NJ, 2002.

51. C. Cherniak, Z. Mokhtarzada, R. Rodriguez-Esteban, and K. Changizi. Global optimization of cerebral cortex layout. *Proceedings of the National Academy of Sciences, USA*, 101(4):1081–1086, 2004.

52. J.B. Kruskal. Nonmetric multidimensional scaling: A numerical method. *Psychometrika*, 29:115–129, 1964.

53. J.B. Kruskal. Multidimensional scaling by optimizing goodness of fit to a nonmetric hypothesis. *Psychometrika*, 29:1–27, 1964.

54. M.P. Young, J.W. Scannell, M.A. O'Neill, C.C. Hilgetag, G.A.P.C. Burns, and C. Blakemore. Non-metric multidimensional scaling in the analysis of neuroanatomical connection data and the organization of the primate visual system. *Philosophical Transactions of the Royal Society of London: Series B*, 348:281–308, 1995.

55. L.G. Ungerleider and J.V. Haxby. 'what' and 'where' in the human brain. *Current Opinion in Neurobiology*, 4:157–165, 1994.

56. C.C. Hilgetag, G.A.P.C. Burns, M.A. O'Neill, and M.P. Young. Cluster structure of cortical systems in mammalian brains. In J.M. Bower, editor, *Computational neuroscience*, pages 41–46. Plenum Press, New York, 1998.

57. G. Tononi, G.M. Edelman, and O. Sporns. Complexity and coherency: Integrating information in the brain. *Trends in Cognitive Science*, 2:474–484, 1998.

58. O. Sporns and G. Tononi. Classes of network connectivity and dynamics. *Complexity*, 7(1):28–38, 2002.

59. S. Zeki. Functional specialization in the visual cortex of the Rhesus monkey. *Nature*, 274:423–428, 1978.

60. S. Zeki. *A vision of the brain*. Blackwell, Oxford, 1993.

61. W. Vanduffel, B.R. Payne, S.G. Lomber, and G.A. Orban. Functional impact of cerebral connections. *Proceedings of the National Academy of Science (USA)*, 94:7617–7620, 1997.

62. A. Papoulis and S.U. Pillai. *Probability, random variables, and stochastic processes*. McGraw-Hill, New York, NY, 2002. 4th edition.

63. D.S. Jones. *Elemenary information theory*. Clarendon Press, 1979.

64. G. Tononi and G.M. Edelman. Consciousness and complexity. *Science*, 282:1846–1851, 1998.

65. G. Tononi, A.R. McIntosh, D.P. Russell, and G.M. Edelman. Functional clustering: identifying strongly interactive brain regions in neuroimaging data. *Neuroimage*, 7:133–149, 1998.

66. S.E. Petersen, P.T. Fox, M.I. Posner, M. Mintun, and M. Raichle. Positron emission tomographic studies of the cortical anatomy of single-word processing. *Nature*, 331:585–589, 1988.

67. R. Hari, S. Levänen, and T. Raij. Timing of human cortical functions during cognition: role of MEG. *Trends in Cognitive Science*, 4(12):455–461, 2000.

68. G. Tononi, O. Sporns, and G.M. Edelman. A complexity measure for selective matching of signals by the brain. *Proceedings of the National Academy of Science (USA)*, 93:3422–3427, 1996.

69. G. Tononi, O. Sporns, and G.M. Edelman. Measures of degeneracy and redundancy in biological networks. *Proceedings of the National Academy of Science (USA)*, 96:3257–3262, 1999.

70. G.M. Edelman and J. Gally. Degeneracy and complexity in biological systems. *Proceedings of the National Academy of Sciences, USA*, 98(24):13763–13768, 2001.

71. A.K. Seth. *On the relations between behaviour, mechanism, and environment: Explorations in artificial evolution*. PhD thesis, University of Sussex, 2000.

72. M. Mitchell. *An introduction to genetic algorithms*. MIT Press, Cambridge, MA, 1997.

73. V. Latora and M. Marchiori. Efficient behavior of small-world networks. *Physical Review Letters*, 87(19):198701–4, 2001.

74. C. Cannings and C. Penman. Random graphs. In C.R. Rao and D.N. Shanbhag, editors, *Stochastic processes: Modelling and simulation*, Vol. 21. Handbook of Statistics series. Elsevier, 2002.

75. M. E. J. Newman Who is the best connected scientist? A study of scientific coauthorship networks *Physical Review E*, 64:016131, 2001.

76. A.R. McIntosh and F. Gonzalez-Lima. Structural equation modeling and its application to network analysis in functional brain imaging. *Human Brain Mapping*, 2:2–22, 1994.

77. A.R. McIntosh, C.L. Grady, L.G. Ungerleider, J.V. Haxby, S.I. Rapoport, and B. Horwitz. Network analysis of cortical visual pathways mapped with PET. *Journal of Neuroscience*, 14:655–666, 1994.

78. M.S. Pinsker. *Information and information stability of random variables and processes.* Holden-Day, San Francisco, 1964.

79. R.M. Gray. *Probability, random processes, and ergodic properties.* Springer-Verlag, Berlin, 1988.

80. M. Doppelmayr, W. Klimesch, J. Schwaiger, and T. Winkler. Theta synchronization in human EEG and episodic retrieval. *Neuroscience Letters*, 257(1):41–4, 1998.

81. J.L. Cantero, M. Atienza, and R.M. Salas. Human alpha oscillations in wakefulness, drowsiness period, and REM sleep: different electroencephalographic phenomena within the alpha band. *Neurophysiol. Clin.*, 32(1):54–71, 2002.

82. W. Singer and C. Gray. Visual feature integration and the temporal correlation hypothesis. *Annual Review of Neuroscience*, 18:555–586, 1995.

83. C. Gray. The temporal correlation hypothesis: still alive and well. *Neuron*, 24:31–47, 1999.

84. L.F. Lago-Fernandez, R. Huerta, F. Corbacho, and J. Siguenza. Fast response and temporal coherent oscillations in small-world networks. *Physical Review Letters*, 84:2758–2761, 2000.

85. G.M. Edelman. Naturalizing consciousness: A theoretical framework. *Proceedings of the National Academy of Sciences, USA*, 100(9):5520–5524, 2003.

86. J.L. Krichmar and G.M. Edelman. Machine psychology: Autonomous behavior, perceptual categorization and conditioning in a brain-based device. *Cerebral Cortex*, 12(8):818–30, 2002.

Index

accelerated growth model 110
acyclic 53, 71, 73, 77
Adamic 33, 103, 110, 124, 161, 368, 371,
    372, 388, 395, 396, 398, 422
adaptation model 121
adaptive 205, 483, 484, 487, 493, 494,
    500–503, 505, 508
adjacency matrix 35–37, 43–45, 47, 165,
    176, 191, 263
Alava 124, 139, 160, 185
Albert 18, 32, 33, 49, 84, 86, 103, 106,
    123–125, 136, 142, 159–161, 169, 185,
    186, 205, 233, 275, 296, 300, 309, 310,
    367, 369, 422, 426, 439, 440, 447, 455,
    456, 459, 480, 508, 509
Aldous 51, 86, 87, 233
amino acid 118, 450
anisotropic magnetic systems 255
antiferromagnetic 280
archaea 111, 112
archive 124, 313, 315, 322, 323, 331, 333,
    340, 341, 343–346, 348, 349, 356, 357,
    359, 360, 362, 363, 366
artificial neural network 484, 493
arXiv.org 117, 313, 314, 316, 323, 324,
    333, 334
assortative 115–117, 140, 145, 147, 192,
    193, 196, 199, 200, 206, 401, 404, 422,
    423
asynchronous 255–257, 272, 273, 480
attack 3, 4, 19–23, 32, 235, 303–306, 308,
    448, 456, 457
automated detection 326
autonomous system 44, 49, 122, 142,
    143, 148, 425, 427, 428, 438
average distance 9, 17, 28, 51, 58, 89, 90,
    98, 99, 102, 119, 159, 357–361, 366
average height 51, 219, 221

backbone 128, 197, 426, 435
bacteria 110
Barabási 49, 50, 103, 186, 300, 309, 426,
    440, 443
Barabási-Albert model 44, 49
Battiston 399, 405, 423
battle-field simulations 255
Bayes rule 5
ben-Avraham 3, 32–34, 137, 310, 440
Ben-Naim 186, 211, 233
Bethe lattice 36, 45, 48
betweenness 31, 107, 121, 303, 337, 347,
    353–356, 360, 364, 366, 369, 371,
    373–375, 378, 396, 397, 422, 426, 427,
    432–435, 438
binary 47, 130, 252, 326, 328, 333, 334,
    341, 364, 438, 461, 462, 480, 485, 498,
    499, 504
bombing algorithm 131, 136
Boolean 459, 461, 462, 465–467, 469,
    479, 480
brain 333, 483–488, 492, 504–510
branching process 4, 30, 71, 133, 278,
    287, 296
Braunstein 127, 128, 130, 132, 134, 136,
    186
breakdown 3, 4, 18–22, 32, 303–305, 451,
    457
Buldyrev 127, 136, 137, 186, 206
burst analysis 322

C. elegans 456
Caldarelli 144, 148, 160–162, 206, 399,
    400, 402, 404, 406, 408, 410, 412, 414,
    416, 418, 420, 422, 423, 439
Capital Asset Pricing model 416
carbohydrate 450
cascades 299–305, 307–309
Catanzaro 399, 402, 423

514   Index

Cayley tree   6, 25, 127, 128, 133, 134,
        136, 197
cell   164, 186, 255, 333, 377, 443–445,
        448, 457, 459–464, 466–472, 474, 475,
        477, 478, 480, 481, 508
centrality   31, 107, 117, 121, 206, 303,
        352, 360, 364, 366, 369, 371, 373,
        376–378, 396, 397, 422, 456
cerebral cortex   483, 485, 507–509, 511
chemical potential   167, 171
Chung   33, 34, 49, 50, 89, 103, 104, 178,
        186, 187
citation network   106, 340
CiteBase   324
clause   241
cleaning   313, 331, 334, 336
cliquishness   449
closeness   337, 360, 361, 366, 377
cluster   3–9, 15, 18–20, 22, 24, 25, 27–31,
        54–57, 67, 73, 75, 119, 122, 128, 133,
        159, 162, 189, 200, 213, 214, 405, 407,
        419, 420, 422, 444, 446, 448, 471, 484,
        486, 490–492, 510
clustering   45, 49, 51, 54, 55, 58, 66, 67,
        89, 90, 139–154, 156–160, 165, 191,
        277, 299–302, 305, 333, 343, 350, 351,
        369, 399, 400, 402–405, 419, 427–430,
        435–438, 444, 445, 447–451, 455, 457,
        486, 491, 510
Cohen   3, 32–34, 125, 127, 136, 137, 186,
        310, 440, 456, 457
collaboration network   353
collaboration network   106, 341, 351,
        356, 366
collapse   20, 153, 180, 205
colouring of graphs   236
communication network   17, 255, 256,
        261
community   102, 144, 145, 206, 313, 318,
        321, 337, 344, 348, 351, 371–376,
        378–380, 386, 395–397, 457
conceptual network of language   301
conditional variance   70
configuration model   116–118, 140, 141,
        144
connectivity   9, 85, 89, 91–93, 102, 128,
        131, 136, 189, 199, 203, 243, 248, 250,
        252, 261, 263, 278, 288, 305, 306, 308,
        393, 399, 404, 409, 425, 426, 428, 430,
        432–437, 444, 449, 483–486, 488,

        490–492, 494, 496, 498, 503, 504,
        507–510
constrained satisfaction   235
contact process   252, 254, 286
convection-diffusion equation   230
Cornell   313, 323, 324, 366
correlation function   29, 106, 115, 193,
        259, 265, 282, 284
correlation length   3, 4, 28, 29, 261, 264,
        265, 267, 270, 272, 282, 283, 285, 290
covariance matrix   488, 489, 491, 493,
        499, 500, 504
critical exponent   3, 24, 25, 28, 180, 183
crossover   29, 32, 113, 123, 278
cumulative distribution   215, 219, 221,
        224, 269, 414

D. melanogaster   443, 448
data mining   325, 336
Davis-Putnam-Logemann-Loveland
        (DPLL)   236
degeneracy   492, 493, 510
degree   3–15, 17, 19–23, 25, 27, 28,
        30–32, 35–37, 39, 41–49, 51, 54, 55,
        57, 58, 66, 69, 83, 86, 89–106, 108–
        110, 113–117, 119–121, 125, 128,
        132–136, 139–155, 157–159, 165,
        169–173, 175–182, 184, 189–192,
        194, 196–198, 200–204, 212, 213,
        225–230, 238–240, 244, 252, 253, 303,
        305, 315, 359, 362, 369, 370, 380, 381,
        383, 384, 388, 389, 397, 399–405,
        408–412, 415, 416, 418, 422, 425–438,
        444, 445, 447, 449, 451, 453, 486, 489
delta correlated Gaussian noise   258
Derényi   50, 163
deterministic model   111
diameter   7, 9, 35, 51, 58, 86, 89–91,
        96–102, 104–107, 118–122, 174, 191,
        367, 491, 508
diffusion   33, 35, 37, 331, 396
Dijkstra algorithm   130, 131, 212
directed   31, 53, 54, 56, 58, 59, 63, 64, 69,
        71, 73, 74, 77, 78, 81, 82, 110, 141,
        163, 412, 413, 417, 461, 465, 469, 470,
        483, 485
dissortative   17, 115–117, 192
DNA   333, 460, 462
Dorogovtsev   33–35, 49, 50, 86, 110, 124,
        125, 136, 162, 166, 176, 184–187, 205,

206, 233, 296, 309, 310, 439, 440, 455, 456
download patterns   323
dynamic channel allocation   255
dynamical annealing   247, 250

Edelman   483, 493, 507–511
edge   54, 56, 57, 59, 63, 64, 66–69, 72–81, 84, 91, 93–97, 106–108, 111, 112, 115, 143, 147, 149, 151, 165, 167–171, 173–177, 180, 195, 203, 236, 237, 244, 341, 352, 355, 373–375, 379, 381, 400, 404, 420, 428, 435, 462, 469, 470
effective medium approximation   35, 40, 45
efficiency   17, 119, 120, 253, 300, 302, 306, 307, 377
eigenvalue   41–45, 47, 49, 178, 187, 262, 295, 408
eigenvector   252, 290
email network   144, 372, 376, 377, 388, 389, 391
embryonic   459, 463, 464, 480
Emericella nidulans   112
energy   47, 118, 129, 140, 145, 164–168, 171–175, 179–183, 185, 203, 280, 285, 314, 315, 317, 322, 324, 340
entropy   164, 186, 187, 193–198, 200, 202–204, 206, 232, 484, 489, 491, 492, 499, 500, 503, 505, 508
enzymes   110, 460
epidemics   102, 382
eukaryotes   110, 111, 443, 452, 453
Euler   163, 185, 269, 339
evolution operator   244, 245, 247, 252
exhaustive search   235
expansion   30, 63, 252, 253, 264, 281, 290–294, 314, 469, 470
extremal   186, 211–214, 221, 222, 225, 231, 252, 269–271
extreme-height fluctuations   259–261, 269, 270

Farkas   50, 163, 164, 166, 168, 170, 172, 174, 176, 178, 180, 182, 184–187
financial network   412
finite-size   8, 28, 40, 42, 47, 114, 201, 259, 260, 264, 266, 268, 293, 294, 432
fitness model   121, 143, 417
forest   213

Fourier   39, 258
fractal   3, 28–30, 111, 274, 297, 423, 456
funneling   353, 356, 366, 367

Garlaschelli   399
gauge symmetry   237
gelation   212, 224, 225
gene expression   331, 459, 460, 462, 464, 473, 475
generating function   7, 19, 22, 24, 25, 30, 37, 42, 223, 230, 380, 381, 401
genomics   162, 333, 443
geographical   278, 427–429, 435
geometric growth model   110
Ghim   105, 124
giant component   13, 58, 104, 128, 134, 169, 177, 183, 212, 224, 225, 243, 343, 349, 350, 356–358, 360, 370
Ginsparg   313, 335, 336, 366, 367
Glauber   256
Godeaux   339
Goh   34, 50, 105, 124, 125, 161, 162, 187, 310, 369, 440
Goltsev   35, 50, 186, 187, 296, 456
graph   4, 6, 8–12, 15, 16, 18, 19, 27, 28, 30, 36, 37, 39, 42, 44–49, 53, 56–58, 63–65, 67, 71–75, 77, 79–82, 84, 86, 89–105, 112, 113, 119, 127, 128, 131–136, 141, 143, 144, 148, 162–171, 173–187, 189, 191, 195–205, 212, 223, 226, 236–245, 248, 251–254, 262, 324, 339, 349–352, 357–361, 363, 364, 366–368, 370, 371, 373–376, 379, 380, 384, 386, 396, 399–401, 404–406, 408–410, 412, 426–430, 435, 436, 439, 445, 447, 460, 461, 469, 484–486, 491, 493, 507
greedy algorithm   93, 387, 389
Greedy heuristic   239–241, 243, 245, 251
Green's function   290, 291, 293
Guclu   255, 256, 258, 260, 262, 264, 266, 268, 270, 272, 274, 275, 296, 297

H.pylori   443, 448
Hamiltonian   145, 146, 173, 258, 259, 279
hard network   288, 291, 293, 295
Harris criterion   278, 284
Hastings   262, 264, 273–275, 277, 296, 297

Havlin   3, 32–34, 125, 127, 136, 137, 186,
    206, 310, 440, 456, 457
height statistics   212, 218, 220
Hermitean   252
hierarchical network   446, 447
hierarchical network   448
hierarchy   11, 159, 239, 351, 364, 376,
    377, 384, 387, 389–391, 393–395, 411,
    412, 425, 427, 434–437, 451, 463
Hilgetag   488, 492, 507–510
HP Labs   371, 373, 376–378, 383–385,
    388–390, 392
hub   11, 200, 201, 400, 404
Huberman   33, 103, 110, 124, 161, 368,
    371, 373, 375, 396–398, 422

impurity averaged perturbation theory
    290
information flow   302, 371, 379, 395
information network   313, 334
interactome   455, 456
Internet   3, 9, 15, 18, 20, 31–33, 35, 36,
    43–45, 47–49, 89, 103, 104, 106, 110,
    112, 113, 115–118, 120–124, 136,
    141–143, 146, 148, 161, 185, 201, 205,
    206, 212, 256, 273, 299, 318, 336, 338,
    367, 368, 399, 425–430, 434, 435,
    437–440, 443, 456, 457, 484, 486
Ising   175, 277, 280, 284, 285

Kac   339, 368
Kahng   34, 50, 105, 124, 125, 161, 162,
    187, 310, 369, 440, 456
Kalisky   3, 4, 6, 8, 10, 12, 14, 16, 18, 20,
    22, 24, 26, 28, 30, 32–34
Kardar-Parisi-Zhang   258, 279
Kauffman   206, 461, 462, 480
Kertész   139
Kim   34, 50, 105, 124, 125, 148, 160–162,
    186, 187, 275, 296, 309, 310, 369, 440,
    456, 457
kinetic roughening   255, 272
Kleinberg   33, 90, 103, 104, 162, 336, 367,
    369, 387, 388, 392, 398, 422
Korniss   186, 255, 273–275, 296, 297
Kosterlitz-Thouless   285
Kozma   262, 264, 273, 274, 277, 296
Krapivsky   86, 124, 162, 185, 186, 211,
    233, 254, 347, 369

laggard   217, 218, 221, 222, 224, 231
Lai   125, 299, 309, 310
Langevin equation   258, 261, 263
Laplacian   37, 42, 258, 262, 289, 290
largest cluster   5, 9, 27, 28, 119, 122
largest component   169, 170, 173, 177,
    179, 212, 304, 343
largest degree   179, 182, 213, 435
layer   3, 11–17, 30, 31, 144, 244, 483
leader   212, 214–218, 221, 223–228, 231
leadership statistics   212, 222, 225, 231
leaf   246, 374
lethal   448, 449
link   4, 7, 14, 19, 21, 22, 24, 116, 127–131,
    133, 146, 147, 150, 152, 156, 164, 170,
    171, 177, 192, 194, 197, 223, 226, 229,
    262, 264, 278, 280, 283–285, 287, 288,
    292, 293, 295, 300, 304, 306, 339, 373,
    378, 386, 404, 406, 409, 412, 414, 428,
    435, 447, 450, 451
lipid   450
load   105–111, 113–117, 122, 123, 261,
    269, 303–309, 369, 426, 428, 432, 435,
    438
loops   5–7, 11, 36, 41, 91, 113, 123, 128,
    129, 140, 143, 159, 165, 176, 351, 462,
    472
Lotka   344, 345, 369
Lu   33, 34, 50, 89, 103, 104, 186, 187, 347,
    365, 456
Lubachevsky   257, 273, 296

macaque   485–487, 492, 507, 508
Markovian   238, 239, 241, 244, 247, 252
Master equation   37, 219
maximal degree   12, 227, 230
Maxwellian   131, 132
mean-field   24–28, 32, 51, 52, 60, 62, 64,
    85, 140, 141, 150, 159, 160, 254, 265,
    266, 272, 273, 277, 278, 280, 281, 283,
    285, 287, 288, 290, 293, 295, 296, 456
Medline database   345, 351
Mendes   33–35, 49, 50, 86, 124, 125, 136,
    162, 185–187, 205, 206, 233, 296, 309,
    310, 439, 440, 455, 456
metabolic network   111–113, 141, 199
Methanococcus jannaschii   113
metric copying   52, 59, 66, 85
Milgram   89, 103, 356, 357, 369, 389, 391,
    397, 456

minimal spanning tree (MST)   406
mixing   17, 115, 145, 147, 148, 192, 193,
        199, 200, 206, 397, 404, 422, 437
modularity   140, 159, 191, 192, 197, 206,
        446, 448–451, 453, 456
module   201, 446, 448, 450
Molloy-Reed construction   10, 12
Monasson   50, 232, 235, 253, 254, 274,
        297
monomer   218, 222, 224
Monte-Carlo   173, 180, 181, 183, 184
motif   198, 452, 453, 457
Motter   299, 309, 310
mRNA   443, 460, 465, 466, 468
multi-fractal   150
multigraph   11
mutation   191, 203, 333, 472, 474, 496
mutual information   189, 193, 195, 206,
        489, 493, 508

navigability   444
NCSTRL database   345, 357, 359
network   3–6, 8–11, 15–20, 22, 25, 28,
        30–32, 35–37, 41, 44–47, 49–51, 57,
        58, 85–87, 90, 91, 95, 102, 105, 106,
        108–113, 115–119, 121, 123, 125, 127,
        128, 131, 132, 135, 139–146, 149, 151,
        158, 159, 163, 164, 167, 168, 170, 173,
        174, 178, 180, 184, 189, 191–196,
        198–205, 207, 225–228, 231, 255, 256,
        261–264, 268, 271–273, 277–280,
        288–296, 299–306, 308, 309, 313, 316,
        318, 324–326, 334, 335, 337–342,
        348–358, 360, 362–364, 366–373,
        376–382, 386–389, 391, 395–397,
        399–402, 404–407, 412–414, 417,
        418, 421, 423, 425, 427–430, 432,
        435, 437–439, 443–453, 455–457,
        459, 461–465, 467–477, 479, 480,
        483–488, 490, 491, 493–506, 510, 513
neuron   333, 483, 509, 511
New York Stock Exchange (NYSE)   414
Newman   6–9, 32, 33, 49, 50, 86, 103, 108,
        116, 124, 141, 143, 159–161, 186, 206,
        232, 274, 297, 309, 310, 335, 337, 354,
        367–370, 373, 387, 396–398, 401, 402,
        422, 423, 440, 457, 508, 510
Nishikawa   299, 309
node   3–5, 9, 11–15, 31, 36, 37, 90, 130,
        131, 133, 134, 139–142, 146, 148–158,

        189, 191, 192, 196, 197, 200, 212, 213,
        225–231, 237–241, 243–246, 248,
        251, 261, 269, 271, 277, 289, 299–306,
        371, 374, 375, 381, 385–387, 389, 400,
        402, 404, 407, 409–412, 420, 444,
        446–448, 450, 451, 461, 462, 465–471,
        473, 475, 485, 486, 491, 495, 496, 500,
        501, 506
noise   118, 189, 193, 196–198, 200,
        202–204, 258, 261, 263–265, 288, 289,
        321, 408, 499
non-metric multidimensional scaling
        (NMDS)   487
Novotny   255, 273–275, 296, 297
NP-complete   235, 236
nucleotide   450

Oh   34, 105, 124
Oltvai   443
optimal distance   127
Othmer   462, 480
outbreak   287, 382, 383, 385
overlap matrix   451
overload failures   299, 303, 308

packet   105–107, 123
Palla   163, 185
parallel discrete-event simulations   255
Pareto's law   417, 423
partition function   167, 168, 280, 281,
        285
Pastor-Satorras   33, 34, 50, 124, 136, 161,
        162, 186, 206, 310, 397, 422, 425, 439,
        440, 457
pathway   105, 107, 108, 112, 119, 465
peer review   316, 317, 319–321, 335
peer-to-peer network   141
peptide   333, 450
percolation   3–6, 15, 18, 20, 21, 23–29,
        31–33, 57, 58, 81, 102, 127, 128, 131,
        133, 169, 170, 177, 223, 233, 241, 277,
        368
phase transition   3, 22, 26, 174, 178, 180,
        181, 184, 285, 404
phenotypic   449, 464, 503
Poisson   6, 14, 42, 43, 47, 52, 53, 55,
        59–61, 67, 70, 71, 105, 128, 135, 136,
        144, 169, 197, 219, 226, 255, 257, 289,
        389, 409, 447
polylogarithm   380

polymerization   212, 223
portfolio   406, 408, 413–421, 423
power law   3, 4, 11, 24, 27, 31, 58, 85,
    89–93, 95, 97–100, 102, 103, 105, 106,
    110, 120, 123, 131, 134, 142, 146, 148,
    159, 169, 177–182, 184, 187, 203, 204,
    218, 230, 250, 270, 282, 344, 345, 349,
    399, 404, 410–412, 430–434, 436, 437,
    449
price correlations   406
processing elements   255, 256, 279, 288
protein   110, 115, 118, 141, 143, 144, 149,
    206, 333, 443, 445, 448–453, 455–457,
    460, 462, 464, 465, 468, 469, 472, 476,
    477, 479
proteome   201, 206, 449, 453–455
proteomics   162, 443, 455–457
proximity   216, 387–389, 391, 395, 405,
    487

Rényi   50
random graph   27, 30, 39, 42, 44, 48, 53,
    57, 58, 63, 64, 71, 81, 82, 84, 86, 89,
    91, 92, 98, 102–105, 119, 127, 128,
    132–136, 141, 144, 163, 168, 169, 171,
    174, 176–183, 187, 197, 203, 223, 238,
    244, 251–254, 262, 351, 358, 359, 366,
    370, 399, 401, 409, 426, 429, 430, 435,
    447
random matrix   35, 36, 406, 407
random tree   216
random tree   224, 225, 412
random walk   30, 36, 140, 150
random-sequential Monte Carlo   255
range-based attacks   299, 305
rate-equation   151, 152
Redner   50, 86, 124, 136, 162, 185, 186,
    211, 233, 348, 367, 369
remaining degree   115, 116, 194, 202
replica trick   280
repository   313
residual graph   6
resilient   18, 20
rich get richer   229
RNA   333
routing   3, 17, 425, 427, 428, 435, 439

S. cerevisiae   110, 443, 448, 449, 453
sabotage   20
Saccharomyces cerevisiae   110, 455

Samukhin   33–35, 50, 86, 124, 125, 162,
    185–187, 233
Santa Fe Institute   189, 205, 207, 274,
    337, 367
satisfiability   253
scalability   255–258, 269, 272, 279
scale-free   3, 4, 8–11, 17, 18, 22, 25, 28,
    30–32, 35, 36, 43, 44, 46–49, 86,
    103–105, 113, 143, 149, 159, 170, 173,
    179, 181, 182, 184, 189–192, 198,
    203–206, 272, 273, 278, 299, 300,
    302–306, 308, 309, 367, 369, 372, 379,
    382, 397, 399, 400, 404, 414, 416, 418,
    425–427, 430, 432, 434, 438, 443,
    445–449, 453, 454, 456, 457, 486
scaling   4, 27, 28, 31, 51, 58, 62, 86, 103,
    109, 132, 141, 142, 148, 149, 157, 189,
    190, 194, 201, 203, 205, 211, 212, 214,
    215, 218, 224, 225, 228, 231, 250,
    258–261, 267, 268, 277, 279–285, 287,
    288, 290, 295–297, 304, 306, 308, 309,
    326, 359, 369, 415, 426, 429, 432–434,
    456, 487, 495, 509
scattering   291–293
scheduling   235
scholarly information network   313
scientific coauthorship networks   337,
    510
search tree   236–239, 243–247, 249–252
segment polarity gene network   463,
    475, 479, 480
selection principle   221
self-consistent Born approximation
    (SCBA)   291
self-organizing processes   443
Seth   483, 493, 506, 508–510
shareholder network   413, 414
shell   12, 30
shortest path   19, 38, 127, 128, 263,
    305–307, 352, 354, 355, 374, 377, 387,
    389, 486
simulated annealing   189, 202–204, 212
SLAC SPIRES-HEP database   324
small-world   36, 104, 119, 169, 189, 205,
    255, 261–263, 266, 267, 271, 277–280,
    283–285, 287–290, 293–296, 299–
    302, 305, 309, 357, 359, 368, 397, 398,
    445, 456, 486, 504, 505, 510, 511
social group   386, 395

social network   143, 145, 337, 339, 348,
    350, 353, 360, 367, 369, 371, 372,
    377–379, 386, 396, 400, 402
soft network   288, 290, 291, 295
software class diagrams   199
Solé   189
specific heat   281, 283
spectral   35–37, 47, 49, 163, 177–179,
    185, 406, 408
SPIRES database   342, 344, 345,
    348–350
Sreenivasan   127
standard deviation   51, 304, 376
Standard Industrial Classification System
    423
Stanley   89, 127, 136, 137, 160, 186, 205,
    206, 274, 297, 309, 310, 347, 356, 368,
    423, 440
star graph   198, 201, 202
static model   106, 109, 110, 114, 115,
    119–121, 123
steady state   259, 261, 267, 468, 472–477,
    479
stock investment network   412
structure factor   258, 259, 264–267
subgraph   36, 58, 71, 72, 74, 77–79, 93,
    94, 96, 100, 102, 173, 182
supervised learning   326
Support Vector Machines   326, 336
survival probability   216–218, 221, 222,
    229, 286
susceptibility   281–283, 285, 286
synchronization   255–257, 261–264, 267,
    269, 271, 272, 279, 288, 289, 333, 507,
    509, 511
Szabó   139

tail   4, 8, 10, 11, 27, 35, 36, 40, 42–45, 48,
    51, 56, 57, 83, 110, 119, 120, 170, 178,
    179, 189, 220, 221, 224, 227, 228, 269,
    270, 324, 325, 345, 346, 388, 389, 414,
    417, 430, 432, 464
text classification   313, 326, 327, 329,
    333, 334
thermodynamic limit   113, 123, 139, 145,
    213, 260, 293, 295
Thesaurus   141, 301, 302
Thomas   461, 462, 480
tomography   3, 4, 11, 492
Toroczkai   186, 255, 274, 275, 296, 297

traceroute   427
train routes   338
transcendental   135, 219, 221, 230
transcription   144, 460, 463–466, 469,
    470, 472, 479
transcriptomics   443
transition matrix   42, 43, 47, 48
translational invariance   257
transmissibility   381, 382
tree   4–6, 25, 36, 43, 45, 47, 61, 82, 111–
    113, 123, 127, 128, 131, 133–136, 148,
    159, 164, 184, 197, 212–216, 218–222,
    224, 225, 231, 236–239, 242–252, 323,
    324, 341, 351, 354, 355, 364, 404, 406,
    407, 409, 412, 427, 428, 450, 451, 513
triad formation model   151
triangle   57, 83, 140, 143, 350

undirected   35, 107, 110, 119, 141, 151,
    165, 427
University of Michigan   161, 337, 342,
    439, 440
utilization   257, 259, 264–268, 271, 272

Vázquez   50
Valverde   161, 189, 206, 207
Vázquez   425
vertex   7, 35, 37, 38, 41, 42, 45, 49, 54, 57,
    59–68, 71, 74, 76, 77, 79, 81, 83–85,
    91, 92, 95–97, 99–101, 105–110, 113–
    116, 118–120, 122, 123, 146–148, 150,
    151, 165, 169–171, 173, 176, 177,
    179–182, 192, 195, 197, 206, 237, 239,
    240, 244, 245, 248, 250, 352, 354–357,
    359, 360, 362–364, 366, 373–375, 377,
    381, 399, 400, 409, 412, 413, 416, 418,
    427–430, 432–437
Vespignani   33, 34, 50, 124, 136, 161, 162,
    186, 206, 296, 310, 397, 422, 425, 439,
    440, 457
viable   374, 449
Vicsek   50, 124, 162, 163, 185, 187, 369,
    457
virtual time horizon   255–258, 260, 263,
    264, 266, 267, 269, 272, 273
virus   26, 32, 382, 386, 395, 430, 448, 455
visual cortex   485–488, 492, 510

Watts   6, 32, 33, 35, 49, 84, 90, 103, 104,
    159–161, 169, 185, 186, 274, 296, 297,

305, 309, 310, 352, 353, 367–370, 386–388, 391, 395, 397, 398, 422, 440, 456, 486, 508, 509
Watts-Strogatz model 300, 308
weight 51, 92, 95, 99, 101, 106, 127–133, 164, 167, 168, 175, 176, 262, 280, 281, 295, 327–329, 331–333, 347, 354, 355, 361–363, 400, 404, 414, 418, 420, 494, 496
Wigner theorem 36
wild type 459, 462, 464, 467, 473, 474, 479

world wide web 3, 103, 149, 339, 346, 347, 445, 508
Wuchty 443, 456, 457

XY model 285

yeast 110, 120, 121, 451–453, 455–457
Young 487, 492, 508–510
Yule formula 74

Ziff 233, 339, 368